T0337819

Multifunctional Agriculture

Multifunctional Agriculture
Achieving Sustainable Development in Africa

Roger R.B. Leakey
International Tree Foundation, Oxford, England, United Kingdom

ACADEMIC PRESS

An imprint of Elsevier
elsevier.com

Academic Press is an imprint of Elsevier
125 London Wall, London EC2Y 5AS, United Kingdom
525 B Street, Suite 1800, San Diego, CA 92101-4495, United States
50 Hampshire Street, 5th Floor, Cambridge, MA 02139, United States
The Boulevard, Langford Lane, Kidlington, Oxford OX5 1GB, United Kingdom

Copyright © 2017 Elsevier Inc. All rights reserved.

No part of this publication may be reproduced or transmitted in any form or by any means, electronic or mechanical,
including photocopying, recording, or any information storage and retrieval system, without permission in writing from
the publisher. Details on how to seek permission, further information about the Publisher's permissions policies and our
arrangements with organizations such as the Copyright Clearance Center and the Copyright Licensing Agency, can be
found at our website: www.elsevier.com/permissions.

This book and the individual contributions contained in it are protected under copyright by the Publisher
(other than as may be noted herein).

Notices
Knowledge and best practice in this field are constantly changing. As new research and experience broaden our
understanding, changes in research methods, professional practices, or medical treatment may become necessary.

Practitioners and researchers must always rely on their own experience and knowledge in evaluating and using any information,
methods, compounds, or experiments described herein. In using such information or methods they should be mindful of their
own safety and the safety of others, including parties for whom they have a professional responsibility.

To the fullest extent of the law, neither the Publisher nor the authors, contributors, or editors, assume any liability for any
injury and/or damage to persons or property as a matter of products liability, negligence or otherwise, or from any use
or operation of any methods, products, instructions, or ideas contained in the material herein.

British Library Cataloging-in-Publication Data
A catalog record for this book is available from the British Library

Library of Congress Cataloging-in-Publication Data
A catalog record for this book is available from the Library of Congress

ISBN: 978-0-12-805356-0

For Information on all Academic Press publications
visit our website at https://www.elsevier.com/books-and-journals

Working together
to grow libraries in
developing countries

www.elsevier.com • www.bookaid.org

Publisher: Andre G. Wolff
Acquisition Editor: Nancy Maragioglio
Editorial Project Manager: Billie Jean Fernandez
Production Project Manager: Nicky Carter
Designer: Victoria Pearson

Typeset by MPS Limited, Chennai, India

Contents

Biography

Roger Leakey has diplomas in practical agriculture (NDA and CDA) and degrees in agricultural science (BSc, PhD, DSc), with truly global experience of field-based research in tropical agriculture, horticulture, and forestry spanning nearly 50 years. He was Director of Research at the International Centre for Research in Agroforestry (ICRAF 1993–1997) and Professor of Agroecology and Sustainable Development of James Cook University, in Cairns, Australia (2001–2006). He has been Vice President of the International Society of Tropical Foresters and is Vice Chairman of the International Tree Foundation, a UK-based charity supporting development projects in Africa. He holds a number of Fellowships in learned societies, universities, and international research centres. He was a Coordinating Lead Author in the International Assessment of Agricultural Science and Technology for Development (IAASTD) which was approved by 58 governments in an Intergovernmental Plenary meeting in Johannesburg, South Africa in April 2008. This Assessment examined the impact of agricultural knowledge, science, and technology on environmentally, socially, and economically sustainable development worldwide over the last 50 years and suggested that to meet these challenges agriculture has to advance from a unifunctional focus on food production and to additionally embrace more environmental, social, and economic goals to become multifunctional.

Preface

This book charts the evolution of a practical and effective means of transforming the lives of farming households in the tropics and subtropics from deprivation due to hunger, malnutrition, and poverty, toward plenty, with food and nutritional security as well as social and economic empowerment. Currently one-third of farm land is severely degraded with staple crops actually yielding only 10–15% of their biological potential and many millions of rural households suffering from food shortages, dietary deficiencies, ill health, and extreme poverty. The emerging story in this book illustrates how this can relatively easily be transformed by rehabilitating soils and agroecosystems, leading to greatly improved actual food crop yields. Crucially, however, the story also includes the creation of new nutritious fruit and nut crops by farmers trained to simply domesticate traditionally important and marketable indigenous food species, which are important for better nutrition and health, as new crops. This process then creates new local businesses and industries to further develop the rural economy and livelihoods.

This approach of combining agroecological restoration with income generation to address poor crop yield is especially important in developing countries where farm land is severely degraded, farmers are trapped in poverty, hunger and malnutrition, and opportunities for improved lifestyles are virtually inaccessible. Past practices of clearing forest as new land for agriculture are no longer acceptable, so, if the global food crisis is to be averted, ways have to be found to rehabilitate degraded farm land and make existing fields more productive. The cultivation of tree crops also helps to mitigate climate change by sequestering carbon above- and belowground.

Trees of many different types offer a range of useful and appropriate functions: from atmospheric nitrogen fixation; the restoration of soil structure and nutrient cycling; niche creation for organisms from the smallest microbe to birds and mammals; as well as producing a wide range of edible products from leaves, fruits, and nuts, as well as a wide array of nonfood products for existing and future markets. In addition, trees protect watersheds and sequester carbon. Interestingly, farmers in the tropics and subtropics are still very much better connected with the role of trees and forests in both the environment and society—connections lost by centuries of so-called civilization associated with modern living in industrialized countries.

With his research partners, the author has been instrumental in developing ideas, strategies, techniques and practices spanning many disciplines, which are now allowing these functions to be harnessed by poor farmers in small-scale, bottom-up, community development projects. In essence, this involves the evolution of a logical approach to "multifunctional agriculture"—food production together with social, economic, and environmental benefits—through the sustainable intensification of farming practices within an integrated approach to rural development.

Over the last decade, many international reports have pointed a critical finger at agriculture for its environmental and social costs to both global health and human livelihoods, calling for a more sustainable approach, saying that "business as usual" in farming is no longer acceptable (e.g., MEA, 2005; GEO, 2007; CAWMA, 2007; IAASTD, 2009; The Royal Society, 2009), especially in the tropics and subtropics. However, these reports do not go on to say how to achieve the desired outcome. This desire for a new and more multifunctional approach to rural development has also been expressed in the seventeen 2030 Sustainable Development Goals. To address these broad-ranging social, economic, and environmental goals, which form a matrix of interacting issues, will require a set of equally interrelated interventions that together form a holistic solution. Furthermore, the current tendency to focus on individual goals misses the opportunity to be as cost effective as a more integrated approach. Thus, it is important that development agencies focus on activities that promote the simultaneous delivery of multiple goals.

In a world increasingly focused on specializing in the detail of small components of a single scientific discipline, this book is innovative and perhaps unique in looking in the other direction. It draws on the experience and output of a single scientist trained in agriculture and botany, and who then over the course of his career became increasing involved in work with scientists in other disciplines. This multi- and transdisciplinary work at the interface of agriculture with forestry, horticulture, plant physiology, genetics, ecology, soil science, food science, economics, social sciences, and environmental science has led to a more holistic way to address the complex issues of today's food crisis.

This book compiles a selection from the 300 + publications that span many disciplines and are the outcome of this research. Organized by core concept, each chapter of this book also includes an update, revealing the progress that has been made as well as some pointers to future work. This will also identify those concerns and limitations that still factor into the adoption and utilization of multifunctional agriculture in maximizing crop production, closing the yield gap in staple food crops, enhancing the livelihoods of subsistence farmers and their communities, and addressing environmental degradation and natural resource mismanagement. From this we can get an insight into how a new multidisciplinary approach to agriculture (agroforestry) has emerged over the last 25 years. It resonates with the complex interacting problems facing the achievement of the new 2030 Sustainable Development Goals. However, agroforestry is often misunderstood by decision makers who consider that trees should be cleared to make way for crops, rather than recognizing their essential role in ecosystem and planetary health.

The challenge now for the unconventional approach to modern agriculture described in this book is to promote the scale-up of these research outputs in development programs to the point where they have regional, and even global, impact. This requires acceptance of the concepts and outcomes by policymakers, agribusiness, and academics, most of whom are currently going in a different—more conventional and less sustainable—direction, perhaps unaware of the business opportunities presented by this environmentally more sustainable and socially acceptable option.

In conclusion, this book provides the research publications underpinning the story presented in Roger Leakey's personal odyssey of field experience from around the world, *Living with the Trees of Life—Towards the Transformation of Tropical Agriculture*. Both have been compiled for those focused on long-term solutions to the environmental, as well as food supply, challenges facing the world, and provide an important resource for understanding how and why trees provide a crucial part of the total solution.

Part I

The Basics

Section 1

Agroecology and the Role of Trees

There are many definitions of agroforestry, one of the delivery mechanisms of multifunctional agriculture. Until the mid-1990s these definitions related to the many ways of integrating trees into farming systems as stand-alone agronomic technologies. This viewpoint overlooks the ecological functions that underpin the ways that allow it to enhance the environmental sustainability, as well as the productivity, of farming systems. Here, agroforestry is seen as applied agroecology in which the planting of trees in many combinations and/or configurations creates niches in farming systems that are colonized by a wide range of wild organisms initiating successive phases of an agroecological succession with increasing ecological integrity. The niche-forming principle within this functional definition also opens the opportunity for farmers to add diversity and fill some niches with plants and animals producing useful and marketable products in ways that increase total productivity and promote the social and economic sustainability of the farm.

Diversified and functional agroecosystems thus rehabilitate degraded agroecosystems, making them more productive: an unconventional approach to agricultural intensification. The processes are becoming better understood—both in the early pioneer and later mature phases of succession. This knowledge emerged when biodiversity studies found that mature cash crop systems, such as cacao and coffee with shade trees, provide habitat for wildlife of importance for conservation. These findings then led to studies on the ways that insectivorous birds and other natural predators reduce the numbers of pests and pathogens, providing information about the complexity of interacting factors that regulate ecological food chains. However, much more research is needed to better understand how these processes can be managed and manipulated so that environmentally friendly farming systems can provide enhanced productivity for food and nutritional security, as well as poverty alleviation, in harmony with wildlife.

Chapter 1

Definition of Agroforestry Revisited

This chapter was previously published in Leakey, R.R.B., 1996. Agroforestry Today, 8 (1), 5−7, with permission of World Agroforestry Center

Agroforestry has been defined in several ways (Nair, 1989). The current definition of the International Center for Research in Agroforestry (ICRAF—now called the World Agroforestry Center) is a collective name for land-use systems and practices in which woody perennials are deliberately integrated with crops and/or animals on the same land-management unit. The integration can be either in a spatial mixture or in a temporal sequence. There are normally both ecological and economic interactions between the woody and nonwoody components in agroforestry. This definition has served well and helped agroforestry to become recognized as a branch of agricultural science in its own right (Sanchez, 1995).

Agroforestry practices come in many forms but fall into two groups: those that are sequential, such as fallows, and those that are simultaneous, such as alley-cropping (Cooper et al., 1996). In all, some 18 different agroforestry practices have been recognized by Nair (1993a,b), although each has an infinite number of variations. Thus, at the moment, agroforestry is viewed as a set of stand-alone technologies that together form various land-use systems in which trees are sequentially or simultaneously integrated with crops and/or livestock. In agroforestry research, practices are often applied after diagnosis and design, participatory research or characterization studies, as appropriate, depending on the social, economic, and environmental problems in an area.

Agroforestry is generally practiced with the intention of developing a more sustainable form of land-use that can improve farm productivity and the welfare of the rural community.

My problem with the current view of agroforestry is that many people still see it as a set of distinct prescriptions for land-use. As a result, it falls far short of its ultimate potential as a way to mitigate deforestation and land degradation and thus alleviate poverty. A different view, however, is that agroforestry practices can be seen as phases in the development of a productive agroecosystem, akin to the normal dynamics of natural ecosystems. Over time, the increasing integration of trees into land-use systems through agroforestry can be seen as the passage toward a mature agroforest of increasing ecological integrity. By the same token, with increasing scale, the integration of various agroforestry practices into a landscape is like the formation of a complex mosaic of patches in an ecosystem, each of which is composed of many niches.

These niches are occupied by different organisms, making the system ecologically stable and biologically diverse. Filling some of these niches with species that provide important environmental services or economically valuable products, or both, should result in land-use that is both more sustainable and productive. Furthermore, the benefits with increasing scale from the farm to the landscape and the region are exponential, since the ecological and social benefits of diversity on a landscape-scale are considerably greater than the sum of the individual farm-scale benefits.

Within this ecological framework, farmers can manipulate and manage their trees to take advantage of the benefits of the processes in ecosystem services or products, by breaking the process of agradation—or ecosystem development—at any point, or by allowing a mature agroforest to develop. Fallowing and relay cropping, e.g., make use of the benefits of early successional stages of ecosystem development, while complex multistrata systems approach a mature vegetation, such as the commercially valuable damar agroforests of Sumatra.

Therefore, I suggest that agroforestry should be reconsidered as *a dynamic, ecologically based, natural resource management system that, through the integration of trees in farm- and rangeland, diversifies and sustains smallholder production for increased social, economic, and environmental benefits.*

Multifunctional Agriculture. DOI: http://dx.doi.org/10.1016/B978-0-12-805356-0.00001-5
© 2017 Elsevier Inc. All rights reserved.

If these concepts are accepted, then agroforestry researchers and extension workers have a challenge—to start the process of integrating a number of the current agroforestry practices into productive and sustainable land-use systems that alleviate poverty. Contrary to the alternative of monocultures, over time and space these land-use systems become more complex, biodiverse and both ecologically and economically resilient to the normal patterns of climatic variability and pest and disease outbreaks. It is worth noting that in some areas of high population density, farmers are already ahead of the game, and are already practising this kind of agroforestry.

Chapter 2

The Role of Trees in Agroecology and Sustainable Agriculture in the Tropics

This chapter was previously published in Leakey, R.R.B., 2014. Annual Review of Phytopathology 52, 113–133, with permission of Annual Reviews

SUMMARY

Shifting agriculture in the tropics has been replaced by sedentary smallholder farming on a few hectares of degraded land. To address both low yields and low income, the soil fertility, the agroecosystem functions, and the source of income can be restored by diversification with nitrogen-fixing trees and the cultivation of indigenous tree species that produce nutritious and marketable products. Biodiversity conservation studies indicate that mature cash crop systems, such as cacao and coffee with shade trees, provide wildlife habitat that supports natural predators, which in turn reduce the numbers of herbivores and pathogens. This review offers suggestions on how to examine these agroecological processes in more detail for the most effective rehabilitation of degraded land. Evidence from agroforestry indicates that in this way productive and environmentally friendly farming systems that provide food and nutritional security, as well as poverty alleviation, can be achieved in harmony with wildlife.

INTRODUCTION

The future of agriculture is the focus of much international debate (Cribb, 2010; Garnett et al., 2013; Giovannucci et al., 2012; IAASTD, 2009). The main area of consensus is that it needs to be more sustainable and more productive, and that business as usual is not an option (GEO, 2007; Hassan et al., 2005; IAASTD, 2009; Molden, 2007; Royal Society, 2009). The problems are numerous and complex, especially in the tropics and subtropics, where interactions between both biophysical and socioeconomic issues are affecting agricultural productivity, mainly through a combination of land degradation driven by population growth and a declining resource of productive land for agriculture. This has led to farmers becoming sedentary smallholders without the financial resources to replenish soil fertility (Leakey, 2012b, 2013).

In his book *The Coming Famine*, Julian Cribb (2010) says that "... the challenge facing the world's 1.8 billion men and women who grow our food is to double their output of food—using far less water, less land, less energy and less fertilizer. They must accomplish this ... amid more red tape, economic disincentives, and corrupted markets, and in the teeth of spreading drought." Adding more detail to this scenario, the International Food Policy Research Institute in Washington has indicated that the world population will rise from the current 7 billion to approximately 9 billion by 2050, and that to support all these people, food production will have to increase by 70% (IFPRI, 2011). However, it is not clear whether this projection is accepting the status quo in terms of the living standards of the poor in least-developed countries or of the effect on the environment. The real need is, of course, to feed marginalized people and also to improve their lives.

These projections make pretty gloomy reading, but is it a lost cause? The context of the problems here is important. We need to understand that hundreds of millions of poor farmers are trying to support their families on a small area of land without access to fertilizers and other technologies. This new sedentary lifestyle is the result of population growth and the lack of new forest to clear for the practice of shifting cultivation. As sedentary farmers, they have to survive,

Multifunctional Agriculture. DOI: http://dx.doi.org/10.1016/B978-0-12-805356-0.00002-7
© 2017 Elsevier Inc. All rights reserved.

feed, and provide for all the daily needs of their families on approximately two to five hectares of cleared land without a source of income or the lifeline of social services if something goes wrong.

Regarding solutions, much of the focus of discussion is polarized between whether to improve productivity through biotechnology and genetic modification or through better soil husbandry and organic farming. A third commonly advocated solution is to pay greater attention to agroecology, which is certainly relevant to addressing the loss of biodiversity and ecosystem function. However, achieving a sustainable future for agriculture will involve more than just improving agroecological functions. Instead, it is necessary to take a more holistic view of a problem that encompasses the need for improved soil fertility and health; reduced risk of pest, disease, and weed outbreaks; improved crop yields; improved rural and urban livelihoods; and opportunities for economic growth.

UNDERSTANDING THE PROBLEM BEFORE SEEKING SOLUTIONS

Current estimates are that land degradation affects 2 billion hectares (38% of world cropping area) (Eswaran et al., 2006), with soil erosion affecting 83% of the global degraded area (Bai et al., 2008). The problems are especially severe in Africa, where more than 80% of countries are nitrogen-deficient, and where nutrient loss has been estimated at $9-58$ kg ha^{-1} year^{-1} in the 28 worst affected countries and $61-88$ kg ha^{-1} year^{-1} in 21 others (Chianu et al., 2012). Nitrogen (N), phosphorus (P), potassium (K) deficiencies occur on 59%, 85%, and 90% of harvested land area, respectively (Chianu et al., 2012).

Understanding the complexity of the numerous and interacting socioeconomic and biophysical causal factors underlying land degradation and the problems of modern agriculture in the tropics and subtropics is difficult. In the past, they have been lumped together as a downward spiral in which poverty drives land degradation and land degradation drives further poverty, but this is a gross simplification (Scherr, 2000). In an attempt to make the concept more meaningful, Leakey (2010) added some steps to a conceptual diagram of the land degradation and social deprivation cycle. In this highly interactive cycle, a desire for security and wealth drives deforestation, overgrazing, and unsustainable use of soils and water, all of which cause agroecosystem degradation. In farmers' fields, this is seen as soil erosion, breakdown of nutrient cycling, and the loss of soil fertility and structure. The consequence of this degradation is the loss of biodiversity and the breakdown of ecosystem functions that regulate food chains, nutrient cycling, and the incidence and severity of pests, diseases, and weeds. All of these things result in lower crop yields, which in turn lead to hunger, malnutrition, and increased health risks, all of which manifest as declining livelihoods and so return the cycle to a desire for security and wealth. It is recognized that at all of the steps within this conceptual diagram, there are a range of socioeconomic and biophysical influences that will determine the speed of the downward progress at any particular site. Such factors include access to markets, land tenure, and local governance; external factors, such as natural disasters and conflict and war; and economic drivers, such as international policy and trade agreements.

The net result highlighted by this conceptual cycle of land degradation and social deprivation is that in developing countries where farmers are poor and must rely on a small piece of land for all their household needs, the benefits expected from existing agricultural technologies, such as improved crop varieties and livestock breeds, are constrained by a need for income to purchase the inputs essential for food production. As a consequence, crop yields decline as the soil fertility is depleted. This leads to what is called the yield gap in food crops, which is the difference between the yield potential of new modern varieties and the yield actually achieved by a farmer (Tittonell and Giller, 2013).

ADDRESSING SOIL FERTILITY

Conventional agriculture would advocate that inorganic fertilizers be used to make up these deficiencies, but unfortunately in many of the least-developed countries where this need is greatest, the farmers are too poor to purchase them. Even when available, they are typically much more expensive than in industrialized countries. An alternative promoted by the organic agriculture movement is the use of manure, compost, and mulching. The problem here is that organic manure is seldom available in the quantities ($10-40$ Mg ha^{-1} year^{-1}) needed for large areas (Mafongoya et al., 2006).

Another alternative is to utilize biological nitrogen-fixing technologies, such as 2-year improved fallows, relay cropping, or Evergreen Agriculture, by which nitrogen-deficient soils are enriched with fast-growing leguminous trees, shrubs, or vines. A study involving 16 tree species found that the rate of nitrogen fixation was higher than the $23-176$ kg N ha^{-1} year^{-1} for food or fodder legumes (Herridge et al., 2008), and in high-density plantings, such as improved fallows or fodder banks, could be as high as $300-650$ kg N ha^{-1} year^{-1} (Nygren et al., 2012). Together, these research findings on crop production conform to the general experience of adopting farmers who report yield increases of two- to threefold. The combination of inorganic fertilizers and biological nitrogen can, however, be synergistic.

Agroforestry has a long and well-documented history in the use of low-cost, biological nitrogen fixation by leguminous trees and shrubs to restore nitrogen deficiencies in the soils across Africa and other areas of the tropics and subtropics (Sanchez, 1976; Young, 1997). A recent meta-analysis of 94 studies in sub-Saharan Africa found that these tree technologies increased maize yields by $0.7-2.5$ Mg ha^{-1} (or $89-318\%$) (Sileshi et al., 2008a). Furthermore, these yield increases were more stable than monocultures treated wih recommended applications of artificial fertilizer (Sileshi et al., 2011). Indirectly, leguminous trees can also recycle other macro- and micronutrients through uptake from their more extensive and deeper root systems (which act as a safety net to absorb deep nutrients) and subsequent leaf litter fall and fine root turnover (Mweta et al., 2007). By comparison with artificial fertilizers, nitrogen of tree origin is less likely to be lost to, or pollute, groundwater.

Environmentally, biological nitrogen fixation by leguminous trees and shrubs is important for soil health, as it increases the organic matter and carbon contents of soil, which improve soil structure (aggregate stability, porosity, and hydraulic conductivity), reduce soil erosion, and promote greater water infiltration. Soil organic matter content is important in all agroecosystems, but it is especially important in areas where rainfall can be a limiting factor. Ecologically, the organic nitrogen not taken up by the crops is stored in the soil organic matter pools (Schroth et al., 2001) and so is available to other organisms, such as mycorrhizal fungi and earthworms (Sileshi et al., 2008b,c). This nitrogen resource is less likely to be leached and to pollute groundwater than is the use of inorganic fertilizers. Leguminous trees have also been shown to considerably improve rain-use efficiency (the ratio of aboveground net primary production to annual rainfall) (Sileshi et al., 2011) Likewise, water-use efficiency was higher in maize intercropped with leguminous trees than in the sole maize (Chirwa et al., 2007). In general, the use of biological nitrogen fixation to improve soil fertility also leads to crops less susceptible to pests and diseases (Schroth et al., 2000). On the negative side, as with inorganic fertilizers, there can be N_2O emissions associated with higher levels of soil nitrogen.

BIODIVERSITY AND AGROECOLOGICAL FUNCTIONS

The role of ecological processes in agricultural sustainability has been studied for many years (Swift and Anderson, 1994), and its importance for the future of global agriculture is well recognized (Tscharntke et al., 2012). However, its application has not occurred on a large-enough scale to have impact on the global problem of land degradation. The challenge, therefore, is to demonstrate conclusively and to the satisfaction of policy makers, agribusiness, and academics that diverse mixed cropping systems can be productive while using natural resources sustainably. This concept is described as sustainable intensification (Garnett et al., 2013). Care, of course, has to be taken not to introduce species that cohost serious pests and diseases of the planted crops (Schroth et al., 2000).

Previously we saw that integrating leguminous trees and shrubs in cropping systems can initiate the process of rebuilding soil fertility and also boost agroecosystem functions. In recognition of this role of trees, Leakey (Leakey, 1996, 1999b) described agroforestry practices as "phases in the development of a productive agroecosystem, akin to the normal dynamics of natural ecosystems ... and the passage toward a mature agroforest of increasing ecological integrity. By the same token, with increasing scale, the integration of various agroforestry practices into the landscape is like the formation of a complex mosaic of patches in an ecosystem, each of which is composed of many niches ... occupied by different organisms, making the system ecologically stable and biologically diverse." Over recent years, there has been a considerable increase in the number of studies examining the biological diversity associated with complex multistrata agroforestry systems, especially cocoa (Clough et al., 2009a) and coffee (Somarriba et al., 2004), and their comparison with other farming systems (Collins and Qualset, 1999; Schroth et al., 2004; Table 2.1). Many of these studies find that agroforestry systems maintain a level of biodiversity that is considerably higher than other agricultural systems but generally a little lower than that of natural forest. Again, a typical finding is that agroforests provide habitat suitable for forest-dependent species and so, in most cases, are important for wildlife conservation.

The belowground ecosystem is out of sight and so often out of mind, but it is very complex and crucially important to the proper functioning of agroecosystems. Lavelle (1996) identified four complex assemblages of organisms with different roles that differ in the size and number of individuals—from the smallest to the largest: the microflora, the micropredators, the litter transformers, and the ecosystem engineers. Agroforestry, as an agricultural system with high aboveground biodiversity and minimum soil disturbance, is recognized to have a diverse rhizosphere associated with the four belowground species assemblages that together form a soil ecosystem typical of a fully functional late successional stage. Tillage, however, disturbs the soil, killing many of the larger organisms and breaking the fungal networks

TABLE 2.1 Published assessments of biodiversity in agroforestry systems (see also Moguel and Toledo, 1999; Perfecto et al., 1996).

Organisms	Country	Crop	Reference
Phytotelmata plants and ceratopogonid pollinators	Brazil	Cocoa	Fish and Soria (1978)
Bat and bird	Brazil	Cocoa	Faria et al. (2006)
Bat	Brazil	Cocoa	Faria and Baumgarten (2007)
Ants	Brazil	Cocoa	Delabie et al. (2007)
Soil and litter fauna	Brazil	Cocoa	Da Silva Moço et al. (2009)
Sloth	Brazil	Cocoa	Cassano et al. (2011)
Plants	Cameroon	Cocoa	Sonwa et al. (2007)
Soil fauna	Cameroon	Cocoa	Norgrove et al. (2009)
Mammals	Cameroon	Cocoa	Arlet and Molleman (2010)
Birds	Costa Rica	Cocoa	Reitsma et al. (2001)
Dung beetles, mammals (including bats), birds	Costa Rica	Cocoa and banana	Harvey et al. (2006a,b), Harvey and González Villalobos (2007)
Sloth	Costa Rica	Cocoa	Vaughan et al. (2007)
Epiphytes	Ecuador	Cocoa	Andersson and Gradstein (2005), Haro-Carrión et al. (2009)
Birds	Indonesia	Cocoa	Abrahamczyk et al. (2008)
Epiphytes	Indonesia	Cocoa	Sporn et al. (2009)
Amphibians, reptiles, and ants	Indonesia	Cocoa	Wanger et al. (2009, 2010a,b)
Ants	Indonesia	Cocoa	Wielgoss et al. (2010)
Rats	Indonesia	Cocoa	Weist et al. (2010)
Monkeys	Mexico	Cocoa	Muñoz et al. (2006)
Birds	Mexico	Cocoa	Greenberg (2000)
Homoptera	Costa Rica	Coffee	Rojas et al. (2001a,b)
Birds	Dominican Republic	Coffee	Wunderle and Latte (1998)
Birds and other seed dispersal agents	Ecuador	Coffee	Lozada et al. (2007)
Mammals	India	Coffee	Bali et al. (2007)
Termites	India	Coffee	Gowda et al. (1995)
Soil coleoptera	Mexico	Coffee	Nestel et al. (1993)
Mammals	Mexico	Coffee	Gallina et al. (1996)
Ants	Mexico	Coffee	Perfecto and Snelling (1995)
Arthropod	Mexico	Coffee	Perfecto et al. (1997)
Birds	Mexico	Coffee	Perfecto et al. (2004)
Birds	Mexico	Coffee	Greenberg et al. (1997)
Ants and birds	Panama	Coffee	Roberts et al. (2000a)

(Continued)

TABLE 2.1 (Continued)

Organisms	Country	Crop	Reference
Ants	Panama	Coffee	Roberts et al. (2000b)
Collembola	Indonesia	Rubber	Deharveng (1992)
Mammals	Indonesia	Rubber, damar, and durian	Sibuea and Herdimansyah (1993)
Plants	Indonesia	Rubber	Michon and de Foresta (1995)
Birds	Indonesia	Rubber, damar, and durian	Thiollay (1995)
Birds, bats, butterflies, and dung beetles	Nicaragua	Fallows and pastures	Harvey et al. (2006a,b)

(Cavigelli et al., 2012). This disturbance switches the soil ecosystem back to an early successional stage in which some functions are destroyed and must be rebuilt (Beare et al., 1997).

Forest clearance rapidly changes the soil microflora from one associated with the trees and understory to one associated with the invading pioneers (Mason et al., 1992), with complete clearance of trees having more serious impacts than partial clearance. Complete clearance makes the establishment of tree plantations more difficult even in the moist tropics, as reduced soil inoculum of the appropriate mycorrhizal fungi makes it difficult for planted tree seedlings to develop the best mycorrhizal associations (Mason et al., 1992) and hence survive and thrive. Once established, however, these trees slowly rebuild the soil inoculum over several decades (Wilson et al., 1992).

Importantly, in the few studies that have looked at the effects of conserving biodiversity and the productivity of these systems, there is little, if any, evidence that including biodiversity conservation within these wildlife-friendly perennial crop agroforests has any negative effects on crop production. Indeed, it seems that these agroforests can often successfully combine wildlife and crop production benefits without the need for further deforestation (Clough et al., 2011). In effect, when these complex agroforests also include trees that produce high-value marketable products, replace the natural fallows of shifting agriculture and become commercial fallows, which can be productive and support the livelihoods of local communities for many decades (Michon and de Foresta, 1996), if not a perennial and sustainable sedentary farming system. McNeely & Schroth (2006) recognize that maintaining diversity in agroforestry systems provides a wide range of options for adapting to changing economic, social, and climatic conditions.

So, although this focus on biodiversity conservation within productive farming systems is highly desirable, and indeed the focus of the Ecoagriculture Initiative, it is unfortunate that it mainly focuses on late/mature succession agroecosystems, typically dominated by large trees, and seldom addresses the need to understand the role of biodiversity in the pioneer stages of agroecosystem succession (Leakey, 1999b). One consequence of the predominant focus on mature agroecosystems as habitat for wildlife is that we still have a very poor understanding of the importance of this species diversity in the control of pests, diseases, and weeds. However, recent research has started to fill this knowledge gap by focusing on the agroecological factors that affect the phytopathology of crop productivity, especially in mature agroecosystems.

Regulation processes are complex, but it is recognized that important biological synergisms are a function of plant biodiversity (Altieri and Nicholls, 1999), which, of course, is variable depending on the climate temperature and rainfall, soil types, and levels of disturbance. One important factor in the prevention of disease, pest epidemics, and weed invasions is the limitation of their dispersal. Other factors are the density of herbivores, predators, and other natural enemies; the distance to, and the diversity of, natural vegetation; and the turnover rate of crops within these agroecosystems and the intensity of their management. Furthermore, ecological functions are probably also subject to variation in species abundance, the density dependence of food chains and life cycles, the rates of colonization, reproductive success, mortality rates, and the ability to locate hosts. In addition, there are more species-specific factors, such as odors, repelling chemicals, feeding inhibitors, synchrony of life cycles, etc. One conclusion from agroecological research investigating these factors has been that species composition is more important than the number of colocated species (Altieri and Nicholls, 1999). Thus, identifying the best species assemblages needs to be the focus of future research to determine ecological principles for application under very different circumstances around the world.

REBUILDING AGROECOLOGICAL FUNCTIONS

In ectomycorrhizal, and perhaps endomycorrhizal, species there are fungi associated with roots of seedling trees, and as the seedling grows, these fungi remain associated with young roots and radiate out away from the tree's stem, allowing late-stage fungi to colonize closer to the stem. In this way, a successional series of fungi develops over time (Last et al., 1987; Peay et al., 2011). Consequently, for land restoration and stimulation of vegetational colonization, it is important for rapid tree establishment to innoculate tree seedlings with early-stage fungi in the tree nursery prior to planting (Lapeyrie and Högberg, 1994). Likewise, in trees with endomycorrhizal symbionts, the rapid development of these symbiotic relations is important, especially on severely degraded sites where good levels of seedling survival are highly dependent on mycorrhizal inoculation with the best symbionts (Wilson et al., 1991); however, any symbiont seems to be better than none. For agricultural production on severely degraded land, tree establishment is a precursor to the reestablishment of the soil inoculum. This inoculum is then beneficial to the infection and growth of associated crops (Hailemariam et al., 2013). Mulch from leguminous tree species has also been found to encourage the development of soil microflora, as evidenced by increased microbial biomass and increased enzyme activity (Mafongoya et al., 1997; Tian et al., 2001). In Zimbabwe, e.g., actinomycete populations were six to nine times greater when biomass of *Vachellia* and *Calliandra* species was applied to the soil surface than when incorporated into the soil (Mafongoya et al., 1997).

Leguminous fertilizer trees have also been shown to promote the return of soil fauna even in highly degraded soils (Sileshi et al., 2008a; Sileshi and Mafongoya, 2006). For example, in Zambia earthworm densities were two to three times higher in maize intercropped with *Vachellia*, *Calliandra*, *Gliricidia*, and *Leucaena* species compared with maize receiving NPK fertilizer (Sileshi et al., 2008a), whereas maize planted after *Sesbania* with *Tephrosia* and pigeon pea fallows had two to three times greater numbers of earthworms than maize alone (Sileshi and Mafongoya, 2006).

Phytopathological knowledge in developing tree-based agroecosystems is substantially based on an understanding of the regulation of pests, pathogens, and weeds in farming systems and on a limited number of case studies (Altieri and Nicholls, 1999; Schroth et al., 2000). In pioneer and early successional stages of agroforestry ecosystems the best-known case studies involve leguminous species. For example, the leguminous shrub *Sesbania sesban* induces suicide germination of the seeds of the parasitic weed *Striga hermonthica*, resulting in reduced infestations on cereals (Khan et al., 2007). In other legumes, *Desmodium* spp. acts as a repellent to stem borers of cereals, whereas Napier grass (*Pennisetum purpureum*) attracts the pests away from the crops. The combination of these agroecological functions has been described as a push-pull technology (repelling and attracting pests) (Cook et al., 2007). *S. sesban* has also been credited with reducing the dispersal of maize rust (Krauss, 2004). Some tree and shrub species (e.g., *Lantana camara*, *Melia azedarach*, *Azadirachta indica*, *Tephrosia* spp.) are known to have insecticidal properties, and some specific insecticidal chemotypes of *Tephrosia vogelii* have been identified (Belmain et al., 2012). However, there has been concern that some leguminous shrubs also harbor insect herbivores. However, when pure and mixed species legume fallows were tested there was no evidence of serious biomass loss due to herbivory (Girma et al., 2006). *Tephrosia* fallows had the lowest population densities of 18 species recognized as pests of fallows.

Predator-prey interactions can also be modified in early-stage agroecosystems by diversifying the habitat around fields by planting trees on the field boundaries. For example, in China, pond cypress (*Taxodium ascendens*) is planted around rice to provide habitat for spiders that control leafhoppers; in Latin America, ants protect cocoa from the arthropod vectors of disease. Trees planted throughout coffee plantations provide habitat for birds, spiders, and ants, all of which contribute to pest control. These properties have been used in farming systems as part of integrated pest management strategies, in which the need for pesticides is reduced by harnessing these natural processes (Dix et al., 1999).

MAINTAINING LATE SUCCESSIONAL OR MATURE AGROECOSYSTEMS

Recently, a number of studies have started to seek agroecological understanding of pest, disease, and weed problems in mature agroforestry systems in order to find practical, simple, and affordable techniques for poor farmers. These phytopathological studies have been based on both observation and the enumeration of the interactions between certain pests and pathogens and likely predator/parasite species (Table 2.2). Very recently, a few studies have taken observation to the next step in acquiring understanding of specific predator-prey relationships by manipulating parts of the agroecosystem. To some extent, there have been three phases of this research.

TABLE 2.2 Published investigations of agroecological functions in agroforestry systems.

Organisms	Country	Crop	Reference
Parasitoid wasps	Brazil	Cocoa	Sperber et al. (2012)
Midges	Costa Rica	Cocoa	Young (1997, 1983)
Ants and beetles	Indonesia	Cocoa	Bos et al. (2007a,b,c, 2008)
Birds	Indonesia	Cocoa	Clough et al. (2011)
Spiders	Indonesia	Cocoa	Stenchly et al. (2011)
Birds and bats	Indonesia	Cocoa	Maas et al. (2013)
Birds	Panama	Cocoa	Van Bael et al. (2007a,b, 2008)
Moniliophthora and *Phytophthora* spp.	Peru	Cocoa	Krauss and Soberanis (2001)
Birds and arthropods	Costa Rica	Coffee	Peters and Greenberg (2013)
Birds	Guatamala	Coffee	Greenberg et al., 2000
Birds, arthropods, and fungi	Jamaica	Coffee	Johnson et al., 2009
Birds, ants, and leaf miners	Mexico	Coffee	Delabie et al. (2007)
Ants and phorid flies	Mexico	Coffee	Pardee and Philpott (2011)
Birds	Mexico	Coffee	Philpott and Bichier (2012)
Birds and caterpillars	Mexico	Coffee	Perfecto et al. (2004)
Mealy bug, coffee rust, and berry blotch	Central America	Coffee	Staver et al. (2001)

Shade Modification

Tree shade is important to provide the best growing environment for some crops, especially those originating from the forest understory. In cocoa, this shade can be controlled to manage the incidence of diseases, such as frosty pod rot, which is caused by the fungus *Moniliophthora roreri* (Krauss and Soberanis, 2001). Likewise, in coffee plantations, shade trees can be managed to provide optimal light conditions to minimize the risks from pests [e.g., *Cercospora coffeicola* (coffee berry and leaf blotch), *Planococcus citri* (citrus mealy bug), *Hemileia vastatrix* (coffee rust)] and maximize conditions for beneficial fauna and microflora, even in areas with different soils and climate. Shade has been found to be more beneficial in the dry season and should be reduced by pruning in the wet season (Staver et al., 2001). Predation of insect pests by canopy birds is greatest when the canopy is not intensively managed, with the richness of shade trees explaining much of the variation in bird diversity (Van Bael et al., 2007a,b). In another study, disease and insect attacks were more prevalent under single species tree canopies than under mixed canopies, supporting the hypothesis that tree diversity minimizes the risks of pest outbreaks (Bos et al., 2007a,b,c). In addition, shade trees also provide breeding sites for beneficial insects, such as midges, which are pollinators of cocoa (Young, 1982, 1983).

Removal of shade trees has been found to lower the abundance and richness of birds of most guilds, including insectivorous species (Philpott and Bichier, 2012); conversely, the abundance of insectivorous birds was greatest when the canopy cover was dense and species-rich, and there was some dead vegetation. In conclusion, the multifunctional role of shade trees for agriculture and biodiversity conservation is now recognized, but their important role in risk avoidance from insect pest outbreaks is inadequately understood. Nevertheless, it is clear that a diversified food-and-cash-crop livelihood strategy is possible (Tscharntke et al., 2011).

Bird Exclusion

A small number of experiments have investigated the importance of bird and bat populations on organisms further down the food chain in multistrata cocoa and coffee agroforestry systems by setting up exclusion experiments (Johnson et al., 2009; Maas et al., 2013; Peters and Greenberg, 2013). An experiment in which caterpillars were placed on coffee plants with and without bird exclusion confirmed that birds do reduce pest outbreaks, but this was only significant in association with high floristic diversity, indicating the importance of species diversity in the habitat (Perfecto et al., 2004). Likewise, Greenberg et al. (2000) found that exclusion of birds also resulted in a 64–80% reduction in large (>5 mm) arthropods and consequently some increased herbivory. Other studies in cocoa and coffee have generally confirmed these findings, showing that exclusion does generally increase the abundance of foliage-dwelling insect herbivores, with knock-on impacts on leaf damage (Van Bael et al., 2007a,b, 2008). In Indonesia, the exclusion of birds and bats resulted in a 31% reduction of crop yield, illustrating the economic importance of these insectivores (Maas et al., 2013).

Food Chain and Life Cycle Studies

Shaded coffee and cocoa agroecosystems with a well-developed canopy of shade trees have high biodiversity, a wide diversity of microclimates and ecological niches (Siebert, 2002), and, as we have seen, fewer pest problems, most likely because of the greater abundance and diversity of the predators of herbivores. However, the species interactions in predator or parasite impacts on prey and crop production can be very complex. For example, without shade phorid flies negatively affect ant populations, and because ants negatively affect coffee berry borer, there is more damage to coffee berries (Pardee and Philpott, 2011). This indicates how shade impacting on phorid flies can enhance the biological control of berry borer by ants. Another example of the shade benefits involves parasitic wasps of the coffee leaf miner. Leaf damage by the miner was lower when twig-nesting ants were abundant, indicating that ant populations control leaf miners in shaded coffee (De la Mora et al., 2008); thus, through the effects of shade, ants provide important ecosystem services in coffee agroecosystems. Perhaps similarly, a study of parasitoid wasps in cocoa agroforests has found that wasp community composition was influenced by season, disturbance, the species richness of the shade trees, and the scale of the cocoa planting (Sperber et al., 2012). However, the impacts of such interactions can be complex and affected by such things as the climate zone and feeding habits of the bird species. For example, large frugivorous and insectivorous birds are more attracted to agroforests than small-to-medium insectivores and nonspecialist feeders. As might be expected, frugivores and nectarivores favored canopies with high species richness (Clough et al., 2009a,b). Thus, studies of bird populations need to take into account species from different functional groups (Sekercioglu, 2012).

Other studies in cocoa agroforests have found that canopy thinning leads to a reduction in the diversity of forest ant species but not of beetles (Bos et al., 2007a,b,c). This contrasting effect of canopy management illustrates the different effects of shade on different organisms and the need for a better understanding of food chains and life cycles in agroecosystem functions. For example, web-building spiders may be important predators. A study by Stenchly et al. (2011) found that web density in cocoa agroforest was positively related to canopy openness and that line and orb webs were the most common type. The presence of *Philidris* ants was also positively associated with orb-web density but not with other web types. The practical importance of these studies is that fruit losses due to pathogens and insect attacks have been found to be greater when the tree diversity of the canopy is low. This supports the hypothesis that diversity in the upper strata of agroforestry systems has beneficial effects on agroecological functions (Bos et al., 2007a,b,c).

Meta-analysis of the biodiversity and ecosystem service benefits of cacao and coffee agroforests has confirmed the wide-scale benefits of shade trees for the provision of habitat for pest control by insectivores in Latin America and Asia, and less-well-studied Africa (De Beenhouwer et al., 2013; Poch and Simonetti, 2013). However, it is quite possible that these studies have only started to explore the ecological interactions of importance in sustainable agriculture and so further ecological research is needed.

To gain a better understanding of the complex life cycle and food chain interactions between species in mixed species plantings, such as agroforests, recent studies have established some long-term experiments involving species mixtures that vary in the diversity of the planned biodiversity (the planted crops and trees), their species densities, and their configurations. This creates very different microclimates (irradiance, temperature, humidity, etc.) within both the ground flora and the canopies, which in turn affects the colonization of the ecological niches so formed (Siebert, 2002). To try to develop some deeper understanding of the importance of all this variability on the incidence, severity, and dynamics of pest, disease, and weed organisms, Leakey (2012b) has proposed two experimental designs. These experiments were designed to explore agroecosystem breakdown with regard to witches' broom disease in cocoa.

These designs seek to answer some multidisciplinary research questions spanning the agroecological successional from pioneer to mature phases, i.e., from planting through to maturity (Table 2.3), including the impacts of the agroecosystem on the health, yield, and sustainability of cocoa production. It is recognized that in the long-term the results of experiments like these will probably be very different depending on the level of land degradation at the planting site and its distance to mature forest. It is also recognized that such experiments are extremely difficult to implement

TABLE 2.3 Research questions aimed at an examination of the interactions between planned and unplanned biodiversity in cocoa agroforests.

Experimental design	Research questions
Nelder Fan	1. What is the ecologically acceptable number of cocoa plants per hectare? What density of cocoa is sustainable? 2. Can an ecologically acceptable density of cocoa be made economically acceptable to farmers by diversification with other cash crops? 3. What combinations of cocoa, shade trees, and other trees/shrubs can create a functioning agroecosystem that is also profitable for farmers? 4. How many trees/shrubs (the planned biodiversity) are required to create sufficient ecological niches aboveground and belowground (the unplanned biodiversity) to ensure that cocoa grows well, remains healthy, and produces beans on a sustainable basis? 5. Do the microclimate and biotic environment of a cocoa agroecosystem influence cocoa/chocolate quality?
Replacement series	(a) What is the effect of replacing cocoa with one to four other cash crops under upper and middle story tree species, and what is the ecologically acceptable mixture of cocoa and other cash crop plants per hectare? (b) Can a mixture of cocoa and other cash crops be made economically acceptable to farmers? (c) What combinations of cocoa, shade trees, and other cash crops can create a functioning agroecosystem that is also profitable for farmers? (d) How many trees/shrubs (the planned biodiversity) are required to create sufficient ecological niches aboveground and belowground (the unplanned biodiversity) to ensure that cocoa grows well, remains healthy, and produces beans on a sustainable basis? (e) What are the impacts of different distances (with and without the physical barrier of other species) between cocoa plants on the incidences of pests and diseases? (f) Do the microclimate and biotic environment of a cocoa agroecosystem influence cocoa/chocolate quality? (g) What are the differences in production and in sustainability between a range of species planted as a mixed agroforest in a 50% cocoa/50% other cash crops mixture (Treatment 3) and that of a monocultural mosaic?

without the confounding effects of competing root systems between juxtaposed treatments; thus individual replicates of each treatment are surrounded by guard rows of similar species composition.

The first design involves a complex multistrata agroforestry experiment laid out as a Nelder Fan (Nelder, 1962), and the second experiment is a complex replacement series (de Wit, 1960). Specifically, these designs seek to examine the relationships in mixed species cocoa agroforests between spacing, microclimate, the planned and unplanned biological diversity, the incidence of pests and diseases in cocoa, and the overall production and economic returns from cocoa and the other companion trees. The replacement series study adds an assessment of the impacts of different configurations of companion species grown in differing proportions within the canopy. Both these experiments are also examining how a range of indigenous companion crops producing agroforestry tree products can additionally provide income generation and livelihood benefits.

The Nelder Fan design aims to examine some ecological questions (Table 2.3) that arise from growing a single crop (cocoa) at different densities (60–2000 trees per hectare) under a constant density of a canopy composed of either one under-story shrub species and one canopy tree species or of 16 understory shrub species and eight canopy tree species (Fig. 2.1).

The second design modifies the replacement series design to test the replacement of cocoa with five other cash crops under a canopy of five tree species and five subcanopy species (Fig. 2.2). The upper-story trees and subcanopy trees/shrubs are at the same spacing (9 m × 9 m), each of which is surrounded by eight cocoa plants, two plants each of four cash crops, or a mixture of four cash crops with four cocoa (3 m × 3 m), depending on the treatment within the replacement series. The modular structure of the canopy ensures that the physical properties of the multistrata agroforest (i.e., the three stories of the canopy) remain constant across the experiment. To act as a control treatment in the replacement series, each of the different species is grown as a monoculture and at the same spacing as it occurs in the 50:50 treatment of the replacement series. This allows the comparison of production and income per unit area from a monoculture with that of the integrated species mixture.

In both of these experiments, the design embraces the idea that the canopy and subcanopy trees should provide marketable products in addition to shade and other environmental/ecological services. This is in recognition that in addition to minimizing the risks of pest, disease, and weed problems in commodity or food crops, farmers want to increase their income and improve their livelihoods. For the past 20 years, the domestication of such trees has been aimed at this poverty-alleviating benefit (Leakey et al., 2012).

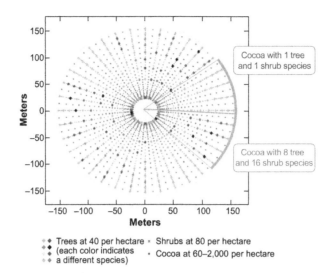

FIGURE 2.1   A Nelder Fan experiment [trees (*diamonds*) at 40 per hectare; shrubs (*crosses*) at 80 per hectare; cocoa (*circles*) at 60–2000 per hectare]. *Figure modified from an image by Ron Smith of the Center for Ecology and Hydrology, Edinburgh, Scotland.*

	Treatment 1	Treatment 2	Treatment 3	Treatment 4	Treatment 5
Cocoa under trees	100%	75%	50%	25%	0%
Other cash crops	0%	25%	50%	75%	100%

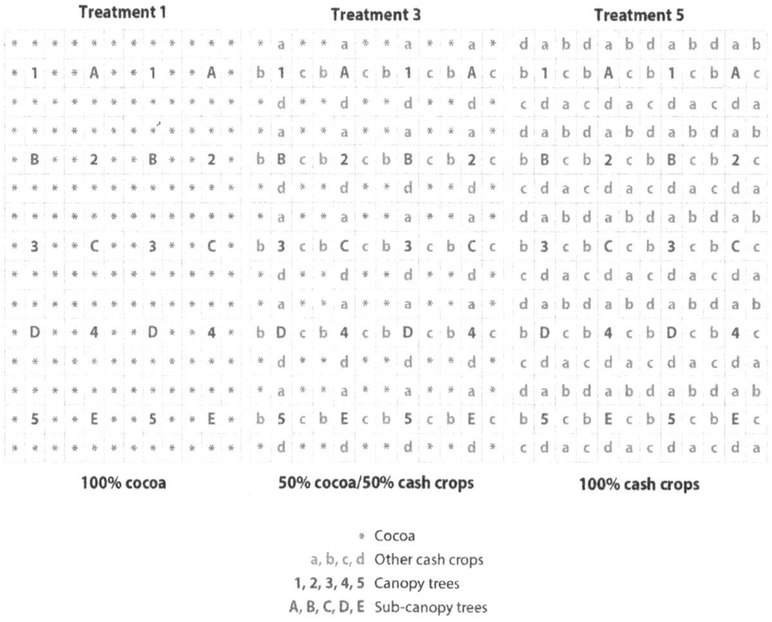

Treatment 1　　　　　　**Treatment 3**　　　　　　**Treatment 5**

100% cocoa　　　**50% cocoa/50% cash crops**　　　**100% cash crops**

* Cocoa
a, b, c, d Other cash crops
1, 2, 3, 4, 5 Canopy trees
A, B, C, D, E Sub-canopy trees

FIGURE 2.2   A Replacement series experiment (Cocoa = green stars; other cash crops = a–d: both 3m × 3m) under trees (9 m × 9 m: Canopy trees = 1–5, Subcanopy trees = A–E).

FILLING THE NICHES BELOW THE CANOPY WITH USEFUL PLANTS

Agroforestry often creates opportunities for shade-adapted species to fill shady niches and increase the benefits derived from mixed cropping systems. In this connection, most existing food crops have been selected and bred for cultivation in full sun, so there are opportunities for plant breeders and domesticators to develop new crops or crop varieties that are better adapted to partial shade (Leakey, 2012b), e.g., Eru (*Gnetum africanum*) in Cameroon. In other initiatives, socially and commercially important herbaceous species are also being domesticated as new crops to meet the needs of local people for traditional foods and medicines as well as other products, and these too can fill niches in complex species mixtures (Okafor, 1980; Schippers, 2000; Smartt and Haq, 1997). In addition to these ecological benefits, this diversification also has social and economic benefits that go a long way toward meeting the goals of multifunctional agriculture.

LANDSCAPE AND SCALING ISSUES

With little new land for the expansion of agriculture without the loss of now scarce forest, it is clear that the agroecology of agroforestry systems is an important area for more research (Siebert, 2002). This need is probably further expanded by the complexity added by scaling up to the level needed for ecological equilibrium in agricultural landscapes. In practice, the achievement of scale may need to depend on the formation of land-use mosaics with corridors between forest patches (Leakey, 1999b). This is illustrated by the finding that the diversity of frugivorous and nectarivorous birds decreased with increasing distance from the forest, whereas granivorous birds increased in diversity with increasing distance from the forest (Clough et al., 2009a,b). This illustrates that species of different functional groups differ in their need for a forest resource in the landscape. This is of importance both for wildlife conservation and almost certainly for the optimization of agroecological interactions.

In Central America, coffee agroforests are important habitat for both sedentary and migratory birds of regional and international importance. However, the role of agroforests on migratory bird populations and of migratory birds on agroecological functions are not well understood. However, one study has found that the effects of bird predation were especially strong when bird diversity was high as a result of the presence of migratory birds. Importantly, however, this period of high bird diversity was associated with greater reductions of arthropod density (Van Bael et al., 2008).

Fortunately, when considering landscape issues there is now increasing information about the tree species richness and composition of smallholder farming systems around the world (Goenster et al., 2011; Kindt et al., 2006a,b; Michon and de Foresta, 1996) and what it means for production, farm management, and farmer livelihood, and vice versa (Carsan et al., 2013). Of course, the farmer is much more interested in the livelihood benefits from agroforestry systems than in the wildlife per se. Thus, when engaging farmers in discussions about agroforestry, it is important to explain the livelihood benefits from complex agroforests and the need to utilize different tree species that produce useful and marketable products as the structure of the mature agroecosystem, i.e., the planned biodiversity (Perfecto and Vanderdeer, 2008). As we have seen, these perennial crops then create niches aboveground and belowground for the wild organisms—unplanned biodiversity—which are vital for the completion of complex food chains and the closure of numerous interactive life cycles at different trophic levels. These functions, together with nutrient, carbon, and water cycling, are the basic processes determining the functioning of ecosystems.

THE BIG PICTURE: THE ROLE OF AGROECOLOGY AND AGROFORESTRY IN TROPICAL AGRICULTURE

As mentioned in the "Introduction" section, deforestation and land degradation are now seriously affecting both agricultural crop yields in the tropics and the availability of productive land. This, together with the abject poverty of many smallholders unable to purchase artificial fertilizers and pesticides, makes the importance of land rehabilitation by agroecosystem restoration a prime objective for rural development (Leakey, 2012b). How can this be done? Two literature reviews concluded that it is possible for productive agriculture to provide both food security and biodiversity using appropriate agricultural practices that support functioning agroecosystems (Chapell and LaValle, 2011; Tscharntke et al., 2012).

One suggestion for how to simultaneously resolve the issues driving food and nutritional insecurity, poverty, and environmental degradation is to take a three-step agroforestry approach (Fig. 2.3). This combines biological nitrogen fixation with the diversification of the farming system for the restoration of agroecological functions and then adds to that the generation of income from the domestication of indigenous tree species that produce useful and marketable products for local and regional trade (Leakey, 2010, 2012e, 2013). These three steps are also aimed at closing the yield gap (the difference between potential yield of a food crop and the actual yield achieved by farmers) in staple food production by increasing soil fertility and generating income for farmers.

Ecological Restoration (Step 1) builds on over 25 years of research experience, led by the ICRAF on the use of a number of leguminous nitrogen-fixing trees and shrubs, in improved fallows, relay cropping, and, more recently, evergreen agriculture. As mentioned earlier in the section on addressing soil fertility, field experience indicates that twofold to threefold increases in crop yield are typical after 2 years of these nitrogen-fixing tree fallows. Even on a small farm,

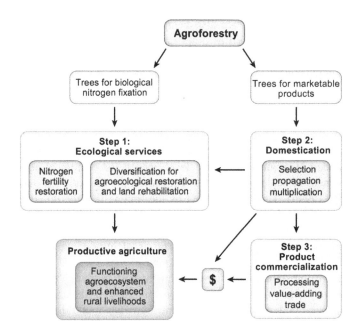

FIGURE 2.3 A diagrammatic representation of the role of trees in agroecology and the development of sustainable agriculture.

this yield increase is enough to free up space to diversify the cropping system with other income-generating crops or farm enterprises and initiate agroecological restoration. Then adding trees which produce marketable products further diversifies the farming system and forms a productive tree canopy that provides shade to cocoa and coffee. This leaves room for other crops in the canopy gaps and in the landscape mosaic.

Tree domestication (Step 2) by selecting and vegetatively propagating their elite trees for planting as superior cultivars to produce culturally and traditionally important and nutritious foods (Asaah et al., 2011; Leakey and Asaah 2013; Tchoundjeu et al., 2006, 2010), has, over the past 20 years, made great progress in assisting poor rural communities.

Product commercialization (Step 3) creates business and employment opportunities in cottage industries engaged in processing, value adding, and marketing of the products of these agroforestry trees.

By combining agroecological restoration with income generation within a participatory integrated rural development program that provides community training and education in a wide range of relevant skills (Degrande et al., 2012), agroforestry becomes a powerful new tool to address the cycle of land degradation and social deprivation. The income-generating component adds critical value to agroecology and may even be an important incentive for farmers to diversify their farming systems. Consequently, a collective action initiative in Cameroon (Gyau et al., 2012) is now being seen as an approach to delivering multifunctional agriculture and sustainable intensification to tackle the big issues of poverty, malnutrition, and hunger. If implemented on a sufficient scale, this approach also has potential to mitigate climate change because the trees, and their soils, sequester substantial quantities of carbon. These lessons for integrated rural development have now been drawn together as a set of 12 Principles and Premises (Leakey, 2014f) based on the agroecological role of trees for upscaling and wider implementation of the sustainable intensification of agriculture.

Chapter 3

Trees: A Keystone Role in Agroecosystem Function: An Update

R.R.B. Leakey

There are many definitions of agroforestry, most of which are based on early ideas presented by Nair (1989, 1993a,b), all relating to the many ways of integrating trees into farming systems as stand-alone agronomic technologies (Atangana et al., 2014); some of which are described in Chapter 4, (Cooper et al., 1996). These can be modified by the use of different species that suit the social, economic, and physical environment of the site. The magnitude of this variability has led to Nair (2014) saying that agroforestry cannot be defined exactly, but that it basically "refers to the purposeful growing of trees and crops in interacting combinations for a range of objectives including a variety of products and commodities and a vast array of environmental and other ecological services."

The difficulty I have with this technological approach to defining agroforestry is that it overlooks the ecological functions that underpin agroforestry in ways that allow it to both enhance the environmental sustainability, as well as the productivity of farming systems. The functional definition of agroforestry, presented here (Chapter 1 [Leakey, 1996]), which is independent of any particular practice, agronomic system or species combination, was accepted and adopted by the ICRAF in its Medium-Term Plan (ICRAF, 1997). This definition sees agroforestry as applied agroecology in which the planting of trees in any combination or configuration initiates phases of an agroecological succession with increasing ecological integrity. Within mixed species plantings, especially if some are tall perennials, niches are formed above- and belowground for natural organisms to colonize. Through their interacting life cycles and food webs, these organisms create, support and maintain the ecological equilibrium between species, and importantly, enhance the function of the nutrient, carbon and water cycles. Furthermore, over time and with increasing scale, the landscape becomes a complex mosaic of patches in different phases of ecological succession, as in natural ecosystems. Both these processes are akin to the normal dynamics of natural ecosystems, processes which are at the heart of ecological and environmental sustainability.

The niche-forming principle within this functional definition also encompasses a mechanism for enhancing the social and economic sustainability of farming systems, as Leakey (1999b) suggested that farmers should enrich these agroecosystems by filling as many niches as possible with trees and other plants producing useful and marketable products (the "planned biodiversity"). Enrichment in this way increases farm total productivity while further creating niches for colonization by natural organisms—the "unplanned biodiversity." Productivity can then be further intensified by the domestication of these useful, marketable and culturally important indigenous species, especially for nutritious food products to support rural livelihoods and local trade, as described in Section 3. This intensification then provides an added incentive for farmers to practice agroforestry. These concepts are examined further in Sections 4 and 5.

At the turn of the millennium relatively little was known about how to harness the ecological benefits by the deliberate enrichment of farming systems through agroforestry at either early or late phases of agroecological succession. So, there were many unanswered questions (Leakey, 1999b):

- What is the extent of unplanned biodiversity in mixed cropping systems?
- How much biodiversity (and the ratio of planned to unplanned biodiversity) is required?
- What are the relationships between agroecological function and biodiversity? And are they always positive?
- What are the relationships between diversification and biodiversity?
- What are the relationships between biodiversity and intensification?
- What are the relationships between ecological factors and economic output?

Multifunctional Agriculture. DOI: http://dx.doi.org/10.1016/B978-0-12-805356-0.00003-9
© 2017 Elsevier Inc. All rights reserved.

- How important is scale and does it vary between trophic levels?
- Can the patch dynamics of mosaics and the use of biodiversity corridors be a useful means of addressing scaling issues?

To address some of these questions, Leakey (1999b) called for three types of research:

1. To develop prototype systems
2. To test hypotheses and hence to define principles
3. To implement, test and monitor specific interactions between crops and taxa in unstructured mixtures.

To date, prototype systems (best-guess species combinations) have mainly been used in soil-plant interaction studies for productivity (Szott et al., 1991) and have not been widely used to examine any of the previous ecological questions, although Matos et al. (2003) have looked at the configuration of mixed species plots to investigate the susceptibility of mahogany to damage by *Hypsipyla* shoot tip moths. To define some ecological principles based on hypotheses to elucidate some of the relationships between planned and unplanned biodiversity, agroecological functions and total productivity, Leakey (2012b; 2014e) has presented experimental designs (Figs. 2.1 and 2.2). However, complex and long-term experiments of this sort require a multidisciplinary team of scientists and consequently are very costly and demanding. This has, so far, been a major obstacle to progress. Thus it has been the third of these approaches where most advance has been made, as seen in Chapter 2 (Leakey, 2014e). This review summarizes work which has thrown light on some key aspects of the agroecological function of early (or pioneer) stages of succession; and late (or mature) stages of succession. In the former, leguminous trees and shrubs importantly improve soil fertility and structure through the symbiotic nitrogen-fixing bacteria on their roots, while initiating the diversification of the agroecosystem both above- and belowground, while in the latter the predominant focus has been on cocoa and coffee agroforests in Latin America and Indonesia in which shade trees provide roosts for insectivorous birds and bats, habitat for ants and other small predators that prey on pests and pathogens.

In early-stage agroecosystems, relatively simple crop combinations between legumes and some grasses have been found to either deter pests (including some parasitic weeds) or to attract the predators of pests themselves. In late-stage phases of agroecological succession, the studies have taken two main forms: enumeration of different organisms (Table 2.1), and a start to unravel the complex interactions of different food webs (Table 2.2). This is beginning to provide evidence that explains the important role of trees in agroecosystem functions. This role is founded on the effects of the perennial above- and belowground structure of trees and tree canopies to create ecological niches for an extraordinary array of organisms, from the minutest microbe to the largest top predators and herbivores (including *Homo sapiens*). It is the level of activity of the food chains/webs and life cycles of these organisms that embodies the ecological balancing trick between organisms, and so determines the sustainability of the system. At all agroecological phases there is growing evidence of the benefits of species diversity on the success of the nutrient, water and carbon cycles that are crucial for healthy agroecosystems. It is the functioning of the carbon cycle resulting from the integration of woody perennials into agricultural and multifunctional landscapes that gives agroforestry the ability to mitigate climate change by both the reduction of greenhouse gas emissions and the sequestration of carbon in standing biomass (especially of perennial plants) and soil organic matter (van Noordwijk et al., 2011).

To update the literature presented here, there has been growing research activity to understand the dynamic interactions of organisms in agroforestry systems (Bennett et al., 2015; Garbach et al., 2014; Lavelle et al., 2014; Perfecto et al., 2014; Leakey, in press) and in landscapes (Scherr et al; 2014). Understanding the roles of this biodiversity is critical to understanding how to manage agroecosystems sustainably. This is needed in order to both capture the benefits of functioning agroecosystems that are important for the production of the food and numerous other useful products that provide the livelihoods of the farming households; and to promote the international public goods and services (biodiversity conservation, watershed protection, mitigation of climate change, etc.) important for humanity. In Chapters 37 and 39 (Leakey, 2014 and Leakey and Prabhu, 2017), we will see further evidence that the trade-offs that some believe to occur between production and biodiversity conservation (Garnett et al., 2013; Godfray and Garnett, 2014) can be avoided (and certainly are not inevitable) to create more sustainable farming systems.

Finally, to conclude, I think we can emphasize that it is the flexibility provided by the diversity of species and practices accorded by the ecological concept of agroforestry that gives it a unique advantage over what are often the rather prescribed land-management practices of conventional modern farming systems—which have been subject to economic boom and bust, environmental degradation and social deprivation. The delivery of a highly adaptable generic model to address and reverse these undesirable outputs of many farming systems will be presented in Chapters 31 and 34 (Leakey and Asaah, 2013; Leakey, 2013). This model provides much needed hope for the future (Leakey, 2014f). Importantly, it recognizes that improved agroecological functions alone are not sufficient to deliver the 2015 Sustainable Development Goals, and that farmers also need a good source of income to allow them to intensify and maximize production and to improve their livelihoods (Fig. 2.3, Leakey, 2012a; 2013). In the final chapter of this book (Leakey, 2017i) we will see that global sustainability has to embrace *Homo sapiens* within the functioning of intensified agroecosystems.

Section 2

Agroforestry Practices and Systems

In the 1990s agroforestry turned from being a practice aimed primarily at soil fertility improvement while supplying poles, wood, and fuel, to one also focused on producing livestock fodder and marketable products, especially foods and medicines, for income generation—an approach which can be intensified by tree domestication. The first paper reviews the considerable progress made with smallholder farmers across sub-Saharan Africa in the development of simple and appropriate agroforestry technologies using fast-growing leguminous trees and shrubs to address the production constraints of declining soil fertility in the absence of accessible external inputs. This practice is important both to improve food security from staple crops, as well as to enhance the sustainability of small-scale farming systems. Many of these same trees and shrubs also provide high-quality livestock fodder supplements. The first paper also points to the need to examine the potential of high-value indigenous and exotic trees to generate income, and to link agroecological benefits to economic and social benefits, so addressing the many disconnects at the whole farm or landscape level between the systems and their expected benefit flows. It concludes that agroforestry systems will need to evolve in both diversity and intensity if they are to remain relevant and effective for tomorrow's Africa, and to be better understood as an integrated land-use strategy.

The second paper introduces the idea of domesticating high-value indigenous trees with commercial potential in local and regional markets, both to meet the needs of farmers and to diversify and intensify agroecosystems for greater and more sustainable production in parallel with income generation. While these tree species, and their products, are well known to local people, they are virtually unknown to science. Thus there is much to do to assess their potential for, and amenability to, genetic improvement as new crops. At this time in the 1990s, the crucial first step was to identify priority species, and to formulate a simple and rapid domestication strategy appropriate for use by smallholder farmers, while capturing the genetic potential of each species. In addition it was recognized that to meet the objective of poverty alleviation, it was crucial to determine which traits were important to markets and the development of a value chain. While some traits that are relatively easy to identify do benefit the farmer, there are undoubtedly others that are important to the food, pharmaceutical or other industries that require more sophisticated evaluation. In later chapters we see how these concepts have been further developed and become a powerful tool in rural development across Africa.

Chapter 4

Agroforestry and the Mitigation of Land Degradation in the Humid and Sub-Humid Tropics of Africa

This chapter was previously published in Cooper, P.J.M., Leakey, R.R.B., Rao, M.R., Reynolds, L., 1996. Experimental Agriculture, 32, 235–290, with permission of Cambridge University Press

SUMMARY

In the last 35 years, the population of sub-Saharan Africa has increased nearly threefold and is expected to reach 681 million by the year 2000, with nearly 50% of the population living in urban centers. Such population pressures, exacerbated by a range of social and political factors, have already resulted in widespread land degradation in areas of high population densities and the expansion of agriculture on to marginal and sloping land. Declining soil fertility and soil erosion are increasingly threatening the sustainability of small-scale farming systems throughout Africa, and affordable external nutrient inputs are seldom available to farmers. In addition, shortages of wood for construction and fuel and high-quality dry season fodder for livestock are widespread and serious constraints to farm productivity.

Agroforestry, the deliberate integration of woody perennials into crop and livestock systems, has the potential to mitigate many of these constraints through both the service and production functions played by trees. In recent decades much agroforestry research has been undertaken in sub-Saharan Africa. In this review we focus specifically on research that addresses the potential of agroforestry systems to enhance soil fertility, prevent soil erosion, provide high-quality dry season fodder or generate much-needed income through the production of high-value goods.

Much emphasis has been placed on a wide range of agroforestry systems for the maintenance of soil fertility and the prevention of soil erosion losses, and encouraging results, both in technical performance and farmer enthusiasm, have occurred. However, it is clear that agroforestry solutions to land degradation are always likely to be location-specific in their relevance, performance, and farmer acceptability. It is essential that farmers are included as research partners to determine what is appropriate for their conditions.

Good progress has also been made on identifying fast-growing leguminous trees and shrubs for high-quality livestock fodder supplements. Where livestock enterprises, such as peri-urban milk production, are market-oriented, the adoption and impact of such systems have been high. Given population and urbanization projections, it is likely that fodder trees and shrubs will have a major role to play in meeting future feed demands for both milk and meat production. Research on the potential of high-value indigenous and exotic trees to generate income has been less extensive in Africa, although the huge potential of this approach has been clearly demonstrated by farmers in Southeast Asia. We suggest that there is a need for increased research emphasis on the domestication of high-value indigenous trees and their integration into more sustainable, diverse and intensive land-use systems.

We conclude that, although good progress has been made in agroforestry research in Africa and farmer adoption is occurring, future population projections pose a clear challenge. Agroforestry systems that provide solutions for today's land degradation problems will need to evolve in both diversity and intensity if they are to remain relevant and effective for tomorrow's Africa.

Multifunctional Agriculture. DOI: http://dx.doi.org/10.1016/B978-0-12-805356-0.00004-0
© 2017 Elsevier Inc. All rights reserved.

LAND DEGRADATION

It is estimated that the population of sub-Saharan Africa has nearly trebled in the last three and a half decades, rising from 210 million in 1960 to 495 million in 1990, and it is predicted that it will reach 681 million by the year 2000 (UNFPA, 1992). In long-term predictions of population growth, Bulatao et al. (1990) estimate that the population of Africa as a whole will continue to increase well into the 21st century, reaching a maximum of around 3000 million by the year 2100. It is true that these figures are based on estimates and projections that exclude the impact of natural disasters, wars, and disease, and hence must be treated with caution. Furthermore, they are not specific to the humid and subhumid tropics of Africa. However, they do clearly illustrate the alarming situation of unprecedented population increases that are likely to occur, and which are already occurring, particularly in the higher-potential humid and subhumid regions.

The impact of the already huge increases in population on the degradation of the natural resource base has been as predictable as it has been devastating. This devastation has arisen from a complex interaction of biophysical, socioeconomic, and political factors. For example, as population densities have increased, traditional and sustainable systems of shifting cultivation have suffered from shorter and shorter fallow periods of natural forest or woodland regeneration (Fig. 4.1). This situation has eventually given way to sedentary agriculture on small-scale land holdings. In addition, continuing population pressure has resulted in further forest clearance for agricultural land. From a study of 40 countries in Africa, Dembner (1991) estimates that 50 million hectares of land was deforested between 1980 and 1990. The possible future impact of 3000 million people in 2100, if steps are not taken to reduce the rate of deforestation and land depletion, is hard to imagine.

High rates of urbanization in Africa have also exacerbated land degradation. The growing numbers of urban dwellers require food, fuelwood, and timber products. These needs result in a steady transport from agricultural land to urban settlements of nutrients that end up as useless and unrecycled waste. In Africa as a whole, the proportion of people living in urban settlements has risen from around 15% in 1950 to an estimated 40% in 1995 and is expected to reach 54% by 2025 (UNEP, 1992). Nutrient export from agricultural land to urban centers is, however, not the only source of soil-nutrient depletion, particularly with regard to nitrogen (N). In analyses of nutrient budgets of cropped fields on small-scale mixed farms in Western Kenya, Shepherd et al. (1994) estimated that in such high-rainfall areas (1600$-$1800 mm a^{-1}), 63 kg N $ha^{-1}a^{-1}$ were lost through leaching, denitrification, and volatilization. This can be compared with 43 kg N $ha^{-1}a^{-1}$ lost through export from the field in crop yield. For the range of contrasting farm management options they examined, net nitrogen balances for cropped land ranged from -39 to -110 kg $ha^{-1}a^{-1}$ and net phosphorus (P) balances from -7 to $+31$ kg $ha^{-1}a^{-1}$. For the sub-Saharan region as a whole, it has been estimated that the annual average loss per hectare is 22 kg N, 2.5 kg P, and 15 kg K (Stoorvogel et al., 1993).

A further result of population increase and the nutrient depletion of potentially productive land has been the expansion of agriculture on to marginal sloping land and into drier areas traditionally reserved for livestock grazing. Such lands not only suffer from nutrient export but unless carefully managed are especially prone to wind and water erosion of the topsoil where the bulk of soil nutrients available to crops is located. It is estimated that steeply sloping land accounts for about 10% of the geographical area of sub-Saharan Africa, although in some countries in east and southern Africa, such lands constitute as much as 40$-$70% (Fig. 4.2). Human-induced water erosion accounts for nearly 50% of land degradation in Africa (WRI, 1992).

FIGURE 4.1 In Cameroon, as in much of the humid tropics, increasing population densities have resulted in shorter and shorter fallow periods in traditional shifting cultivation systems. *Photo R.B. Leakey.*

FIGURE 4.2 In Burundi, and elsewhere in Africa, population growth is forcing farmers to cultivate steep slopes. Without proper soil conservation measures, soil erosion is inevitable. *Photo A. Njenga.*

FIGURE 4.3 Continuous cultivation, lack of affordable external inputs and a complex mosaic of social and political factors have caused widespread land degradation in humid and subhumid Africa. *Photo A. Njenga.*

Few small-scale farmers can afford to replenish this steady loss of soil nutrients with external inputs of organic or inorganic fertilizers. Often fertilizer is simply not available or too expensive and in many instances farmers are forced to use animal manure as fuel rather than as fertilizer. They cannot afford to implement long-term soil conservation and improvement strategies that do not provide short- or medium-term benefits. Many of these problems have been exacerbated by, e.g., insecure land tenure, land fragmentation, lack of access to credit, pricing policies that favor urban consumers, inappropriate trading policies, and poor access to both local and international markets. These problems have all fueled the downward spiral of land degradation (Fig. 4.3). The results of these combined forces have been the development of a range of major production constraints now facing small-scale farmers.

Raintree (1987) examined the relative importance of these constraints by summarizing the output of ICRAF's (the International Center for Research in Agroforestry, now called the World Agroforestry Center) diagnosis and design (D&D) exercises (equivalent to participatory rapid rural appraisal) undertaken in 47 land-use systems in 9 countries of sub-Saharan Africa. These D&D exercises asked farmers to list the major constraints they faced (Fig. 4.4). What is immediately apparent is that farmers were aware of both declining soil fertility and soil erosion as important constraints to their welfare. Further, it is interesting that shortages of building materials and fuelwood, a direct consequence of forest and woodland clearing and the denudation of the landscape of trees, appear to be of more concern in the subhumid than in the humid regions. This is associated with both the slower regenerative capacity of woody vegetation in the drier areas and greater population densities. Fodder shortages are also an important constraint, particularly in the subhumid regions. This reflects the greater livestock numbers, the more severe dry season and the continuing destruction of natural grazing lands for agriculture. Given declining soil fertility and hence declining crop yields, it is perhaps surprising that farmers did not mention food shortages and cash shortages more frequently, but this probably reflects the subsistence nature of many small-scale farms in sub-Saharan Africa and their ability to find other sources of family income through off-farm employment.

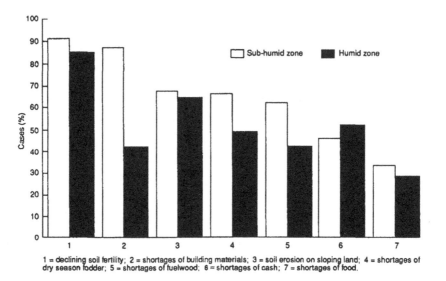

1 = declining soil fertility; 2 = shortages of building materials; 3 = soil erosion on sloping land; 4 = shortages of dry season fodder; 5 = shortages of fuelwood; 6 = shortages of cash; 7 = shortages of food.

FIGURE 4.4 Frequency distribution of problems affecting land-use systems in the humid and subhumid tropics (Raintree, 1987).

In conclusion, if we accept the postulated population growth estimates, the prospects for future generations of the urban and rural people of Africa are indeed bleak. What can be done to alleviate this crisis? This chapter attempts to draw together evidence from biophysical, social, and environmental studies to illustrate that agroforestry, if applied on a large enough scale, does have the potential to help alleviate the problems. It is, however, self-evident that, unless backed by supportive national and international policies and social motivation, farmers will continue to struggle to make a living in ways that lead to land degradation, rather than adopting improved land management techniques.

THE ROLE OF AGROFORESTRY

In order to place agroforestry in a context with other land-use systems, ICRAF has adopted the following definition:

Agroforestry is a collective name for land-use systems and practices in which woody perennials are deliberately integrated with crops and/or animals on the same land management unit. The integration can be either in a spatial mixture or in a temporal sequence. There are normally both ecological and economic interactions between the woody and nonwoody components in agroforestry.

ICRAF (1993).

Raintree (1987) noted that a narrow interpretation of this definition would require the actual physical interaction of trees with herbaceous crops, pastures, or livestock in space or time. However, from a problem-solving point of view a broader and more useful interpretation allows trees to be grown on land management units ranging from a field, to a farm, a community, or a watershed. This more liberal interpretation, which is increasingly being adopted, only requires that there be interactions between the components. For example, at the farm level, a block planting of a woodlot, fodder bank, or a discrete home garden can each be considered as an agroforestry system. This is by virtue of their economic interaction with other components of the farm in drawing on the same pool of family labor, capital and management resources.

At the farm level, trees have the potential service functions of soil fertility improvement and/or maintenance, the reduction of wind and water erosion, enhanced microclimate, livestock containment through living fences, and demarcation of internal and external boundaries. On a larger-scale, trees can improve the hydrological cycles in watersheds, and play an important role in the maintenance of soil, insect, plant, and wildlife biodiversity. Trees in agroforestry systems also have a potential role in the sequestration of carbon in above- and belowground biomass and enhancing levels of soil organic carbon. Unruh et al. (1993) estimated that 1550×10^6 ha of land are suitable for some kind of agroforestry intervention in sub-Saharan Africa. If this potential were fully realized, they calculated that between 8 and 54×10^{15} g of carbon would be sequestered, enough to offset global emissions of carbon from fossil fuels for only 1.7–9 years. They concluded that the secondary effects of agroforestry in reducing rates of deforestation and hence emission of carbon may be more important than the primary effects of carbon sequestration.

From a production perspective, trees provide fuelwood, building materials, and high-quality dry season fodder. They are also an important source of income generation through the provision of high-value products such as poles, timber, fruits, medicines, resins, and gums. There is increasing awareness of farmers' indigenous knowledge and uses of trees. For example, ICRAF (1992a) recently presented detailed information of 32 distinct service and production functions of over 170 indigenous and exotic species found in Kenya alone. Among scientists, there is also greater awareness of the potential to domesticate these species (Leakey and Newton, 1994a,b). Agroforestry, through both the service and production functions of trees, therefore has the potential to alleviate or, in some instances, remove completely many of the principal constraints recognized by farmers which result directly from land degradation.

In this review, we shall deliberately focus on the role of trees in mitigating land degradation at the farm and field levels. It is at this level that the bulk of current information is available, and it is farmers who are and will be the prime instigators of improved land management. Farms are the most important building blocks from which larger-scale impacts will be achieved. As previously indicated, there are many interlinked and interactive causes and processes involved in the downward spiral of land degradation (Fig. 4.5). We shall structure our review accordingly and provide evidence on the capacity of agroforestry to break this downward spiral through soil fertility improvement (intervention point A), the reduction of soil erosion (point B), the provision of dry season fodder (point C), and generating income (point D).

SOIL FERTILITY IMPROVEMENT

It has long been recognized that trees have the potential to increase nutrient availability to crops, a fact which is self-evident given the historical sustainability of shifting cultivation throughout the humid and subhumid tropics. Some forms of agroforestry attempt to mimic the role of secondary forest or woodland fallows through the deliberate inclusion of fast-growing leguminous trees into the farming system for the maintenance or improvement of soil fertility. Various systems are being evaluated which include trees grown either in spatial mixtures or in temporal sequences with crops. Several processes have been identified by which trees can enhance the chemical and physical properties of soils (Ingram, 1990). They include:

1. Nitrogen input into the system through biological nitrogen fixation by rhizobial bacteria in root nodules of certain tree species.
2. Deep nutrient uptake by trees from below the crop rooting zone and subsequent surface deposition as litter, biomass burning, or prunings. In addition, belowground litter of decaying root systems may be important.
3. A more closed nutrient cycle through the capture of nutrients which would otherwise be leached from beyond the tree/crop rooting zone.
4. Improved soil physical conditions (e.g., better aggregation, reduced bulk density, improved infiltration) resulting from higher levels of organic matter, old tree root channels, and increased macrofaunal activity.
5. Reduced aluminum (Al) toxicity and low pH through enhanced cycling of bases and the production of metabolic compounds that temporarily complex Al.
6. Increased availability of P through mycorrhizal associations.
7. Improved activity of soil organisms (such as fungi, arthropods, termites, and worms) through a cooler and moister microclimate.

Early research in agroforestry largely concentrated on evaluating "systems" for improving soil fertility with less emphasis on trying to assess the relative importance of the possible biophysical processes involved. Such research has in many instances been invaluable in defining "what works where" but has been less successful in explaining "why." Given the considerable complexity of above- and belowground interactions, and the technological challenge of measuring discrete processes in agroforestry systems, there is still some way to go. Recent reviews of process-oriented agroforestry research have identified some hard data (Ingram, 1990; Anderson and Sinclair, 1993) but Sanchez (1996), in discussing the evolution of agroforestry as a science, concluded: "Agroforestry is not there yet. Its underlying principles are yet to be developed in a sufficiently rigorous manner that assures predictive understanding."

Given the ongoing development of these principles, we will concentrate on presenting examples where rigorous "systems-oriented" research, both on-station and in farmers' fields, has made a start in demonstrating what works, or does not, and where.

Hedgerow Intercropping

Hedgerow intercropping (synonymous with alley cropping) is perhaps the most widely researched agroforestry system in sub-Saharan Africa (Kang et al., 1990). Trees are grown as spatial mixtures in hedges, typically spaced 4—6 m apart with a within-hedge spacing of 0.25—0.5 m. Crops are grown between the hedges and tree biomass, obtained through

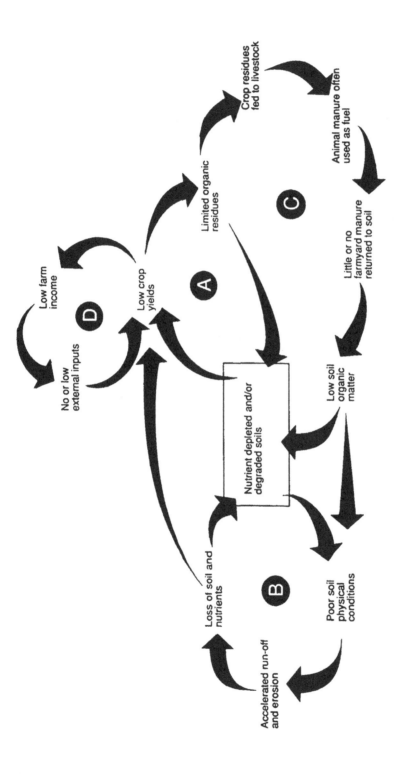

FIGURE 4.5 Breaking the downward spiral of land degradation and rural poverty through agroforestry interventions: (A) Soil fertility improvement, (B) Soil conservation, (C) Fodder supply, (D) Income generation.

periodic pruning during the cropping period, is added to the soil as green manure. Early work on this system was pioneered on Alfisols and Entisols by the International Institute for Tropical Agriculture (IITA), and appeared very promising (Kang et al., 1981). Since that time scientists throughout sub-Saharan Africa have evaluated this system under a wide range of soil and climatic conditions. Results have been mixed and in general disappointing. Recent reviews of hedgerow intercropping trials reported in the literature (e.g., Akyeampong et al., 1995) have highlighted some of the following important factors.

First, many of the earlier trials suffered from experimental design faults, which could have resulted in an upward bias of the yields reported in hedgerow intercropped plots and reduced yields in the control plots. These have been summarized by Coe (1994), who identified 10 problems commonly encountered in research reported by ICRAF as well as other institutes. Principal among these were:

1. Experimental plot size too small, resulting in above- and belowground interference between treatments. For example, one experiment in Kenya which compared three tree species had tree roots of all three species in every plot after 4 years. Such an effect is perhaps even more important when tree roots invade adjacent "no tree" controls. The trees get extra soil resources from an area perhaps twice that of the hedgerow plot, and these soil nutrients are then concentrated as green manure within the hedgerow treatment area. Not only does this result in unrealistically high amounts of nutrients added as green manure but also control plots are depleted of nutrients and water. Coe quotes an example where nearby yields from sole maize plots outside the hedgerow intercropping experimental area were two to three times higher than the sole maize control within the experiment, even though the management of both plots was identical. More recent experiments have to a large extent overcome these problems through the use of much larger plots and the installation of root barriers or root trenches around control plots. However, such barriers must be reinstalled frequently since tree roots have been shown to grow under them and up into the control plots within 1 year (Singh et al., 1989).
2. Many experiments have no control or inappropriate control treatments such as continuous maize with zero inputs. In assessing whether or not hedgerow intercropping can provide a sustainable and more productive alternative to farmers' current practices, it is essential that those practices be represented as control treatments within the experimental design. Few farmers grow maize continuously without any inputs at all.

While these acknowledged flaws in design are not peculiar to hedgerow intercropping trials, they certainly add to the complexity of reviewing the literature in order to ascertain under what biophysical conditions hedgerow intercropping will work. This task is already challenging enough given the wide range of tree species, hedgerow management and spacings, crops and local modifications (livestock inclusion, fallow periods) which have been evaluated by scientists throughout sub-Saharan Africa. Added to this complexity are the crucial factors affecting the adoption of this technology by farmers and its long-term impact on their welfare. Several such reviews have been undertaken in recent years and a consensus is emerging concerning the biophysical and socioeconomic conditions under which hedgerow intercropping will benefit farmers (Nair, 1990; Kang, 1993; Rao, 1994; Akyeampong et al., 1995; Sanchez, 1996; Reynolds, 1994; Woomer et al., 1995; Carter, 1995; Whittome, 1994). The biophysical parameters required for success appear to be:

- Soils of moderate pH (greater than 5.5) with a high base status for nutrient recycling in biomass
- Declining soil fertility (recognized as a serious problem by farmers)
- Scarcity of trees in the landscape (thus interest in tree products)
- Rainfall of more than 1000 mm a^{-1} (adequate for crops and trees)
- Cropping systems dominated by maize (in general, maize responds better to hedgerow intercropping than other crops such as cassava, upland rice, soyabean, and groundnuts)
- Well adapted tree species/provenances for high biomass production
- Timely pruning, efficient utilization of prunings, and sustained hedge growth
- Absence of serious termite attack on the hedge during establishment for certain susceptible species.

Many socioeconomic parameters have also been identified through extensive participatory on-farm research which appear important both in terms of farmer interest and capacity to meet the management requirements (establishment, pruning, and biomass distribution) of the technology. Principal among these are:

- High land pressure associated with declining or absent fallow periods
- Knowledge or tradition of producing seedlings for hedge establishment and filling gaps
- Available labor at pruning time, preferably male as the task is often considered too onerous for women
- Individual secure tenure of discrete parcels of land

FIGURE 4.6 Adoption of hedgerow intercropping in sub-Saharan Africa has been low. Here, at Maseno, Kenya, farmers found that labor requirements for hedgerow pruning conflicted with other demands, and crop yield responses were disappointing. *Photo A. Njenga.*

- Confinement of livestock to avoid dry season browsing of hedges
- Farm revenue as a major source of household income
- Infrastructural arrangements that promote extension-farmer and farmer-farmer linkages
- A clear perception by farmers that they are substantially benefiting from the technology.

This last point is of overriding importance and is, of course, clearly linked with the biophysical performance of the technology. Two observations have perhaps encapsulated the current status of the acceptance of hedgerow intercropping. First, reviews of long-term and well-designed trials have identified biophysical situations where the technology has failed to work but have been less successful in finding examples where substantial and sustained increases in crop yields have been achieved (Sanchez, 1996). Second, studies of farmers' reaction to and acceptance of hedgerow intercropping have concluded that, in spite of more than a decade of research, development and promotion, adoption of this technology in sub-Saharan Africa is very low (Fig. 4.6; Carter, 1995; Whittome, 1994; Reynolds, 1994).

Is hedgerow intercropping therefore merely a scientific curiosity, destined to become labeled as one more example of scientists' "technology push"? Certainly at the moment the evidence is not encouraging, but it still may well be a technology for the future. Why? It is reasonable to imagine that almost all the required socioeconomic criteria identified here will become more conducive to the adoption of hedgerow intercropping as populations continue to increase, agriculture becomes more intensive and farming and its infrastructural support mechanisms become more developed. But what of the essential improvements required in the biophysical performance of the technology? Ong (1994) developed an equation to describe the component effects of hedgerow intercropping, which in its simplest form can be written as:

$$I = F - C$$

where I is the overall interaction as a percentage increase in yield achieved over sole crops, free from hedge influence; F the percentage crop yield increase due to fertility improvements (including soil physical improvements and changes in microclimate) from the hedge prunings and roots; and C the percentage decrease in yield due to crop competition with trees for light, water, and nutrients. The challenge is clear: can we maximize F (i.e., increase biomass production and its efficiency as a green manure) and minimize the competition effects (C)?

There are many published examples of contrasting pruning regimes, biomass management strategies and species comparisons, all studied with the objective of optimizing biomass production and its efficiency of use. Very positive results have been reported and yet in many cases the crucial issues of design criteria cast some doubt on the validity of the results. For example, apparently useful modifications to pruning time and biomass management were assessed in a trial at Chitedze Research Station, near Lilongwe in Malawi, which receives an average of 900 mm unimodal rainfall per season (Table 4.1). At the time of the third pruning during the cropping cycle (a time at which the crop is too mature to benefit from additional nutrients), the hedge prunings were sun-dried and stored and then applied as a handful of leaves buried around each plant in the next season 1–2 weeks after maize planting. This practice conserved about two-thirds of the annual biomass produced and utilized it in the most temporally and spatially efficient manner possible. There is plenty of evidence (see Table 4.5) to show that the greatest responses of maize to green manure are achieved when the green manure is incorporated into the soil near the time of maize planting. When this practice was combined with increasingly timely hedge pruning to minimize hedge/tree competition, substantial responses to hedgerow intercropping were achieved in 1991/92, a drought year when competition for moisture was likely to be severe (Table 4.1). It is interesting to note that

TABLE 4.1 Maize grain yields (t ha^{-1}) in an alley cropping hedge screening trial at Chitedze Research Station, Malawi.

	1989/90[a]	1990/91[b]	1991/92[c]
Sole crop maize			
0 kg N ha^{-1}	1.11	0.82	0.40
100 kg N ha^{-1}	2.08	4.53	1.93
Maize + *Senna spectabilis* hedge			
Rows next to hedge	0.35	0.89	2.66
Center rows	1.11	1.02	2.48
Mean	0.73	0.96	2.57
Maize + *Gliricidia sepium* hedge			
Rows next to hedge	1.12	1.01	1.37
Center rows	1.48	0.96	0.98
Mean	1.30	0.99	1.18

[a]*The trial was established in the 1987/88 season; in the 1989/90 season there were two primings, the first at 4 weeks after planting (WAP), the second after harvest.*
[b]*In the 1990/91 season, the first pruning was three WAP and the second after harvest.*
[c]*In the 1991/92 season there were three prunings: one WAP, midseason, and after harvest.*
Source: From Bundersen (1992)

FIGURE 4.7 When tree pruning management, spatial arrangement and efficient use of green manure are all optimized, mixed intercropping of *Gliricidia sepium* in maize in Malawi results in large yield increases. *Photo M. R. Rao.*

as the hedge pruning regime improved between 1989 and 1992, the competition effects on the maize adjacent to the hedge declined. It is also impressive that maize yields in the *Senna spectabilis* hedgerow system were considerably greater than those of maize in the control plot, which received 100 kg N ha^{-1}. However, in this trial the control treatments were separated from the "hedge" treatments by only four guard rows of maize (3.6 m). C. K. Ong (personal communication) has found that in the first year of establishment, senna roots spread 4 m laterally at a semiarid location in Kenya; thus it seems highly probable that they invaded and mined the control plot at Chitedze after 4 years.

However, other interesting evidence is also provided by a second trial in Malawi at Makoka Research Station near Zomba (Maghembe, 1994), which confirmed that minimizing tree-crop competition coupled with efficient management of the prunings of spatially mixed trees enhances crop yields. In this trial, *Gliricidia sepium* was established in a maize crop at a similar density to that commonly used in hedgerows, but at a dispersed spacing of 0.9 × 1.5 m rather than as a hedge. During the growing season, competition was minimized by cutting back the gliricidia to a height of 10 cm at the time of land preparation in October and twice more during the growing season in December and February. All the biomass was incorporated into the ridges on which the maize was grown (Fig. 4.7). The trial was established in 1992 and

TABLE 4.2 Maize grain yields (t ha^{-1}) in sole maize and hedgerow intercropping plots at Yaoundé, Cameroon.

	1990		1991		1992		1993		1994	
	Season		Season		Season		Season		Season	
	1	2	1	2	1	2	1	2	1	2
Sole maize	1.52	NF[a]	2.98	NF	3.54	NF	2.54	NF	2.17	NF
Hedgerow intercropping	2.13	TF[b]	3.70	TF	4.79	TF	5.09	TF	4.55	TF
SED	0.38	–	0.29	–	0.35	–	0.44	–	0.14	–

[a]NF, *natural fallow*.
[b]TF, *tree fallow*.
Source: From Duguma, B., Mollet, M., Tiki Manga, T., 1994. Annual Progress Report. Institute of Agronomic Research (IRA), Cameroon and International Center for Research in Agroforestry (ICRAF) AFRENA Report.

the cycle of management was repeated in 1993 and 1994. Root barriers were installed at the start of each season to help eliminate belowground interference between plots.

No major effects were observed in the season of establishment but in the second season, 1993, the maize demonstrated a substantial response to the accumulated additions of green manure, giving 2.5 t grain ha^{-1} compared with 1.2 t ha^{-1} from sole maize control plots receiving no N or P and 2.4 t ha^{-1} from plots receiving 50% of the recommended rates of N and P. In addition, 2.4 t wood ha^{-1} was produced, currently valued at US\$25 t^{-1}. Very similar results were obtained in the third season, 1994. Here is early but solid evidence that minimizing competition and optimizing the efficiency of use of biomass can give positive benefits in spatially mixed systems. This system is now being evaluated with farmers in Zambia.

Other modifications to hedgerow intercropping have also given positive results, in particular the inclusion of short fallow periods during which hedgerows are allowed to grow unchecked and cropping between the hedges is temporarily halted. Reynolds (1994) recently concluded that "alley farming cannot maintain soil fertility without external inputs, or the inclusion of fallow periods." A good example of this has been observed from a trial, free of known design faults, at Yaoundé in Cameroon which receives a bimodal rainfall of 1600 mm a^{-1} (Duguma et al., 1994). Mixtures of *Leucaena leucocephala* and *G. sepium* were established in 1988 as hedgerows 4 m apart in large plots of 11 × 16 m. Maize was grown in one season each year only, in both hedgerow plots and control plots of sole unfertilized maize. The hedges were pruned back at the time of maize planting and twice more during the growing season to minimize competition for light. The biomass was incorporated into the soil. In the second season of each year, plots were left either as a tree fallow in hedgerow plots or as a natural fallow in control plots (Table 4.2). The inclusion of a natural fallow period in the control treatment resulted in only a slow decline in maize yield on this moderately fertile Oxisol. However, the inclusion of a tree fallow period in the hedgerow intercropping plot has maintained yields at a high level. Soil analyses (0–15 cm) undertaken in 1994 showed significantly higher organic matter, total N and pH levels in the hedgerow plots. Here again is evidence that a modified hedgerow intercropping system can sustain maize yields at high levels over time. Similar results have been reported by Jabbar et al. (1994) from a bimodal rainfall location (1250 mm a^{-1}) near Ibadan in Nigeria.

A third success story for hedgerows has been their use in contour plantings on moderately and steeply sloping land for soil erosion control. Here the evidence of benefits is concrete, as discussed further in the section on soil conservation.

In conclusion, as far as hedgerow intercropping is concerned, we suggest that on relatively fertile soils with sufficient rainfall there is emerging evidence that the performance of this technology can be made sufficiently impressive for it to be of interest to farmers. It is important to remember that much of the research conducted on hedgerow intercropping has focused on the nitrogen response of the system. As such, other major plant nutrients were often applied as a blanket dressing. Even under conditions where the system maintains nitrogen levels, it will almost certainly be necessary to add other nutrients, especially phosphorus, for the long-term sustainability of system production.

We also believe that, as agriculture intensifies in such high potential conditions, variants of hedgerow intercropping will become increasingly acceptable through their integration with other systems. In conditions where moisture is limiting, and on inherently infertile and strongly acid soils, the evidence suggests that this is an inappropriate technology.

Improved Fallows

As indicated earlier, trees are a component of naturally regenerating fallows in many traditional shifting cultivation systems throughout Africa and the rest of the tropics. Farmers in many humid tropical regions have recognized certain tree species as being associated with improved soil fertility and therefore deliberately promote such species in fallows either by planting them toward the end of the cropping phase or by selectively controlling those species that compete with the preferred species. Such tree fallows are known as enriched fallows. In southeastern Nigeria farmers encourage species such as *Dactyladenia (Acioa) barteri*, *Alchornea cordifolia*, *Anthonata macrophylla*, *Crestis ferrugina*, *Dialium guineense*, and *Harungana madagascariensis* in the fallow period (Kang et al., 1990). Similarly, farmers in Latin America are known to encourage certain economic and soil improving indigenous trees such as *Stryphnodendron excelsum*, *Dalbergia tucurensis*, *Dipteryx panamensis*, *Vochysia ferruginea*, and *Tabebuia rosea* (Montagnini and Sancho, 1990).

With increasing population pressure the inevitable need for shortening the fallow period and increasing cropping intensity has been well recognized. However, 3–6 years of natural tree fallows cannot maintain soil fertility at similar levels to those achieved in traditional shifting cultivation systems (Aweto et al., 1992). Because of this, a concept of improved fallows is now being pursued which involves the planting of fast-growing, nitrogen-fixing, and deep-rooting trees to enhance and maintain soil fertility over a shorter span of time.

Unlike spatial mixtures such as hedgerow intercropping, there is no competition between the tree and the crop component in improved fallow systems (except between adjacent fallow and cropland, which could be important on small farms) as they follow one another in temporal sequence. However, the same principles of experimental design as highlighted in the previous section are still important. Above- and belowground "mining" of resources can still occur in experiments where tree and crop plots (controls) are adjacent. It will also occur when fallows of different lengths are being compared and the fallow treatments are incorrectly phased. For example, crops following a 1-year fallow will suffer interference from adjacent 2- or 3-year fallow plots still in the ground (Rao et al., 1990).

Research on improved fallows in Africa has not been as extensive as that on hedgerow intercropping, but it is gathering momentum and impressive results are being obtained. A good example of such research comes from the Miombo ecozone of southern Africa where shifting cultivation systems such as *Chitemene* are no longer sustainable. If tree crowns are lopped, 20–30 years' regeneration is required to sustain the *Chitemene* type of shifting cultivation in northern Zambia (Mansfield et al., 1975). If trees are coppiced, Stomgaard (1985) estimated that 43 years are required for miombo regeneration. However, recent reports indicate that the commonly practiced fallow period in northern Zambia is now less than 10 years (Kwesiga and Kamau, 1989) and on the plateau in eastern Zambia bush and grass fallows of 1–5 years are common (Kwesiga and Chisumpa, 1990).

Early research with improved fallows at Msekera Research Station at Chipata in Zambia, which receives between 800 and 1000 mm unimodal rainfall, has been encouraging but suffered from design faults (Kwesiga and Coe, 1994). Two-year fallows of *Sesbania sesban*, planted at a density of 10,000 trees ha^{-1}, resulted in large increases in yields compared with unfertilized maize in the first 2 years after fallow clearance. Yields of 5.0 and 5.6 t ha^{-1} were obtained in 1990 and 1991 compared with 4.9 and 4.3 and 1.2 and 1.9 t ha^{-1} from continuously cropped maize with (112 kg N ha^{-1}) and without fertilizer, respectively. In addition, approximately 10 t fuelwood ha^{-1} were harvested at fallow clearance (Fig. 4.8). However, as already indicated, incorrect phasing in the experimental design and small plot size almost

FIGURE 4.8 Substantial litter accumulates under *Sesbania sesban* fallows in Zambia which, when incorporated into the soil, results in high maize yields. Between 10 and 15 t ha^{-1} of fuel wood is also produced. *Photo F. Kwesiga.*

TABLE 4.3 The effect of 1- and 2-year fallows of different species on maize yield in the first year after clearance at Chipata, Zambia, 1994.

	Maize yield (t ha^{-1})	
	Grain	Stover
Sesbania sesban		
1-year fallow	3.43	4.60
2-year fallow	5.47	7.01
Tephrosia vogelii		
1-year fallow	2.80	3.49
2-year fallow	3.65	5.34
Sesbania macrantha		
1-year fallow	2.08	4.56
2-year fallow	3.00	4.22
Cajanus cajan		
1-year fallow	2.40	4.05
2-year fallow	2.78	4.39
Grass fallow		
2-year fallow	2.12	3.69
Continuous maize		
With fertilizer	3.40	5.09
Without fertilizer	1.09	2.33
Groundnut/maize rotation	2.21	3.52
SED	0.44	0.63

certainly resulted in above- and belowground interference between tree and crop plots in this trial (Kwesiga and Coe, 1994). Research on improved fallows in Zambia has subsequently expanded, with the use of larger plots, root barriers or trenches surrounding plots, and properly phased designs. Furthermore, a wider range of potential fallow species are now being evaluated as 1-, 2- and 3-year fallows.

S. sesban continues to show its superiority as a 1- or 2-year fallow, and *Cajanus cajan* and *Tephrosia vogelii* are also showing promise (Table 4.3). These species are currently being jointly evaluated with over 150 farmers in eastern Zambia, and increasingly farmers are beginning to establish spontaneously their own fallow experiments with sesbania. Farmers who have already harvested their first crop of maize after fallow clearance are observing similar yield increases to those reported in Table 4.3 (Fig. 4.9). Similar maize responses have been observed on acid Ultisols in Onne, Nigeria (Gichuru, 1991) where a 2-year fallow of *Tephrosia candida* gave maize yields of 2.46 t ha^{-1} compared with 0.95 t ha^{-1} following a natural bush fallow. On the basis of a survey of farms under *G. sepium* fallow in southern Nigeria, Adejuwon and Adesina (1990) concluded that soil fertility improved progressively with the length of fallow. The most significant improvements were in terms of organic matter, nitrate-nitrogen, and potassium. Unfortunately, no test crop data from this study are available.

Such improved fallows clearly work and, as is the case in Zambia, farmers are interested and wish to try them. Initially it was thought that this practice would be most relevant to farming systems where land is not a constraint and farmers are setting, or could afford to set, land aside for fallowing. However, recent surveys in western Kenya, in a high population density zone (300−1000 people km^{-1} with a bimodal rainfall of 1800 mm a^{-1}, have shown that nearly 50% of farmers with farms of between 0.5 and 5.0 ha leave their land under fallow for one or two seasons). Severe land depletion, with maize yields falling below 0.5 t ha^{-1}, and an inability to purchase inputs, are the principal causes, but insufficient labor to cultivate crops was also cited as a reason (ICRAF, 1994, p. 27). These farmers have expressed interest in short-duration improved fallows for both soil fertility improvement and wood production.

FIGURE 4.9 ICRAF is evaluating sesbania fallows with over 200 farmers in Zambia. This farmer harvested an equivalent of 6.9 t ha^{-1} of maize following 2 years of sesbania (on the left) compared with only 2.9 t ha^{-1} following a grass fallow (on the right). *Photo A. J. Simons.*

FIGURE 4.10 In Malawi, relay planted sesbania grows rapidly in the dry season after the maize crop matures. The incorporation of sesbania litter into the soil at land preparation results in improved yields of the subsequent maize crop. *Photo A. Njenga.*

In the Shire highlands of Malawi, a unimodal rainfall zone, farm sizes are even smaller, ranging from 0.2–0.5 ha, and preclude the setting aside of land for fallow. As expected, soil fertility is declining, with many farmers unable to afford fertilizer. Many farmers currently intercrop (or relay crop) their maize with *C. cajan*, both as a source of income from the local sale of seed and also as a means of adding organic nitrogen through above- and belowground litter. Cajanus is known to fix about 90% of its nitrogen requirements from the atmosphere (Kumar Rao et al., 1987) and other studies (Poth et al., 1986) have shown residual benefits to subsequent maize crops equivalent to 40 kg N ha^{-1}. An innovative form of dry season fallow (relay cropping) is being investigated in which no land is taken out of maize production. *S. sesban* is established as seedlings at the same time as maize, which is planted and managed according to the recommended practice. Sesbania is planted at a spacing of 90 × 150 cm and maize at 30 × 75 cm. Superimposed are three levels of fertilizer application to the maize, 0%, 50% and 100% of the recommended amount for Malawi (14 kg N ha^{-1} and 10 kg P ha^{-1} at planting, plus 42 kg N ha^{-1} as a top dressing) (ICRAF, 1995).

After the harvest of maize, the sesbania is allowed to continue growing throughout the dry season on residual moisture (Fig. 4.10). Prior to the sowing of maize in the rainy season of the following year, the sesbania is harvested and the leaves, pods, and small branches incorporated into the soil. Material suitable for fuelwood is separated and removed. This cycle has been repeated on the same plots for five consecutive seasons (Table 4.4).

In the season of establishment (1989/90), as expected, the relay cropped plots within any given fertilizer subtreatment showed no benefit of the technology, although significant responses to fertilizer were found and continued to be observed throughout the life of the trial. In the second season, the effect of incorporating sesbania residue into the soil prior to maize sowing gave significant increases in the unfertilized maize subtreatment, and this effect continued to be observed in subsequent seasons, both in the unfertilized maize crop and in maize receiving 50% of the recommended

TABLE 4.4 Maize yields (t ha^{-1}) obtained in a relay cropping system with *Sesbania sesban* at Makoka, Malawi.

| | Fertilizer (% of recommended amounts for Malawi) | | | | | | |
| | 0 | | 50 | | 100 | | |
	Sole maize	Relay cropping	Sole maize	Relay cropping	Sole maize	Relay cropping	SED[a]
1989/90	1.00	1.02	3.75	3.27	4.48	3.77	0.40
1990/91	1.33	2.41*	6.21	6.13	7.12	7.84	0.60
1991/92	0.67	2.57**	2.99	3.75*	3.47	4.32	0.38
1992/93	1.02	2.33**	4.79	5.99*	6.06	7.40**	0.51
1993/94	0.51	1.16*	2.41	3.29**	3.23	3.49	0.36

[a]SED provided is for comparison of yield differences between sole and relay cropped maize within fertilizer levels.
*Significant difference at 10%. **Significant difference at 5%.
Source: From Maghembe, J.A., 1994. Out of the forest: indigenous fruit trees in southern Africa. Agrofor. Today 6 (2), 4–6.

rate of fertilizer. A similar trend was also observed in the maize receiving 100% of the recommended rate of fertilizer but the effect was only significant in the climatically favorable season of 1992/93, when rainfall was high and well distributed.

Fuelwood is in very short supply in the Shire Highlands and is currently valued at US$25 t^{-1}. The added benefit of the fuelwood produced from the dry season sesbania fallow (mean of 2.08 t ha^{-1} across treatments and years) has aroused considerable interest among farmers who have seen the performance of this technology. Again, however, there is some concern that the small plot size used in this experiment may have resulted in unrepresentative exploitation of the belowground soil resources during the dry season. Thus in the 1994/95 season, relay cropping with *S. sesban* is being evaluated in large plots directly with farmers in the Zomba district.

In conclusion, with respect to improved fallows, various forms appear to have considerable potential as a strategy to restore soil fertility and improve crop yields, and, as far as farmers are concerned, do not suffer the complexity of managing tree-crop competition. In addition, short-term fallows have the potential to produce useful quantities of firewood which appears to be attractive for farmers in many areas (see Fig. 4.1). Although yet to be proven, the technology may be widely acceptable since it builds on a widespread tradition and knowledge of fallowing as currently practiced by farmers when crop yields fall below a certain economic threshold.

Appropriate fallow tree species have been identified for a wide range of soil and climatic conditions (Rao, 1994). This is important because care must be taken not to overpromote a single species, such as *S. sesban*, which is known to be susceptible to attack by the Mesoplatys beetle. In addition, sesbania can lead to a build-up of soil nematode populations, so is unsuitable for use prior to susceptible cash crops such as tobacco.

Research on improved fallows is likely to expand in coming years and it is essential that new experiments pay careful attention to experimental design issues (Rao et al., 1990; Coe, 1994) in order to avoid the pitfalls which may have befallen much of the previous research on hedgerow intercropping. It is also important that long-term trials be established to assess how sustainable such improved fallows are over a longer period. Nitrogen-fixing trees will certainly maintain soil nitrogen at high levels, but other major plant nutrients such as phosphorus may well become limiting after several cycles of fallowing.

As with any intervention, the adoption of improved fallows will depend upon farmers' perception of the benefits, particularly when additional labor is required in fallow establishment and management. Clearly, farmers recognize an economic threshold below which they do not consider cropping to be worthwhile and this is the underlying cause for land abandonment and natural fallows. Research must demonstrate that the returns to labor for improved fallows are superior to those achieved through natural fallows. Such analysis (ICRAF, 1994, p. 111–114) demonstrated substantially greater returns to labor when a 2-year *S. sesban* fallow was compared with continuous maize cultivation over a 6-year period. Data for a more appropriate comparison with a natural fallow were not available in the data set studied. Long-term trials with appropriate controls are required to address this issue but ultimately it will be the farmers who decide. For this reason, participatory research with farmers, as is currently being undertaken in Zambia, will be the most relevant approach.

TABLE 4.5 The effect of species, method of application and time of application of mulch on maize yields at Chipata, Zambia, 1993 (main effects from a factorial trial).

	Maize yield (t ha^{-1})
Species	
Senna siamea	2.89
Piliostigma thonningii	1.80
SED	0.13
Method of application	
Applied on surface	1.60
Incorporated in soil	3.02
SED	0.13
Time of application	
14 days before planting	2.79
At planting	2.57
14 days after planting	1.84
28 days after planting	2.18
SED	0.18

Source: From ICRAF, 1994. Annual Report 1993. International Center for Research in Agroforestry, Nairobi, Kenya, p. 118.

Biomass Transfer

The transfer of tree or shrub biomass from distant areas to food crop production fields to maintain soil fertility and crop yields is a well-known practice in certain countries. For instance, such a practice has been used for decades in Asia for irrigated rice production. *Gliricidia maculata* is grown on the paddy field bunds and incorporated into the soil before rice transplanting (Singh et al., 1991). In southern Africa, collection and use of miombo woodland litter as a source of plant nutrients is a common practice. Farmers use litter directly as green manure or they dry and store it for later use, especially if the material is collected during the noncropping period when the demand for labor is low. In addition, it may be cured in cattle pens and mixed with manure (Nyathi and Campbell, 1993). The traditional *Chitemene* and *fundikila* shifting cultivation systems in northern Zambia and its surroundings also heavily depend on exploiting miombo litter (Matthews et al., 1992). Recent research at Domboshawa in Zimbabwe has demonstrated that high-quality litter gives substantially greater maize yields than miombo leaf litter (Dzowela et al., 1994). Yields of 5.7 and 5.6 t ha^{-1} were obtained following the incorporation of 5 t litter ha^{-1} from *L. leucocephala* and *C. cajan*, respectively, at planting compared with only 2.5 t ha^{-1} achieved through the incorporation of miombo litter at the same rate.

Several factors affect the response of crops to the application of transferred biomass and have been reviewed by Rao (1994). Major factors are the chemical composition of the litter of different species and the method and timing of application. The impact of these factors is illustrated by results obtained in Zambia (Table 4.5). These results confirm that incorporation of the litter into the soil close to the time of maize planting produces the greatest maize yield response. However, with species that decompose rapidly (such as gliricidia and sesbania) the timing is less critical and incorporation may be delayed for 3 or 4 weeks after planting (Read et al., 1985). Palm (1996), from an extensive review of the literature, concluded that, as a general guideline, if the concentration of nitrogen in the green manure is less than 1.74%, then net nitrogen immobilization will occur and will possibly continue for the bulk of the cropping season. If the concentration of nitrogen is greater than 1.74%, then net mineralization will occur, but will decrease as a function of the ratio of lignins + polyphenols to nitrogen.

Many studies have, however, confirmed that the efficiency of organic nitrogen is lower than that of inorganic nitrogen (Gutteridge, 1992; Read et al., 1985). Several reasons seem to be evident, such as slow decomposition of residues, lack of synchrony in the mineralization of nutrients and crop uptake, unfavorable soil conditions (moisture, temperature, biological activity), and loss of nutrients through leaching, volatilization, and immobilization (Palm, 1996). The need for synchrony between nutrient mineralization and crop demand increases with rainfall and leaching intensity and shallow-rooted crops, especially on acid soils in the humid tropics (van Noordwijk et al., 1991). The less efficient use of organic nitrogen is typified by results obtained from a trial conducted over a 4-year period at Chipata in Zambia. Five rates of inorganic nitrogen (0, 30, 60, 90, and 120 kg ha^{-1}) and organic nitrogen from *G. sepium* or *L. leucocephala* prunings (0, 5, 10, 15, and 20 t fresh weight ha^{-1}) were applied in factorial combination on the same plots for four seasons. Basal dressings of phosphorus and potassium were supplied to maize each year to focus on the nitrogen response. Consistent and linear responses of the maize to the different rates of both sources of nitrogen were observed in all seasons, with the response in any given season always being greater to inorganic nitrogen. The results are typified by the regression equations obtained in the 1989/90 season (ICRAF, 1994, pp. 114−117). For gliricidia:

$$Y = 1844 + 16.0\, N_i + 10.1\, N_o (r^2 = 0.96)$$

and for leucaena:

$$Y = 1922 + 16.7\, N_i + 5.1\, N_o (r^2 = 0.96)$$

where Y is maize yield (kg ha^{-1}) and N_i and N_o are kg added nitrogen ha^{-1} in inorganic and organic form, respectively.

These results confirm that nitrogen from gliricidia litter is used more efficiently than that from leucaena. Several researchers have shown that nitrogen release from gliricidia is rapid compared with leucaena (e.g., Constantinides and Fownes, 1994).

However, we must emphasize that in this trial, as in many others, the response to the different sources of nitrogen application was measured on an annual basis. There is a need to examine residual effects in more detail to assess the longer-term relative efficiencies of organic and inorganic nitrogen. It has also been observed that the combination of organic and inorganic nitrogen fertilizers improves the efficiency of nitrogen uptake. Soil microbial activity may be increased and nitrogen capture by the microbial pool from inorganic sources may be improved when fresh organic matter, which can supply soluble carbon as a source of energy, is used. This would reduce leaching and volatization losses, and enhance nitrogen cycling rates and thus nitrogen availability (Buchanan and King, 1992; Snapp, 1995).

There are several other key research issues that remain to be addressed. First, how feasible will it be to find suitable niches on-farm to establish high-quality litter banks? A rough generalization is that approximately 0.5 ha litter bank of an adapted tree species would be required to produce useful amounts of biomass (around 5 t dry matter ha^{-1}) for 1.0 ha of cropped land. Second, how sustainable would a litter bank be before nutrient mining of the soil reduced its productivity? Last, the labor requirements and the opportunity cost of the land set aside for litter banks need to be carefully assessed. Research in Zimbabwe and Kenya (Dzowela et al., 1994; A. Niang, ICRAF, personal communication) is currently addressing these issues.

SOIL CONSERVATION

As populations expand, both in numbers and in terms of area settled, more marginal and sloping land inevitably comes under crop production. The consequence of farming fragile soils on sloping land in areas receiving intense rainfall is as inevitable as it has been damaging in the humid and subhumid tropics of Africa. Even in 1938 Hailey concluded that water erosion of soils "is now one of the most serious problems of Africa." Nearly 60 years later examples of the impact of Hailey's timely warning can still be seen in the widespread existence of contour bunds in Zimbabwe, the commonly practiced contour ridging in Malawi and the bench terraces of southwestern Uganda (Fig. 4.11). Since that time, researchers and farmers have been acutely aware of the continuing gravity of the problem (Young, 1989; Kiepe and Rao, 1994) and yet mechanical erosion control measures, involving some form of earth movement, are seldom adopted by small scale farmers. Without incentives, as often provided in the developed world, farmers do not perceive the short-term benefits achieved as being worth the cost of installation and maintenance (Mwakalogho, 1986). Young (1989), after an extensive review, concluded that "conservation is likely to be most effective where it is conducted with the active cooperation of farmers, in their perceived interests, and integrated with other measures for agricultural improvement." Agroforestry can achieve this through its capacity to combine short-term production with longer-term soil conservation functions.

Three broadly defined agroforestry systems have potential in the humid and subhumid tropics of Africa: barrier hedges planted on the contour, tree/shrub combinations to stabilize existing conservation structures and multistrata

FIGURE 4.11 Gradients in sorghum yields are common across scoured bench terraces in southwest Uganda. Fast-growing leguminous trees can restore fertility in the upper parts of the terrace, and provide much-needed fuelwood. *Photo R. B. Leakey.*

systems that simulate natural forests and woodlands. In the semiarid tropics of Africa, linear plantings of trees as windbreaks have shown considerable potential to reduce wind erosion but fall outside the scope of this review. Their potential is reviewed elsewhere (FAO, 1986; Kiepe and Rao, 1994).

Barrier Hedges

Barrier hedges, planted on the contour line, have been widely evaluated throughout the humid and subhumid tropics. Optimum spacing between hedges is dependent on the degree of slope but within-hedge spacing is commonly between 10 and 25 cm. Hedges are pruned frequently to minimize competition effects with adjacent crops in a similar manner to hedgerow intercropping. Since erosion control is the principal objective, hedgerow prunings are usually applied as a surface mulch to provide added soil protection, a factor which can be critical during the early growth stages of crops. Long-term and properly designed trials, in which surface water run-off and soil erosion losses from land under barrier hedges are actually recorded, are not frequently reported in the literature, but the results that are available are encouraging and often dramatic (Table 4.6). Hedges spaced 4−5 m apart appear effective at controlling erosion on moderately sloping land (10−25%) regardless of the annual rainfall or soil type. Other results, however (Anecksamphant et al., 1990; Nyamulinda, 1991), have shown that on severely sloping land (35−60%) in Thailand and Rwanda such a spacing is too great to control soil loss effectively. In contrast, on a slope of 44% at Ntcheu in Malawi Banda et al. (1994) demonstrated that a spacing of 0.9 m almost completely eliminated soil loss (Table 4.6) and also maintained maize yields at about 2 t ha^{-1} compared with the control where yields declined to 0.2 t ha^{-1} over 6 years.

However, this dramatic impact of erosion control on crop yields is not noted in the other trials reported in Table 4.6. Njoroge and Rao (1994) concluded that short-term yield increases cannot be expected where the soils are deep and fertile and soil erosion losses are only moderately high. Positive crop responses are more common in the short term on steep slopes with shallow soils which experience high-rainfall and where soil loss and hence nutrient depletion rates are rapid.

Experiments conducted in the Philippines and Indonesia through the Sloping Lands Network of the International Board for Soil Research and Management (IBSRAM) demonstrated substantial declines in crop yields under farmers' management where no soil conservation was practiced. In contrast, barrier hedgerows maintained yields at about twice those achieved by farmers' conventional practice over a 3-year period (IBSRAM, 1994).

In many instances soil loss measurements, such as those reported in Table 4.6, have not been recorded but evidence for soil conservation is inferred through the rapid accumulation of soil on the upslope side of barrier hedges, resulting in the natural formation of "biological terraces" (Fig. 4.12). For example, a drop of between 0.5 and 0.6 m in soil levels between the upslope and downslope side of barrier hedges was reported in the Philippines on slopes of 20−30% after only 3 years (Maclean et al., 1992; Garrity, 1994). Such trapping of soil and the progressive formation of terraces result in the reduction of slope length and angle. Although the natural formation of "biological terraces" is a positive attribute of barrier hedges, it occurs because soil movement is still taking place between hedgerows. The displacement of soil and nutrients from the upper parts of the alley, and their deposition on the lower portion results in rapid formation of a soil fertility gradient across the terrace. Many researchers (Bannister and Nair, 1990; Maclean et al., 1992, Solera, 1993; Garrity, 1994) have observed the rapid development of this phenomenon, and an associated gradient of crop yields. Typically,

TABLE 4.6 Soil loss and surface water run-off from land under barrier.

	Rainfall (mm a^{-1})	Slope (%)	Soil type (USDA)	Hedge spacing (m)	Period reported (years)	Mean soil loss (t ha^{-1} a^{-1})		Mean run-off (mm a^{-1})		Hedge species
						Barrier hedge	Control	Barrier hedge	Control	
Yurimaguam, Peru[1]	2200	16	Paleudult	4	4.5	6	79	60	347	*Inga edulis*
Citayan, Indonesia[2]	2044	12	Haplusdox	4	1	11	103	56	232	*Flemingia congesta*
Los Banos, Philippines[3]	2074	17	Trapudult	5	1	3	127	75	347	*Desmanthus virgalus*
Ibadan, Nigeria[4]	1358	7	Paleustalf	4	6	<1	5	72	252	*Leucaena leucocephala*
Butare, Rwanda[5]	1279	28	Hapludox	5	4	7	303	28	111	*Calliandra calothyrsus*
				10	4	4	303	35	111	*Leucaena leucocephala*
Ntcheu, Malawi[6]	1125	44	UStropept	0.9	6	2	44	na	na	*Leucaena leucocephala*
Machakos, Kenya[7]	750	14	Rhodustalf	4	4	<1	19	4	21	*Senna siamea*
			Rhodustalf	4	4	3	19	6	21	*Leucaena leucocephala*

Source: Adapted from Njoroge, M., Rao, M.R., 1994. Barrier hedgerow intercropping for soil and water conservation on sloping lands. Paper Presented at the 8th International Soil Conservation Organization Conference. Soil and Water Conservation: Challenges and Opportunities. 4–8 December, 1994, New Delhi, India. From [1]ICRAF, 1994. Annual Report 1993. Nairobi, Kenya: International Center for Research in Agroforestry, [2]Hawkins et al. (1990), [3]Paningbatan (1990), [4]Lal (1989), [5]Konig (1992), [6]Banda et al. (1994), [7]ICRAF, 1993. ICRAF: The Way Ahead. Strategic Plan. ICRAF, Nairobi, Kenya.

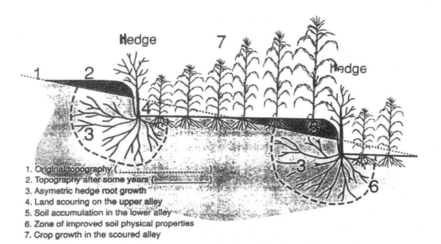

1. Original topography (............................)
2. Topography after some years (...............)
3. Asymetric hedge root growth
4. Land scouring on the upper alley
5. Soil accumulation in the lower alley
6. Zone of improved soil physical properties
7. Crop growth in the scoured alley

FIGURE 4.12   A schematic diagram of barrier hedgerow technology showing various processes and their effects on crop growth.

FIGURE 4.13   Barrier hedgerows planted on contours have been successfully adopted in many parts of the world, as here in the Philippines. Adoption in Africa has been less widespread. *Photo A. Young.*

crop yields are 50−75% less on the scoured upper parts of the terrace than on the lower parts. Such "terrace scouring" is not restricted to barrier hedges. In southwest Uganda, near Kabale, a farm survey was undertaken of sorghum yields across bench terraces constructed 30 years ago on slopes of around 60%. It was observed that terrace scouring was widespread and, averaged across all farms surveyed, sorghum yields were 0.75, 1.10, and 2.85 t ha^{-1} for the upper, middle, and lower parts of the terrace, respectively (ICRAF, 1994, p. 101) (Fig. 4.11). Little research has been reported that examines ways to prevent or cure the impact of such terrace scouring. Some work is underway in the Philippines which suggests that a combination of contour ridge land preparation and barrier hedges may be effective as a preventative measure (ICRAF, 1995) and early results from southwest Uganda indicate that leguminous trees, planted as fodder banks or woodlots, could play an important role in reclaiming the upper portions of scoured terraces (ICRAF, 1994, pp. 104−106). However, the issue is largely underresearched and remains a priority for future research (Sanchez, 1996).

Little detailed research has been undertaken of the processes leading to the effectiveness of barrier hedges but from what evidence is available. Njoroge and Rao (1994) concluded that several probable mechanisms are involved: hedgerows are semipermeable and allow the disposal of excess water without ponding and stagnation; hedge roots stabilize terraces as they form, and improve infiltration rates immediately adjacent to the hedge; surface mulch, when applied, reduces rainfall impact and soil erodability, improves soil fertility and soil physical conditions, and also conserves moisture by reducing soil evaporative loss during early crop growth (Fig. 4.12).

Barrier hedges are not, however, a new concept and have been adopted by farmers in several parts of the world. Kerkhoven first advocated their use in 1913 and widespread planting occurred in the 1930s and 1940s in Indonesia (Metzner, 1976), and more recently in the Philippines and Haiti (Tacio, 1991; Pelleck, 1992) (Fig. 4.13). Currently, adoption in sub-Saharan Africa is low, although the practice is widely recognized as being effective, and is being actively promoted by several development schemes, e.g., in Malawi (Franks, 1992).

In recent adoption studies in the Philippines it was found that farmers had modified the technology by replacing shrubs for mulch production with trees and shrubs with a cash value (Fujisaka, 1993). Analogous to this is the widespread practice of contour planting fodder grasses in Rwanda (Niang et al., 1996), Kenya (O'Neill and Muriithi, 1994), and Burundi (Akyeampong, 1996). The message from farmers seems clear: barrier hedges are an acceptable innovation but in the absence of incentives or short-term crop yield advantages they must produce a valued product rather than green manure. This is corroborated by an economic analysis of alternative uses of leucaena hedge prunings (ICRAF, 1993, pp. 27–29). Prunings used as a fodder supplement for milk production gave economic returns between three and seven times those achieved when they were used as green manure on maize.

However, the use of hedge prunings as fodder, or the use of alternative high-value trees or shrubs, begs the question as to what extent erosion control will be reduced if the surface mulch is not applied to the soil between the hedges. Under mildly erosive conditions at Machakos, Kenya, it was found that, when the mulch from *Senna siamea* barrier hedges was not applied, the hedges alone reduced annual soil erosion losses from 36 t ha^{-1} in the control plots to 2 t ha^{-1} in 1990, and from 12 t ha^{-1} to 1.5 t ha^{-1} in 1992 (ICRAF, 1993, p. 63). However, in the same trial in 1994, a highly erosive rainfall of 115 mm occurred in a 24-h period and in this case erosion was only reduced from 46 t ha^{-1} in the control to 17 t ha^{-1} in the "hedge alone" treatment. In the "hedge plus mulch" treatment, soil losses were only 1 t ha^{-1} (C. K. Ong, ICRAF, personal communication). As far as we are aware, studies such as this one are rare. The results indicate that under mildly erosive conditions barrier hedges alone are effective but that, as rainfall intensity (or slope) increases, the added benefit of surface mulching becomes important. However, as noted earlier, it is under such conditions that farmers are likely to observe short-term crop yield benefits (IBSRAM, 1994).

An innovative approach comprising a modification of barrier hedges is being assessed in Mindanao, Philippines, which may well have potential in sub-Saharan Africa (ICRAF, 1994, pp. 140–141). Narrow contour strips of the field area are left uncultivated and allowed to vegetate naturally. The natural vegetative strips capture run-off sediment and form biological terraces in a similar manner to barrier hedges. Farmers in the Philippines have indicated that this low investment form of erosion control is attractive, and that the biological terraces can provide the foundation for agroforestry. They are planting income-generating perennials and fodder species along the terrace risers.

In conclusion, barrier hedges have proved to be effective in controlling soil erosion on gentle to moderate slopes. Under such conditions, hedges alone are sufficient. The adoption of this technology in sub-Saharan Africa could well be enhanced if the hedgerow prunings of leguminous shrubs were used by farmers as a high-quality fodder supplement (see subsequent section) rather than as a surface mulch. Under conditions of more severe soil erosion, the added soil protection of the surface mulch appears important but short-term crop yield benefits which are likely to accrue under such conditions may be a sufficient incentive to farmers. Alternatively, even on steep slopes the application of surface mulch may be avoided if other proven soil conservation practices such as contour ridging and the surface application of crop residues are utilized (Young, 1989, pp. 35–39). Even though such practices will control soil loss from the slope as a whole, it appears that soil movement between barrier hedges is still likely to occur, leading to the terrace scouring effect. More research is required to find ways to prevent this happening and to reclaim degraded soils where scouring has already occurred.

Trees and Shrubs on Conservation Structures

Where conservation structures such as earth bunds, terraces, or conservation grass strips are already in place, trees and shrubs can play an important role in both stabilizing such structures and increasing the diversity of this potential farm niche through the production of fodder, wood, or fruits. Although such niches usually occupy only a small proportion of the land area, they are often fertile due to the accumulation of nutrient-rich soil from terrace scouring.

The inclusion of trees and shrubs on conservation structures is not an erosion control research issue as such since it is the conservation structure itself that is the main focus of research interest. There are nevertheless important research issues which need to be considered. To exploit fully the productivity of conservation structures, the priority needs of farmers must be assessed and potential markets for high-value products ascertained before choices are made from a wide range of possible trees. In identifying appropriate species, spacing and establishment techniques must be known. For example, timber-producing trees with a tall and linear growth habit (i.e., *Grevillea robusta*, *Markhamia lutea*, and *Alnus acuminata*) may be spaced 3–5 m apart but fruit trees such as citrus, mango, avocado, and jackfruit require a spacing of not less than 10 m (Kiepe and Rao, 1994). Where fodder production is a high priority, species such as *Calliandra calothyrsus* and *L. leucocephala* (Akyeampong, 1996; Niang et al., 1996) can be established as closely spaced pure hedges or in combination with existing grass strips. In many instances, farmers may wish to plant a mixture of upper-story and understory trees to provide a range of products. For example, mixtures of upper-story trees for poles (*Casuarina equisetifolia* and *G. robusta*) and understory fodder trees (*C. calothyrsus*) are being evaluated on terrace risers at Kachwakano in southwest Uganda by ICRAF.

FIGURE 4.14 Existing contour bunds in farmers' fields in Zimbabwe are a possible niche to grow dry season fodder for draft oxen. *Photo B. Dzowela.*

TABLE 4.7 Maize production adjacent to fodder bunds planted with different hedgerow species, Dombashawa, Zimbabwe, 1994.

	Maize production (kg (5 m)$^{-1}$)			
	Acacia angustissima	*Leucaena leucocephala*	*Cajanus cajan*	*Calliandra calohyrsus*
Row[a]				
1	0.77	0.93	0.97	0.99
2	1.20	1.43	1.64	1.67
3	1.63	1.56	1.92	2.28
4	2.02	1.70	2.06	2.26
5	2.13	1.84	2.07	2.38
6	2.29	1.95	2.03	2.34
7	2.19	1.97	2.05	2.34
8	2.11	2.29	2.16	2.20
9	2.18	2.19	2.16	2.25
10	2.18	2.38	2.27	2.34
Maize yield loss (t ha^{-1})[b]				
	0.59	0.70	0.41	0.33

[a]*Row 1 adjacent to contour bund, Row 10 furthest away.*
[b]*Maize yield loss was calculated utilizing the mean yields of Rows 7, 8, and 9 as the control, and assuming 500 m bund length ha^{-1}.*
SED for maize yield loss = 0.138.
Source: From Dzowela, B.H., Mafongoya, P.L. Hove, L., 1994. SADC-ICRAF Agroforestry project, Zimbabwe. 1994 Progress Report. ICRAF, Nairobi.

The competition for resources between trees and crops has been discussed in a previous section on hedgerow inter-cropping, and the same principles hold when trees are planted on conservation structures. For example, studies in Zimbabwe (Dzowela et al., 1994) are evaluating the potential of leguminous hedgerows planted on old earth contour bunds to supply dry season fodder for draft oxen. In addition, the impact of these hedges on the performance of maize grown adjacent to these bunds is being evaluated. The results have shown that such bunds (assuming 500 m bund per farm) do have the potential to provide sufficient fodder for the needs of a 300-kg draft animal, but at the cost of lost maize yield due, in this case, to belowground competition for water (Fig. 4.14). There were differences in the competitive effects of the species evaluated and in the case of leucaena this extended for up to five rows (4.5 m) away from the bund (Table 4.7). In another study in Uganda, Okorio et al. (1994) demonstrated similar competitive effects of linear contour plantings of a range of trees being evaluated for pole production. In this case, they demonstrated that both above- and belowground competition for resources were important and for the most competitive species (*Maesopsis eminii*) this effect extended for more than 3 m either side of the linear planting. It therefore appears that on steeply sloping land, where conservation structures are closely spaced, the competition between trees and crops will be more important than on more gently sloping land where earth bunds may be spaced up to 20 m apart, as is typical, for instance, in many small-scale farms in Zimbabwe.

In conclusion, high-value trees and shrubs do have an important role to play in both stabilizing conservation structures and increasing their productivity and diversity of output. However, this will almost always be at some cost in terms of reduced productivity of associated crops. Research priorities are those associated with the determination of local community needs and opportunities, and the identification of appropriate species and their management, rather than the impact of trees and shrubs on soil erosion control per se.

Multistrata Systems

Multistrata agroforestry systems have evolved primarily to generate income and provide and diversify basic family food needs. Their function and productivity in these roles is discussed in a subsequent section (see section on high-value trees). However, because of their structural resemblance to secondary forests, they exhibit a profound capacity to prevent soil erosion (Kiepe and Rao, 1994).

Little research has been undertaken to investigate the detailed mechanisms involved but from available evidence it is clear that it is the understory and litter which play the primary role in soil conservation, and that the upper-story canopy has little, if any, beneficial effect. Raindrops reach 95% of their terminal velocity in 8 m free fall from an upper-story canopy, frequently with increased drop size and greater soil erosivity (Soemwarto, 1987). The relative importance of the different layers in multistrata systems was illustrated by a study of erosion under a 5-year-old *Acacia auriculiformis* stand in Java. The removal of the canopy and understory, but the retention of the litter alone, reduced soil erosion by 95% compared with that experienced from bare soil. The retention of the canopy alone increased the erosive power of raindrops by 24% (Wiersum, 1985). Other evidence for the importance of litter was provided in a comprehensive review by Wiersum (1984). Erosion losses reported in undisturbed forest plantings in the humid and subhumid tropics range from 0.02 to 6.20 t ha^{-1} a^{-1} compared with 5.92 to 104.8 t ha^{-1} a^{-1} where the understory was burned and the litter removed. These values can be compared with erosion in multistory gardens, which ranged from 0.01 to 0.14 t ha^{-1} a^{-1}. Indirect evidence was also recently obtained from Uganda where *C. calothyrsus* and *L. leucocephala* were established at high populations (1 × 0.5 m) on 60% sloping land. After 3 years a dense canopy of about 3 m in height and a thick layer of litter had formed. Observations indicated that, even on such steep slopes, soil erosion had been completely eliminated (ICRAF, 1994, pp. 101−108).

MULTIPURPOSE TREES AND SHRUBS AS FODDER SUPPLEMENTS

Feed Quality

Shortages of high-quality dry season fodder supply have been widely recognized as a constraint to ruminant production in the tropics of Africa, both for small and large ruminants (Fig. 4.4). Crop residues are utilized as fodder and seldom returned to the soil, and animal manure is often used as a fuel. Such practices exacerbate the negative nutrient budgets of crop production fields and contribute substantially to the steady decline in soil organic matter (Fig. 4.5). In addition, natural pastures and crop residues are of low feed quality, and animal condition and performance almost always declines during the dry season. Already a serious constraint to many small-scale farmers, this situation is projected to worsen dramatically in the future as population pressure continues to build.

A recent extensive review of animal agriculture in sub-Saharan Africa concluded that dwindling feed resources will be the greatest constraint to the productivity of an expanding livestock population (Winrock International, 1992). It is forecast that unless feed production trends improve dramatically, the current 140,000 t meat imported annually into the region (costing US$200 × 10^6) will rise from the present level of 3% of the total requirements to about 35% of the region's needs, or 7 × 10^6 t meat by the year 2025. Similarly, milk product imports are predicted to rise from the current level of 11% of total requirements (costing US$500 × 10^6 annually) to about 25% by the year 2025. The combined cost of these projected imports of meat and milk products is estimated at a staggering US$16 × 10^9 annually, beyond the realms of economic feasibility.

Given the inevitable decline in availability of already poor quality grazing areas and increased pressure on-farm feed resources, a major research and development initiative is required to improve the supply of quality feed at the farm level. Without such an initiative, the projected pressure on the natural resource base to provide livestock feed will result in further widespread land degradation.

However, most small-scale farmers cannot afford or do not have access to concentrated feed which, even when available, can be of variable quality. Legume forage, however, which has a similar effect to commercial concentrates on feed utilization and animal performance, has proved to be adoptable on smallholder farms. Leguminous trees and shrubs have advantages over herbaceous legumes inasmuch as they are persistent, produce more edible dry matter and usually retain their leaves during the crucial dry season period; indeed, they can be deliberately managed to optimize leaf dry matter production during this period. However, many browse species contain antinutritive factors, such as saponins, alkaloids, and tannins, that can significantly reduce their nutritive value and may affect palatability. Condensed tannins have a significant effect on fermentability in the rumen, but have little effect on palatability. For example, *C. calothyrsus* contains a high level of condensed tannin and is highly palatable when fresh, but has a low level of fermentability in the rumen and low digestibility in the postrumen intestinal tract. Drying and wilting reduces total phenolics and condensed tannin levels, but depresses palatability, degradability and digestibility (Palmer and Schink, 1992; Ahn et al., 1989). *G. sepium*, on the other hand, which has a low level of tannin, is more palatable after wilting to animals unaccustomed to the feed, but both fresh and wilted material is highly palatable to animals that are used to the feed (Ash, 1989; Smith et al., 1994).

There is also wide variation in the site and extent of breakdown of protein from browse species (Table 4.8). For example, drying increases the rumen degradability of protein and the proportion of undegradable protein that is catabolized in the small intestine in albizzia, erythrina and leucaena, but decreases it in gliricidia and calliandra. Time spent in the rumen is affected by the level of food intake and the rate of breakdown of particles in the rumen.

For all browse species studied, an increase in the time spent in the rumen raises the level of rumen degradability and decreases the level of subsequent catabolism in the small intestine. The proportion of dietary nitrogen excreted in dung or lost in urine from livestock on a forage diet will depend upon the site of catabolism in the digestive tract (Reynolds and de Leeuw, 1994). It is not biologically efficient to include more rumen-degradable protein in a diet than can be trapped by the rumen microbes. This is particularly true in farming systems where manure is essential for soil fertility maintenance (e.g., those in the highlands of Rwanda) since nitrogen in dung is more easily managed than nitrogen in urine. Urine cannot be easily handled by farmers, and there is a high loss of nitrogen from urine through volatilization and leaching. It would be more efficient (in terms of nitrogen utilization) to compost surplus high-nitrogen vegetation than to pass it through animals. However, as previously indicated, the economics of green manure versus fodder supplementation need to be carefully assessed.

In a study of three browse species using cattle, sheep and goats, Smith and colleagues (1994) have found highly significant browse x animal species interaction for the rate and extent of degradation of the slowly degradable dry matter fraction in browse. Since there is no difference in the capacity of sheep and goats to digest most conventional diets (grasses, crop residues, concentrates), except when dietary protein levels are very low (Tolkamp and Brouwer, 1993), this suggests specific adaptation by different animal species to browse species. Care must therefore be taken in extrapolating nutritive values for browse species from one animal species to another.

Effects on Animal Performance

The addition of leguminous browse species to a basal grass ration can lead to a marked improvement in animal performance. High-quality browse can be grazed in situ (e.g., as practiced for beef cattle in Australia) or it can be offered as a cut-and-carry supplement to stall-fed animals (e.g., dairy cattle on smallholder farms in Kenya). Cattle growth rates, e.g., improved from -20 g d^{-1} on natural pasture in Indonesia to $+540$ g d^{-1} when leucaena comprised 40% of the ration (Wahyuni et al., 1982). Milk production of dairy cattle on smallholder dairy farms in Kenya increased by 0.5 kg milk per kg leucaena added to a diet of *Pennisetum purpureum* (Fig. 4.15). During early lactation the response in milk production is high, particularly in the dry season when grass quality is poor (Fig. 4.16), but in mid to late lactation the milk response is more muted as a greater proportion of additional nutrients are directed toward body tissue (Muinga et al. 1995). In Embu, in Kenya, the replacement value of *C. calothyrsus* as a fodder supplement has been found to be in the region of 3 kg fresh material ($=1$ kg dry matter), equivalent to 1 kg of dairy meal with 16% crude protein for grade Friesian and Ayrshire milking cattle (R. T. Paterson, unpublished data). However, for local Ankole cattle in Rwanda with low milk productivity, it has been shown that there is little economic gain to be had through the use of high-quality tree fodder as a supplement (Niang et al., 1996). The use of tree fodder for increased milk production must be judged according to the productivity and hence nutritional needs of the animals involved. An improvement in body condition should increase reproductive performance by raising conception rates and reducing calving intervals, although direct evidence from long-term trials with legume forage is lacking.

TABLE 4.8 Concentrations of protein (CP) and coefficients of rumen degraded protein (RDP), intestinally digested protein (IDP) and excreted dietary protein (undegraded and undigested, ExDP) for various browse species, as affected by forage preparation, forage regrowth period, and length of time the material remains in the rumen.

	Proportions of			
	CP	RDP	IDP	ExDP
	$(g \ (kg \ DM)^{-1})$			
Forage preparation[a]				
Erythrina verigata				
Fresh	253	0.64	0.25	0.11
Sun-dried	253	0.74	0.20	0.05
Gliricidia sepium				
Fresh	261	0.69	0.25	0.06
Sun-dried	261	0.70	0.19	0.11
Leucaena leucocephala				
Fresh	246	0.39	0.37	0.24
Sun-dried	246	0.48	0.34	0.18
Calliandra calothyrsus				
Fresh	256	0.71	na	na
Dried 25°C	256	0.39	na	na
Dried 65°C	256	0.32	na	na
Forage regrowth time before harvesting[b]				
Callianda calothyrsus				
12 weeks	282	0.59	0.32	0.09
24 weeks	253	0.58	0.26	0.16
48 weeks	231	0.51	0.23	0.26
Forage time in the rumen[c]				
Sesbania sesban				
24 h	267	0.77	0.10	0.13
48 h	267	0.84	0.06	0.10
Leucaena leucocephala				
24 h	257	0.44	0.32	0.24
48 h	257	0.57	0.21	0.22
Calliandra calothyrsus				
24 h	217	0.24	0.31	0.45
48 h	217	0.31	0.25	0.44

[a]*Perera et al. (1992), Palmer and Minson (unpublished).*
[b]*Kaitho et al. (1993).*
[c]*Kamatali et al. (1992).*

FIGURE 4.15   Small-scale milk producers at Embu and elsewhere in Kenya are increasing their milk yields by growing *Calliandra calothyrsus* and *Leucaena leucocephala* (shown here) as a high-quality feed supplement to napier grass. *Photo A. Njenga.*

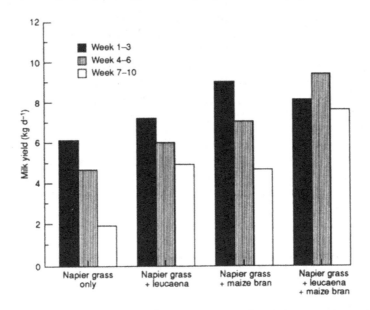

FIGURE 4.16   Effect of leucaena supplement on milk production of crossbred dairy cows in early lactation during the dry season (Muinga et al. 1995).

Small ruminants also benefit from the inclusion of leguminous browse in their diet, with faster growth rates and improved reproductive performance. The most significant change following supplementation of pregnant and lactating dams is in the survival rates of their offspring, which rises from 45% on a grass-only diet to over 95% when browse comprises 40% of the ration (Reynolds and Jabbar, 1994). Significant health and nutrition interactions were observed, increasing survival rates, when browse supplements were offered to sheep infected with trypanosomiasis (Reynolds and Ekurukwe, 1988). This is a nutritional effect rather than a direct consequence of browse per se but for many farmers browse is the most available option for improving the quality of the ration in the dry season. Several studies have confirmed that the supplementation of native browse with high-quality fodder also enhances the dry matter intake and live-weight gains of young goats and sheep. In Rwanda, *Mimosa scabrella* fed to young goats as a supplement to a basal diet of *Setaria splendida* increased their live-weight gain from 31 to 50 g d^{-1} (ICRAF, 1994, pp. 98−101). Similar results have been reported from Zimbabwe when *Acacia angustissima* was used as a fodder supplement for local goats browsing native dry season pastures. Live-weight gains were dramatically increased from −20 to +12 g d^{-1} during the 50-day period of the trial (Dzowela et al., 1994).

FIGURE 4.17 In the highlands of East Africa, many farmers have planted the grasses *Pennisetum purpureum* (shown here) and *Trypsacum laxum* for erosion control and fodder production. *Photo A. Njenga.*

Fodder Production Systems

Three issues are of particular concern in the development of improved fodder production systems for small-scale farmers. The first is the identification of the available niches on farms for the introduction of fodder trees. Many exist and the choices farmers make will determine both the spatial arrangement and, to some extent, the management of the trees. The second issue is that of tree management to optimize fodder production at critical times of the year, which in most instances occurs during the dry season. The third is the choice of species.

In many instances, opportunities exist for linear planting of fodder trees on farms as internal hedges, in combination with upper-story trees as boundary plantings, along terrace risers on sloping land or along permanent contour bunds. Fodder trees can also be planted in blocks either as pure stands or interplanted in existing grass fodder banks (O'Neill, 1993). Furthermore, trees can be planted in cropland either as scattered trees (ICRAF, 1992b, pp. 83−84) or in a hedgerow intercropping arrangement (Reynolds and Jabbar, 1994).

The management of trees to optimize fodder production will be specific to species, spatial arrangement, environment, and interaction with associated trees or shrubs. Where trees are planted in linear arrangements, a within-row tree spacing of between 0.25 and 0.5 m is recommended and experience suggests that pruning fodder trees to a height of between 60 and 100 cm is the most appropriate with regard to ease of labor, quantity of fodder produced, and competition with nearby crops (Niang et al., 1996). At Embu, in Kenya, in a bimodal rainfall regime, when *L. leucocephala* was pruned at intervals of about 3 months to a consistent height of 100 cm, initial leaf yields were highest when side branches were pruned back to the main stem. Annual cumulative yields were increased, however, by cutting the side branches at about 20 cm from the stem. The yields could be further increased by hand-stripping the remaining leaves after the pruning but the extra labor this involved may not be justified by the small increase in dry matter production (ICRAF, 1994, pp. 109−111). Production is also enhanced if trees are allowed to develop strong root systems before pruning.

Studies have shown that fine-root die-back occurs when fodder trees are pruned (e.g., Reynolds and Jabbar, 1994), so that belowground root regrowth is needed in addition to aboveground fodder production. Pruning too frequently over a prolonged period will reduce fodder production.

In the highlands of East Africa, the contour planting of fodder grasses (*P. purpureum* below 2000 m altitude and *Trypsacum laxum* above 2000 m) has been widely adopted by farmers as a dual-purpose erosion control and fodder production system (Fig. 4.17). High-quality fodder-producing trees can easily be established at the same time as the grass strips or within established lines. Research has shown that the inclusion of such trees in combination with grass strips dramatically increases the production of digestible protein without affecting total dry matter production (Niang et al., 1996; ICRAF, 1995). For example, in Burundi the productivity of two contour-planted rows of trypsacum grass was compared with a single row of trypsacum combined either with a row of *C. calothyrsus* or *Leucaena diversifolia* (Table 4.9).

From these results it can be calculated that a Burundian farmer would only need to plant 60 m of a calliandra/trypsacum hedge to meet fully the annual maintenance and digestible protein requirements of a local 20-kg goat. Such an approach is being widely adopted by milk producers near Embu in Kenya where farmers are planting calliandra in established contour plantings of pennisetum and are reporting significant increases in milk yield. Other research in Zimbabwe has demonstrated that *A. angustissima*, grown on existing 2-m-wide contour bunds which average 500 m

TABLE 4.9 Fodder production from different combinations of grasses and shrubs in Burundi.

	Mean annual production 1991–93 (kg DM m^{-1})	Mean annual production of digestible protein (g m^{-1})
Two rows trypsacum	2.22	89
One row trypsacum + one row calliandra	2.06	179
One row trypsacum + one row leucaena	1.83	119

Source: From ICRAF (1995).

bund length per hectare of cropped land, can provide the crucial supplementation needs of draft oxen during the dry season (Dzowela et al., 1994). Fodder production on these bunds is now being evaluated with local farmers for a range of adapted fodder species (Fig. 4.14).

Fodder trees can also be grown as linear hedgerows in crop fields in an identical arrangement to that described for hedgerow intercropping. Where such hedgerows are planted on the contour for erosion control, hedgerow prunings can be most economically utilized for fodder under mildly erosive conditions (see previous section on barrier hedges) and this is an attractive option for farmers. However, on flat land such an approach has little to offer. Removal of prunings for fodder instead of their use as a green manure will inevitably result in a reduction in associated crop yield. This could largely be avoided if the same number of trees were planted as a block in part of the field, since this arrangement results in a reduced zone of tree/crop interface. In such block plantings, where trees are commonly spaced at approximately 0.5 × 1.0 m, fodder yields range from 0.5 to 1.0 kg tree" a^{-1} in the humid and subhumid tropics. The recommended level of fodder supplementation for a grade dairy cow is about 2 kg dry matter d^{-1} or 700 kg a^{-1}. Between 700 and 1400 trees are therefore required per lactating animal, or a land area of between 0.035 and 0.07 ha only.

In many areas, particularly those with long dry seasons, farmers need to adopt a tree pruning strategy that maximizes fodder production at periods of greatest shortage. Under such conditions moisture stress and leaf senescence can lead to the loss of fodder material unless cutting times are chosen judiciously. If, e.g., the aim is to maximize the yield of fodder at the height of the dry season in August, as in Burundi, research has shown that the final wet-season cut should be made 6 months earlier in February (Akyeampong and Muzinga, 1994). Earlier cutting resulted in increased leaf senescence while later cutting gave a suboptimal period for recovery before the onset of the dry season.

Cultivation of fodder trees on grassland is only feasible where grazing access can be controlled. For example, over 200 km^2 of natural pasture on commercial farms in Queensland has been strengthened with leucaena for use by beef or dairy herds (Wildin, 1994). On communal land or where livestock graze freely after crops have been harvested, young fodder trees can be killed by free-roaming animals. In some parts of Africa, however, farmers collectively impose control over special areas set aside for dry season grazing. An example is the *Ngitiri* system practiced by the Sukuma peoples in the Shinyanga district of Tanzania (Otsyina and Asenga, 1993). Collectively or individually owned areas are protected during the rainy season and are then grazed with careful regulation during the dry season by draft oxen. However, farmers recognize the generally poor quality of the dry season grazing and have expressed considerable interest in establishing adapted and productive browse species in their *Ngitiri*. They have indicated that linear plantings of such species, to subdivide *Ngitiri* areas into paddocks, would further assist in dry season grazing control. *L. leucocephala*, *G. sepium*, *A. angustissima*, *C. calothyrsus*, and *Atelia herbertsmithii* are currently being evaluated together with farmers.

The choice of species will depend not only on their feed quality, as discussed earlier, but also on their specific adaptation and their capacity to remain productive under continuous pruning. *S. sesban*, e.g., is a widely adapted and fast-growing species with high-quality fodder. However, experience has shown that its regrowth after pruning is poor and early death can result (Niang et al., 1996).

In general, it has been found that adapted exotic material shows greater vigor and fodder productivity than indigenous species. In many parts of sub-Saharan Africa with relatively neutral soils, *L. leucocephala* was a most productive fodder species and was becoming popular with farmers until the arrival of the leucaena psyllid (*Heteropsylla cubana*) into the region. This pest first appeared on the African mainland just north of Mombasa (Kenya) in August 1992 (Reynolds and Bimbuzi, 1993). It has recently appeared in Zimbabwe and its presence in neighboring countries would suggest that it will become endemic in the whole region before long. It remains to be seen if natural or, possibly, introduced predators will be able to control it in the long run. In Asia, the impact of the psyllid was much reduced 4–6 years after its initial invasion and yields of leucaena biomass are now almost equal to preinfestation levels (van den Beldt and Napompeth, 1992). Other species such as *L. diversifolia*, *Leucocasia esculenta*, and *Lambula pallida*, together with a number of composites and interspecific hybrids, have shown some degree of tolerance to the psyllid (Dzowela et al., 1994; Otsyina et al., 1994a), but work with them is of recent inception and more local experience will have to be gained before these species and lines can be recommended to farmers. Of other exotic leguminous species tested, those with widest general potential across the region would appear to be *C. calothyrsus* where the annual rainfall exceeds 1100 mm and *G. sepium* where the annual rainfall is less than 1100 mm but above 700 mm. On acid soils in Zimbabwe the cold tolerance of *A. angustissima* makes it a very promising species (Dzowela, 1994) while at altitudes above 2300 m on extremely acid soils (pH below 4) in Rwanda *M. scabrella* has been shown to outperform most other species (ICRAF, 1994, pp. 98–101).

Selection work with indigenous fodder trees is not as well advanced as with exotic species. Trees which are valued by farmers as sources of fodder include both legumes and nonleguminous species. In the bimodal rainfall highlands of Kenya, farmers value the local species *Trema orientalis* and *Sapium elipticum* while the naturalized species *Morus alba*, which is commonly recognized as a source of feed for silkworms, also shows potential for the feeding of ruminant livestock (Thijssen et al., 1993). In eastern Zambia the local species *Zizyphus abyssinica* and *Diplorynchus condylocarpon* have been shown to be useful supplements for goats fed on a basal diet of poor quality roughage (Phiri et al., 1992) while in Tanzania *Margaritaria discoides* is showing similar promise for goats (Otsyina et al., 1994b).

In conclusion, considerable information is available on the potential of trees to provide high-quality fodder, their adaptation to specific environmental conditions, the range of farm niches where they can be planted and appropriate management regimes to optimize their production and impact. In spite of such information, widespread adoption is currently limited to peri-urban milk producers where market forces provide a clear incentive and farmers are already aware of the impact of feed supplementation on milk yields. Where oxen are used for draft power and provide a source of income for farmers (as in Zambia and Tanzania), farmers are interested in feed supplementation during the dry season but widespread adoption has yet to occur.

In the future, as urban populations increase and natural grazing areas are unable to support expanding herds of beef cattle, it is probable that peri-urban beef industries will develop, as is already being observed for poultry and egg production. The rate at which such systems develop and their economic viability will depend upon a reliable and cheap source of high-quality feed supplement. Fodder trees will have an important potential role (Winrock International, 1992). However, research will need to address the issue of the sustainability of such tree-based fodder production systems since they will result in high rates of nutrient export (Shepherd et al., 1994). Efficient nutrient recycling of animal waste is likely to be important in this respect.

HIGH-VALUE TREES FOR INCOME GENERATION

In the last few decades, since emerging from being just a traditional form of land-use, agroforestry has been viewed as a panacea for the achievement of sustainability and land amelioration. More recently, however, agroforestry has reemerged as a means of producing traditionally important and valuable forest products such as timber, fruits, medicines, and extractives on farmland. In this way agroforestry can now be viewed as an intervention point to break the downward spiral of land degradation and rural poverty (Fig. 4.5). The production of forest products can be used either to generate cash with which to buy fertilizers to increase the yields of staple crops or as a profit-motivating incentive to promote the establishment of more trees on-farm to ameliorate soil depletion and land degradation. Thus the vision now is of agroforestry as an integrated land-use policy that combines increases in productivity and income generation with environmental rehabilitation and the diversification of agroecosystems. Such a vision can be fitted to the range of situations found in the major ecoregions of the tropics. The realization of this vision, however, is going to depend on three factors in particular. The first is an appreciation by the international community and donors of the importance of high-value indigenous species in the lives and welfare of local people, as well as an incentive (or the removal of disincentives) for local people to plant trees on their farms. The second is domestication of commercially important indigenous tree species producing forest products, and the third is the development of a processing and marketing infrastructure.

Naturally Regenerating Trees in Farmland

In Africa, there are still relatively few examples where trees are being deliberately established on farms to generate income. While examples of farmers growing trees for this purpose are known to the authors, there is very little published evidence of the scale of such activity in different areas of Africa. It is, however, quite common for farmers to

FIGURE 4.18 In local markets, such as this one in Kumba, Cameroon, one finds a wide variety of indigenous fruit tree products on sale, such as the kernels of *Ricinodendron heudelotii* (left foreground), used as a spice, and *Irvingia gabonensis* (right) used to thicken food. *Photo R. B. Leakey.*

protect and retain naturally regenerating trees on farmland, both for the production of a regular income from fruits and as a capital investment in the form of quality timber which can be liquidated to pay school fees or wedding/medical/funeral expenses.

In the Sahelian parklands, large trees of *Butyrospermum paradoxum* and *Parkia biglobosa* are grown as scattered trees in the fields, and the fruits of these trees are harvested and marketed. Despite a reduction in the production of crops under the trees, the overall returns per hectare are increased by 6500 FCFA ha^{-1} (250 FCFA = US\$1) with one *B. paradoxum* and one *P. biglobosa* per hectare and by 13,600 FCFA ha^{-1} if the number of *B. paradoxum* is increased to eight (Bonkoungou, 1995). Although these economic benefits are not large, they do represent an important cash income to relatively poor farmers.

In southern Africa the indigenous fruits of the miombo woodlands are also important and are beginning to contribute to commercial trade. Data are scarce on the value of this trade but, e.g., the US\$103 made by a marketeer in Zambia in September 1993 from the sale of *Ziziphus mauritiana*, one of the locally marketed fruits (Kwesiga et al., 1994), is significant when the average annual per capita income for Zambia is only \$272.

In the humid forests of West Africa there are also numerous species the products of which are traded (Fig. 4.18). For example, cola nuts and the kernels of the bush mango (dika nut) are sold on a substantial regional scale. Again the economic value of this trade is not well documented. The cola nut trade in 1981 was reported to be 22,500 t (Nkongmeneck, 1985) of which 20,400 t entered commerce, mostly being traded to the north where it is the only stimulant allowed to Muslims. Data on bush mango kernels from Cameroon, however, show that the price per kilogram fluctuates seasonally from a low in August (535 FCFA) to a high in February (1670 FCFA), with farmers receiving 50−72% of these urban prices in July−August but only 18−39% in August−October (Ndoye and Tchamou, unpublished). The trade in some other products, such as the chewing sticks from *Garcinia* species, has been quantified. In this case the street value from Kumasi market in Ghana was US\$9 × 10^6 a^{-1} (Falconer, 1992). In all these examples the marketing has developed to meet local and regional demands and is based on traditional practice.

Another example, and one involved in international trade, is "pygeum" (*Prunus africana*) which grows in discontinuous populations throughout sub-Saharan Africa in the montane areas above 1200 m and where rainfall exceeds 1000 mm. The bark of this tree is harvested for pharmaceutical products used in the treatment of benign prostatic hyperplasia and prostate gland hypertrophy, ailments suffered by 60% of men in Europe and the United States (Fig. 4.19). The market value of the commercial product has been estimated at US\$150 × 10^6 a^{-1} (Cunningham and Mbenkum, 1993). The overexploitation of this tree's bark is seriously threatening the resources of the species, with implications for the future source of the pharmaceutical product, as it is apparently unlikely that this drug will be synthesized, as well as for the montane ecosystems of sub-Saharan Africa. This tree, which also produces good timber, is beginning to be grown by farmers in Cameroon and has potential for agroforestry.

Planted Trees in Farmland

The best example of farmers deliberately growing indigenous trees for cash generation is in Southeast Asia, not in Africa. Some Asian examples are included in this review to illustrate the potential of the technologies, from which we believe Africa can learn.

FIGURE 4.19 The bark of many indigenous trees, such as *Prunus africana*, is highly valued, both locally and internationally, for its medicinal properties. *Photo R. B. Leakey.*

FIGURE 4.20 In Sumatra, farmers have developed over 10,000 ha of multistrata agroforests, dominated by *Shorea javanica*, a resin-producing species. In 1993, the profits from damar resin exceeded US$7 million. *Photo R. B. Leakey.*

In southwest Sumatra, near Krui, farmers have for more than a century been planting damar trees (*Shorea javanica*) for resin production, in mixtures with a range of indigenous fruit tree species. The area now exceeds 10,000 ha and 65−80% of the households are involved in damar production (Fig. 4.20). In 1994, the production was expected to reach 10,000 t (Dupain, 1994) at a value of US$0.3−0.4 kg^{-1}. These trees are planted among rice, coffee, and other crops, so that at all stages of their growth the farm is productive—food crops in years 1−3, coffee and bananas in years 3−8, fruits and fuelwood in years 2−20, and resins, fruits and timber from year 20 onwards (De Foresta and Michon, 1994). In the case of damar, the resins are utilized by industries in Indonesia or exported worldwide. In 1984, the export

FIGURE 4.21 *Grevillia robusta* is widely grown on-farm and field boundaries on the slopes of Mount Kenya for fuelwood and timber. Yields of adjacent maize can be reduced by tree/crop competition for resources. *Photo A. Njenga.*

TABLE 4.10 Cost/benefit analysis ($ ha^{-1}) at net present value of 18-year-old boundary plantings of *Grevillea robusta*.

Added benefits				Added costs				Net discounted benefit
Fuelwood	Timber	Saved inputs	Total	Lost maize	Potential yield lost	Nursery cost	Total	
47	29	11	87	47	28	4	79	8

From Tyndall, B., 1996. The Anatomy of Innovative Adoption: The Case of Successful Agroforestry in East Africa. Ph.D. Thesis, Colorado State University, 212pp.

market represented one-third of the harvested volume, and trade increased from 250 to 400 t a^{-1} between 1972 and 1983 (Michon and de Foresta, 1995). Most of this trade (80%) is met by the damar agroforests. The economic value of the damar trade and its associated activities is of major significance to the villages around Krui. In 1993, the profits from damar production were US$7.25 × 10^6 (US$3.25 × 10^6 from sales, US$2.65 × 10^6 from added value and US $1.35 × 10^6 from wages). To this should be added US$271,000 profit made by Krui traders (Michon and de Foresta, 1995). This analysis excludes the locally consumed products from these agroforests such as fruits, vegetables, spices, fuelwood, timber, palm thatching, rattan, bamboo, and fibers.

Systems similar to the damar agroforest are also practiced in Sumatra with rubber and cinnamon. The "jungle rubber" agroforests covered more than 2.5 × 10^6 ha while there were 42,600 ha cinnamon in 1989 (Aumeeruddy, 1994). These damar, cinnamon and jungle rubber multistrata agroforests are probably the ultimate example of an alternative to slash-and-burn agriculture, being highly productive, biologically diverse and very similar in structure to natural tropical high forest. In short, they are the nearest thing to a sustainable manmade ecosystem.

Probably the best example of the economic value of planted trees in an agroforestry system on a small farm in Africa is the cost/benefit analysis of 18-year-old *G. robusta* trees in boundary plantings in Embu, Kenya (Tyndall, 1996). Here at 1500 m altitude, with a bimodal rainfall pattern of 1300 mm a^{-1} and on deep nitisol soils, *G. robusta* is grown on farms producing maize, beans, coffee and milk and averaging 1.5 ha in area. Boundary plantings resulted in 85% of households being self-sufficient for fuelwood (Fig. 4.21). The trees are grown mainly for firewood and timber in a single row along the edge of a field at a spacing of 2.7 m. This cost/benefit study examined the inputs and outputs over an 18-year period and discounted these at 20% in a 1-ha plot of maize in which 37 trees were grown down one side of the plot. This was compared with a 1-ha plot without trees. In the 13 m adjacent to this row of trees the maize suffered from competition, resulting in an average loss of 5% over the whole hectare. The *G. robusta* plots were found to have added benefits that totaled US$87 ha^{-1}, while the added costs totaled US$79 (Table 4.10). Thus the discounted net benefit was US$8 ha^{-1} (10% greater than added costs). This, like the indigenous fruits example, highlights the fact that in purely economic terms the benefits of even high-value trees only just exceed the associated costs and yield losses. Despite this, farmers are planting grevillea widely. Indeed, although higher elevation divisions of Embu (Manyata and Runyenjes) have some of the highest human population densities in Africa, they also have rapidly growing tree populations as farmers are adopting agroforestry. The unquantified benefits probably include reduced risk, reduced pressures on neighboring forests, erosion control on sloping land, shade and amenity, freedom from boundary disputes and enhanced climate effects (e.g., a windbreak effect).

Data on the yields of crops planted adjacent to rows of tree crops are relatively scarce but in one study in Uganda 17 upper-story tree species were compared for their variability in competition with associated crops (Okorio et al., 1994). The relatively fast-growing trees with broad and dense canopies (*M. eminii* and *Cordia abyssinica*) or tall canopies (*M. lutea* and *Melia azedarach)* appear to be more competitive than those that are slow growing (*A. acuminata*) or those with narrow and lax canopies, suggesting that shading is the major source of competition. Relatively little is known, however, about the belowground competition, although species like *M. eminii*, in contrast to *A. acuminata*, have large amounts of fine feeder roots in the top 30 cm of the soil. In addition to this species variation in belowground architecture, it is likely that there is variation within-species. The identification and development of methods to utilize this variation is an area for future research. In the Ugandan study *A. acuminata* was the only species that had an apparent positive effect on crop performance. This was attributed to nitrogen fixation by its Frankia symbionts, since soils sampled from under this species were significantly richer in nitrogen, soil organic matter, calcium, and potassium, but conclusive evidence of the mechanism involved is lacking.

Home Gardens

The intimate association of trees, crops, and animals in home gardens makes them a good example of an agroforestry system. The outputs from home gardens are diverse and production provides farmers with extra income. Unfortunately, in Africa home gardens are not well developed although they exist in many areas (Fig. 4.22). In Nigeria, e.g., they are important in areas with high population densities (1000 people km^{-2}). In such areas up to 29% of the cultivated area can be in compound gardens (Okafor and Fernandes, 1987) and these produce 59% of the crop output (Watson, 1990). In monetary terms the output of these gardens is 5−10 times greater than that of crop fields, with returns to labor 4−8 times greater. Despite this, most of the examples quoted in this review are again from Southeast Asia or from Latin America. Many of the biophysical interactions taking place in home gardens are complex, difficult to research and poorly understood. Despite this, it is probably fair to claim that home gardens are among the most sustainable of land-use systems and that they have the following characteristics (Torquebiau, 1992).

Home gardens are generally accepted as having a high biological diversity. For example, in Java 270 plant species have been recorded in the 41 home gardens of one village while 338 plant species were reported in one area of Mexico.

FIGURE 4.22 Multistrata home gardens, like this one in northern Tanzania, are highly productive and sustainable systems. They provide a diverse range of food and cash products and high returns to labor. *Photo A. Njenga.*

In addition, it is common to find local varieties of tree crops established in home gardens, e.g., in Micronesia 21−37 varieties of each of three species (coconuts, breadfruit, and bananas) have been recorded. In Mexico 15% of the woody plants were locally wild species. A correlation has been found between the diversity of home gardens in Mexico and the available labor force. These home gardens are also biologically diverse in terms of wildlife. In Java, e.g., out of 121 different bird species identified in home gardens in four villages 15 were endangered species.

From the socioeconomic viewpoint food production from home gardens gives good returns on labor. In Indonesia 7% of people's time is spent in the home garden but this effort produces 44% and 32% of their total intake of carbohydrates and proteins, respectively. This production can be achieved at low cost. For example, in Java production costs only represent about 10% of the output or 15% of the gross income, while in paddy rice equivalent values are between 30−50% and 56%, respectively. Nutritionally, home gardens are also very important. In Hawaii, e.g., a 35-m^2 home garden can provide 100% of the vitamin A and C requirements, 50% of the iron requirements and 18% of the protein requirement of a family of five. Thus, as in Java, the home garden species supply the nutrients in shortest supply from the staple diet of rice. Reported values for total calorie and total protein intakes provided by home gardens are variable, with values as high as 44% and 32%, respectively. Interestingly, during 1975−83 fruit prices rose 10-fold in Indonesia, while the price of rice rose by only 2.4 times, indicating that the market for diverse tree-crop products may be increasing relative to that for staple commodity crops.

Production from home gardens is not, however, limited to food crops. Levels of wood production of 7−9 m ha^{-1} a^{-1} have been recorded in Java, which compare well with growth rates obtained in plantation forestry. In Java 51−90% of fuelwood comes from home gardens, while in Bangladesh it has been noted that 70% of the sawlogs and 90% of the harvested fuelwood and bamboo come from the homestead. In addition the diversity of products from home gardens also provides opportunities for the development of cottage industries. This creates jobs and off-farm employment.

In conclusion, home gardens are both highly productive and sustainable. However, it must be noted that this is, at least in part, due to regular inputs of household, animal and human waste which supplement nutrient uptake by deep rooted trees.

Domestication

Extractive harvesting or gathering, retention of natural trees on the farm and the planting of unimproved germplasm are the first three steps of the domestication of a tree species. Accepting the argument that there is an important niche in agroforestry systems for high-value wild species, a strong case can be made for the further domestication of these "Cinderella" species (Leakey and Newton, 1994a,b; Leakey et al., 1994; Newton et al., 1994). The domestication process is multifaceted (Leakey and Newton, 1994c) and will vary depending on the value of the species, the duration of its reproductive cycle and the length of the agroforestry/forestry rotation to which it is suited. In all cases it is important that a domestication strategy be developed which ensures the maintenance of genetic diversity and the continuing introduction of new cultivars, so that the risks of a narrowing genetic base and stagnating commercial production are avoided (Leakey, 1991; Simons et al., 1994). The development of genetic homeostasis between cultivars, furthermore, allows a deployment strategy in which cultivars can fill different niches within a farm to take advantage of differences in optimal environmental conditions and in competitiveness (Foster and Bertolucci, 1994).

In recent years a number of new tree products have appeared on the supermarket shelves. Kiwi fruit, pecan nuts and macadamia nuts are perhaps the best examples. In the case of macadamia nuts domestication arose from the combination of a breeding program in Hawaii and the clonal propagation of cultivars selected for tree habit, yield and kernel recovery, which are grafted onto canker resistant rootstocks (Cannell, 1989). The newest clones are reputed to yield four times as many nuts per tree and to have 10% more first-grade kernels than the earlier clones.

In the tropics a number of indigenous fruit trees are in the early stages of domestication (Okafor and Lamb, 1994; Clement and Villachica, 1994), with the selection of superior individuals for grafting and the collection of germplasm. These approaches need to be linked to studies to maximize harvest index (Cannell, 1989) and to the integration of trees and crops into food crop systems (Watson, 1983). At ICRAF, the domestication of some indigenous fruit trees has started in Malawi and in Cameroon (ICRAF, 1995; Leakey and Maghembe, 1994). In Malawi field trials have demonstrated that, contrary to expectation, the growth of indigenous fruit trees is quite fast and that *Azanza garckeana*, *Annona senegalensis*, *Bridelia micrantha*, *Bridelia cathatica*, *Z. abyssinica*, *Z. mauritiana*, and *Ficus vallis-choudea* can start to produce flowers and fruits within 2 years (Maghembe, 1994). The indications are that there are substantial differences between trees in fruit size, color, and flavor, and current work is directed toward capturing this variation by vegetative propagation.

A similar approach with indigenous fruits is in progress in Cameroon but in this instance the first step has been to develop and apply guidelines to determine the species preferences of farmers in Ghana, Nigeria, Cameroon, and Gabon in order to prioritize species for domestication on a regional basis (Jaenicke et al., 1995). In both southern and west Africa, these domestication initiatives have also involved workshops for scientists to gather information on the priority species (Maghembe et al., 1995; Boland and Ladipo, unpublished). The result in west Africa has been the choice of *Irvingia gabonensis* as the first priority and consequently germplasm collections have been made in each country for the establishment of living ex situ germplasm banks in Nigeria and Cameroon (ICRAF, 1995). Coupled with the establishment of these germplasm collections has been the identification by farmers of the best trees in each village. Scions and cuttings from these trees will be used to establish local cultivars that can be rapidly reintroduced into the villages and established for on-farm testing.

In conclusion, while tree products are one of the main factors determining the inclusion of high-value trees on farms, they are probably the least well studied and understood reasons for farmers' acceptance of agroforestry. This is particularly the case in Africa, but the examples from Asia and Latin America provide hope for the future.

In many instances the demands of growing human populations, coupled with the rates of deforestation, are leading to the exhaustion of the natural resource and an uncertain future. Agroforestry has the potential to reverse these trends, especially if the incentives to plant can be generated by the genetic improvement of the products, so that farmers can increase their returns on investment, and by the expansion of the markets. Unfortunately, both these precursors of change must occur simultaneously and international agricultural research and commerce are not well integrated. However, the market constraints have recently been receiving considerable attention (Vosti and Witcover, 1995) so there is hope that the desired progress can be made. Given the potential of recently modified horticultural techniques applied to the domestication of the trees which traditionally have only been important as sources of extractive reserves, there is now a new impetus for examining the market potential of the most promising commercial species. Ways will have to be found to promote the marketing of species prior to their domestication as the costs of domestication are high. However, since the greatest potential of these species will be in local urban and regional markets where they are already well known, these costs may be very much less than those associated with breaking into an international market. There are therefore many areas for both biophysical and socioeconomic research for each potential domesticated tree, spanning the full range of processes associated with domestication.

CONCLUSION

To date, research in sub-Saharan Africa has placed emphasis on the development of agroforestry systems which directly address the maintenance of the soil resource base through enhanced soil fertility and soil erosion control, and good progress has been made. Where the development of such systems are built on indigenous concepts such as fallowing, the chances of adoption and impact appear high. We believe, however, that agroforestry solutions to land degradation are always likely to be location-specific in their relevance and acceptability. For this reason it is imperative that research increasingly includes farmers as partners in the research process.

We introduced this review with a brief overview of projected population increases and their probable impact on land-use and land degradation. The inference is clear. Agroforestry systems that provide solutions for today's land degradation problems will need to evolve in both diversity and intensity if they are to remain relevant and effective for tomorrow's Africa. The intensification and diversification of agroforestry systems should be equated to an ecological succession and agroforestry should therefore be seen as a dynamic rather than a static form of land-use. In this connection it is interesting to note that in many areas of high population density, particularly in Southeast Asia, the pathway of evolution of land management following forest clearance has nearly gone full circle. Land originally cleared for shifting cultivation and crop production is now increasingly coming under complex manmade agroforests which closely resemble, both in structure and environmental function, the indigenous forests they have replaced. This phenomenon offers great hope for the deforested and degraded areas of the tropics, especially in Africa, but it is clear that the political and policy environment has to be favorable for this type of change to occur—land tenure in particular is a probable requirement. Increased research emphasis needs to be placed on the domestication of high-value indigenous trees and their integration into more sustainable, diverse, and intense land-use systems. Such research is both long-term and complex, but should be embarked upon now if the research community is to be in a position to meet the needs of future generations.

ACKNOWLEDGMENTS

The authors acknowledge, with gratitude, all those who contributed to the preparation of this paper, in particular Dr. K. Shepherd, Dr. C. K. Ong and Dr. R. Coe of ICRAF for their review and useful comments on early drafts, Miss M. Mwangi for all her secretarial input and Mr. A. Njenga who provided many of the photographs.

Chapter 5

The Domestication and Commercialization of Indigenous Trees in Agroforestry for the Alleviation of Poverty

This chapter was previously published in Leakey, R.R.B., Simons, A.J., 1998. Agroforestry Systems, 38, 165–176, with permission of Springer

SUMMARY

New initiatives in agroforestry are seeking to integrate into tropical farming systems indigenous trees whose products have traditionally been gathered from natural forests. This is being done in order to provide marketable products from farms that will generate cash for resource-poor rural and peri-urban households. This poverty-alleviating agroforestry strategy is at the same time linked to one in which perennial, biologically diverse and complex mature stage agroecosystems are developed as sustainable alternatives to slash-and-burn agriculture.

One important component of this approach is the domestication of the local tree species that have commercial potential in local, regional or even international markets. Because of the number of potential candidate species for domestication, one crucial first step is the identification of priority species and the formulation of a domestication strategy that is appropriate to the use, marketability and genetic potential of each species.

For most of these hitherto wild species, little or no formal research has been carried out to assess their food value, potential for genetic improvement or reproductive biology. To date, their marketability can only be assessed by their position in the local rural and urban marketplaces, since few have attracted international commercial interest. To meet the objective of poverty alleviation, however, it is crucial that market expansion and creation are possible; hence, for example, it is important to determine which marketable traits are amenable to genetic improvement. While some traits that are relatively easy to identify do benefit the farmer, there are undoubtedly others that are important to the food, pharmaceutical or other industries that require more sophisticated evaluation.

This chapter presents the current thinking and strategies of ICRAF in this new area of work and draws on examples from our program.

INTRODUCTION

Agroforestry practices come in many forms but have traditionally been categorized into two groups—those that are sequential, such as fallows, and those that are simultaneous, such as alley cropping (Cooper et al., 1996). In all, some 18 different agroforestry practices have been recognized by Nair (1993), although each has an infinite number of variations.

Leakey (1996), however, has suggested that agroforestry practices should be seen as stages in the development of an agroecosystem such that the increasing integration of trees into land-use systems can be seen as the passage toward a mature agroforest of increasing ecological integrity. In this way, with increasing scale, the integration of various agroforestry practices into the landscape is like the formation of a complex mosaic of patches in an ecosystem, each of which is composed of many niches. These niches are occupied by different organisms, making the system ecologically stable and biologically diverse. Filling some of these niches with indigenous species that provide important environmental services or economically valuable products traditionally obtained from natural forest, or both, should result in land-use that is both sustainable and productive.

Multifunctional Agriculture. DOI: http://dx.doi.org/10.1016/B978-0-12-805356-0.00005-2
© 2017 Elsevier Inc. All rights reserved.

Increasing the quality, number and diversity of domesticated trees that provide a wide array of nontimber forest products (NTFPs) to fill these niches should enhance agroforestry's capacity to fulfill its ultimate potential as a way to alleviate poverty and to mitigate deforestation and land depletion. In the humid tropics this could lead to the development of viable alternatives to slash-and-burn agriculture. The large-scale adoption of such an approach should be especially beneficial, since the ecological and social benefits of diversity on a landscape scale are considerably greater than the sum of the individual farm-scale benefits. There is great urgency to achieve these benefits if the severity of the current manmade episode of species extinction—the so-called "Sixth Extinction"—which was so dramatically described by Leakey and Lewin (1996), is to be defused.

In this respect, Sanchez and Leakey (1997) see domestication as one of the three determinants for balancing food security with natural resource utilization in sub-Saharan Africa. The other two determinants are the need for an enabling policy environment that favors smallholder rural development, and the means to reduce soil fertility depletion.

Discussion of the role of domestication cannot, however, be divorced from that of commercialization, since without an expanded or a new market, the incentives to domesticate intensively for self-use are insufficient. Conversely, if the market explodes, the incentive for large-scale producers to establish monocultural plantations may sweep away the benefits that agroforestry could deliver to small-scale, resource-poor farmers around the tropics (Leakey and Izac, 1996). Thus policies that promote the linkages between the domestication and commercialization of NTFPs are one of the important areas in which policy scientists need "to stretch their conceptual framework... and to consider more carefully the links between markets, the environment, household production and household welfare" (Dewees and Scherr, 1996).

DOMESTICATION STRATEGIES

The term "Cinderella trees" (Leakey and Newton, 1994a,c) is now widely accepted as a phrase applicable to traditionally important indigenous species that have been overlooked by science for agroforestry and forestry, as evidenced by the term's use in numerous articles and conference proceedings (see Leakey et al., 1996). Similarly, the need to rapidly domesticate the Cinderella trees has been accepted, and is now one of the three pillars of ICRAF's research program. In genetic terms, domestication is accelerated and human-induced evolution. Domestication, however, is not only about selection, as domestication integrates the four key processes of the identification, production, management, and adoption of agroforestry tree genetic resources (ICRAF, 1995).

The Working Group on "Product Domestication and Adoption by Farmers" in the recent ICRAF Conference on the "Domestication and Commercialization of Nontimber Forest Products in Agroforestry" (Leakey et al., 1996) defined the domestication of trees producing NTFPs as "a progression from collection and utilization of products, through protection, management and cultivation, which culminates with genetic manipulation." Two extreme strategies (Fig. 5.1) were envisaged:

1. making incremental improvement through management on-farm and
2. making major leaps in improvement by genetic selection and breeding.

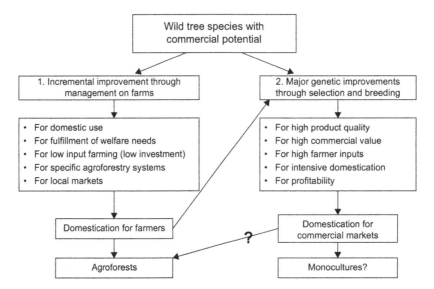

FIGURE 5.1 Two extreme pathways in the domestication and commercialization of nontimber forest products. *Modified from Leakey et al. (1996).*

The first strategy is farmer-oriented, while the second is market-oriented. The question is whether the market-oriented strategy will automatically lead to monocultures. History would suggest that this is likely, but history is not always the precedent people expect. It seems likely that between these extremes there are many scenarios that could benefit farmers and the environment. For example, the two strategies are not mutually exclusive: farmer-domesticated trees can be inputs into more formal genetic improvement programs, just as selected cultivars from genetic improvement programs can be grown on smallholder farms (Fig. 5.1). Furthermore, there is the precedent of "jungle rubber" production in Indonesia, where rubber produced in complex agroforestry systems is overtaking that from monocultural, clonal plantations (Michon and de Foresta, 1996). To encourage this development for other tree species producing NTFPs, incipient markets require good information flows so that domestication may attract a price premium for high-quality products and better marketing practices. In this way, small-scale farmers will be able to respond to, and benefit from, domestication opportunities, reducing the risk of displacement by monoculture plantations. With regard to market information, Ndoye (1995) has called for price surveillance systems for minor NTFPs, similar to those for international commodity crops, so that farmers will be more aware of marketing opportunities and the benefits of domestication will be maximized.

The Working Group on Policy and Institutional Aspects of Domestication at the conference referred to earlier (Leakey et al., 1996) considered the economic and social implications of domestication and commercialization on indigenous peoples, and the trade-off, if any, between the objectives of raising farmers' incomes and biodiversity. It was concluded that the domestication of trees to produce NTFPs for agroforestry must benefit both the farmer and the environment, and that there must be no policy barriers that discriminate against smallholder farmers. The recommendations charged ICRAF with developing guidelines for the domestication of NTFPs that would ensure benefits to the farmer. It was recognized that local traders and processors, and hence the community, would also benefit. It was felt that a farmer-oriented approach to domestication would also benefit the environment and biodiversity. It should enhance biodiversity: (1) within the domesticated species by deliberate genetic conservation, so ensuring the maximization of the genetic base of new cultivars and (2) within the land-use system, by rehabilitating abandoned farmland through agroforestry with cultivars of locally important trees (Leakey, 1998a,b).

ICRAF's philosophy of domestication and commercialization of trees for agroforestry is closely allied to the African proverb:

If many little people,
in many little places,
do many little things,
they can change the face of the earth.

Consequently, ICRAF has started domestication programs with a number of tree species in each of the six ecoregions in which ICRAF operates. These programs, which are described in following sections, involve farmer-driven, genetic selection, and improvement.

RESEARCH IN PROGRESS

To date, more than 2500 tree species have been documented for their use on farms (Burley and von Carlowitz, 1984). It is essential for ICRAF to develop priorities to select the species most suitable for domestication. There is a move away from domesticating only fast-growing, nitrogen-fixing trees toward wild 168 indigenous trees with potential to generate cash for farmers. The germplasm of the new priority species is, almost without exception, unimproved. Given the restrictions on funds and research manpower, the emphasis of ICRAF's approach to domestication is on developing decision-making frameworks that first ascertain whether domestication should proceed, and then determine its direction and intensity (Simons, 1996b).

In the domestication of "Cinderella trees" for agroforestry, ICRAF is dealing with a very imperfect knowledge base, since these species have often been virtually overlooked by science, and are little known commercially, except in their local area. The situation is further complicated by the fact that the most appropriate domestication methods will vary depending on the uses of the trees, the value of their products or environmental services, the length of time they will be retained in agroforestry systems, and their biological traits, such as genetic variability.

The majority of domestication methods under consideration have been adapted from those used for agricultural/horticultural crops or industrial forest species. It is now recognized that the Northern-temperate bias to tropical tree improvement needs to be redressed by the development of novel approaches that take into consideration the requirements of small-scale, resource-poor farmers and their farming systems. Given that improvement is as much a social and political challenge as a biological one, it will only be through experimental implementation of a range of approaches

TABLE 5.1 Priority tree species selected for domestication by ICRAF and partners through implementation of farmer preference surveys and priority setting guidelines.

	West Africa		Latin America
Priority order	Humid lowlands	Semiarid lowlands[a]	Peruvian Amazon[a]
1	Irvingia gabonensis/I.wombolu	Adansonia digitata	Bactris gasipaes
2	Dacryodes edulis/D. klaineana	Vitellaria paradoxa	Cedrelinga catenaeformis
3	Ricinodendron heudelottii	Parkia biglobosa	Inga edulis
4	Chrysophyllum albidum	Tamarindus indica	Calycophyllum spruceanum
5	Garcinia kola/G. afzelii	Zizyphus mauritiana	Guazuma crinita[a]

[a]Preliminary, awaiting full economic evaluation. Source: Franzel et al. (1996).

TABLE 5.2 Priority fruit tree species selected for domestication by ICRAF and partners through market and ethnobotanical surveys.

Priority order	Southern Africa plateau
1	Uapaca kirkiana
2	Sclerocarya birrea
3	Zizyphus mauritiana
4	Vangueria infausta
5	Azanza garckeana

that methods and strategies will progress. Consequently, ICRAF uses a few priority species as models, hoping that others will then utilize these models to domesticate a much wider array of species.

The first two steps in the domestication process are common to most tree species; thereafter species will vary in the methods applied (Simons, 1996a). The first step is the determination of which species should receive priority for domestication. This step of priority setting has already been widely reported (Leakey and Jaenicke, 1995; Jaenicke et al., 1995; Franzel et al., 1996; Simons, 1996b). It involves household interviews to determine farmer preferences, the assessment of market potential and, finally, the inputs of researchers on technical points relating to genetic variability. A shortlist of priority species for a region can then be assembled. The second step is the likely need to make rangewide germplasm collections of the chosen species. These collections must conform to the requirements of the Convention on Biological Diversity and the rights of sovereign states and of the rural people. The implementation of this priority setting process by ICRAF and her partners has already led to the preliminary selection of 18 priority species from the humid and semiarid lowlands of West Africa, as well as in the Peruvian Amazon (Table 5.1).

Less rigorous means than those described here have been used to select other species for domestication in the southern Africa plateau or Miombo woodland and for the East African Highlands (Table 5.2). The application of systematic priority setting has been beneficial in some cases and less than satisfactory in others. For example, although it has clearly identified a number of species of great interest to farmers, it failed to identify any fodder trees in the Sahelian region. In an earlier Diagnosis and Design exercise in this region, farmers clearly identified the need for dry season fodder as one of their major constraints to livestock farming (Djimde, 1991). Why did farmers not identify this need the second time? Perhaps at the time of the interview, fodder was not in short supply. Alternatively, the interviewers might have shown more interest in other species and forgotten to ask specifically about fodder. A different problem emerged in Amazonia, where many farmers of migrant origin did not know the pool of local species from which they could choose.

Currently, germplasm collection by ICRAF is focused strongly, but not exclusively, on the species identified through farmer input and the priority setting exercise. However, in the Sahel, work is also in progress with *Pterocarpus erinaceus*, *Bauhinia rufescens*, and *Combretum aculeatum*, all of which have fodder potential but were added to the list of priority species by researchers aware of the need for dry season fodder. Similarly, in Cameroon, two commercially important medicinal species, *Prunus africana* and *Pausinystalia johimbe*, were added to the priority list for the humid lowlands of West Africa because of opportunities for farmer income, fears about the future of the resource for industry, and the perceived needs for conservation of the species and their habitats.

Rangewide germplasm collections, the second step in the domestication process, are both expensive and time consuming. Furthermore, once germplasm has been collected, especially for recalcitrant species such as *Irvingia gabonensis* and *Uapaca kirkiana*, there is a large commitment to nursery work and the establishment and maintenance of living genebanks. Consequently, the germplasm of only a few priority species has so far been extensively collected by ICRAF and its partners, the National Agricultural Research Systems (NARS) of countries in the different regions.

Prior to the present phase of ICRAF's domestication program, seeds were collected mainly from species used primarily for wood. These included *Markhamia lutea* in Kenya and Uganda, *Sesbania sesban* (and a few related species) in southern Africa, and *Faidherbia albida*, *C. aculeatum*, *Balanites aegyptiaca*, and *Prosopis africana* in the Sahelian zone. The collections of *Ses. sesban* were the first to be done by ICRAF in a fully participatory manner with the NARS (Ndungu and Boland, 1994). This involved detailed planning and extensive training of field teams in each country and then, through Material Transfer Agreements, the exchange of germplasm between participating countries for storage in national and regional seed banks.

More recently, seeds of three indigenous fruit tree species were collected using similar collaborative procedures. These collections include *Irvingia gabonensis* (and some related species, notably *Irvingia wombolu*) in West Africa, and *U. kirkiana* and *Sclerocarya birrea* in southern Africa. For the *Irvingia* species, genebanks have been established at Onne and Ibadan in Nigeria, and at Mbalmayo in Cameroon. Each genebank contains approximately 60 accessions (single-tree progenies) of each species, from a wide range of sites in Ghana, Nigeria, Cameroon, and Gabon. The collection strategy used for *Irvingia* involved farmers in the targeting of superior trees. In each village visited, farmers selected 20–30 trees with desirable traits (fruit shape, fruit size, kernel size, tree form, etc.). Further selection by the collection team (NARS, NGOs, and ICRAF) narrowed these down to two trees per village. During this period of exploration and collection, studies were made of the phenotypic variation in tree form, phenology, and fruit characteristics such as shape, color, and sweetness (Ladipo et al., 1996). Variation in fruit characteristics is very extensive in wild populations, as has been reported for *Dacryodes edulis*, the species of second priority in humid West Africa (Leakey and Ladipo, 1996).

In addition to studies on the morphological variation of these collections, work is in progress to characterize genetic variation using polymerase chain reaction (PCR) techniques. Aspects of genetic structure, amount of variation, ecogeographic partitioning, and hybridization are being examined. Molecular characterization is also in progress for *Prunus africana* populations.

Building upon the experience of collecting West African fruit species, ICRAF's latest collection efforts with national partners focused on two indigenous fruit species (*S. birrea* and *U. kirkiana*) of the Miombo woodlands of southern Africa. Although conservation and utilization of the genetic resources are intimately linked, collection missions usually assign one of these objectives higher priority than the other by carrying out either random or targeted collections, respectively. In the absence of knowledge on the efficacy of phenotypic selection, it was decided to carry out collections from the same populations, using both random and targeted selective procedures.

For *U. kirkiana*, seeds of 25 trees have been collected from each of 30 provenances by ICRAF and its partners in five countries, through close collaboration with the Southern Africa Development Community (SADC) Tree Seed Center Network. These have been exchanged between Malawi, Mozambique, Tanzania, Zambia and Zimbabwe, and nurseries are overflowing in advance of setting up field genebanks. A similar exercise has been conducted for *S. birrea* with collections in Botswana, Malawi, Mozambique, Namibia, Swaziland, Tanzania, Zambia, and Zimbabwe.

In the next phase of the work on indigenous fruit trees, ICRAF plans to use the standard horticultural techniques of grafting, air-layering, and rooted cuttings, to capture the additive and nonadditive genetic variation between individual plants (Leakey and Jaenicke, 1995), a strategy that is highly appropriate for the genetic improvement of long-lived, high-value tree species. Interestingly, *U. kirkiana* and *S. birrea* are both diecious. Therefore, female trees (fruit-bearing) can be preferentially multiplied, while recognizing the need to retain a certain number of male trees.

COMMERCIALIZATION

Throughout the tropics the products of indigenous trees are marketed locally on a small-scale, as a means of generating cash to supplement a subsistence lifestyle (Falconer, 1990; Arnold, 1995, 1996; Lamien et al., 1996; Melnyk, 1996). Frequently these products are collected in natural forest or from wild trees retained on farmland. Some of these products enter regional trade (Ndoye, 1995) and, increasingly, a few are marketed in the United States and Europe. It has been

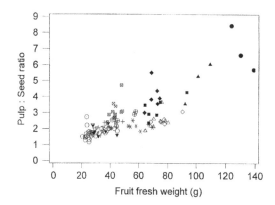

FIGURE 5.2   Genetic variation in fruit characteristics of *Dacryodes edulis* in a Yaoundé market, Cameroon. Different symbols represent different fruitlots. Solid symbols represent fruits selling for more than 500 CFA (US $1.00) per kg of pulp. *After Leakey and Ladipo (1996).*

recognized that expanded markets for these products would increase the value of natural forests and benefit forest dwellers (Peters et al., 1989). Similarly, markets for NTFPs produced on areas already deforested could improve the income of subsistence farmers and provide an alternative to slash-and-burn agriculture, one of the major causes of deforestation.

There is, however, a problem. The market for most of these NTFPs is fairly small, and the supply is often seasonal. Furthermore, the wild, unimproved product may not have enough market appeal to encourage greater commercial interest. As would be expected, there is great genetic diversity in the products of each species (see Ladipo et al., 1996) affecting, e.g., fruit size, shape, color, flavor, and pulp:seed ratio. It was interesting to note in *D. edulis* that large size and high pulp:seed ratios were not the only factors determining the prices of different fruitlots in Yaoundé market. Presumably, therefore, flavor and other nonvisible attributes are important contributors to price determination (Fig. 5.2). These genetic traits provide the opportunity for selection to enhance the appeal to both the market and the growers. In the case of fruits, traditional horticulture provides the methodology for domestication. For other products (e.g., food additives and spices), the selection of appropriate traits is more difficult to determine. For instance, what characteristics need to be improved to make a better food thickening agent? Here there is a need for dialog between the food industry and the field scientists—a dialog that does not exist at present. Similar problems relate to the pharmaceutical and cosmetic industries, where again appropriate domestication could lead to higher quality natural products if the desired traits were known to field scientists. Furthermore, there are regulations affecting the food, pharmaceutical, and cosmetic industries that determine the toxicological acceptability of products. Scientists involved in the domestication of species producing potential new products need to be aware of the limits and opportunities imposed by these regulations.

Another commercial problem is the seasonality of production, especially for perishable products such as fresh fruits. Domestication can almost certainly have considerable impact on the seasonality of production. For example, in the fruit trees *I. gabonensis* and *U. kirkiana*, there is considerable within-species variation in phenology, with some trees flowering and fruiting outside the main season or fruiting several times per year instead of the normal once per year. Since kernel extraction is a very labor-demanding process, which potentially reduces market appeal, a remarkable discovery for *I. wombolu* in Gabon was a self-cracking seed. Assuming that this is a genetically determined variation, these trees are strong candidates for selection to become cultivars through cloning (Leakey and Jaenicke, 1995).

Returning to the problem of market size, one is confronted with the question of which comes first, the demand or the supply? This is a "chicken or egg" type of problem, which is further complicated by the price elasticity of supply and demand.

In one sense, the relatively small market for these products is perhaps an asset, at least at this stage in the development of domestication activities. If demand for the products of improved cultivars were too high, large-scale company interests might swamp the production before small-scale farmers practicing agroforestry had developed marketing infrastructures. Ideally, for the purpose of poverty alleviation, and incidentally for environmental benefits through agroforestry, small-scale operations are preferable. However, a small, seasonal supply of goods is not attractive to traders. Thus, it would be more desirable commercially to have a year-round supply of products from a number of species coming in and out of production at different times of the year. It is clear that such developments require two conditions: an appropriate policy environment (Leakey and Izac, 1996) and commercial interests sympathetic to small-scale production (e.g., cooperatives of farmers), such as those associated with organic farming. Fortunately, there are already examples of highly successful international marketing of produce grown by small-scale farmers, e.g., French beans grown in Kenya (see also Sanchez and Leakey, 1997). Furthermore, while integrating a range of tree crops into farming systems at the landscape scale, there is scope for farmers to consider segregating these from staple food crops (van Noordwijk et al., 1995). The contiguous blocks of damar agroforests in Sumatra are a good example of this approach (Michon and de Foresta, 1996). Numerous other configurations of trees and crops are, however, possible and at a later stage of the domestication process there will be a need to match tree cultivars with crop varieties, in a way already well advanced for linear tree/crop plantings (Ong and Huxley, 1996).

CONCLUSION

The concepts, strategies, and policies associated with agroforestry are rapidly evolving, and need to continue to evolve, if the promises of agroforestry to alleviate poverty and mitigate environmental degradation are to be met. In this paper, we have developed the philosophy of integrating the domestication and commercialization of trees that produce valuable NTFPs into agroecosystems involving a number of agroforestry practices. This strategy gives hope of improving the well-being of the rural poor, while also promoting better land-use practices.

Chapter 6

Trees: Delivering Enhanced Crop Production and Income: An Update

R.R.B. Leakey

Chapters 4 and 5 (Cooper et al., 1996, and Leakey and Simons, 1998) illustrate the turning point from agroforestry as a practice or system primarily aimed at soil fertility improvement and a domestic source of poles, wood fuel, and timber, to one where enrichment with domesticated indigenous trees producing marketable products initiates income generation for the intensification of functioning agroecosystems.

The first paper focuses specifically on farmer-managed research which addresses the potential of agroforestry systems to enhance soil fertility and prevent soil erosion. It then presents, for the first time, the concept of combining this soil fertility restoration with the planting of domesticated high-value, indigenous food/fodder trees to generate income. The paper concludes that agroforestry systems will need to evolve in both diversity and intensity if they are to remain relevant and effective for tomorrow's Africa.

Interestingly, there does not seem to have been much recent new research on the agronomic aspects of agroforestry practices, although there has been much synthesis (e.g., Sileshi et al., 2008) and focus on their adoption. In addition, there have been some revealing assessments to quantify the extent and nature of the "planned" biodiversity in some areas (e.g., the Kenyan highlands: Kindt et al., 2006a,b) and some practices (e.g., upland coffee systems). One fairly consistent finding in several locations has been a roughly 50:50 mix of exotic and indigenous tree species in agroforestry landscapes. As we saw in Chapter 2 (Leakey, 2014), coffee and cocoa tree shade systems have also been the prime focus of assessments of the ecological role of the "unplanned biodiversity." See also Chapter 3 (Leakey, 2017a).

Looking to the future, it is quite common for farmers to practice more than one or two agroforestry systems within a single farm (Fig. 6.1). However, interestingly, there has been little "whole farm" research to examine the interactions between different agroforestry practices either for maximized production, or for their agroecological benefits, including the mitigation of climate change. Such work could examine how mosaics and boundary plantings (cf. "biodiversity corridors") affect the habitat and populations of predators, parasites, and defoliators at different trophic scales. Thus, more systems-based quantification of either "planned" or "unplanned" biodiversity is needed if progress is to be made toward "whole farm" agroecological understanding in which different practices form components of functioning landscape mosaics. In particular, there needs to be better understanding of the density and/or configuration of these species in different agroforestry systems within a landscape (Fig. 6.2) and with their expected agroecological benefits.

Likewise, there has been little work to examine the interactions between agroforestry systems linking their agroecological benefits to income or domestic/social benefits. In other words, there are many disconnects at the whole farm or landscape level between the systems and their expected benefits flows. Thus, it can be concluded that agroforestry systems research still has some way to go before it can be better understood as an integrated land-use strategy. This becomes even more important, and complex, when adding the next layer of management—the intensification of the tree component by domestication through the selection and deployment of tree cultivars in farming systems.

The second paper introduces the idea of domesticating high-value indigenous trees to meet both the needs of farmers and as a means of both diversifying and intensifying agroecosystems for greater and more sustainable production and for income generation. This idea was first raised in 1982 (see Chapter 11 [Leakey et al., 1982]) and then explored in more detail during a conference in 1992 in Edinburgh (Leakey and Newton, 1994a,b). Now, these concepts have been further developed (See Chapters 14 and 16: [Leakey et al., 2012 and Leakey and Akinnifesi, 2008]) and have emerged as a powerful concept for rural development in Africa and indeed other parts of the tropics and subtropics (see Chapter 28 [Leakey, 2014]; and Chapter 37 [Leakey, 2014f]).

Multifunctional Agriculture. DOI: http://dx.doi.org/10.1016/B978-0-12-805356-0.00006-4
© 2017 Elsevier Inc. All rights reserved.

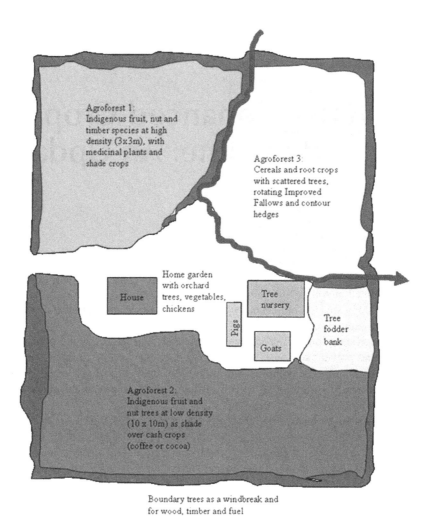

Agroforest 1:
Indigenous fruit, nut and
timber species at high
density (3x3m), with
medicinal plants and
shade crops

Agroforest 3:
Cereals and root crops
with scattered trees,
rotating Improved
Fallows and contour
hedges

Home garden
with orchard
trees, vegetables,
chickens

House

Pigs

Tree
nursery

Tree
fodder
bank

Goats

Agroforest 2:
Indigenous fruit and
nut trees at low density
(10 x 10m) as shade
over cash crops
(coffee or cocoa)

Boundary trees as a windbreak and
for wood, timber and fuel

FIGURE 6.1   A diagrammatic representation of "whole farm" agroforestry in which different practices have come together as an integrated approach to maximize economic, social, and agroecological benefits.

FIGURE 6.2   An agricultural landscape mosaic providing diverse niches for "unplanned biodiversity."

As we saw in Chapter 5 (Leakey and Simons, 1998), the domestication of trees producing useful products was defined as "a progression from collection and utilization of products, through protection, management and cultivation, which culminates with genetic manipulation" which can either be farmer-oriented or market-oriented (Fig. 5.1), although the orientations are not mutually exclusive. They do, however, call for a clear and wise strategy for genetic improvement that protects the diversity of the genetic resource (Chapter 16 [Leakey and Akinnifesi, 2008]; Dawson et al., 2014). By 2004, a much more complex definition of domestication was proposed. It was summarized as "tree domestication in agroforestry is farmer-driven and about matching the intra-specific diversity of many locally important tree species to the needs of subsistence farmers, the markets for a wide range of products and the diversity of agricultural environment" (Simons and Leakey, 2004). This was linked to the concept of a decentralized approach to developing horticultural cultivars using vegetative propagation (see Chapters 17 and 20 [Leakey et al., 1990; Leakey, 2014]) and simple characterization of the very extensive tree-to-tree variation (Chapters 22 and 23 [Leakey et al., 2005b,c]) of many underutilized indigenous fruits, nuts, and leaves (Chapter 7 [Leakey, 1999]). Typically, these wild species are little known to science, but have great cultural value as traditional foods (Chapters 7 and 10 [Leakey, 1999; Leakey, 2017c]) important for nutrition and local livelihoods (Schreckenberg et al., 2006; Degrande et al., 2006; Jamnadass et al., 2011). They are now attracting growing interest in some industrial countries. In addition, many of these species are traditionally important as medicines extracted from bark (Chapters 12 and 14 [Leakey and Simons, 2004; Leakey et al., 2012]) and numerous other wood-derived products of interest to pharmaceutical or other industries (Page et al., 2010a).

The decentralized, bottom-up community approach has been called "participatory domestication" (Chapters 27 and 28 [Leakey et al., 2003; Leakey, 2014]; Tchoundjeu et al., 2002, 2006, 2010; Leakey et al., 2003; Asaah et al., 2011). It has focused on the priority species identified by local communities and it has gone on to deliver many social and economic benefits (Chapter 29 [Lombard and Leakey, 2010]) arising from linkages with the value chain (Leakey and Izac, 1996; Leakey and van Damme, 2014; Lombard and Leakey, 2010). Today, it is clear that a focus on the development of local markets holds great promise for the development of cottage industries in rural and peri-urban areas providing business and job opportunities for local people (Chapter 28 [Leakey, 2014]). These are opportunities that have not existed previously and that allow some members of the rural community to leave subsistence agriculture and climb the ladder into the cash economy. This is an area for great future expansion creating new markets and new local industries, based on very specific trait combinations or "ideotypes" (Chapter 25 [Leakey and Page, 2006]; Figs. 16.4 and 28.5). However, these opportunities raise a new set of challenges about how to protect innovative farmers and their families from commercial exploitation (Chapter 29 [Lombard and Leakey, 2010]; Santilli, 2015).

Since the early days of tree domestication in Cameroon, the concept has spread to other countries, as e.g., in the forest region of West Africa (Peprah et al., 2009; Onyekwelu et al., 2014), and to the Sahel (Abasse et al., 2011; de Smedt et al., 2011; Assogbadjo et al., 2005, 2006; Ræbild et al., 2011; Sotelo Montes and Weber, 2009), East Africa (Thiong'o et al., 2002; Goenster et al., 2011; Jamnadass et al., 2010), Southern Africa (Akinnifesi et al., 2006; Mng'omba et al., 2015) and to Latin America (Cornelius et al., 2010; Weber et al., 2001; Parker et al., 2010), Southeast Asia (Roshetko and Evans, 1999; Narendra, et al., 2013; Mulyoutami et al., 2015), and to Oceania: Solomon Islands (Pauku et al., 2010), Vanuatu (Page et al., 2010a,b), and Papua New Guinea (Nevenimo et al., 2008; Leakey et al., 2008)—see Figs. 14.3 and 14.4.

Section 3

Importance of Tree Products

Many thousands of edible plants are wild and undomesticated, and many of them have been regularly consumed and highly appreciated by local people. As common property natural resources from the wild, the products of these species are said to be nontimber or nonwood forest products. Typically these species are much richer in nutrients than the few cereals and root crops that have been domesticated as staple foods on account of their carbohydrate content. The first paper reviews what is known about the nutritive values of the 17 species identified by African farmers as those of top priority for domestication in four ecoregions of Africa. Typically the published data are the mean values for each species, with little if any information about the extent of genetic variation in these values to guide the domestication process. Furthermore, there is seldom any information of the ranges of size, flavor, or indeed any traits that may be important to the trade or commercialization of the products.

The second paper presents data for two species (152 trees of *Irvingia gabonensis* and 293 trees of *Dacryodes edulis*) gathered from six villages in Cameroon and Nigeria. In addition to quantifying the extent of tree-to-tree morphological variation in a number of fruit and kernel traits, the frequency distribution of this data is examined with regard to a hypothesis that the range and frequency of variation in the different populations can be used to identify five stages of domestication. On the basis of the results of this study, *D. edulis* can be said to be virtually wild in Nigeria but semidomesticated in Cameroon, while *I. gabonensis* is wild in Cameroon and semidomesticated in Nigeria. With regard to the domestication of these species, this study clearly indicates both the perceived importance by local people of these species and the potential for phenotypic selection of different traits at the village level. The paper also recognizes the need for collaboration between agroforesters, farmers, and the food industry in initiatives to domesticate these species, both for local consumption and for larger-scale trade and for value-addition in wider-scale commercialization. The extension of these concepts is reported in later chapters.

In recognition of the role these species can play as new crops in the development of many tropical and subtropical countries through the actions of local farmers, the third paper proposes that the products should be known as "agroforestry tree products". This name change reflects their origin as private property when cultivated on-farm, a change which should therefore exclude them from legislation governing products gathered from natural forests and woodlands. This name change will also allow their production to be incorporated into agricultural statistics.

Chapter 7

Potential for Novel Food Products From Agroforestry Trees

This chapter was previously published in Leakey, R.R.B., 1999. Food Chemistry, 64, 1—14, with permission of Elsevier

SUMMARY

The domestication of trees for agroforestry approaches to poverty alleviation and environmental rehabilitation in the tropics depends on the expansion of the market demand for their non-timber forest products (NTFPs). This paper reviews published data on the nutritive values of the flesh, kernels, and seed-oils of the 17 fruit tree species that have been identified, in 4 ecoregions of the tropics, by subsistence farmers as their top priorities for domestication.

In some species, genetic variation in nutritive value has been reported, but in most species there is still inadequate information on which to base programs for the genetic improvement of these species. Farmers and agroforesters have identified many of the biological constraints relevant to their viewpoint on production, but there is a need for inputs from the food industry into the identification of the desirable traits and characteristics of potentially novel food products. This paper calls for greater collaboration between agroforesters and the food industry in the effort to promote the domestication and commercialization of underutilized tree products.

INTRODUCTION

New initiatives in agroforestry are seeking to promote poverty alleviation and environmental rehabilitation in developing countries, through the integration of indigenous trees, whose products have traditionally been gathered from natural forests, into tropical farming systems (ICRAF, 1997). This is being done in order to provide marketable products from farms that will generate cash for resource-poor rural and peri-urban households. One important component of this approach is the domestication of the local tree species that have commercial potential in local, regional or even international markets (Leakey and Simons, 1997). Consequently, in collaboration with ICRAF (formerly International Center for Research in Agroforestry, now the World Agroforestry Center), farmers in four ecoregions of the tropics (the humid and dry zones of West Africa, Amazonia, and southern Africa) have identified their priority indigenous trees for "domestication," from among the many that have been, and are still being, used traditionally, to provide people's needs for food and nutritional security (Fig. 7.1). For most of these hitherto wild species, little attention has been paid to seek market opportunities for the products within the international food industry, although in some instances research has been carried out to assess their food value and potential for domestication. To meet the objective of market creation and expansion, it is important to identify potential market niches and then to determine whether there are important product characteristics, which should be improved through genetic selection. While some traits that are relatively easy to identify do benefit the farmer, there are undoubtedly others that are important to the food industry, but that require more sophisticated evaluation in collaboration with the private sector. This review seeks to draw together, in priority order for each region, all the existing information on the characteristics of the products from the tree species that farmers have identified as being their preferred choice for domestication.

Multifunctional Agriculture. DOI: http://dx.doi.org/10.1016/B978-0-12-805356-0.00007-6
© 2017 Elsevier Inc. All rights reserved.

FIGURE 7.1   Tree products on a market stall in Kumba, Cameroon.

FIGURE 7.2   Fresh fruits of *Irvingia gabonensis* in southern Cameroon. Note variation in color and size.

HUMID LOWLANDS OF WEST AFRICA

Irvingia gabonensis (O'Rorke) Baill. and Related Species (Bush Mango or Dika Nut)

This fruit is like that of a small, cultivated mango in appearance (Fig. 7.2), although the two are unrelated. The pulp of this fruit is eaten fresh and the kernel of the nut is a food additive. The flesh is juicy and varies between sweet and bitter. The sweeter form is generally considered to be *I. gabonensis* var. *gabonensis*, while the bitter form is var. *excelsa*, now called *I. wombolu*. Trees are being selected for the sweetness of their fruits, fruit size, color, and other desirable traits (Ladipo et al., 1996), but not so far for any kernel traits. The pulp can be used for the preparation of juice, jelly, and jam. The extraction rate of juice from the fruit pulp was 75% and the sugar concentration of this juice is comparable with pineapples and oranges (Akubor, 1996), but with a higher ascorbic acid content (67 mg/100 ml). This concentration of ascorbic acid is also nearly three times that of *Dacryodes edulis* and *Chrysophyllum albidum* (Achinewhu, 1983). These fruits are therefore a good local source of vitamin A.

Evaluation of the winemaking potential of the juice (Akubor, 1996) found that wine produced after 28 days fermentation had 8.12% alcohol content. Sensory evaluation showed no significant difference in color, mouthfeel, sweetness, flavor, and general acceptability from a German reference wine.

FIGURE 7.3 Prepared and unprepared kernels of *Irvingia gabonensis* in Cameroon.

A study of fruit ripening and storage (Joseph and Aworh, 1991) has shown that fruits harvested at the mature green stage and ripened at 26–29°C were preferred to tree-ripened fruits in color and texture, although they were both comparable in composition. Fruits held at 12–15°C developed symptoms of chilling injury. In a separate study (Aina, 1990), ripening fruits were found to increase in soluble solids and carotenoid content, decrease in acidity and to undergo starch hydrolysis.

The most important product from these species (especially *I. wombolu*) is, however, the kernel of the nut, which is extracted, dried, and can be stored for long periods (Fig. 7.3). These kernels are traded on both a local and a regional scale in West Africa (£1 – 3 kg^{-1} depending on season). Uzo (1980) considered that the fruits from a single tree could generate income of US$300 per annum. The composition of *I. gabonensis* var. *excelsa* (now *I. wombolu*) kernels at 88.1% dry matter has been reported by Ejiofor et al. (1987) to be 51.3% fat, 26.0% total carbohydrate, 2.5% ash, 7.4% crude protein, 0.9% crude fiber, 9.2 mg/100 g vitamin C, and 0.6 mg 100 g^{-1} vitamin A.

Other reports (e.g., Oke and Umoh, 1978) have quoted values of 54–67%, and even 72%, for fat content and 38.8% for carbohydrate (Ejiofor, unpublished). Okolo (unpublished) reports that the fat has an absence of volatile oils, a melting point at 37–42°C, saponification value of 233–250 and an iodine value of 2–9. He also quotes reports from 1929–39, that the myristic acid and lauric acid content of *Irvingia* kernels vary depending on the source of the fruits (Nigeria: 50.6% and 38.8%; Sierra Leone: 33.5% and 58.6%, respectively). Unpublished data (Hellyer, 1997) has given myristic acid and lauric acid values of 39.2% and 51.1% from *I. wombolu* kernels from Cameroon. The amino acid composition of kernels has been reported by Amubode and Fetuga (1984).

A comparison of kernel composition between *I. gabonensis* and *I. wombolu* has shown that *I. wombolu* has less fat, more crude protein, less crude fiber, and less vitamin C than *I. gabonensis* (Ejiofor et al., 1987). The fat from *I. wombolu* has lower iodine and saponification values (Joseph, 1995).

Kernels are processed by grinding and separating the residue from the fat. The residue is used as a food additive to thicken soups and stews, as it produces a viscous consistency when added a few minutes before serving. A rheological study of the polysaccharides in dika nut found that the variation of "zero-shear" specific viscosity was broadly similar to the general form of disordered polysaccharides, although with some specific attributes consistent with it having a compact molecular geometry rather than a "random coil" conformation (Ndjouenkeu et al., 1996). Joseph (1995) reports that the viscosity of mucilaginous solutions is lower at high temperatures and at high shear rates, making it appropriate as a thickening agent. The residue can be made into cubes/pellets with enhanced storage life (Ejiofor et al., 1987). Okolo (unpublished) has calculated that a pilot plant, with a capacity of 100 kg per hour, would require 256 tonnes of kernels per year. Calculated on crude protein basis, dika nut meal shows comparatively better water and fat absorption properties than raw soy meal and hence it may have useful applications in processed foods, such as bakery products and minced meat formations (Giami et al., 1994).

Dacryodes edulis (G. Don) H.J. Lam and Related Species (African Plum, African Pear, or Safoutier)

A recent workshop in Cameroon (Kengue and Nya-Ngatchou, 1994) reviewed knowledge of this species in view of new initiatives for its domestication. The flesh of the fruit has good nutritional value and has been reported by Umoro

FIGURE 7.4   Fruits of *Dacryodes edulis* from Yaoundé market. Note variation in color, size, and price.

Umati and Okiy (1987) to contain, as a percentage of dry matter (dm), 31.9% oil, 25.9% protein, 17.9% fiber. The main fatty acids in the lipid fraction are palmitic acid (36.5%), oleic acid (33.9%), and linoleic acid (24%), giving a profile similar to palm oil (*Elaeis guineensis*). The main essential amino acids are leucine (9.57%) and lysine (6.3%), while others are glutamic (17.0%), aspartic (15.1%), and alanine (7.7%) acids. The ascorbic acid content of the flesh is 24.5%, but this is lost by some forms of cooking (Achinewhu, 1983). Many of the nutrients are, however, in the skin of the fruit, which is usually discarded.

The seeds of *D. edulis* are usually discarded, but analysis shows them to have considerable nutritional value and a lack of toxins that makes them at least useful as a supplement to animal feed (Obasi and Okolie, 1993).

Okafor (1983) defined two varieties (*D. edulis* var. *edulis* and *D. edulis* var. *parvicarpa*) in Nigeria on the basis of their size and the relationship between their longitudinal and midtransverse circumferences. In the Congo, on the other hand, Silou (1996) characterized four fruit types that vary in size and shape. However, a small survey of three markets in Yaoundé (Leakey and Ladipo, 1996) determined that there was four- to fivefold variation between fruitlots in fruit weight, pulp:seed ratio and price per kilogram of pulp (Fig. 7.4). While large fruits with a high pulp:seed ratio were usually highly priced, it was clear that some small fruits also commanded a high price, presumably because flavor, quality, and other variables were important in the marketplace. Although Leakey and Ladipo (1996) report continuous variation in fruit size, pulp:seed ratio and other fruit characteristics between different fruitlots of different origins, Youmbi et al. (1989) indicate that there are two morphological types on markets in Cameroon: a large fruit with a large seed and a small (short) fruit with a well-developed mesocarp. They further indicate that these two types vary in their chemical composition, with the large type characterized by a higher lipid content in the mesocarp than in the seed, and the converse in small fruits. Fatty acid content was, however, not significantly different in two contrasting fruit types (Kapseu and Tchiegang, 1996). Nonstructural carbohydrates are higher in the seed than in the mesocarp of both types. There is also much variation in taste and some variation in protein content (Kapseu and Tchiegang, 1996). The further characterization of these differences is important in the domestication of the species and their orientation to different markets.

Tests have determined that storage life of fruits can be prolonged beyond 8 days by refrigeration (Emebiri and Nwufo, 1990). At 15°C, storage life was 2 weeks, although some fruit types did deteriorate over this period. The causes of this variation in storage life need to be determined. A palm oil dip, or enclosure in a polythene bag, enhanced storage life at 15°C. At 5°C, susceptible fruit types remained firm, but they deteriorated before day 25. The apparent genetic variation in shelf life is a trait that should be included in the selection of cultivars.

Ricinodendron heudelottii (Baill.) Heckel (Peanut Tree, Essessang, or Nyangsang)

The kernels of the nut are widely traded in Cameroon and used as a flavoring in food dishes with the oil used in cooking (Fig. 7.5). The paste of ground kernels is said to have a better taste than groundnut sauce (Ndoye, 1995). However, remarkably little is known about the products of this species. In Ivory Coast the kernels are used as a condiment (Téhé,

FIGURE 7.5 Kernels from fruits of *Ricinodendron heudelottii*.

FIGURE 7.6 Fruits of *Chrysoplyllum albidum* from Nigeria. *Photo by D.O. Ladipo.*

1986). The nutritive value of kernels is recorded in Pélé and Berre (1967), but the data has not been seen by the author. Kapseu and Parmentier (1997), however, report that the polyunsaturated acids are high (79.4%) and that the unsaponifiable matter is low (1.6%). The kernels can be stored for long periods.

Chrysophyllum albidum G. Don (White or African Star-apple)

Achinewhu (1983) has reported that fruit pulp (Fig. 7.6) contains 21.8 mg 100 g^{-1} ascorbic acid, while the skin contains 75 mg 100 g^{-1}, while Edem et al. (1984) report 446.1 and 239.1 mg/100 g for pulp and skin, respectively. The latter authors also indicate that proximate analysis of fruit pulp was protein (8.8%), lipid (15.1%), ash (3.4%), carbohydrate (68.7%), and crude fiber (4.0%), with only minor differences between pulp and skin. With the exception of calcium (100 vs 250 mg 100 g^{-1}) and iron (10 vs 200 mg 100 g^{-1}) in pulp and skin, respectively, the mineral content of these components of the fruit were also very similar. According to Achinewhu (1983), the levels of toxic substances in both the mesocarp and the pericarp were low, although the juice was highly acidic. Edem et al. (1984), on the other hand, identified high levels of tannins in pulp (627 mg 100 g^{-1}) and lower levels in peel (264 mg 100 g^{-1}). Fruit storage was best at 10°C, while for the kernel the traditional method of storing in layers of red clay was best.

The juice of fruits has potential as an ingredient of soft drinks and can be fermented for wine or other alcohol production (Ajewole and Adeyeye, 1991).

The seeds of this species are not particularly rich in lipids (3.2%), but linoleic (38.4%) and oleic (29.6%) acids are the main fatty acids present (Essien et al., 1995). Ajewole and Adeyeye (1991) have, however, reported higher lipid content (16.6%) and confirmed that unsaturated fatty acids are the main component of the oil (74%) and hence it is desirable in the context of heart disease risk reduction. The residual cake also has potential for animal feed.

Garcinia kola Heckel and Related Species (Bitter Cola)

The flesh of the fruit is edible and has medicinal uses (Fig. 7.7). Comparison of the nutritive value of the pericarp and mesocarp of fresh fruits from Nigeria (Dosunmu and Johnson, 1995) shows that crude protein was higher in the mesocarp than in the pericarp (7.8% vs 3.9%), while the pericarp was richer in crude fiber (16.5% vs 13.9%) and macro elements (e.g., K: 990 vs 499; Fe: 150 vs 4.2; Ca 200 vs 100 mg 100 g^{-1}). The mesocarp was richer in N (1248 vs 624 mg 100 g^{-1}) and P (720 vs 520 mg 100 g^{-1}). The mesocarp was also richer in crude lipid (8.7 vs 6.9%) and ascorbic acid (127 vs 93 mg 100 g^{-1}).

The kernels of the nuts are widely traded and eaten as a stimulant (Fig. 7.8). Unsaturated fatty acids (linoleic acid: 40.5%, oleic acid: 30.8%) are the main components of the lipids (4.5%) found in the seeds of this species (Essien et al., 1995; Omode et al., 1995). The low kernel oil content of this species, however, probably eliminates it as a commercial source of oil (Foma and Abdala, 1985).

The chemical, brewing, and antimicrobial properties of *Garcinia kola* seeds have been compared with hops in lager beer brewing, because of their similarity in flavor and greater availability in West Africa (Aniche and Uwakwe, 1990). Treatment of *G. kola* with methanolic lead acetate produced a yellow precipitate from which organic acids (alpha acids) were confirmed by thin-layer chromotography. Hops, however, had a higher concentration of organic acids than *G. kola*. Laboratory brewing trials with both products gave beers with similar chemical properties. Organoleptically, *G. kola* beer was as acceptable to tasters as hopped beer, but with an improved bitterness. *G. kola* and hop extracts exerted similar antimicrobial effects on two beer spoilage microorganisms (*Candida vini* and *Lactobacillus delbruckii*).

The products of three *Garcinia* species (*G. kola* = 36%), are widely used in Ghana and 70% of this use is as chewing sticks. These are bought in urban markets as an alternative to toothpaste and brush (Adu-Tutu et al., 1979). The

FIGURE 7.7 Flowers and fruits of *Garcinia kola* in Cameroon.

FIGURE 7.8 Kernels from the nuts of *Garcinia kola*.

good dental health is attributed to these chewing sticks, despite the shortage of dentists (1 per 150,000 people) by comparison with the United Kingdom (1 per 3000 people), although it has to be remembered that there are also dietary differences between these countries.

SEMI-ARID LOWLANDS OF WEST AFRICA

Adansonia digitata Linn. (Baobab)

The young tender leaves of baobab are used as green or dried vegetables (Fig. 7.9), rich in vitamin A and calcium, while the white powdery pulp of the fruit capsule is extracted and used as a flavoring in a variety of cool and hot drinks. The fruit are rich in pectins and have a vitamin C content of 169 mg 100 g^{-1} (Agbessi Dos-Santos 1987), at least 10-fold greater than that of oranges (Booth and Wickens, 1988). The seed kernels contain 12−15% edible oils, more protein than groundnuts and are rich in lysine, thiamine, calcium, and iron (Booth and Wickens, 1988).

In the Sahel four types of baobab are recognized: black-bark, red-bark, gray-bark, and dark-leaf. The dark-leaf baobab is preferred for use as a leafy vegetable, while the black- and red-bark baobabs are preferred for their fruits (Fig. 7.10). Baobab leaf is an excellent source of calcium, iron, potassium, magnesium, manganese, molybdenum,

FIGURE 7.9 Dried and ground leaves of *Adansonia digitata* in Mali. *Photo by J. Baxter.*

FIGURE 7.10 Fruits of *Adansonia digitata* in Malawi.

phosphorus, and zinc (Yazzie et al., 1994), and has an amino acid composition that compares favorably to that of an "ideal" protein. Dried leaves are rich in β carotene (Nordeide et al., 1996). The leaves also contain an important amount of mucilage (Gaiwe et al., 1989).

Recently, Sidibé et al. (1996) assessed the tree-to-tree variation in vitamin C content of fruits from the black-, red-, and gray-bark types in 2−3 trees from 4−5 villages in three areas of Mali spanning a range of rainfall zones (450−500 mm; 600−700 mm; 750−850 mm). The vitamin C content varied threefold between trees, but there were no consistent differences in vitamin C content between zones or tree types. The powders from fruits are added to drinks and to gruel as it cools after cooking, so preserving the vitamins. Healthy, nonsmoking adults need 23 g per day of baobab powder to meet their vitamin C requirements, while convalescents or nursing mothers require 90 g.

In Senegal, Becker (1983) reported that fruit with 91.3% dry matter contained 73.7% carbohydrates, 8.9% fiber, 0.2% fat, 2.7% crude protein, and 209 mg per 100 g vitamin C.

In Malawi, fruit pulp of baobab at 86.8% dry matter was found to contain: 79.4% total carbohydrate, 8.3% fiber, 4.3% fat, and 3.1% crude protein, and high levels of several minerals, including 28.4 mg g^{-1} K and 1156 μg g^{-1} Ca (Saka et al., 1994). Ascorbic acid content is 179.1 mg 100 g^{-1} fresh weight (fw) (Saka, 1995). Proximate analysis of the seed kernel at 92.12% dry matter indicates that it is 29.6% fat, 28.7% crude protein, and 7.3% crude fiber, while K and Ca are 1186 and 456 mg 100 g^{-1}, respectively. Fatty acid composition is 31.7% oleic acid, 30.8% palmitic acid, and 25.2% linoleic acid.

In Nigeria, similar results were obtained (Odetokun, 1996), although carbohydrate contents were lower and protein content higher than in Malawi. Carbohydrates were higher in the pulp than in the seed and vice versa for protein. Eromosele et al. (1991) observed that fruits were rich in Mg (209 mg 100 g^{-1}) and ascorbic acid (337 mg 100 g^{-1}).

Traditionally, baobab seeds and pulp are sun-dried, roasted, or fermented to extend shelf life and enhance nutritive value. When fruits from Maiduguri (Nigeria) were treated experimentally (Obizoba and Amaechi, 1993), it was found that fermentation for 6 days was better than roasting with regard to the value of crude protein (36.4 vs 32.7%), fat (34.1 vs 32.0%), and carbohydrate (30.0 vs 23.5%). Fruit pulp and aqueous extracts stored over a period of 8 months, with and without sodium metabisulphite, was found to deteriorate rapidly during periods of high humidity, unless treated with an antioxidant (Ibiyemi et al., 1988). Storage of pulp was also prolonged by use of airtight containers, while juice could be stored at 10°C.

Vitellaria paradoxa Gaertn. syn. *Butyrospermum paradoxum* (Sheanut or Karité)

This tree is one of the most common components of the Sahelian Parklands and occurs over very large areas of Africa. The nuts (Fig. 7.11) are used for oil extraction for cooking, soap, and cosmetics. Nut production is about 3−6 kg of dry kernels per tree, but varies considerably between trees and years (Hilal, 1993; Boffa et al., 1996). One hundred kilograms of fruits give about 10 kg of dried kernels. These will yield about 5 kg of butter, with an oil content of 46.3−51.6% (33% nonsaturated and 67% saturated).

After removing the fruit pulp, the seeds are dipped in boiling water, dried or smoked and stored. After shelling and grounding, the butter is extracted. The wet extraction process uses either boiling water or churning in cold water. The dry method uses heat and pressing. Another method uses organic solvents. The butter itself consists of a saponifiable

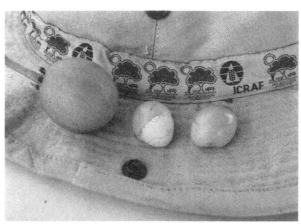

FIGURE 7.11   Fruit, nut, and exposed kernel of *Vitellaria paradoxa* in Mali.

fraction, containing triglycerides rich in vitamin F, and an unsaponifiable fraction, consisting of karitens, triterpenic alcohols, phytosterols, and vitamins A, D, and E, which give the butter its cosmetic hydrating, protecting, revitalizing, and curative qualities (Hilal, 1993).

Many analyses have been done of shea butter (see Booth and Wickens, 1988), but according to Sawadogo and Bezard (1982), it contains 45.6% oleic acid and 44.3% stearic acid. Oleic acid was found preferentially esterified in the 2-position (60%). The total triacylglycerols were fractionated and the fractions were analyzed for fatty acid and triglyceride compositions: the monounsaturated fraction accounted for 50% and the di-monounsaturated fraction for 27.3% of the fat. The proportion of 30 possible isomers could be determined. Only 11 isomers could be found at over 1%. Two isomers accounted for 60% of the shea butter.

According to Badifu (1989), the nonpolar lipid components of shea butter were sterols, diglycerides, free fatty acids, and triglycerides. The main components of nonpolar lipids were triglycerides. The major fatty acids of the triglyceride were stearic acid (about 46%) and oleic acid (about 41%). Others present in relatively small quantities were 4% palmitic, 7% linoleic, and 1% linolenic acids. The free sterols were 11% campesterol, 20% stigmasterol, and 68% beta-sitosterol. The polar lipid components in phospholipids were phosphatidylcholine (lecithin), phosphatidylserine, and phosphatidylethanolamine (cephalin). The glycolipid component was digalactosyldiglyceride and the main sugar moieties were galactose (about 32%) and glucose (about 66%). The predominant fatty acids in phospho- and glycolipids were stearic (36−50%), oleic (41−50%), and linoleic (6−11%).

Chavelier (1943), however, reports that of the glycerides 7.0% are saturated (tributyrine 3.1%, dibutyrostearirine 3.1%, arachidodipalmitine 1.0%), while 93% were nonsaturated (dipalmitsoleine 19%, dibutyrsoleine 54%, and palmitodioleine 19%).

After refining, traditionally prepared shea butter is tasteless and odorless. It has been sold as baking fat, margarine, and other fatty spreads and finds increasing use in edible products (Booth and Wickens, 1988). The fat is useful in patisserie and confectionary, the latex in the fat giving pliability to the dough. It is also used to formulate a cocoa butter substitute, which is unnoticeable in the final product.

Trees of *Vitellaria paradoxa* also produce a latex which can be tapped. No literature has been found giving the properties of this latex.

Parkia biglobosa (Jacq.) R. Br. ex G. Don (Néré or Locust Bean)

The seeds of Néré are fermented to make soumbala or dawadawa, a black, strong-smelling, flavorsome, tasty, proteinaceous food that is eaten for 50−90% of the year (Fig. 7.12). This keeps without further treatment for long periods and

FIGURE 7.12 Soumbala made from *Parkia biglobosa* fruits in Mali. *Photo by J. Baxter.*

is eaten in small quantities with sorghum or millet dumplings or porridge (Booth and Wickens, 1988). It is rich in protein (40%), lipids (35%), linoleic acid, and vitamin B_2 (0.4−0.9 mg 100 g^{-1}) and widely traded in urban markets. Soumbala is deficient in the amino acids methionine, cystine, and tryptophan, like other legume seeds, but the cereals in the diet compensate for this deficiency.

The yellow, floury pulp around the seeds in the seedpod is a high-energy food with up to 60% sugar (20% reducing sugars and 10−24% sucrose) and 291 mg vitamin C 100 g^{-1} dm (Campbell-Platt, 1980). This pulp can be eaten raw, pressed into a cake or made into a refreshing drink with water, or fermented into an alcoholic beverage. The pods and leaves can also be eaten. Dried flour, unlike dried fermented seeds, were rich in α and β carotene (Nordeide et al., 1996).

Tamarindus indica Linn. (Tamarind)

Tamarind products are highly developed and widely used in Asia and so far little used in Africa, although syrup, and jam are made from fruits (Fig. 7.13). In India and Thailand especially, cultivars are grown and the food industry is active. Tamarind gum (or hydrocolloid) is a polysaccharide polymer (D-galactose, D-xylose, and D-glucose) obtained from the endosperm of the seeds. It is extracted, purified, and refined, and used as a thickening, stabilizing, and gelling agent in foods, especially in Japan where Dainippon Pharmaceutical Co conducted 2 years of feeding toxicity tests (Glicksman, 1996). In India it is the chief acidifying agent in curries, chutneys, and sauces. The gum can also be used as a binder in pharmaceutical tablets, as a humectant and emulsifier (Hulse, 1996). Proximate analysis of seed kernels shows that 65.1−72.2% is nonfiber carbohydrate, 15.4−22.7% is protein, 3.9−7.4% is oil, and 0.7−8.2% is crude fiber.

Two main products are used by the food industry: (1) tamarind kernel powder (TKP), which contains about 50% gum and (2) tamarind gum polysaccharide (TGP), the purified product that is virtually 100% pure. These two products have different specifications (see Glicksman, 1996) and uses. TKP hydrates quickly in cold water, but reaches maximum viscosity if heated for 20−30 minutes. TGP is more soluble but still requires some heat. A typical 1.5% gum solution will yield a viscosity of 500−800 cps at 25°C. TGP has excellent stability over a range of pH, with electrolytes (e.g., 20% salt) and at temperatures below 65°C and degrades rapidly at higher temperatures and low pH. TSP has the ability to form gels in the presence of sugar or alcohol and can be used to form pectin-like gels in jams, jellies, and other preserves (Glicksman, 1996). The xyloglucan from tamarind seeds offers no chemical advantage over guar gum as a viscosifier, but tamarind flour is cheaper, indicating that a bioprocess to upgrade the tamarind polysaccharide might be commercially viable (Reid and Edwards, 1995).

In Nigeria fruits have been analyzed for their ascorbic acid content (Eromosele et al., 1991), but found not to be particularly rich in this vitamin.

In Malawi, tamarind fruits with 73.1% dry matter were found to contain: 85.0% total carbohydrate, 5.9% fiber, 1.6% fat, 4.1% crude protein (Saka et al., 1994). Ascorbic acid content is 19.7 mg 100 g^{-1} fw (Saka, 1995), but α and β carotene are absent from both dried leaves and dried fruits (Nordeide et al., 1996).

FIGURE 7.13 Syrup and jam made from fruits of *Tamarindus indica*.

Zizyphus mauritiana Lam. (Jujube or Ber)

The fruits of jujube are one of the best edible wild fruits and some cultivars are planted. The fruits, which vary in size, are sweet and rather dry with a comparatively large stone and the larger fruits are often eaten raw (Booth and Wickens, 1988). The fruits are also boiled with rice and millet and stewed or baked. Alternatively they are made into jellies, jams, chutneys, or pickles. They can also be candied or sun-dried.

Great variation has been recorded in the fruit's nutritional value (Becker, 1983; Geurts, 1982), but they are generally rich in sugars (5.4−23%), vitamin C (96−500 mg/100 g), vitamin A and carotene (21−81 mg 100 g^{-1}). Eromosele et al. (1991) have reported that the fruits are rich in Ca (712.5 mg 100 g^{-1}) and Mg (227 mg 100 g^{-1}).

The potential of these fruits is virtually untapped in Africa, but is commercially exploited in India and Pakistan where cultivars are well-developed—see also the section on this species under southern Africa.

SOUTHERN AFRICAN PLATEAU—MIOMBO WOODLANDS

Sclerocarya birrea (A. Rich) Hochst. (Marula)

The products of this pan-African, dry forest tree (fruit, nuts, oils, juice, gums, etc.) have been extensively characterized in South Africa and the findings reviewed by Weinert et al. (1990). The fruits and nuts, in particular, have considerable commercial value. The fruits (Fig. 7.14), which vary 30-fold in their reported flesh:stone ratio, are described as having exotic flavor and high nutritive value (e.g., vitamin C is two to three times that of orange), with a few trees yielding 1.5 tonnes of fruit per tree. The strong aroma of the fruits has also been characterized by freon 11 or 12 extraction and over 100 components have been identified.

The nuts too have been described as a delicacy and yield an oil with a quality (fatty acid composition) comparable with olive oil, but with a stability that is 10 times greater. This stability is explained in terms of its tocopherol/sterol composition (Δ^5-avenasterol and α-tocopherol). The amino acid content, with the exception of lysine which is deficient, has been likened to human milk and whole hens' eggs. It has been concluded that the oil could be of value to the food industry where it could be used as coating of dried fruit, as a frying oil or as a substitute for high-oleic safflower oil in baby foods.

Proximate analyses performed on fruits from different areas of southern Africa reveal some variation that may be either genetic or environmental. The causes of this variation need to be investigated, as genetic variation of this magnitude would be of importance to domestication programs. Similarly, the ascorbic acid content of Marula fruits in Nigeria (403 mg 100 g^{-1}) has been reported to be twice that found in Botswana, although it is said to vary considerably depending on the stage of ripening, being highest in ripe fruits (Eromosele et al., 1991).

The gum of marula is acidic and has a low intrinsic viscosity, low molecular weight and high methoxyl content. The main sugar in the gum is galactose (63%), without any rhamnose (Weinert et al., 1990).

White edible flesh surrounds a large nut, which contains three edible kernels. The nut represents about 50% of the weight of the fruit (Taylor and Kwerepe, 1995). In some trees the fruit pulp is sweet and in others very sour. Fruits are rich in vitamin C (194.0 mg 100 g^{-1} at 85% moisture). They are very popular in Botswana and are used to make a local

FIGURE 7.14 Fruits and leaves from *Sclerocarya birrea*.

FIGURE 7.15   Amarula liqueur and Marulam wine made from fruits of *Sclerocarya birrea.*

beer. They can vary between 10.4 and 16.0 degrees Brix (sweetness). In South Africa, the Mitsubishi Corporation have also brewed a beer, Afreeka, which has been undergoing market trials in the United Kingdom in 1997. Also in South Africa, the internationally popular liqueur Amarula is marketed by Distillers Corporation. In Zambia a wine called Marulam is also marketed commercially (Fig. 7.15). A pasteurized juice has also been marketed in Botswana and early "browning" problems overcome (Taylor and Kwerepe, 1995). Juice flavor has been evaluated by a tasting panel, who quantified 19 characteristics of flavor, odor, mouthfeel, and aftertaste. The prominent trait identified was sourness, although one of five juices was much sweeter than the others (Schäfer and McGill, 1986). General experience suggests that there is considerable variation between trees within the species in sweetness, and that trees from drier environments are sweeter than those from wetter areas. Numerous small enterprises in the countries of southern Africa produce Marula jam and jellies.

The nut kernels are nutritious and widely eaten. Kernel oil is also highly prized both for cooking and for cosmetics. At 96% dry matter, the kernel is 57.3% fat, 28.3% protein, 3.7% carbohydrate, and 2.9% fiber (Taylor and Kwerepe, 1995) and rich in phosphorus and magnesium. Ogbobe (1992) has, however, reported from Nigeria that the kernels contain 11% crude oil, 17.2% carbohydrate, 37% crude protein, 3% fiber, and 1% saponins. He also reported that the oil contains nine fatty acids of which palmitic, stearic, and arachidonic acids are the most dominant.

Marula fruits fall from the tree before they are ripe, while still green, and turn yellow as they ripen on the ground. After gathering, the ripening process in storage involves concurrent changes in pH, total acids and total soluble solids. The process reaches a climax after 7 days (Weinert et al., 1990). Fallen fruit have a storage life of 14 days at 12.8°C. After 21 days, 89% of fruits have rotted. There are contradictory results from low-temperature storage, with one report that fruits can be kept for 16 days at 4°C and another indicating that temperatures of 9°C cause damage.

Figures about the current level of production seem to be unavailable, but in 1985 it was estimated that 600 tonnes of juice were processed in South Africa. Since then there has been the introduction of several alcoholic beverages onto the international market, suggesting that the current figure must now be considerably higher. It is assumed that further growth is constrained by the fact that harvesting is restricted to the collection of fruits from wild trees, although domestication programs have been initiated in South Africa, Botswana, and Malawi.

Much research has been done on juice, extraction, product characterization, and processing (see Weinert et al., 1990). Variation in total soluble solids of puree and juices varied between 7.5 and 15.5 degrees Brix over three seasons, the lower value coinciding with a drought and the higher value with a wet year. Total titratable acidity was similarly affected, altering the sugar:acid ratio, an index for sensory quality in fruit juices. Sensory characteristics have also been evaluated by 14 descriptor terms, including odor, flavor, mouthfeel, aftertaste, etc. The combination of these traits and those of yield are currently being used in South Africa to register cultivars for testing horticulturally for commercial fruit production.

FIGURE 7.16 Fruits growing on *Uapaca kirkiana* in Zimbabwe.

Uapaca kirkiana Muell. Arg. (Masuku or Mahobohobo)

The fleshy pulp of the masuku fruit (Fig. 7.16) is eaten fresh or processed into a variety of products: juices, squashes, wines, sweet beer, porridge, jams, and cakes (Ngulube, 1995). In Zambia, popular brands of wine are Masau and Mulunguzi. They are produced commercially and sold in supermarkets. A beer called Napolo Ukana and a gin called Kachasu are produced.

In Zambia the fruits of masuku are mostly (80%) cream-colored, but others are rufous (18%) and a few brown (2%) (Mwamba, 1995), with trees bearing cream-colored fruits having the greatest fruit load. The pulp forms only about 45% of the fruit, the skin being 38% and the seed 17%.

In Malawi, Uapaca fruits with 27.4% dry matter were found to contain: 86.5% total carbohydrate, 8.4% fiber, 1.1% fat, and 1.8% crude protein (Saka et al., 1994). Ascorbic acid content is 16.8 mg 100 g^{-1} fw (Saka, 1995).

The total free sugar content of masuku fruit juice from Zambia is 8.5% (Sufi and Kaputo, 1977) as determined by paper chromatography and confirmed by ultraviolet absorption spectrophotometry. It contains glucose (4.1%), fructose (2.7%), sucrose (1.5%), and xylose (0.2%).

Zizyphus mauritiana Lam. (Jujube or Ber)

In Malawi, Zizyphus fruits at 14.8% dry matter were found to contain: 73.0% total carbohydrate, 3.4% fiber, 9.5% fat, 4.1% crude protein (Saka et al., 1994). Ascorbic acid content is 13.6 mg 100 g^{-1} fw (Saka, 1995).

See also the previous section on jujube in the semi-arid lowlands of West Africa.

Vangueria infausta Burch. (Wild Medlar)

Vangueria fruits in Malawi have been found to contain at 26.5% dry matter: 78.1% total carbohydrate, 10.2% fiber, 2.6% fat, 5.7% crude protein (Saka et al., 1994). Ascorbic acid content is 16.8 mg 100 g^{-1} fw (Saka, 1995). In Botswana, ascorbic acid content of 4.7 mg 100 g^{-1} has been reported for fruits with 64.4% moisture.

Azanza garckeana (F. Hoffm.) Exell & Hillcoat (Snotapple)

In Malawi, Azanza fruit pulp at 52.8% dry matter was found to contain: 35.2% total carbohydrate, 45.3% fiber, 1.1% fat, 12.0% crude protein (Saka et al., 1994). Ascorbic acid content is 20.5 mg 100 g^{-1} fw (Saka, 1995).

WESTERN AMAZONIA

Inga edulis Mart. (Inga or Guaba)

The pulp around the seeds in the pod (Fig. 7.17) is sweet and tender and is widely marketed and eaten as a fresh fruit in Amazonia (Villachica, 1996). There is great variability within the species and potential to create cultivars, but there is

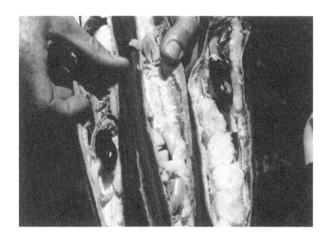

FIGURE 7.17 Exposed seeds and pulp in pods of *Inga edulis*.

FIGURE 7.18 Fruits of *Bactris gasipaes*.

not much published information on the nutritional aspects of these fruit. The pulp, which is over 80% water, is rich in carbohydrates and has high-energy value. The nutritive value of fresh pulp is low, as reported by Villachica (1996). The fruits, which can normally be kept for only 3–4 days, can be stored in a refrigerator for 3 weeks. The embryos of this and other *Inga* species are cooked and are more nutritious than the fruit pulp (Pennington and Robinson, 1998). Boiled embryos of *Inga ilta*, e.g., contain 57.7% moisture, 13.5% protein, 0.2% fat, 1.2% crude protein, 23.2% starch, and 4.2% soluble carbohydrates. The cooking probably degrades trypsin inhibitors and enhances palatability.

Bactris gasipaes H.B.K. (Peach Palm or Pejibaye)

The two major products from peach palm are the fruit (mesocarp) and the "heart of palm" (Figs. 7.18 and 7.19), although the oil, wood, and fiber are also valuable. The main markets for the fruit are as a delicacy for direct human consumption, as an animal feed, and as a starchy ingredient in bread and cakes. The fruit of peach palm, which varies in flavor and texture, is always eaten cooked, as boiling breaks down a trypsin inhibitor that would otherwise have negative effects on human/animal growth. Considerable variation has, however, been reported in the presence of this inhibitor among different samples. Fruits are already marketed in jars and cans and can also be sold dehydrated.

FIGURE 7.19   Stems of *Bactris gasipaes* being prepared as "heart of palm."

Domestication of peach palm toward the different products and uses has arisen from farmer selection within Amerindian communities in tropical America (Clement, 1988). Consequently, eight or nine landraces can be identified, which are suited for different uses. Classification is based on fruit size (Clement, 1990): small fruits are generally more oily and fibrous (two landraces described); large fruits are starchy, low in oil and have a high pulp:seed ratio (two to three landraces). The last four landraces are intermediate in size.

Proximate analysis of mesocarp samples has not taken into account the differences between landraces, but big variation has been reported (see review by Clement, 1990); e.g., oil (8.3−23.0%, with one sample of 61.7%), protein (6.1−9.8%, with one sample of 17.5%), N-free extract (59.5−79.9%), fiber (2.8−9.3%). Analyses of the composition of mesocarp protein have shown that all the essential amino acids are present, although at lower levels than in maize. Arginine (7.3−9.2%) and glutamic acid (4.7−6.3%) are the most abundant. The mesocarp is frequently extremely rich in β carotene, although there is big variation in the presence of this provitamin (Arkcoll and Aguiar, 1984).

Mesocarp oil quality has been studied in more detail than protein quality (e.g., Silva and Amelotti, 1983) and contains both saturated (29.6−46.3%) and unsaturated (53.3−69.9%) fatty acids, with palmitic acid (29.6−44.8%) and oleic acid (41.0−50.3%) the most abundant, respectively. It seems that the triglyceride structure is extremely variable, even within samples. This should allow opportunities for genetic selection at the clonal level. A study of the tocopherols and tocotrienols showed a strong predominance of α-tocopherol (Lubrano et al., 1994). Although the more primitive landraces are apparently rich in oil, there is a problem of extraction as the oil, starch, and water form an emulsion that has to be solvent extracted (Clement and Arkcoll, 1985).

As an animal ration, peach palm fruit flour can substitute for maize, sorghum, or wheat and has been widely tested as a meal for chickens, usually as a partial substitute for cereals (e.g., 50%), especially for older birds (Clement, 1990). For pigs, silaging fruits has been reported to be an excellent means of storing the fruits, which may also be acceptable to cattle. Animal feeds would usually be based on the starchy fruit varieties with low oil content.

Peach palm flour has also been used in bread baking and at 10% substitution for wheat gives dough with excellent baking quality (Tracy, 1996), slightly less protein, more energy (from the oil) and more vitamin A (β-carotene). The flour can also be used in cakes.

Palmito or heart of palm is already grown commercially with more than 2000 ha in Costa Rica by 1990. It is, however, in competition with palmito of acai (*Euterpe oleracea*), that has lower overheads as it is exploiting natural populations, but has lower quality control. Processing technology has been developed in Costa Rica and Brazil (Fig. 7.20). Uses such as deep-fried chips are also being found for some of the residue from palmito preparation.

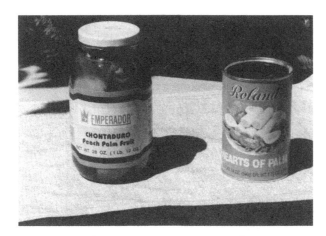

FIGURE 7.20   Bottled fruits and canned "heart of palm" from *Bactris gasipaes*.

CONCLUSIONS

Farmers working with ICRAF throughout the tropics have identified the indigenous trees that they would like to see domesticated. They have also identified the traits, which from their perspective should be improved. For example, they would like to see the trees coming into production at an earlier age; the length of the productive season increased; the tree height reduced; and, of course, the yield and quality of the products increased. The success of this initiative to domesticate fruit trees is however closely linked to commercial (Leakey and Izac, 1996), economic, and policy issues (Cannell, 1989, Leakey and Tomich, 1999), but overridingly there is the need to develop and expand markets to provide the incentive to plant and manage trees in farmland. It is therefore important to examine what is known about the products and to identify ways in which they could be utilized and improved.

This review has indicated that there have been a few studies to characterize the products with commercial potential from the farmer-identified priority species. Very few of these, however, have looked at the range and origin of intraspecific genetic variation and the opportunities it presents to improve the yield and quality of the products. Furthermore, few if any of these studies have made any recommendations as to which components of the products should be improved to enhance their value to the food industry. Agroforestry researchers working in tree domestication need information from members of the food industry about the traits that they would like to see improved by genetic selection. The need is for information about characteristics that would make the products more competitive in the market, ensure their certification as a food additive, or enhance either processing or storage. Clearly dialog and collaboration between agroforestry researchers and food scientists is needed to ensure that progress towards tree domestication is coordinated and steered in a direction that is most likely to result in the significant adoption of novel products by the food industry.

Chapter 8

Evidence that Subsistence Farmers have Domesticated Indigenous Fruits (*Dacryodes edulis* and *Irvingia gabonensis*) in Cameroon and Nigeria

This chapter was previously published in Leakey, R.R.B., Tchoundjeu, Z., Smith, R.I., Munro, R.C., Fondoun, J-.M., Kengue, J., Anegbeh, P.O., Atangana, A.R., Waruhiu, A.N., Asaah, E., Usoro, C. and Ukafor, V., 2004. Agroforestry Systems, 60, 101–111, with permission of Springer

SUMMARY

Ten fruit and kernel traits were measured in 152 *Irvingia gabonensis* and 293 *Dacryodes edulis* trees from six villages in Cameroon and Nigeria. Frequency distribution curves were used to examine the range of variation of each trait of each species in each village and aggregate them into national and regional populations. There were differences between the village subpopulations, with regard to the normality (e.g., mean kernel mass of *D. edulis*) or skewness (e.g., mean flesh depth of D. edulis) of the distribution curves and in the degree of separation between the individual village populations along the x axis, resulting in the development of a bimodal distribution in the regional population. For all traits, populations of both species differed significantly between countries, but only in *D. edulis* were there significant differences between the Cameroon populations. On the basis of the results of this study, *D. edulis* can be said to be virtually wild in Nigeria but semidomesticated in Cameroon, while *I. gabonensis* is wild in Cameroon and semidomesticated in Nigeria. These results are discussed with regard to a hypothesis that the range and frequency of variation in the different populations can be used to identify five stages of domestication. From a comparison of the frequency distribution curves of desirable versus undesirable traits, and statistically identifiable changes in skewness and kurtosis, it is concluded that as a result of the farmers' own efforts by truncated selection, *D. edulis* is between Stages 2 and 3 of domestication (with a 67% relative gain in flesh depth) in Cameroon, while *I. gabonensis* in Nigeria is at Stage 2 (with a 44% relative gain in flesh depth). In this study, genetic diversity seems to have been increased, and not reduced, by domestication.

INTRODUCTION

The present study is part of a wider project, which seeks to determine the socioeconomic constraints and opportunities for the domestication and cultivation of these two indigenous fruit trees in Cameroon and Nigeria (Schreckenberg et al., 2002). Earlier studies in the project have quantitatively assessed the tree-to-tree variation in *I. gabonensis* in Cameroon (Leakey et al., 2000; Atangana et al., 2001) and Nigeria (Anegbeh et al., 2003; 2005), and examined opportunities for multitrait selection, based on the definition of ideotypes for *I. gabonensis* (Leakey et al., 2000; Atangana et al., 2002) and for *D. edulis* (Waruhiu 1999; Leakey et al., 2002). The relationships between various traits and the price received by farmers is being examined.

The domestication of new species is a major undertaking, one that is a continuous process of improvement and one that has to be justified by the benefits that accrue to the producers and consumers of the products. There is currently

Multifunctional Agriculture. DOI: http://dx.doi.org/10.1016/B978-0-12-805356-0.00008-8
© 2017 Elsevier Inc. All rights reserved.

some debate among development organizations focused on poverty alleviation, sustainable livelihoods, and food security, about the direction for future research: some favor biotechnology and an expansion of the Green Revolution (Serageldin and Persley, 2000), while others see potential for broadening the basket of crops (McNeely and Scherr, 2001) and a "Really Green Revolution" based on farming systems diversified with new tree crops (Leakey, 2001b). In the case of a number of agroforestry trees, including *I. gabonensis* (Aubry Lecomte ex O'Rorke) Baillon. and *D. edulis* (G. Don) H.J. Lam, it has been argued that their domestication will enhance farmer livelihoods, reduce poverty, and promote economic development (Leakey, 2001b). At the same time, this domestication of agroforestry trees should encourage the development of sustainable land-use practices that will rehabilitate degraded farmland, sequester carbon, and other greenhouse gases and enhance both biodiversity and the functioning of agroecosystems (Leakey, 2001a). This study contributes to the crop diversification versus biotechnology debate by seeking evidence that farmers have already demonstrated their interest in diversification by initiating the domestication process in indigenous fruit trees.

Typically, trees are outbreeding and genetically very diverse due to the contribution of large numbers of individuals to a shared genepool and the free segregation of alleles during meiosis (Zobel and Talbert, 1984). Consequently, the means of different wild subpopulations for any given trait can differ significantly in response to strong selection pressures or, if these pressures are weak in the absence of geneflow, the range of variation in these subpopulations overlaps. In plant breeding, cycles of selecting and crossing between only the best individuals in the population (truncated selection), results in new progenies, which outperform their parents in the selected trait (Futuyma 1998). The degree of improvement depends on the narrow sense heritability (Stearns and Hockstra, 2000). The domestication of a species must therefore result in changes in the frequency distribution of the values of the selected trait among the members of the population. During the course of several generations of truncated selection, the frequency distribution of the trait can be expected to change through a progression of stages that ultimately lead to the formation of a variety. This is hypothetically portrayed in Fig. 8.1.

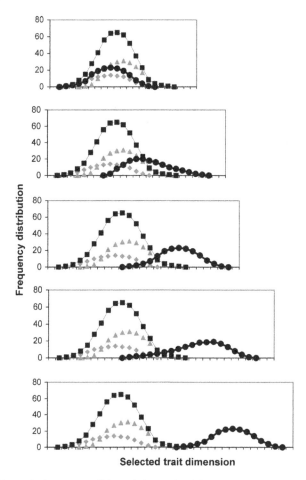

FIGURE 8.1 Hypothetical stages of domestication. 1 = A wild population made of three subpopulations. In stages 2–5 one of these subpopulations (●) is domesticated by truncated selection to form a variety.

In genetic terms, domestication has been defined as "accelerated and human-induced evolution" (Simons, 1996), but will the resulting changes in the frequency distribution patterns of the selected traits be the same during plant breeding and evolution? Perhaps one consideration is that during evolution the species is either introduced by chance into an environment that differs from that of the natural range, or the environment changes. Chance introductions usually occur in small numbers, have a relatively narrow genetic base, and may depend on selection of a rare gene for survival. Consequently, the introduced population is exposed to severe selection pressure for a new or different set of traits. This situation can be viewed as similar to truncated selection for a commercially desirable trait during plant breeding. Environmental change, on the other hand, results in changes in gene expression over longer periods of time. Consequently, collections of fruits at a single time from a well-established wild population may not reveal subpopulation structure differentiation that displays the sort of differences found between bred varieties and their wild population. Most of the species that are cultivated in agriculture and horticulture have been subjected to truncated selection and breeding and may bear little resemblance to their wild origins. Additionally, the wild relatives of these crop plants may have become extinct either through genetic pollution by crossing with the selected varieties, or by being eliminated from the environment by clearance of the natural vegetation. By contrast, in forestry, because of the long generation time of most trees, at least the early stages of domestication have often been by provenance selection. Provenance selection in forestry involves the transfer of geographically discrete subpopulations to new environments. These environments are generally similar to the natural range in many respects, but can differ in ways that enhance either the growth rate, or tree form (e.g., lodgepole pine from Queen Charlotte's Island off the Vancouver coast of Canada, which is stunted in its natural environment, performs much better in Scotland. Domestication through provenance selection thus appears to equate much more closely to evolution through chance introduction to a new environment.

In contrast to the widely cultivated agricultural and horticultural crops of the world that have been domesticated for millennia, the initiatives to domesticate some of the indigenous fruit trees of different ecoregions of the tropics (Leakey and Simons, 1997) are starting now with wild, or virtually wild, genepools. This imposes responsibilities on the scientists involved to develop an understanding of the potential of the species, to ensure that domestication proceeds wisely, efficiently and within the constraints imposed by the Convention on Biological Diversity, and to maintain and protect the diversity of the genetic resource. To do this, it is necessary to obtain a good understanding of the diversity of the genetic base of the species.

This study examines the hypothesis that the stage of domestication in populations of *I. gabonensis* and *D. edulis*, tree species with unknown domestication history, can be determined by an evaluation of the patterns in the frequency distribution of different traits between populations (Fig. 8.1). Ideally, cultivated plants should be compared with wild ones at the same site. Unfortunately, this was not possible in this study and the frequency distribution of traits before any artificial selection pressures were applied are unknown. To circumvent this we have compared frequency distributions of neutral characters with those probably the target of selection by farmers.

MATERIALS AND METHODS

This study is based on data collected from six villages (Table 8.1), four in Cameroon (Atangana et al., 2001, 2002; Waruhiu, 1999; Leakey et al., 2002) and two in Nigeria (Anegbeh et al., 2003, 2005), which measured the following nine fruit traits in 24 ripe fruits of up to 100 trees per village of *I. gabonensis* and *D. edulis*, using kitchen scales accurate to 2 g and calipers accurate to 0.1 mm:

1. Fresh fruit mass (g),
2. Nut mass (g)—for *I. gabonensis* only—after drying the residue flesh,
3. Fresh kernel mass (g),
4. Fruit length (mm),
5. Fruit width (mm),
6. Flesh depth in the fruit thickness dimension (mm),
7. Taste score (1 [bitter] − 5 [sweet])
8. Fibrosity score (1 [low fiber]−5 [high fiber])—for *I. gabonensis* only, and
9. Oiliness score (1 [low oil]−5 [high oil])—for *D. edulis* only.

These data were used to derive: Shell mass (g) = (Nut mass—kernel mass, for *I. gabonensis* only), and fresh Flesh mass (g) (Fruit mass—Nut mass, in *I. gabonensis*, and Fruit mass—Kernel mass, in *D. edulis*).

The data were analyzed for differences between countries, sites and land uses using restricted maximum likelihood (Patterson and Thompson 1971) in Genstat 5 (1993), a technique appropriate to the unequal numbers of fruits and trees

TABLE 8.1 The location of study villages in Cameroon and Nigeria and the numbers of trees for which fruit trait data were collected.

	Latitude °N	Longitude °E	Altitude (m)	No. of trees harvested
I. gabonensis				
Cameroon				
Elig Nkouma	4°06′	11°24′	460	31
Nko′ovos II	2°55′	11°21′	610	21
Nigeria				
Ugwuaji	6°25′	7°32′	175	100
D. edulis				
Cameroon				
Makenene	4°52′	10°48′	580	100
Elig Nkouma	4°06′	11°24′	460	57
Nko′ovos II	2°55′	11°21′	610	12
Chop Farm	3°57′	9°15′	11	31
Nigeria				
Ilile	5°19′	6°55′	54	100

in each category. The equivalent to the F test in these analyses is an asymptotic Wald test (Wald, 1940), which does not give exact significance levels for the size of the data set analyzed in this study. Frequency distributions of the mean values per tree per site for each of the measured and derived traits were plotted using Excel 97.

RESULTS

For both species, using restricted maximum likelihood (Patterson and Thompson, 1971), there were significant differences between countries in the mean values of all traits.

D. edulis (293 trees)

In Cameroon, there were significant differences in all traits between the four within-country populations of *D. edulis* (Table 8.2). Only one population was studied in Nigeria. Significantly skewed frequency distributions were identified for fruit mass in two Cameroon (Chop Farm, Makenene) populations and one Nigerian (Ilile) population (Table 8.2 and Figs. 8.2 and 8.4). Similarly, significantly skewed distributions were found for flesh mass in three of the Cameroon (Chop Farm, Elig Nkouma, Makenene) populations and in the Nigerian population. The Nigerian population was the only one to have a skewed distribution for kernel mass and fruit length. The other traits and populations were all normally distributed. Significant kurtosis was found for flesh depth in Elig Nkouma, flesh mass in Makenene and Ilile, and in kernel mass and fruit length in Ilile.

I. gabonensis (152 trees)

There were no significant differences between the two Cameroon populations (Table 8.3). Fruit mass, flesh mass, nut mass, shell mass, kernel mass, and fruit width were all significantly skewed in the Nigerian population (Ugwuaji) and normally distributed for fruit length and flesh depth (Table 8.3 and Figs. 8.3 and 8.4), while in the Cameroon populations all traits were normally distributed. Significant kurtosis was found in the Nigerian population for fruit mass, flesh mass, and fruit width and in the Elig Nkouma population of Cameroon for nut and shell mass (Table 8.3).

TABLE 8.2 The mean, range, and frequency distribution (skewness and kurtosis) of fruit and kernel characteristics of *D. edulis* in Cameroon and Nigeria.

Trait	Site		*Dacryodes edulis*			
	Country	Village	Mean ± SE	Range	Skewness	Kurtosis
Fruit mass	Cameroon	Chop Farm	36.2 ± 3.0	10.0–81.8	1.05 #	0.78
		Elig Nkouma	59.2 ± 2.6	20.7–114.0	0.61	0.76
		Makenene	52.1 ± 1.5	22.8–100.5	0.72 #	0.93
		Nko'ovos II	49.5 ± 8.7	16.8–106.3	0.95	−0.15
	Nigeria	Ilile	31.6 ± 1.1	10.2–71.4	0.83 #	0.86
Flesh mass	Cameroon	Chop Farm	25.9 ± 2.7	6.8–65.9	1.34 #	1.18
		Elig Nkouma	47.3 ± 2.4	12.5–106.0	0.85 #	1.19
		Makenene	40.2 ± 1.4	14.5–89.8	0.89 #	1.23 #
		Nko'ovos II	39.2 ± 7.8	10.2–92.0	1.03	0.06
	Nigeria	Ilile	22.1 ± 1.0	6.8–62.2	1.20 #	2.15 #
Kernel mass	Cameroon	Chop Farm	10.5 ± 0.8	0.0–16.2	−0.78	0.30
		Elig Nkouma	12.1 ± 0.4	8.0–18.3	0.48	−0.55
		Makenene	12.0 ± 0.3	3.8–20.8	0.25	0.46
		Nko'ovos II	10.6 ± 1.0	6.6–17.9	1.24	0.71
	Nigeria	Ilile	9.5 ± 0.3	0.8–16.1	−0.64 #	1.62 #
Fruit length	Cameroon	Chop Farm	56.5 ± 2.3	33.5–83.5	0.46	−0.44
		Elig Nkouma	74.7 ± 1.5	46.7–105.7	0.02	0.29
		Makenene	68.7 ± 1.0	42.6–94.9	0.01	0.16
		Nko'ovos II	69.8 ± 6.9	41.3–122.4	1.29	1.60
	Nigeria	Ilile	59.4 ± 1.3	39.0–95.1	1.53 #	3.28 #
Fruit width	Cameroon	Chop Farm	32.8 ± 1.0	23.3–35.9	0.49	0.26
		Elig Nkouma	38.9 ± 0.8	24.0–53.5	0.12	0.23
		Makenene	37.7 ± 0.4	28.3–47.7	0.08	−0.45
		Nko'ovos II	36.5 ± 2.0	26.0–45.5	−0.18	−0.04
	Nigeria	Ilile	31.1 ± 0.4	21.8–43.8	0.37	0.06
Flesh depth	Cameroon	Chop Farm	4.8 ± 0.3	1.9–8.0	0.64	0.06
		Elig Nkouma	6.5 ± 0.2	0.6–11.1	−0.20	1.33 #
		Makenene	6.2 ± 0.1	3.6–10.3	0.32	−0.05
		Nko'ovos II	6.7 ± 0.5	4.4–9.6	0.40	−0.84
	Nigeria	Ilile	3.8 ± 0.1	1.8–6.4	0.18	0.00

(# – Skewness/kurtosis >2 x SE of skewness/kurtosis).

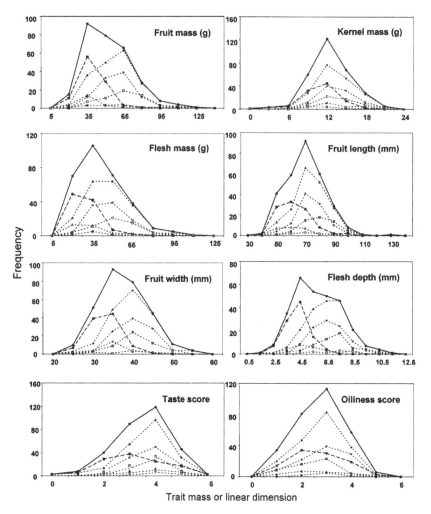

FIGURE 8.2 Frequency distributions of fruit, nut, and kernel characteristics of *D. edulis* from Cameroon and Nigeria (● = region; ■ = Nigeria; = Cameroon. Villages in Cameroon: ○ = Nko'ovos II; ◇ = Makenene; □ = Elig Nkouma; △ = Chop Farm).

DISCUSSION

We are unaware of any other data sets of this sort that have been analyzed in this way, although somewhat similar raw data has been collected for other species (e.g., *Argania spinose*, Bani-Aameur et al., 1999). The quantification of tree-to-tree variation in this way goes beyond the descriptors of genetic variation used by Ladipo et al., 1996 for *I. gabonensis*, or for example, by IPGRI for avocado (*Persea* spp.) and banana (*Musa* spp.).

Significant differences between populations for a variety of traits of economic importance is normal and is the basis of genetic selection programs. Likewise, there is typically significantly different variation between trees within a population. However, in this study, in agreement with the hypothesis presented in Fig. 8.1, there was evidence in certain traits (those likely to be selected by farmers seeking crop improvement) that these differences between populations may be the result of domestication activities. This is illustrated by the degree of separation between the populations along the *x* axis, leading to bimodal distributions in the overall population. For example, for flesh thickness, fruit mass, and fruit width, there were considerable differences in the position of the peak on the *x* axis, while the peak mean kernel mass of all populations was found in the same position on the *x* axis (Fig. 8.2). This separation is particularly marked between the Nigerian and Cameroon populations, although the Chop Farm population in Cameroon seems to be very similar to the Nigerian population. Since the kernel of *D. edulis* is usually discarded, as being of little value, while the mass, and size of the fruit and the depth of the flesh are all indicators of a good fruit, it seems likely that over time there has been intentional selection by Cameroon farmers for these desirable traits. Thus the frequency distribution data (Fig. 8.2), especially when presented as percentage frequency (Fig. 8.4), support the hypothesis that the current stage of

TABLE 8.3 The mean, range, and frequency distribution (skewness and kurtosis) of fruit and kernel characteristics of *I. gabonensis* in Cameroon and Nigeria.

Trait	Site		*Irvingia gabonensis*			
	Country	Village	Mean ± SE	Range	Skewness	Kurtosis
Fruit mass	Cameroon	Elig Nkouma	99.0 ± 4.4	58.8–161.3	0.39	−0.21
		Nko'ovos II	107.5 ± 7.9	44.5–195.4	0.54	0.25
	Nigeria	Ugwuaji	165.3 ± 5.8	69.0–419.8	1.44 #	3.40 #
Flesh mass	Cameroon	Elig Nkouma	85.7 ± 4.0	49.5–142.3	0.40	−0.37
		Nko'ovos II	92.7 ± 7.5	38.5–179.1	0.68	0.51
	Nigeria	Ugwuaji	145.3 ± 5.3	59.5–388.8	1.49 #	3.77 #
Nut mass	Cameroon	Elig Nkouma	13.3 ± 0.6	9.0–23.6	0.54	1.98 #
		Nko'ovos II	14.9 ± 0.8	5.9–22.8	−0.33	0.49
	Nigeria	Ugwuaji	20.0 ± 0.6	9.5–40.6	0.99 #	0.87
Shell mass	Cameroon	Elig Nkouma	9.3 ± 0.5	5.8–18.6	0.71	2.41 #
		Nko'ovos II	10.8 ± 0.5	5.4–15.9	−0.15	−0.17
	Nigeria	Ugwuaji	14.1 ± 0.5	4.9–30.9	0.91 #	0.59
Kernel mass	Cameroon	Elig Nkouma	4.1 ± 0.1	2.8–6.3	0.69	0.21
		Nko'ovos II	4.1 ± 0.4	0.5–6.9	−0.25	−0.44
	Nigeria	Ugwuaji	5.9 ± 0.1	0.4–10.0	0.56 #	0.34
Fruit length	Cameroon	Elig Nkouma	60.3 ± 1.0	50.6–75.7	0.57	0.11
		Nko'ovos II	61.6 ± 1.7	46.2–77.3	0.20	−0.26
	Nigeria	Ugwuaji	66.3 ± 0.8	49.2–89.3	0.36	−0.12
Fruit width	Cameroon	Elig Nkouma	57.8 ± 0.9	46.3–66.1	−0.33	−0.44
		Nko'ovos II	59.7 ± 1.5	45.1–72.9	−0.39	−0.09
	Nigeria	Ugwuaji	63.3 ± 0.8	46.2–105.0	0.58 #	0.99 #
Flesh depth	Cameroon	Elig Nkouma	15.6 ± 0.3	12.2–19.2	0.17	−1.04
		Nko'ovos II	15.5 ± 0.6	11.2–21.8	0.68	−0.21
	Nigeria	Ugwuaji	19.7 ± 0.3	12.9–31.4	0.46	−0.41

(# = Skewness/kurtosis >2 × SE of skewness/kurtosis).

domestication can be traced in populations by the emergence of varieties with the selected traits. Consequently, it is possible to conclude that in Cameroon, especially in Makenene and Elig Nkouma, *D. edulis* is between Stages 2 and 3 of domestication (Fig. 8.1). In this case, farmers in Makenene confirmed that their parents' generation had selected for large fruit size, while the current generation has added selection for flavor. This recent selection for flavor is, perhaps, supported by the separation emerging for fruit taste between the Cameroon and Nigerian populations (Fig. 8.2).

Further evidence in support of farmers' interest in *D. edulis* in Cameroon has been obtained during participatory household surveys (Mbosso, 1999). The species is very widely planted in all four communities and constitutes 42% of all fruit trees in farmers' fields (Schreckenberg et al., 2002). In terms of food value, it is ranked as the most important fruit tree in Chop Farm and Elig Nkouma, and as the second most important in Nko'ovos II and Makenene. It is among the top three species for commercial value in all four communities (Mbosso, 1999). The apparent lack of domestication

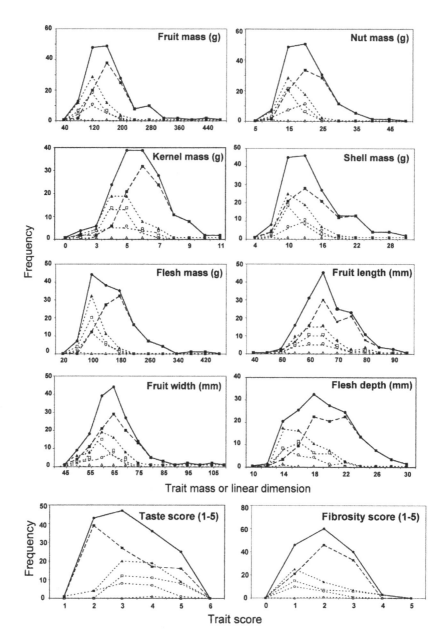

FIGURE 8.3 Frequency distributions of fruit, nut, and kernel characteristics of *I. gabonensis* from Cameroon and Nigeria (● = region; ■ = Nigeria; = Cameroon. Villages in Cameroon: ○ = Nko'ovos II; □ = Elig Nkouma; △ = Chop Farm).

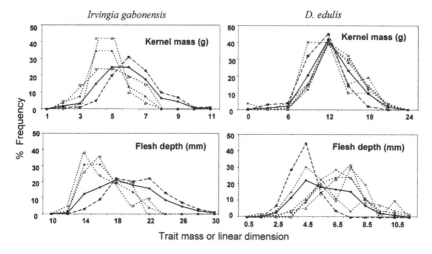

FIGURE 8.4 Frequency distributions (%) of fruit, nut, and kernel characteristics of *I. gabonensis* and *D. edulis* from Cameroon and Nigeria (● = region; ■ = Nigeria; = Cameroon. Villages in Cameroon: ○ = Nko'ovos II; □ = Elig Nkouma; △ = Chop Farm).

of *D. edulis* in Chop Farm may be due to a number of factors. The community consists of immigrants with relatively insecure land tenure, while the landlords are mostly fishermen without great interest in trees (Mbosso, 1999). Second, Chop Farm has a low frequency of all tree species per farm, possibly because of the availability of a large adjacent forest resource. Third, interest in the commercial value of *D. edulis* in this area is very recent, dating only to 1996 when traders from neighboring Gabon first started coming to the village to buy the fruit (Schreckenberg et al., 2002).

In *I. gabonensis*, we obtain a very similar result, but in contrast with *D. edulis*, the evidence for domestication is in Ugwuaji, Nigeria, where all the trees had been planted by farmers, rather than in Cameroon, where wild trees had been retained on-farm when the forest was cleared for shifting cultivation. In this species, domestication does not seem to be so advanced as in *D. edulis*, and probably conforms to Stage 2. There was no evidence for differences in the stage of domestication between subpopulations in Cameroon. This again concurs with the results of the participatory household surveys where only farmers in Nko'ovos II ranked *I. gabonensis* as the most important species for food and commercial value (Mbosso, 1999). However, even in Nko'ovos, farmers have not yet begun to plant it, in spite of greatly increased interest from traders in recent years (Schreckenberg et al., 2002). As in *D. edulis*, the evidence from this study for domestication was strongest for the fruit traits (flesh mass and flesh depth), but none of the frequency curves of traits measured in *I. gabonensis* were found to fall at the same position on the *x* axis for all the populations assessed. This suggests that selection by Nigerian farmers has affected all the traits measured, including kernel mass. This is not surprising, given that the kernel is the commercially important food product.

It is not clear exactly how the differences in skewness and kurtosis in the frequency distributions (Tables 8.2, 8.3) come about, although they are evident in the graphs (Figs. 8.2, 8.3 and 8.4). It can be hypothesized that the first response to selection of trees with large fruits is an increase of big fruited trees in the population. Thus, the first stage is a shift from a normal distribution to a positively skewed distribution (Fig. 8.1). A logical progression based on this argument would then lead to a population with platykurtosis and a normal distribution again, before the development of negative skewness and finally the return to normality (Fig. 8.1). This hypothesis seems to be supported by the data (Tables 8.2, 8.3), which show significant positive and negative skewness and significant kurtosis. It is also interesting to note that, within the individual traits themselves, there is no evidence at the population level for a reduction in diversity as a result of domestication; indeed, there seems to be an increase. This is contrary to the general expectation that reduced intraspecific variation is a consequence of domestication.

In the case of *I. gabonensis* the existence of a few large fruits was associated with a proportional increase in the nut and shell mass. Unfortunately, this did not greatly enhance kernel mass, due to the weak relationship between nut mass and kernel mass ($r^2 = 0.32-0.65$; Atangana et al., 2002). Consequently, it appears that selection for fruit size, while enhancing flesh mass, has unintentionally resulted in selection for shell mass. This is undesirable, as large shelled nuts are generally difficult for women to crack and so difficult for kernel extraction. This negative selection arising from selecting for large fruits is due to the strong positive correlation between shell mass and nut mass ($r^2 = 0.86-0.95$: $p = 0.01$ [Cameroon] and $r^2 = 0.89-0.92$: $p = 0.01$ [Nigeria]) and between nut mass and fruit mass (Atangana et al., 2001). This finding supports the concept that the domestication of *I. gabonensis* needs to be focused on the selection of cultivars based on either fruit or kernel ideotypes (Leakey et al., 2000). It is encouraging, however, that trees with light-shelled, easily cracked nuts can be found in Cameroon (Leakey et al., 2000) and in Gabon (Ladipo et al., 1996), although these seem to be uncommon. Unfortunately, however, these light-shelled nuts have, so far, been associated with relatively small kernels (Atangana et al., 2002).

Overall, this analysis of fruit trait data in *I. gabonensis* and *D. edulis* has identified evidence that farmers in Nigeria and Cameroon, respectively, have by their own efforts achieved at least Stage 2 in the domestication process. This has resulted in a 44% relative gain in flesh depth for *I. gabonensis* in Ugwuaji, Nigeria, and 67% gain in flesh depth for *D. edulis* in Elig Nkouma and Nko'ovos II. This result emphasizes the importance that subsistence farmers attribute to indigenous fruits for their own consumption and for trade. This importance is also emerging from market studies, which indicate that, e.g., with *D. edulis*, wholesale traders in Gabon travel to markets like Makenene to buy fruits for importation into Gabon. Similar evidence of regional trade has recently also been documented for *I. gabonensis* and *Ricinodendron heudelotii* kernels, and *Cola* spp. (Ruiz Pérez et al., 1999).

The recognition that farmers in Cameroon and Nigeria have initiated the domestication of two of their indigenous fruit trees emphasizes the relevance of the current activities of ICRAF and others to further domesticate the indigenous trees that provide marketable NTFPs of importance to local people for food security and income generation. The direction of this research differs from that promoted by biotechnologists and many international donors (Serageldin and Persley, 2000), but apparently is more in tune with the wishes and practices of Third World farmers. The need now is to go to the next stage of domestication in which cultivars are developed, using vegetative propagation techniques (Shiembo et al., 1996), from the very best trees available in each village. This is starting in the ICRAF program

(Tchoundjeu et al., 1998) with the intention of using these cultivars in cocoa and other agroforests to diversify the agroecosystem, and so to create land-use systems that: (1) enhance the income and livelihoods of poor subsistence farmers, (2) provide novel food and other products, which can become export commodities, (3) diversify farming systems, so promoting agroecosystem function, and (4) produce the international public goods and services (carbon sequestration, biodiversity, etc.) that are demanded by the global community (Leakey, 2001a,b).

ACKNOWLEDGMENTS

This publication is an output from a research project partly funded by the UK Department for International Development (DFID) for the benefit of developing countries. The views expressed are not necessarily those of DFID (DFID project code R7190: Forestry Research Program).

Chapter 9

Non-Timber Forest Products — A Misnomer?

This chapter was previously published in Leakey, R.R.B., 2012. Journal of Tropical Forest Science, 24, 145–146, with permission of Forest Research Institute of Malaysia

In the past, it was common practice for hunter-gatherers to collect tree and other products from forests and woodlands. These common property resources were nutritious and often also important as medicinal products, as well as being useful for wood products for crafts, construction, and for tools. As such they were rightly called non-timber forest products (NTFPs). These days, I think we need to recognize that NTFP is sometimes a misnomer as much of the forest has gone or has been severely degraded by logging and shifting cultivations. As a consequence, many of the species which used to provide NTFPs are now more commonly found in farmland. Indeed, many of the traditionally important and useful tree species are now found in places far removed from forest—sometimes even outside the area of their natural distribution. The distribution of these species is now a consequence of anthropogenic activity and they have become components of the agricultural landscape. For example, marula (*Sclerocarya birrea*) trees in southern Africa are typically scattered trees in farmers' fields. Likewise, in the humid belt of West and Central Africa, safou (*Dacryodes edulis*) is often grown as shade tree in cocoa farms. Safou is thought to have a small natural range in southwest Cameroon (Vivien and Faure, 1985), but is now common throughout southern Cameroon and in large areas of central Africa.

In Asia and Latin America there are similar examples such as damar (*Shorea javanica*) which is cultivated in complex agroforests in Indonesia, longan (*Dimocarpus longan*) in Vietnam and peach palm (*Bactris gasipaes*) in Brazil and Costa Rica.

The importance of indigenous fruit and nut tree species as the source of both nutrition and income to rural households was recognized in the early 1990s (Leakey and Newton, 1994a). As a result, a program of tree domestication was initiated (Leakey and Simons, 1997). This has grown to become an international initiative now entering its third decade (Leakey et al., 2012). This initiative was focused on the promotion of indigenous trees for the production of marketable and domestically important products to improve the livelihoods of poor smallholder farmers in the tropics and subtropics. The approach employed was to develop horticultural cultivars by vegetative propagation of elite trees selected for a range of different attributes, thus meeting the needs of different market opportunities. Additionally, the approach involved the active engagement of local communities in participatory domestication (Tchoundjeu et al., 2006, 2010, Asaah et al., 2011), as this was seen as a way to empower the villagers, and to ensure that the social and economic benefits flow to the communities involved (Lombard and Leakey, 2010).

To recognize these new crops, the term *agroforestry tree products* (AFTPs) was proposed (Simons and Leakey, 2004) to distinguish them from NTFPs or nonwood forest product (NWFPs). As such, we can consider these species to be a new generation of agricultural crops. The importance of these new tree crops is not well recognized yet by agricultural scientists and policymakers, who are more focused on the small number of herbaceous plants, often highly domesticated by the Green Revolution. However, I believe that in the future they will play an important role in the fight against poverty, malnutrition, hunger, and environmental degradation, making tropical agriculture much more sustainable and productive (Leakey, 2010). In this respect, the domestication of agroforestry trees can be seen as a new second wave of crop domestication (Leakey, 2012b,c), which is aimed at improving the livelihoods of poor, smallholder farmers in the tropics and tropical nations rather than contributing to the further enrichment of the economies of developed countries.

Multifunctional Agriculture. DOI: http://dx.doi.org/10.1016/B978-0-12-805356-0.00009-X
© 2017 Elsevier Inc. All rights reserved.

In conclusion, I accept that the terms NTFP and NWFP are still highly appropriate for common property resources being collected from forests and woodlands. However, I think we now need to be aware that many of these products come from farms. Therefore, I believe that in these circumstances these important tree products need to be recognized as AFTPs—products from new crops with a vital role to play in the development of many tropical and subtropical countries. This recognition is needed if the statistics of trade and consumption are to influence policymakers and donors of the social, economic, and environmental values of more diversified forms of agriculture based on perennial tree crops (Simons and Leakey, 2004) (Fig. 6.1).

Chapter 10

Trees: An Important Source of Food and Non-Food Products for Farmers: An Update

R.R.B. Leakey

Thousands of wild species are recognized as having edible products, as well as being useful in many other ways (e.g., Elevitch, 2006; Abbiw, 1990; Bekele-Tesemma et al., 1993). Most if not all of these edible fruits, nuts, and leaves have much higher nutrient content (vitamins, minerals, etc., as well as amino and fatty acids, protein, fiber, etc.) than the major cereal and root crops that dominate global diets. In the past, many of the species were derogatively labeled as "famine foods," but this misnomer belies the point that many of these foods are culturally important as highly appreciated traditional foods (Figs. 7.1–7.20). The potential for the diversification and enrichment of diets with the products of these wild and culturally and traditionally important species has now started to be internationally recognized. Furthermore, they are now being seen as critically important food resources (Fig. 7.20; Burchi et al., 2011; Jamnadass et al., 2011; 2015; Stadlmayr et al., 2013). The potential for postharvest transformation of some of the products has still not been widely recognized, but appropriate techniques and strategies for the identification of desirable trait combinations have been developed (see Chapter 25 [Leakey and Page, 2006]).

Chapter 7 (Leakey, 1999a) introduced 16 of these highly nutritious indigenous food tree species (Figs. 7.1–7.19) and promoted them as new crops to counter malnutrition and poverty, as well as to create novel foods (Fig. 7.20) for global industries. This has been highly successful as the number now being studied, cultivated, and/or domesticated, has risen to over 60. These include:

1. Food species: *Allanblackia floribunda, Allanblackia parviflora, Allanblackia stuhmannii, Annona cherimoya, Annona senegalensis, Argania spinose, Artocarpus altilis, Balanites aegyptiaca, Barringtonia procera, Baillonella toxisperma, Berchemia discolor, Blighia sapida, Canarium indicum, Carapa procera, Chrysophyllum cainito, Cola acuminate, Cola nitida, Dyera polyphylla, Garcinia kola, Griselinia lucida, Gnetum africanum, Hyphaene thebaica, Inocarpus fagifer, Moringa oleifera, Morinda citrifolia, Parinari curatellifolia, Pentadesma butyracea, Persea schiedeana, Pometia pinnata, Spondias purpurea, Strychnos cocculoides, Saccharopolyspora spinosa, Terminalia kaernbachii, Terminalia catappa, Treculia africana, Vitex fischeri,* and *Vitex payos*.
2. Nonfood species (medicinal, extractives, fodder, and wood): *Calocophyllum spruceanum, Guazuma crinita, Pausinystalia johimbe, Prosopis africana, Prunus africana, Pterocarpus erinaceus, Santalum austrocaledonicum, S. lanceolata,* and *Warburgia ugandensis*.

In addition, the ethnobotany of many other species is being studied within the context of their potential for domestication (e.g., van Damme and Termote, 2008; Cheikhyoussef and Embashu, 2013; Guissou et al., 2015; Chivenge et al., 2015; Chivandi et al., 2015) to enhance human nutrition and/or grow in severe environments (Ofori et al., 2014). There are literally hundreds of other species which could play a role in agroforestry, and many of these have been listed (Clark and Sunderland, 2004, Elevitch, 2011; Atangana et al., 2014; Toensmeier, 2016). Thus, it is clear that the ideas and initiatives of the early 1990s (Leakey and Newton, 1994a) to promote the domestication of wild, indigenous, and traditionally important plant species as new crops has been widely accepted and adopted around the world, focused on over 50 indigenous fruit trees (Leakey et al., 2012; Awodoyin et al., 2015). The work to date has been acclaimed as an African initiative to resolve the problems of African agriculture (Leakey, 2014d). Nevertheless, there is a need for

Multifunctional Agriculture. DOI: http://dx.doi.org/10.1016/B978-0-12-805356-0.00010-6
© 2017 Elsevier Inc. All rights reserved.

much work to domesticate these species. In parallel, there has also been a call for a stronger emphasis on the genetic quality of these indigenous trees being planted for their AFTPs (Dawson et al., 2014).

Chapter 7, (Leakey, 1999a) presented the wide range of variation in the nutrient content from different research samples. For example, in *Irvingia gabonensis* kernels the fat content has been reported to range between 51% and 72%. This suggests that either the environment has big effects on fat content, or that there is genetic variation. Later studies in Cameroon and Nigeria found tree-to-tree variation in fat content from 37.5 to 75.5 in different trees from one village in Cameroon (Leakey et al., 2005d). This suggests that this variation has more to do with genetics than the environment. Similar findings in *Sclerocarya birrea* and other species will be presented later in more detail (see Chapters 22 and 23 [Leakey et al., 2005a,b]). For now, the "take-home" message is that a mean value for a species conceals much important information with regard to the nutritive value, and/or other traits of interest, in some of these products. Later, in Chapter 25 (Leakey and Page, 2006), we will see how this information can be used when developing elite cultivars for cultivation in agroforestry systems, and also when subsequently developing postharvest product processing and marketing.

Chapter 8, (Leakey et al., 2004) presented the use of the frequency distribution of tree-to-tree variation in different product traits data to test and quantify the extent of farmer domestication in traditionally important species such as *D. edulis* and *I. gabonensis*. This test has found evidence that rural householders have initiated a domestication process based on the slow process of tree breeding. The test has subsequently been applied to *S. birrea* fruits and kernels in southern Africa (see Figs. 22.4, 23.5, and 23.6) Leakey et al., 2005a,b,c; Leakey, 2005), *Tamarindus indica* in the Sahel (van den Bilcke et al., 2014), as well as on *Barringtonia procera* kernels in Oceanea (Pauku, 2005), and *C. cainito* fruits in Latin America (Parker et al., 2010). This clearly demonstrates the importance that farmers attribute to these species, and offers opportunities to unravel the anthropogenic impacts of species distribution. Interestingly, in contrast to Africa, evidence from Amazonia suggests that 83 indigenous species had been locally domesticated to some degree by 1500—before the European conquest (Clement et al., 2015). However, there is no such evidence from the Congo Basin in Africa, although *D. edulis* is commonly found in Nigeria, Congo, Zaire, and Central African Republic, well beyond the area in SW Cameroon considered to be its natural range.

The term AFTPs as cultivated tree products distinct from common property NTFPs has been absorbed into common usage (e.g., Akinnifesi et al., 2008; Atangana et al., 2014) and is proving to be useful in avoiding policy misconceptions (Foundjem-Tita et al., 2012).

Part II

Genetic Selection for Added-Value and New Opportunities

Section 4

Tree Domestication

A key challenge for the domestication of agroforestry trees for fruits and nuts centers on the longevity of trees and their outbreeding reproductive system, both of which make tree breeding a slow process. To get around these difficulties, this chapter follows the development of concepts and practices of tree domestication in agroforestry that use the horticultural techniques of vegetative propagation and cultivar selection. This approach has been gaining momentum since its emergence in the 1980s. Using a decentralized farmer-driven and market-led process, this approach seeks to match the intraspecific diversity of locally important trees to the needs of subsistence farmers, product markets, and agricultural environments. The steps of such a domestication process are: selection of priority species; definition of an appropriate domestication strategy; the deployment of germplasm via appropriate low-technology pathways; and the implementation of a participatory rural development. The special focus placed on the use of bottom-up practices is to ensure that benefits flow to the farmers. Nevertheless, this grassroots approach is implemented in parallel with research to understand the extent and patterns of variation that underpin the wise and appropriate use of genetic diversity. Interestingly and importantly, research has found that about 80% of the morphological variation in tree products is found in every village population. This finding is fundamentally important for the success of the decentralized tree domestication strategy, as well as its close affiliation to a commercialization and value-adding strategy that promotes the formation of a local value chain that seeks to maximize income generation in local and regional markets, as part of a strategy for sustainable rural development.

Over the first two decades of this program, work has progressed on more than 50 tree species around the world with an expanding research agenda characterizing traits important for domestic fresh-food consumption as well as on the chemical and physical composition of products for food and nonfood industries. Looking to the future, suggestions are made for further development and expansion of both the science to underpin agroforestry tree domestication, and for applied research in support of development programs to enhance the livelihoods of poor smallholder farmers worldwide. In parallel with grassroots cultivar development, modern molecular techniques provide understanding of the genetic structure of variation in a number of traditionally important indigenous food trees. Together they open up opportunities for novel trait combinations to meet a wide range of human needs. Thus, at one end, this program has a highly effective and tested, decentralized approach to indigenous tree domestication for the relief of food and nutritional insecurity, poverty, social injustice, environmental degradation, and climate change at the local level; while at the other end, genomic studies ensure that the genetic resource management of indigenous species is based on the latest understanding of the fundamentals of genetic diversity. Together, this approach paves the way for the diversification of domesticated species in tropical agriculture to ensure that agriculture satisfies the needs of billions of individual people as well as the global needs of humanity for the wise management of our planet. In the latter context, the domestication of these tree species also has a key role in the "carbonization" of agriculture and hence in the mitigation of climate change, as well as in the conservation of natural resources and the complex series of interactions between them and living things.

Chapter 11

Domestication of Forest Trees: A Process to Secure the Productivity and Future Diversity of Tropical Ecosystems

This chapter was previously published in Leakey, R.R.B., Last, F.T., Longman, K.A., 1982. Commonwealth Forestry Review, 61, 33−42, with permission of Commonwealth Forestry Association

SUMMARY

In recent years, progress has been made with domestication of *Triplochiton scleroxylon*, an important timber tree of the moist West African forests. Seed viability has been extended from a few weeks to many months by appropriate drying and cold storage. Seed supply is irregular, but the successful development of vegetative propagation methods has provided an alternative, regular supply of planting stock. Rooted cuttings have the further advantage that they can readily be tested to identify promising clones of good form. To this end, experiments on the physiology of branching have allowed the development of a screening technique that can predict later branching habits from tests on small plants. The occurrence of precocious flowering in glasshouse conditions has allowed progress towards control over reproduction, and cross-pollinations with deep frozen pollen have produced viable seeds from clones only 2−5 years old.

The possibility of similarly domesticating many other tree species is discussed in relation to the need for improved selections for use in diverse managed ecosystems. Obtaining sustained yields of food and wood products on land formerly under moist tropical forest clearly depends on learning how to combine increased output from currently underutilized species with soil improvement and ex situ conservation.

INTRODUCTION

The forests of the humid tropics currently amount to around 9 million km^2. If the current rate of destruction, approaching 50 ha min^{-1} (Myers, 1980), were to continue unchecked, it can be projected that all but the most inaccessible parts will disappear within the next 60 years, with the concurrent loss of thousands of plant and animal species. The situation is already serious, but attitudes are fortunately changing through the actions of governments and a variety of international agencies. As a result, a nominal 1.5% of the world's tropical forests has already been designated for conservation, and it is the intention that this should be raised to 10% under the Man and the Biosphere program. Essential as this in situ conservation is, it needs to be augmented by ex situ conservation (see FAO, 1975; Frankel and Hawkes, 1975), as part of the overall process of domestication in the management of the remaining 90% of the forest land.

EXPLOITATION OF NATURAL FOREST AND THE NEEDS OF THE FUTURE

The exploitation of tropical timbers has been and still is highly selective, with characteristically only about 5% of the trees being extracted at any one time (Myers, 1980). This process is also very wasteful, for although some of the remaining trees are used for fuel, many are burnt on site or left to rot. The consequent sudden release of nutrients is rarely efficiently utilized in such disturbed sites before they are lost from the soil.

In some areas, particularly in West Africa, Central America, and parts of India and Southeast Asia, the demand for timber has greatly exceeded natural production and this deficit has stimulated interest both in the establishment of

Multifunctional Agriculture. DOI: http://dx.doi.org/10.1016/B978-0-12-805356-0.00011-8
© 2017 Elsevier Inc. All rights reserved.

exotic species in plantations and a reconsideration of the value of indigenous species previously regarded as of little or no potential. Thus in West Africa where only 10 species were exploited in the 1950s (contributing 70% of timber exports) more than 40 are utilized today, as for instance in Nigeria (Spears, 1980). Similar reappraisals could be made in: (1) Malaysia, where there are at least 4100 species of trees (Whitmore, 1972), and (2) Amazonia, where only 50 of 2500 tree species are used, although about 400 are thought to have some direct commercial value (Myers, 1980).

Although difficult to quantify, there is little doubt that fruits (including nuts), gums, resins, oils, tannins, fibers, latex, dyes, and medicinal products produced by the different components of moist forests could be utilized to a great extent (Robbins and Matthews, 1974; Jong et al., 1973; Okafor, 1980). Many of these forest products are of increasing economic importance (Grainger, 1980), although movement to international markets is often restricted by inadequate storage, transport and infrastructure. Unlike the exracting of timber trees, the mainly nondestructive harvesting of these forest products has little or no detrimental effect on tropical forests. Indeed, increased interest in these products could contribute positively to the mixed assemblages of plants grown in the tropics.

DOMESTICATION OF UNDERUTILIZED TIMBER SPECIES

Domestication implies the collection of seeds or plants, ideally from the entire natural range of the species, and in time, the selection, propagation, and breeding of variants best suited to the needs of man. Because this is a lengthy process, commercial cultivation usually starts before the crop is fully domesticated. This is particularly so in trees because of their long life cycles, where it is the selection processes that often put a constraint on the rate of improvement achieved during domestication.

For forestry plantations, the first steps are usually taken through provenance testing, where seeds collected from several locations are compared, often outside their natural range.

Many of the candidate species for domestication, however, have seed problems which hinder both provenance testing and commercial forestry, these problems ranging from short periods of seed viability, damage from pests and pathogens to irregular, and infrequent flowering. In all these instances, the lack of a dependable supply of seed can be overcome through vegetative propagation. This technique, which has been widely and successfully used to domesticate horticultural crops, has the added advantage that by producing genetically identical trees it allows promising genotypes to be easily identified. However, with a few notable exceptions (*Cryptomeria* and *Populus*), it has only recently been considered for timber species (see Longman, 1976), in spite of the obvious potentials.

Experience with *Triplochiton scleroxylon*

In 1971 the UK Overseas Development Administration, having been persuaded of the significance of vegetative propagation, initiated a cooperative program with staff of the Forestry Research Institute of Nigeria (FRIN), Ibadan, aiming to investigate and develop techniques for conserving and improving indigenous West African hardwoods, and in particular *Triplochiton scleroxylon* K. Schum. (Sterculiaceae), the source of the lightweight hardwood "Obeche." In 1974, a second project was established at the Institute of Terrestrial Ecology near Edinburgh, UK. In concentrating on physiological principles, the latter project complemented the former more practically orientated project, the two together providing a model probably suitable for the conservation and domestication of other tropical trees.

Until recently, *T. scleroxylon* was one of the commonest high forest trees in many of the moist lowlands of West Africa, accounting for up to 13% of the trees present (Hall and Bada, 1979). Although a pioneer species, it was able to maintain itself in established forests in areas with annual rainfall between 1100 and 1800 mm (Leakey et al., 1980). In the 1950s and 1960s "Obeche" formed 60% of Nigeria's roundwood exports, its good peeling properties being favored by plywood manufacturers.

Despite the demand, few plantations were established before 1975 because the inherently short-lived seeds were generally scarce, frequently attacked by weevils (*Apion ghanaense*) and parasitized by smut fungus *(Mycosyrinx* spp.). Since 1975, the development of techniques to extend the viability of sound seeds by dry storage at −18°C (Bowen et al., 1977), together with the modification of standard horticultural techniques in order to root leafy cuttings taken from young trees (Howland, 1975a), have dramatically changed the availability of planting stock. Together these techniques have enabled gene banks and clonal trials to be established at eight different sites in Nigeria, with a concentration at Onigambari Forest Reserve (Howland and Bowen, 1977; Longman et al., 1979a). The gene banks at Ore, Nimbia, and Afaka include material from seed collections from Sierra Leone to Cameroon spanning a large part of the natural range of *T. scleroxylon*.

Initially stem cuttings of *T. scleroxylon* were rooted at FRIN on mist propagators, but subsequently shaded polythene frames were used, with the cuttings sprayed twice a day to maintain humid conditions (Howland, 1975a). Leafy single-node cuttings from seedlings, coppice shoots, and managed stockplants have been successfully rooted with or without applying auxins. However, rates of rooting and numbers of roots per cutting were increased by applying the auxin IBA (indole-5-butyric acid) (Howland, 1975b). More recently, different clones have been found to have different auxin requirements (Leakey et al., 1982a). As a result, and to achieve at least 60% success within 6−8 weeks, it is recommended that cuttings be dipped quickly in 0.2% alcoholic solution or treated with 40 μl of a 50:50 mixture of IBA with a second auxin NAA (α naphthalene acetic acid). After evaporating the alcohol with the help of a fan, cuttings are set in coarse sand with temperatures at the base of the cutting preferably being maintained at about 30°C. To maximize rooting, the leaf of each cutting should be trimmed to about 50 cm^2, this presumably optimizing the balance between photosynthesis and transpiration (Leakey et al., 1982a). The ability of cuttings to root is additionally dependent upon the condition of the stockplants, the most easily rooted cuttings being those from lateral shoots of plants cut back to about 10−20 cm from the ground (Howland, 1975c; Leakey, 1983). The rooting ability of cuttings from such stockplants was greatest when subsequent regrowth was restricted to two shoots each yielding 5−6 single-node cuttings. This number, possibly because of competition for nutrients (Leakey, 1983), could probably be increased by irrigating and applying fertilizers (Howland, 1975c). Cuttings from shaded shoots seem to root better than those from exposed shoots (Leakey, 1983).

Now that *T. scleroxylon* can be readily propagated by cuttings, fuller advantage can be taken of within-species variation by identifying and multiplying genetically superior individuals, as has happened for many years in horticultural and agricultural crops. In field trials now being managed and observed in Nigeria, attempts are being made to analyze the components of growth of promising clones (Howland et al., 1978; Ladipo et al., 1980). Interestingly, branching habit and height growth seem strongly related; the tallest clones have fewest primary branches per meter of mainstem. From one experiment, data taken 18 months after planting suggests that selection of the 10 tallest out of 100 clones might give a genetic gain of 16.5% (Ladipo et al., 1980), whereas the choice of the best provenance might give a 9.3% gain. Clearly the potential for improvement is considerable and warrants the screening of many more clones. As a supplement to field trials which are expensive and time consuming, it may become feasible to predict inferior and superior clones at an early age by a simple test based on an understanding of the processes determining branching habit. Patterns of branching depend on at least two physiological processes: (1) apical dominance, the ability of a terminal bud to inhibit the growth of axillary buds on the current year's growth, and (2) apical control, the influence of one or more shoots on the extension growth of other, usually more proximal, shoots. Experiments with *T. scleroxylon* indicate that these processes operate to different extents between clones and that genetic and environmental influences on apical dominance can be separated (Leakey and Longman, 1986). Current studies (Ladipo et al., 1991) are investigating the apical dominance relationships in small decapitated plants, and comparing these with growth and branching habits of the same clones in plantations. If the fairly strong correlations found so far are confirmed, then it should be possible to screen large numbers of clones in nurseries when 3−6 months old. Experience with *Terminalia superba* Engl. & Diels, another West African hardwood, suggests that it would also be wise to screen clones for wood quality as the genetic differences in various wood characteristics are not always closely related to morphology and growth (Longman et al., 1979b).

Having developed methods of vegetative propagation and made considerable progress with techniques to allow early selection of superior clones, the next phase of domesticating *T. scleroxylon* is to develop the capacity for easily managed controlled breeding. This necessitates methods of inducing flowering when required in small and preferably young plants. In the wild, *T. scleroxylon* does not normally flower until 15 or more years old when the occasional occurrence of severe "short dry" seasons is thought to stimulate flowering (Jones, 1974; Howland and Bowen, 1977). Three approaches have been made to the study of flowering and reproductive biology: (1) phenological and development of natural flowering (Jones, 1974, 1975); (2) establishment of adult clonal material (in the nursery and field) by grafting/budding, and also by rooting adult cuttings (Howland and Bowen, 1977); and (3) stimulation of precocious flowering (Leakey et al., 1981).

When self-pollinated, flowers of *T. scleroxylon* only produce fruits very occasionally and in these instances their seeds do not germinate. By contrast, viable seeds have been produced by cross-pollination (Howland and Bowen, 1977), even when the parent trees were only 2−5 years old and less than 1 m tall (Leakey et al., 1981). Pollinations were equally successful when done with pollen stored dry at −25°C, conditions which extended viability from less than 1 week to at least 25 weeks. The progeny of these controlled pollinations between precociously flowering clones will be used to study the factors influencing floral initiation in *T. scleroxylon*. It is hoped this will lead to the development of reliable techniques for flower induction, which can be applied to clones with superior vegetative growth. The folly of selecting precocious and heavily flowering variants for commercial plantings, as seems to have happened with teak, must be avoided because of the likely adverse effects on form and timber productivity.

The results of the complementary projects discussed in this paper illustrate how the constraints retarding the domestication and use of species like *T. scleroxylon* can be minimized. The stage has now been reached where it is possible to recommend selected clones for more extensive trials.

Experience with other timber producing species

Thirty-three species of tropical trees with known potential for forestry have been vegetatively propagated at I.T.E. using the techniques developed for *T. scleroxylon* (Table 11.1). Generally they were easy to root, although the Dipterocarps (*Shorea* spp. and *Vateria seychellarum*) proved to be more difficult than the others.

With some of the species, the source of the cutting greatly affected its growth after rooting, in addition to altering its rooting ability. For example, lateral shoot cuttings of *Agathis* and *Araucaria* grew plagiotropically, whereas mainstem cuttings grew erectly. In *Cordia alliodora* the early growth of rooted cuttings from lateral branches was similar to that of mainstem cuttings, but later decreased in vigor. These differences highlight the need for (1) good stockplant management and (2) a thorough knowledge of the characteristics of individual species before extensive plantings of clonal material are made. Additionally it should be remembered that cuttings from mature shoots of many tree species are not only difficult to root but have the tendency to grow plagiotropically and are unsuitable for commercial forestry.

If genotypes tolerant to insect attack could be identified, the use of vegetative propagation could be extended to the important species of *Khaya, Entandrophragma,* and *Chlorophora*. Moreover, it should not be restricted to those species with problems preventing their extensive use in plantations, for it is clear that, for example, *Gmelina arborea* and *Tectona grandis* could both benefit from selection for improved form, removing their marked tendency for heavy branching and precocious flowering, respectively. Similarly the use of the techniques described previously for *T. scleroxylon* could also be extended to arid-zone species (Leakey and Last, 1980).

DOMESTICATION OF SPECIES FOR FRUIT AND OTHER FOREST PRODUCTS

The domestication of forest species for fruit and other products is not a new concept. Indeed many plantation crops (coffee, cocoa, tea, citrus, coconut, oil palm, and rubber) have been or are being improved by selection, but the potential of others is still largely untapped (National Academy of Sciences, 1975, 1979). In addition to trees that produce fruits, resins, latex etc., there are many which have medicinal properties (Irvine, 1961), only a few of which have been utilized commercially. In some instances variants may exist in plants occupying similar geographical ranges, as for example with the acridone alkaloids found in *Oricia suaveolens* and *Teclea verdoorniana* and other members of the West African subfamily Toddalioidea (Rutaceae) (Waterman et al., 1978; Fish et al., 1978). If such variation becomes useful pharmaceutically, there is little doubt that selected clones could be multiplied vegetatively (Table 11.2).

Much basic information remains to be acquired regarding the developmental and reproductive biology of many tree species that could be domesticated as producers of a wide range of forest products. In many instances the approach developed for *T. scleroxylon* could be adapted to meet the needs of particular species. In other groups, notably perhaps the fruit trees, budding, and grafting may be a more suitable method of propagating desirable clones, as illustrated by Okafor (1978, 1980) who has studied some 150 Nigerian species spanning 103 genera and 48 families.

PLANTATIONS IN THE TROPICS

As areas of natural forest in the tropics decrease, and the tendency toward semipermanent agriculture increases, both the diversity and indeed the functioning of these ecosystems are at risk. Moreover, the world's timber demand by 2025 AD is expected to be of the order of 300 million m^3 annum^{-1}. For this to be one quarter satisfied, current planting programs will need to be increased by 300% (Spears, 1980). If such an increase were achieved, the pressure on the surviving natural forest would be lessened, but the extensive reliance on plantations will itself represent a substantial loss of diversity. This loss of diversity may increase the risks of attack by pests and pathogens, but to a considerable extent this could be minimized by increasing the number of species grown commercially. In particular the domestication of more indigenous species could usefully supplement the plantations of *Pinus, Eucalyptus, Tectona,* and *Gmelina* which have been widely planted over the last 20 years. In addition to growing a wider range of species, foresters should probably also learn from the experience of tropical farmers who traditionally practice mixed cropping. There are several examples from the temperate zone where even-aged monocultures have been severely attacked by pests and pathogens, and if such plantations are inserted into complex tropical ecosystems, the risks are likely to be greater. Already in Nigeria some stands of *T. scleroxylon*, established since 1975, have been defoliated by the grasshopper *Zonocerus variegatus* while in Ghana caterpillars of *Lamprosema lateralis* have repeatedly attacked stands of *Pericopsis elata*, and other caterpillars have defoliated *Mansonia altissima*.

The arguments presented by foresters and the timber industry when advocating pure plantations are strong. Monocultures tend to give large yields; they are easier to manage and harvest, and produce uniform products

TABLE 11.1 Tropical tree species vegetatively propagated at the Institute of Terrestrial Ecology, Edinburgh, with actual, or potential, value as timber species.

Moist forest

West Africa

Ceiba pentandra (L.) Gaertn.

Chlorophora excelsa (Welw.) Benth. & Hook.

Nauclea diderrichii (De Wild. & Duv.) Merr.

Terminalia ivorensis A. Chev.

Terminalia superba Engl. & Diels

Triplochiton scleroxylon K. Schum.

East & Central Africa

Dalbergia melanoxylon Guill & Perr.

Vateria seychellarum Dyer

Central & South America

Albizia caribaea (Urb.) Britton & Rose

Cedrela odorata L.

Cordia alliodora Cham.

Ochroma pyramidale Urb.

Swietenia mahogani Jacq.

Tipuana tipu (Benth.) Kuntze

Toona ciliata M. Roem.

S.E. Asia & Australasia

Agathis australis (D. Don.) Salisb.

Agathis dammara (A.B. Lam) L.C. Rich.

Agathis macrophylla (Lindl.) Mast.

Agathis obtusa (Lindl.) Mast.

Agathis robusta (Moore) Bailey.

Agathis vitiensis (Seeman) Benth. & Hook.

Araucaria hunsteinii K. Schum.

Gmelina arborea Roxb.

Shorea albida Sym.

Shorea almon Foxw.

Shorea contorta Vidal

Shorea curtisii Dyer ex. King.

Shorea leprosula Miq

Shorea macrophylla (De Vries) Ashton

Dry forest

Acacia senegal L.

Afzelia africana Smith

Azadiracta indica A. Juss.

Prosopis juliflora (Swartz) D.C.

TABLE 11.2 Tropical tree species vegetatively propagated at the Institute of Terrestrial Ecology, Edinburgh, with potential for producing minor forest products or conferring amenity.

Fruit
Casimiroa edulis La Llave
Chrysophyllum cainito L.
Citrus halimii B.C. Stone
Shorea macrophylla (De Vries) Ashton
Tamarindus indica L.
Pharmaceutical
Teclea verdoorniana Exell &Mendonca
Multipurpose (gum/fodder/tannins, etc.)
Acacia senegal L.
Ceiba pentandra L. Gaertn (Kapok)
Prosopis juliflora (Swartz) D.C.
Amenity
Delonix regia Raf.
Caesalpinia spinosa (Mol.) Kuntze
Tabebuia pal/ida (Lindl.) Miers

for processing (Dyson, 1965). However, there is increasing evidence that, particularly in moist tropical forest, they have many disadvantages in relation not only to pests, but in nutrient cycling, accelerated erosion, etc. Additionally, the advocation of monocultures assumes that forestry is only to be practiced by large enterprises. With the increasing interest in agroforestry, however, it is clear that the farmer, without need for great capital investment or large forces of paid labor, could become a significant producer of timber and other forest products, especially if he were supplied by a cooperative nursery with selected material of a wide range of useful indigenous and exotic species. By combining the cultivation of these trees with agricultural crops (Okigbo, 1981) it is probable that the production of both can be extended into areas with infertile and degraded soils. The area suitable for such land use may thus prove greater than that foreseen by Spears (1980). A greater diversity of produce would undoubtedly help many rural communities.

In conclusion, a more enlightened attitude to the untapped resources of the rapidly disappearing tropical forests of the world is necessary if supplies of food, timber, fuel and other forest products are to be available for future generations. A range of techniques discussed here is facilitating the domestication of *T. scleroxylon*. Additionally, the prospects for other tree species appear hopeful, with some sorts of the problems restricting their use now being easier to resolve. The usefulness of clones must not, however, be abused; variation should be maintained by the judicious selection of numbers of superior clones, but perhaps even more importantly, their utilization should be sought within integrated systems of land use which recognize the need to conserve both soil fertility and biological variation within and between species.

ACKNOWLEDGMENTS

We wish to thank: (1) the UK Overseas Development Administration and the Federal Government of Nigeria for funding these projects; (2) the successive Directors of the Forestry Research Institute of Nigeria for their support and interest; (3) Messrs. N. Jones, P. Howland and D. O. Ladipo and Dr. M.R. Bowen for their contribution to the research program; and (4) Mrs. R. Wilkinson, Mrs. M. Gardiner, and Mrs. M. Ferguson for their technical assistance.

Chapter 12

Tree Domestication in Tropical Agroforestry

This chapter was previously published in Simons, A.J., Leakey, R.R.B., 2004. Agroforestry Systems, 61, 167–181, with permission of Springer

SUMMARY

We execute tree "domestication" as a farmer-driven and market-led process, which matches the intraspecific diversity of locally important trees to the needs of subsistence farmers, product markets, and agricultural environments. We propose that the products of such domesticated trees are called Agroforestry Tree Products (AFTPs) to distinguish them from the extractive tree resources commonly referred to as nontimber forest products (NTFPs). The steps of such a domestication process are: selection of priority species based on their expected products or services; definition of an appropriate domestication strategy considering the farmer-, market-, and landscape needs; sourcing, documentation, and deployment of germplasm (seed, seedlings, or clonal material); and tree improvement research (tree breeding or cultivar selection pathways). The research phase may involve research institutions on their own or in participatory mode with the stakeholders such as farmers or communities. Working directly with the end-users is advantageous toward economic, social, and environmental goals, especially in developing countries. Two case studies (*Prunus africana* and *Dacryodes edulis*) are presented to highlight the approaches used for medicinal and fruit-producing species. Issues for future development include the expansion of the program to a wider range of species and their products and the strengthening of the links between product commercialization and domestication. It is important to involve the food industry in this process, while protecting the intellectual property rights of farmers to their germplasm.

INTRODUCTION

Tree domestication is an umbrella term that is often applied erroneously to a subset of activities such as provenance testing or to a narrow application such as industrial forestry. Simply put, tree domestication refers to how humans select, manage, and propagate trees where the humans involved may be scientists, civic authorities, commercial companies, forest dwellers or farmers. In the tropics, the trees involved in tree domestication occur in natural forest, secondary forest, communal fallow lands, plantations, and farms. These trees in turn provide both products (timber, fruit, fodder, etc.) and services (shade, soil improvement, erosion control, etc.). Of course, human history is interwoven with forests and trees before agriculture, urbanization, and commerce began, but until relatively recently the interaction was about extraction of tree products from natural forests.

While much is written about natural forest management, commercial tree improvement, and forest analog systems, relatively less is published on the largest group of people who are the rural population in developing countries and their interactions with the large group of tree species in agricultural landscapes. In Asia, Latin America, Africa, and Oceania this is where the greatest potential exists for tree domestication to contribute to sustainable development. Agriculture and forestry are no longer thought of as mutually exclusive activities, yet national and international statistics are only kept on the differentiated land cover of these systems and the data on the extent of integrated agroforestry systems are not available.

Pantropically, there has been deliberate selection and management of trees by humans in forests to provide what are now collectively termed nontimber forest products (NTFPs). In several cases, the extraction, and often overextraction of

© 2017 Elsevier Inc. All rights reserved.

NTFPs has led to market expansion and supply shortages, which in turn has led to cultivation of trees for the same products. The ambiguity arises, however, in that such products are no longer from the forest and some of the species also provide valuable timber (e.g., the medicinal species *Prunus africana*). The authors therefore suggest that a new term is required to define tree products that are sourced from trees cultivated outside of forests. Most logically these should be referred to as agroforestry tree products (AFTPs). This paper describes the beginnings, current status, and future directions of domestication of trees for the agroforestry systems of the tropics, with a focus on those providing AFTPs.

ORIGINS AND CONCEPTS OF DOMESTICATION

The word *domestication* has had several definitions and interpretations since its first appearance in the English language in 1639 (OED 1989). When applied to animals it refers quite narrowly to taming wild subjects and bringing them into the homestead. With respect to plants, there is a spectrum of meaning from nurturing wild plants through to plant breeding through to genetic modification in vitro. Most commonly the word is used with reference to annual food crop plants that have undergone selection, breeding and adaptation in agricultural systems. Archeologists concur that annual crop domestication began with wheat 10,000 years ago in Eurasia at a time of rising human populations and overexploitation of local resources (Simmonds, 1979). Cereal domestication surely ranks as one of the greatest technological advances in human history since not only does wheat (*Triticum aestivum*) still provide 20% of food calories consumed globally but domestication of other cereal crops (e.g., rice [*Oryza sativa*], barley [*Hordeum vulgare*], maize [*Zea mays*]) has seen human populations increase 1000-fold since domestication started (Diamond, 2002).

Tree domestication is a far more recent phenomenon than annual crop domestication. One of the earliest records of tree domestication is that of manipulating pollination in *Ficus* trees 2800 years ago by the prophet Amos (Dafni, 1992). More important though than the date of onset of domestication in trees is the scale of activity. In terms of conventional improvement, tree species are far more neglected today than agricultural crops, with the exception of temperate fruit trees (Janick and Moore, 1996). In commercial forestry, fewer than 40 taxa have genetic improvement programs underway and most of these are less than 60 years old (Barnes and Simons, 1994). Attempts at improving the N_2-fixing agroforestry species (e.g., *Leucaena*) started even later, in the 1980s, coinciding with concerns about soil fertility management, a tropical fuelwood crisis, and renewed interest in social forestry. Domestication of other agroforestry trees has received substantial recent interest following a number of articles and conferences, most significantly the 1992 IUFRO Conference in Edinburgh, UK (Leakey and Newton, 1994b,c). Much of the progress in tree domestication has been informed by the well-described home-garden systems of the Amazon, Southeast Asia, and Africa (Kumar and Nair, 2004). Domestication of trees for the provision of AFTPs has, however, been more frequently equated to conventional timber tree improvement and horticultural improvement than to home-garden domestication.

Three striking differences between conventional timber tree improvement and agroforestry tree improvement exist. These are the number of taxa involved, the industrial rather than subsistence use, and the number of stakeholders involved. Commercial plantations typically handle one or a few species and one company may control all the operations from planning, germplasm sourcing, tree improvement, nursery management, planting, tree husbandry to harvesting. These operations are all carried out at a scale to maximize profit. In contrast, agroforestry is concerned with thousands of tree species and millions of subsistence farmer clients influenced by a mixture of government, private sector, community, and international partners, each engaged in different and largely uncoordinated activities. In most cases, agroforestry tree improvement has been concerned with on-farm use of firewood, fodder, fruit, live fence, medicinal, and fallow trees. The next large change in agroforestry worldwide, which has already started (Franzel et al., 2004), will probably come from a greater focus on cultivating trees for cash, and most likely for fruit, timber, and medicines. Thus it is inappropriate to simply equate agroforestry tree domestication with industrial-tree improvement since aspects of species prioritization, indigenous knowledge, farming systems improvement, adoption, and marketing are as important as selection and multiplication. For these reasons, the World Agroforestry Center (ICRAF, formerly called the International Council for Research in Agroforestry) introduced a wider concept of tree domestication:

Domesticating agroforestry trees involves bringing species into wider cultivation through a farmer driven and market-led process. This is a science based and iterative procedure involving the identification, production, management and adoption of high quality germplasm. High quality germplasm in agroforestry incorporates dimensions of productivity, fitness of purpose, viability and diversity. Strategies for individual species vary according to their functional use, biology, management alternatives and target environments. Domestication can occur at any point along the continuum from the wild to the genetically transformed state. The intensity of domestication activities warranted for a single species will be dictated by a combination of biological, scientific, policy, economic and social factors. In tandem with species strategies are approaches to domesticate landscapes by investigating and modifying the uses, values, interspecific diversity, ecological functions, numbers and niches of both planted and naturally regenerated trees.

Simons (2003)

This rather wordy concept can be simplified to: tree domestication in agroforestry is farmer-driven and about matching the intraspecific diversity of many locally important tree species to the needs of subsistence farmers, the markets for a wide range of products, and the diversity of agricultural environment.

Traditionally utilized extractive resources, such as fruits, medicines, and fibers from forests have been collectively described as NTFPs or nonwood forest products (NWFPs). Much research has been done on these products in the hope of finding better ways of managing and conserving natural forests while also benefiting the people living in or near the forests. Now there is confusion in the statistics and literature as people describe and discuss these same products as new crops from farmland (Belcher, 2003). It is therefore proposed that the products of domesticated agroforestry trees, including timber, should be called AFTPs to distinguish between the wild and the domesticated products.

OBJECTIVES OF TREE DOMESTICATION

Trees occur naturally in forests and rangelands and can be grown in commercial plantations and on farms. Within the tropics, natural forests cover 35% of the land; commercial tree plantations account for 1% of the land cover; tree crop plantations [such as cacao (*Theobroma cacao*), rubber (*Hevea brasiliensis*), tea (*Camelia sinensis*), coffee (*Coffea* spp.), citrus (*Citrus* spp.), mango (*Mangifera indica*), and oil palm (*Elaeis guineensis*)] account for a further 1.5%; and agricultural land accounts for 40% of the land area (FAO, 2002). With current tropical deforestation rates at around 1%, principally for expansion of agricultural areas, and even with the most optimistic increases in timber-plantation estates (FAO, 2003), the largest scope for future tree planting in the tropics will be on agricultural land (Simons et al., 2000a).

In agroforestry, the objective of domestication is to enhance the performance of trees in terms of improved tree products, such as timber, fruits, and medicines, and/or improved environmental services, such as the amelioration of soil fertility. In the former case, improvements will usually be for yield and/or quality with a specific market opportunity as the driver of genetic selection. The demands for these outputs of tree domestication in agroforestry trees are sure to attract increasing interest and resources over the coming decades, and to have an increasing market orientation.

The scale and direction of tree domestication in agroforestry is dependent upon the varying objectives of different stakeholders. As already mentioned, in most situations in the past, the objective has been for subsistence use and less frequently for income generation. However, to date, implementation of tree "needs assessment" and the consideration of actual or potential contributions of trees to household budgets has been inadequate (Njenga and Wesseler, 1999), and in many locations the relative importance of trees, let alone opportunities for tree domestication, has not been established.

Nevertheless, exercises in tree species prioritization have been carried out quite extensively, as described in the following sections.

Selection of Tree Species

While species of only three genera (*Acacia, Pinus*, and *Eucalyptus*) account for more than 50% of all tropical tree plantations (FAO, 2003), more than 3000 tree species have been documented in agroforestry systems (Burley and von Carlowitz, 1984). Until the 1990s, research priorities among this vast array of agroforestry trees were determined arbitrarily, often based on individual interests of researchers. Recognition of the importance of understanding of user needs and preferences, technological opportunities and systematic methods for ranking species emerged with publication of guidelines for tree species priority-setting procedures (Franzel et al., 1996). Regional surveys of agroforestry tree species prioritization have subsequently been completed for the Sahel, southern Africa, and West Africa (Jaenicke et al. 1995; Sigaud et al. 1998; Maghembe et al., 1998), as well as for individual countries including Bangladesh, Brazil, Ghana, India, Indonesia, Peru, Philippines, and Sri Lanka (c.g., Sotelo-Montes and Weber, 1997; Lovett and Haq, 2000). Species priorities depend on the objectives of domestication, and will differ if it is for income generation, satisfying farm household needs, germplasm conservation through use, forest conservation through enrichment planting, or farm diversification.

Trees found on farms may originate from forest remnants, natural regeneration, or deliberate planting. Much of the deliberate planting of indigenous trees on-farm [such as the Indonesian damar (*Shorea javanica*), cinnamon (*Cinnamomum* spp.), and rubber agroforests (Michon and de Foresta, 1996)] arose because of spontaneous farmer initiatives, while government and donor projects tended to promote exotic species (see Shanks and Carter, 1994). Recent studies of the frequency and abundance of trees on farms in Cameroon, Kenya, Nigeria, and Uganda show the balance between indigenous and exotic tree species in these new plantings (Kindt, 2002; Schreckenberg et al., 2002b).

While indigenous taxa may account for the majority of species on-farm, introduced exotic taxa account for many of the trees on-farm, especially in Africa. Exotics can have some advantages such as superior growth, although they also have several risks including weediness and aggressive use of natural resources. The existence of exotics clearly demonstrates deliberate planting. Certain exotics have been planted for decades or centuries such that local communities consider the naturalized populations as indigenous species. Examples include *Grevillea robusta* in Kenya or *Gliricidia sepium* in Sri Lanka (Harwood, 1992; Stewart et al., 1996). Whether exotic, naturalized, or indigenous, it is clear that many agricultural landscapes may be tree rich, but species poor (frequently within-species diversity, especially in exotic species, is dangerously poor: Lengkeek et al., 2005). For this reason improvement of the landscape by examining and ameliorating tree species diversity within and between functional uses (e.g., boundary, fodder, firewood, and fruit) can be as important as improvement of a single species. Thus the concept of domesticating the landscape becomes relevant. Four points of intervention are relevant here: replacement, substitution, expansion, and better management of trees.

In conventional forestry, a classic step in tree improvement is derivation of a short list of fast growing and productive species, which is often arrived at from species elimination trials. Early work in agroforestry included such trials, either as unreplicated arboreta or as replicated trial series (Stewart et al., 1992). Sadly, much of this research was undocumented and/or discontinued, and this has resulted in agroforestry tree plantings relying on fewer commonly known species. Another common shortcoming of this work, even when it was written up, is that seed sources that were used to represent species are not reported. This means that many of the results of species trials are confounded by the seed source used and species ranking may have been different had other or mixed sources been used. The solutions to this shortcoming include: (1) clear documentation of germplasm used; (2) use of multiple provenances per species, where possible; (3) or, when (2) is not possible, inclusion of more than one provenance to consider a provenance mixture in species elimination and species proving trials.

TREE DOMESTICATION STRATEGIES

The determinants of a sound domestication strategy for an individual species can be grouped under 14 headings. These are:

- Reasons for domestication (home use, market, conservation of the species, agroecosystem diversification, improved livelihood strategies)
- Tree uses required (products [AFTPs] and services)
- History and scale of cultivation (as native and exotic)
- Natural distribution, intraspecific variation and ecogeographic survey information
- Species biology (reproductive botany, ecology, invasiveness)
- Scale and profile of target groups and recommendation domains (biophysical, market, cultural)
- Collection, procurement or production of germplasm and knowledge (including ownership, attribution, benefit sharing, access, and use)
- Propagule types (including symbionts) envisaged
- Nursery production and multiplication
- Tree productivity (biomass, timing, economics, risks)
- Evaluation: scientific and farmer participatory
- Pests and diseases
- Genetic gain and selection opportunities, methods and intensities
- Dissemination, scaling up, adoption, and diffusion.

The problem with most agroforestry tree species is that information is incomplete, which has led to suboptimal tree domestication strategies. While tree domestication work has increased within agroforestry, the documentation of the logic and the approach has been generally scant. Even when results are shared or published, it is typically the outcomes that are reported and not the processes. A few case studies of tree domestication strategies are available (e.g., *P. africana*; Simons et al., 2000b) as well as a generic tree domestigram. What is most needed and lacking, however, is a generalized domestication decision framework, which uses the elements of the domestigram. Case studies of various tree categories (product type, mode of propagation, generation interval) are currently being used to construct such a framework.

Beyond the individual tree species domestication work to meet the needs of the farmer and the market (with and without processing to enhance shelf life and market value), there are several elements to consider for landscape-level domestication:

- Likelihood of interaction of planted trees with natural populations and consequences of introducing external populations;
- Primary and multiple uses of the species since individual species differ in the number of total uses and their primary and secondary uses to different clients, and lack of attention to this has led to the miracle species concepts associated with species such as *Leucaena leucocephala*;
- Combined value of all species (economic, social, biological) in the landscape;
- Diversity (within and between species) in each functional use group since some groups, e.g., fodder, may be dominated by a single species and substitution of species in one use group may be more useful than replacement of existing species with better material of the same species;
- Number of trees per unit area and per farm as informed by considerations of species richness and abundance;
- Niche integration on-farm and within the landscape as it will affect the viability of populations.

A wider consideration of other elements of agrobiodiversity, such as soil biota, is also pertinent and may need attention.

GERMPLASM SOURCING, DOCUMENTATION, AND DEPLOYMENT

One of the most fundamental elements of a tree domestication program is the sourcing and deployment of germplasm. This is especially true in agroforestry since little or no formal selection and breeding are generally carried out. The introductions made on-farm (native and exotic) are in many cases a once-off exercise with subsequent generations derived by the farmer from the original parent stock. While subsequent introductions by farmers may broaden the genetic material, considerations of the diversity in founder stock are rarely considered. It is routinely reported from various authors that seed availability is a significant bottleneck to tree planting. Perhaps these statements could be better phrased as "seed supply and demand in agroforestry are poorly understood and poorly matched." In essence, the key missing information is the quantification of the various flows (Fig. 12.1) of each species between sources, suppliers, and users of germplasm.

The collection of germplasm from wild stands and the distribution of seed from national tree seed centers and scientists (FAO, 2001; Harwood, 1997; Fig. 12.1) have been best described. One feature evident from the literature and

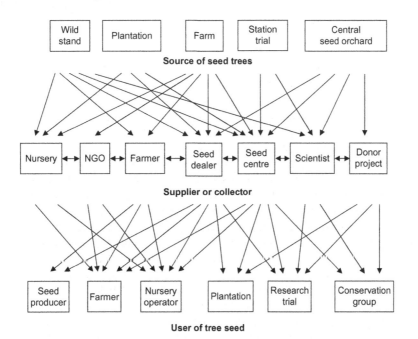

FIGURE 12.1 Diagram showing flows between sources, suppliers, and users of agroforestry tree germplasm.

agroforestry practice is that most people differentiate between seeds and seedlings. To generalize, seeds are seen as the domain of centralized seed centers, and seedlings are seen as the mandate of decentralized small-scale nursery operators.

Herein lies the heart of the "disconnect" in agroforestry. The *Tree Seed Suppliers Directory* (Kindt et al., 2003) seeks to contribute to understanding of the availability from a wider set of international seed dealers, but the companion volumes of small-scale seed producers and nursery operators are missing. For most species, little information is available and the fundamental gap in completing the summary in Fig. 12.1 is the lack of quantitative data to place against each flow, as well as details on the number of actors and volumes of germplasm—whether for a single species, a species group or trees in general. Against this background, a new approach in which farmers or village communities develop cultivars from the superior trees in their own land is being developed by the ICRAF within its Participatory Domestication model (Tchoundjeu et al., 1998; Leakey et al., 2003). This therefore involves a much more localized/internalized selection of germplasm.

As with all other plant species, trees became subject to new international regulations with the coming into force of the Convention on Biological Diversity in December 1993. Most importantly, national legislation stipulates how access to and benefit sharing from plant genetic resources should be handled. In the main, trees have not undergone the same level of scrutiny from advocacy groups or government watchdogs as commercial crop or medicinal plants. This is partly explained by the lack of commercial interest from seed multinationals, and also by the lack of harmonization of national laws on genetic resources of agricultural species and those on forest genetic resources. To be able to regulate germplasm it needs to be well-described. Descriptors provided for germplasm of various annual crops allow plant breeders and others to value and morphologically characterize the materials developed. Such descriptors for agroforestry trees are nonexistent. Interestingly, molecular characterization has been more frequent for agroforestry species and in particular for the leguminous taxa, including those within the genera: *Acacia*, *Calliandra*, *Gliricidia*, *Inga*, *Leucaena*, and *Sesbania*, although this approach is gathering momentum for some timber (e.g., *Swietenia macrophylla*: Novick et al., 2003) and indigenous fruit trees (*Irvingia gabonensis*: Lowe et al., 1998, 2000).

Two contrasting foci have emerged in the agroforestry tree seed industry. First, there has been the need to satisfy immediate seed requirements for donor, nongovernmental organization (NGO), and government-led projects. This relies on sizeable procurement activities and in a few cases large-scale seed production, and the seed is typically given away to farmers and communities for free. Two negative consequences of this approach have been that farmers have undervalued the seed as it has been given out free, and small-scale seed producers have not been able to expand, as they cannot compete against a free good. However, a case could be made that if tree cultivation is considered desirable for society, then tree establishment should be subsidized. The second force has been initiatives by the Danish Forest Seed Center and others to diagnose and establish sustainable seed supply systems. These initiatives are not well suited to project formats where indicators of success can be reported after 2−3 years. However, if more work is not put into sustainable seed systems, the situation of start-stop tree-seed supply will continue due to its association with projects. One way to increase the functionality of seed systems may be with improved material. Often users of tree germplasm do not differentiate within a species and labeling improved provenances, varieties, families or clones would assist.

Whether seed will come from project start-up activities or sustainable supply systems, there is a need for accurate demand forecasting of what type, where, when and how much seed will be needed. A "tree seed forecaster" has been developed by Simons that, for a given species, examines geographic scope for cultivation, potential target population of farmers, testing, and adoption behavior of early and late adopters, average farm size, number of trees per farm, reproductive generation interval and scope for on-farm germplasm production and domestication.

One peculiarity in agroforestry is that germplasm is typically viewed as relating to seed. Yet few farmers plant seed since few agroforestry species are direct sown. The few species that are direct sown are those that tend to provide services rather than products, such as live fences, nutrient replenishment, shade, windbreaks, and erosion control. Thus, to a farmer, germplasm is more about seedlings rather than seed, since the majority of agroforestry trees are planted from nursery-raised seedlings originating from either seed or vegetative propagules. Germplasm supply is, therefore, as much about seedling supply as seed supply. While commercial forestry operations have undertaken substantial research on nurseries and propagation, far less work has been undertaken for agroforestry trees. In agroforestry there has mainly been a focus on the quantity rather than quality of seedlings (Wightman, 1999), although it is now recognized that quality encompasses both physiological and genetic components. Both Jaenicke (1999) and Wightman (1999) describe good practices for research and community nurseries, while Boehringer et al. (2003) describe the importance of the social and organizational aspects of nurseries in addition to the technical ones.

Tree genetic resources used in agriculture should ideally receive commensurate funding for conservation in tune with annual crop plants. Sadly, at the international level, trees are given a back seat compared to herbaceous and graminaceous crops, as evidenced by their exclusion at the 1996 Leipzig, Germany, Conference on Global Plan of Action for Genetic Resources for Food and Agriculture. Furthermore, the recently negotiated FAO International Treaty on Plant Genetic Resources for Food and Agriculture (www.fao.org) lists only two genera for a multilateral conservation system (*Artocarpus* and *Prosopis*).

TREE IMPROVEMENT RESEARCH

Tree improvement research has generally been based on two methods of propagation: sexual or vegetative. Improved tropical fruit trees are typically vegetatively propagated using buds, grafts, marcots (air-layers) or cuttings. Here, elite clones of species such as mango, avocado (*Persea americana*), tamarind (*Tamarindus indica*), and citrus are selected at high intensities for mass propagation as cultivars. Improved timber trees, such as eucalypts (*Eucalyptus* spp.) and poplars (*Populus* spp.), may be propagated by cuttings, or from seed (e.g., *S. macrophylla*, *Cordia alliodora*), in which case cloning may have been used in a seed orchard phase. Nearly all other agroforestry trees are propagated by seed, either direct sown or as nursery-grown seedlings. Given this orientation, tree improvement research has focused both on the propagation method and on the genetic gains from selection among clones or sexual progeny.

The most advanced tree improvement research that has been undertaken on any agroforestry tree is that carried out on *Leucaena* species by Brewbaker and colleagues at the University of Hawaii (Brewbaker and Sorensson, 1994) during the 1970s to late 1990s. *Leucaena leucocephala* is atypical with respect to its self-mating system, thus making it easy to fix elite traits. The backcrossing and interspecific hybridization carried out by Brewbaker and colleagues saw rapid progress in traits such as growth, vigor, palatability to livestock, cold tolerance and insect resistance. Interestingly, this is the agroforestry species that is most widely planted throughout the tropics, and this is presumed to be due both to availability of seed and of improved varieties.

With the exception of *Leucaena*, there has been little formal breeding of agroforestry trees. To date, tree improvement in agroforestry has largely come about through species and provenance screening trials for sexually propagated species, and through clonal selection for vegetatively propagated species. It is worth contrasting this situation with the tree improvement program of a single developed country such as Sweden, which has more than 1500-plus trees and 3500 clones in seed orchards in 22 separate breeding populations for a single species, *Picea abies* (Norway spruce) (FAO Reforgen Database: www.fao.org).

In the absence of formal tree improvement in agroforestry, informal selections by farmers, mostly with trees on their farms, dominate rural areas. Researchers are beginning to understand better the techniques used and traits sought by farmers (e.g., Lovett and Haq, 2000; Schreckenberg et al., 2002b). It is worth noting, however, that the poor results from some farmer selections are due largely to narrow founder populations and inbreeding through propagation from a limited number of parents. This observation is borne out by the superiority of many wild populations that were recollected and compared with exotic landraces (e.g., *Gliricidia*: Simons and Stewart, 1994). With fruit trees and some other trees that are clonally propagated, the issue does not arise, although the optimal number of clones to use is a consideration.

PARTICIPATORY DOMESTICATION

In recent years international aid to developing countries has developed a strong focus on poverty reduction. In parallel with this, the ICRAF initiated its tree domestication program in the mid-1990s with a new focus on products with market potential from mainly indigenous species (Simons, 1996). With this came a shift from on-station formal tree improvement toward more active involvement of subsistence farmers in the selection of priority species for domestication and the implementation of the tree improvement process. In many ecoregions in which ICRAF is active, farmers selected indigenous fruit trees as their top five priorities. Consequently, over the last decade a strategy for the domestication of indigenous trees producing high-value products of traditional and cultural significance has been developed (Tchoundjeu et al., 1998; Leakey et al., 2003). This approach to improving the trees planted by farmers has a number of advantages:

- It has a clear poverty reduction focus, which has been endorsed by a review on behalf of UK Department for International Development (DFID) (Poulton and Poole, 2001). The income derived from tree products (AFTPs) is often of great importance to women and children, for example, to meet the demand for school fees and new uniforms.

- It has immediate impact by going straight into implementation at the village level, so avoiding delays arising from constraints to the transfer of technology from the field station to the field that can be due to technical, financial, dissemination, and political difficulties.
- The approach being developed is focused on simple, low-cost, appropriate technology yielding rapid improvements in planting stock quality based on selection and multiplication of superior trees in ways that create new plants, which also produce fruits within a few years and at heights that are easily harvested.
- It builds on traditional and cultural uses of tree products (AFTPs) of domestic and local commercial importance, and meets local demand for traditional products.
- It promotes food and nutritional security in ways that local people understand, including promoting the immune system, which is especially important in populations suffering from AIDS.
- It can promote local level processing and entrepreneurism, hence employment and off-farm economic development. These benefits can stimulate a self-help approach to development and allows poor people the opportunity to empower themselves.
- It can be adapted to different labor demands, market opportunities, land tenure systems, and is appropriate to a wide range of environments.
- It builds on the rights conferred on indigenous knowledge and the use of indigenous species by the Convention on Biological Diversity and is a model for "best practice," in contrast to biopiracy.
- It builds on the commonly adopted farmer-to-farmer exchange of indigenous fruit tree germplasm as practiced in West and Central Africa, for example *Dacryodes edulis*, which although native to southeast Nigeria and southwest Cameroon, is now found across much of the humid tropics of Central Africa (Cameroon, Congo, Central African Republic, Zaire, Gabon, etc.).
- It builds on the practice of subsistence farmers to plant, select and improve indigenous fruits (Leakey et al., 2004), such as marula (*Sclerocarya birrea*) in South Africa, where the yields of cultivated trees are increased up to 12-fold and average fruit size is 29 g, while trees in natural woodland are 21 g (Shackleton et al., 2003a).
- The domestication of new local cash crops provides the incentive for farmers to diversify their income and the sustainability of their farming systems (Leakey, 2001a,b).

Against these advantages there are possibly disadvantages, such as reduced genetic diversity in the wild population as it is replaced by a domesticated population. However, the implementation of some in situ or ex situ conservation (McNeely, 2004) of wild germplasm, together with the deliberate selection of relatively large numbers of unrelated cultivars can minimize these risks. Indeed, the current strategy, whereby each village develops its own set of cultivars, should ensure that at a national/regional scale the level of intraspecific diversity remains acceptable for the foreseeable future.

Concerns about genetic diversity and how to help farmers to maximize their gains through plus-tree selection for multiple traits have raised questions about how farmers perceive variation. Consequently, quantitative studies of tree-to-tree variation in fruit and nut traits have been implemented at the village level, and it is clear that there is very considerable intraspecific variation in each trait and that many of these traits are unrelated (e.g., *I. gabonensis*: Atangana et al., 2001, 2002; Leakey et al., 2006). Thus, the current approach to selection of trees meeting various market-oriented ideotypes ensures that cultivars are almost certainly highly variable in many other desirable traits, such as resistance to pests and diseases. It is also interesting that in terms of the level of variability in the measured traits, the current semidomesticated on-farm populations are more variable than the wild populations (Leakey et al., 2004), suggesting farmer selections may have multiple population origins.

To maximize the economic, social, and environmental benefits from participatory domestication it is crucial to develop postharvest techniques for the extension of shelf life of the raw products, and processing technologies to add value to them. Without this parallel preparation for increased commercialization, domestication will not provide all the above-listed benefits. The combination, however, has potential applications that extend beyond subsistence agriculture to agricultural diversification of farming systems. In tropical North Queensland, Australia, for example, this is linked, at least in part, to the development of an Australian "bush tucker" industry supplying restaurants and supermarkets worldwide.

In his review of how agroforestry fits the Millennium Development Goals, Garrity (2004) concludes that agroforestry needs "a research and development strategy to reduce dependency on primary agricultural commodities, and to establish production of added value products based on raw agricultural materials (with traditional and cultural values and locally recognized importance and markets), with links to growing and emerging markets." In agreement with much of the previous information on agroforestry tree domestication, Garrity affirms that agricultural R&D institutions must develop new skills in the domestication of indigenous species and the processing/storage of their products, in market analysis and market linkages. This would help focus development on poverty in ways that are of interest and importance to subsistence farmers.

CASE STUDIES

Two case studies are presented for agroforestry species that are used primarily for timber, fruit, and medicinal products.

Prunus africana

Prunus africana (Rosaceae)—"pygeum"—is an afromontane forest species found only above 1000 m altitude and confined to isolated populations forming a wide but disjunct distribution (Cunningham and Mbenkum, 1993). The genus *Prunus* contains more than 200 species (e.g., peach [*Prunus persica*] and plum [*Prunus domestica*]) of which many have undergone intensive domestication through selection and breeding. *Prunus africana* is, however, the only species native to Africa and is an important medicinal tree (Hall et al., 2000). Currently, the commercial harvesting of the bark (approximately 4000 Mg of bark per year) for commercial and domestic medicinal use (treatment of benign prostatic hyperplasia [BPH]) has led to extinction of some populations and the listing of the species on the CITES Appendix II. This is a list of plants requiring protection through extraction permits and monitoring of international trade, which currently has an over-the-counter value for *P. africana* products of $220 million per year (Cunningham et al., 1997).

The combination of the intense conservation interest, the considerable commercial and human health importance of the medicinal product, and the potential of the species to be cultivated by small-scale farmers has resulted in the development of a domestication strategy (Leakey, 1997; Simons et al., 1998) aimed at the restoration of the resource through diverse agroforestry plantings. This strategy is expected to enhance farmer livelihoods, meet future needs of the industry and, to some extent, have positive ecological and conservation benefits.

The objectives of domesticating *P. africana* are:

1. To conserve wild populations through reducing pressure on the natural resource base by encouraging cultivation of trees by small-scale farmers. Cultivated material also serves a useful *circa situ* conservation function if attention is paid to genetic diversity issues.
2. To locate unique and diverse natural populations that warrant specific in situ conservation measures.
3. To identify productive genetic diversity to demonstrate the growth potential of the species.
4. To quantify the genetic control of traits of economic interest (timber volume, chemical profile, bark yield) and determine appropriate selection methods.
5. To undertake participatory domestication with small-scale farmers to determine their preferences and perspectives on cultivating and improving the species.
6. To establish seed production stands and develop appropriate management techniques to be able to deliver sufficient propagules to farmers.
7. To improve propagation methods of the species to encourage wider adoption.
8. To undertake marketing studies in order to better monitor and predict demand and supply, and evaluate prospects for green-labeling and the establishment of premiums for small-scale producers.
9. To use *P. africana* as a case study for the domestication of other medicinal and high-value trees.

The first step before developing a domestication strategy for any species is to collate all available information on the species including: botanic descriptions, geographic distribution, ecology, forest inventories, farmer surveys, harvesting techniques, trade figures, and conservation status. For *P. africana*, the key information gaps identified were details on market intelligence, growth data, reproductive ecology, pests and diseases, genetic variation, and propagation methods. These knowledge gaps have been the subject of recent studies.

Market intelligence projections suggested that, given the increase in the aging male population in Europe and America and the increase in consumer confidence in herbal remedies, demand could rise two to three-fold (to 8000–12,000 Mg per annum). It is clear that natural forests will not be able to meet such demands and thus that cultivation is required. This raises issues about how to achieve this cultivated resource, which are reflected by the two unlikely extremes of a single plantation of 8000 ha (at 4 × 4 m spacing) benefiting a very limited number of producers, or many smallholders each producing a few trees through agroforestry (one farmer growing 5 million trees or a million farmers each producing five trees).

Studies of the few existing plantations of *P. africana* in Cameroon and Kenya found that trees over 12 years of age produced acceptable yields and concentrations of active constituents, although yields of bark-extract increased as trees aged further (Kimani, 2002; Cunningham et al., 2002). The highest bark-extract yield (1.2%) was found in 55-year-old trees, although trees of similar diameter class can have differences in mean extract yield from 0.8% to 1.33%. In addition, the major sterol component, B-sitosterol, varied significantly between provenances ($101-150$ µg/g), while individual tree yields varied from 50 to 191 µg/g. It is not known to what extent this variation is under genetic control. However, molecular studies of neutral genetic variation have revealed that differences between populations across the range account for 58% to 73% of total variation (Muchugi, 2001) reflecting the disjunct nature of afromontane populations. Elite trees of different provenances are now being mixed in a seed orchard, with trees from Kibale, Rwenzori, and Bwindi populations in Rwanda. This approach is comparable to industrial forestry with combining populations in breeding seedling orchards (Barnes and Simons, 1994).

Studies of the storage of *P. africana* seed have found that it is recalcitrant to intermediate in its behavior. Greatest germination ($40-70\%$) was found when purple-colored seeds were collected and depulped. Extracted seeds could be stored at 4°C for up to one year at 15% moisture content, although this decreased viability by 50%. Surveys of seed price in Kenya and Cameroon show it to be $8 to $25 per kilogram, which at $3000-5000$ seeds per kilogram makes it relatively expensive. The infrequent nature of fruiting, high price of seed, and recalcitrant seed behavior indicate that sourcing sufficient seedlings on a regular basis may be problematic.

Vegetative propagation studies for *P. africana* in Kenya and Cameroon have found that juvenile tissues root well ($75-90\%$) as leafy cuttings (Nzilani, 1999; Tchoundjeu et al., 2002b), opening the way for clonal approaches to producing medicinal extracts with high yield, quality and uniformity. On-station trials have been established in Muguga and Kakamega (Kenya), Kabale (Uganda), and Buea (Cameroon) to examine growth and survival, as well as family and provenance variation. Species prioritization studies with farmers in Cameroon, Uganda, and Kenya have confirmed the use and popularity of the species. On-farm planting with *Prunus* has now taken place in Cameroon, Kenya, Madagascar, and Uganda, all with unimproved material. By the mid-1990s these had occurred on several thousand small-scale farms in Cameroon (Cunningham et al., 2002), although to date these immature plantings have not taken pressure off the trees in natural forests.

It is foreseen that the on-farm cultivation of *P. africana* would result in the following outcomes: Trees grown on field boundaries have more spreading crowns than in closely spaced plantations, but are a good source of bark and timber.

- Before felling for timber, nondestructive bark paneling can be undertaken at 15, 23, and 31 years—providing 15, 25, and 60 kg of bark per tree, respectively.
- A tree could conceivably be felled for timber at 40 years of age, when it would yield approximately 200 kg of bark.
- This equates to a production of 7.5 kg of bark per tree per year over a 40-year period.
- For a 3000 Mg market, 400,000 such trees will be needed.
- For a 10,000 Mg market, 1.3 million trees will be needed.

Dacryodes edulis

The domestication of *D. edulis*—Safou or African plum—has been the developing model for Participatory Domestication approaches in Cameroon and Nigeria described previously (see papers in Schreckenberg et al., 2002a), since its identification as the top priority species for agroforestry in West and Central Africa.

It is widely grown in mixed farming systems across many countries of Central Africa ($28-57\%$ of all fruit trees), especially as the shade tree for cacao or coffee in the forest-savanna transition zone, as well as a middle-stratum in cacao farms under secondary humid forest and as a common constituent of home gardens. Mean numbers of trees per farm range from 20 to 200 in some villages in Cameroon (Schreckenberg et al., 2002b), with tree density being greatest in farms under 2 ha. In 1997, the trade of Safou in Cameroon alone was worth $7.5 million, excluding domestic consumption. Of this trade, $2.5 million is in exports (Awono et al., 2002). The fruits are typically roasted and eaten as a nutritious staple; it is rich in fat (64%), protein (24%), and carbohydrate (9%), and there is the potential for vegetable oil extraction on a commercial scale.

There is considerable farmer-to-farmer exchange of germplasm of this species, which almost certainly explains its common occurrence in farmland even outside its natural range, regardless of social features such as the wealth of the farmers, prevailing land tenure regime, labor availability, and level of education. The multiinstitutional and multidisciplinary domestication program is built on earlier tree improvement studies examining reproductive biology, provenance variation, etc. initiated by the Institut de la Recherche Agricole pour le Développement (IRAD) (Kengue, 1998; Kengue and Singa, 1998). Now it is being focused at the village level on the development of cultivars from trees with superior fruit size, color, and taste. Vegetative propagation techniques, especially marcotting, have been used to capture the phenotype of selected trees and so to produce cultivars. To increase the multiplication rate, initial problems with the rooting of leafy stem cuttings have been overcome using simple, low-tech propagators established in a central nursery and replicated in all pilot villages (Leakey et al., 1990). With the help of NGOs, the villagers have been trained in nursery and vegetative propagation skills and become the nursery managers and the owners of the germplasm developed from their local population.

Some 10-fold variation in mean fruit mass between different trees, and additional variation in other commercially important traits, indicates considerable opportunity to increase the size and quality of fruits for market. To this is being added a better understanding of consumer preference and assessment of the genetic variation in sensory traits (taste, oiliness, acidity, smell, etc.), and studies to relate these to other visual characteristics are underway. Selection criteria based on the identification of market-oriented ideotypes have been identified (Leakey et al., 2002; Anegbeh et al., 2005) so that the improved prices of desirable fruit types currently only recognized in retail urban markets can be acquired by the producer dealing with traders at the farm-gate. Currently the diversity in fruit size and quality from the virtually wild trees on most farms, each phenotypically different from other trees, precludes wholesale buyers from recognizing superior fruits. Hopefully, however, when a vehicle can be loaded with the uniform fruits of recognized cultivars, farmers will be financially rewarded for their efforts. In Cameroon, women have indicated that the coincidence of schooling expenditures with income flows from the sale of fruits is one of the attributes of *D. edulis* that they appreciate (Schreckenberg et al., 2002b).

Currently, Safou fruits are perishable and difficult to store for more than a few days. However, studies are in progress to improve shelf life by various forms of processing and value-adding. Once this is achieved, it is anticipated that the demand will increase and the fruits, or fruit products, will become available outside the current four-to-five-month production season. In parallel with these developments are studies to understand more about the potential for the production of different oils for the food industry, involving the relationships between oil quality and other chemical and morphological traits. Safou fruits are already traded on a small scale into Europe and America; it is hoped that the combination of targeted genetic selection and processing will open up market opportunities, which can further benefit subsistence farmers. Much work remains to be done to ensure that this is achieved, with particular recognition that the intellectual property of the farmers developing the cultivars needs to be assured.

RECOMMENDATIONS: FUTURE DEVELOPMENTS

From the issues discussed in this paper, there is evidently much work to be done to build on the work of the last decade. There are very large numbers of species in all ecoregions, which potentially could be domesticated to produce marketable AFTPs and environmental services. The extension of the previously mentioned strategies, principles, and techniques to even a small proportion of these species represents an enormous challenge to countries and to development agencies with limited resources. Probably the biggest constraint to achieving this on the scale required is the serious lack of people in NGOs and national agricultural research systems (NARS) with adequate knowledge of vegetative propagation techniques for trees. There is also a need for better methods of propagating mature trees (Leakey, 2004). Resolving the technical and implementation issues surrounding the expansion of this "grass-roots" revolution will be enormous, with strategies on issues such as how to avoid a loss of intraspecific genetic diversity needing to be resolved, especially when recognizing that crop domestication is an ongoing iterative process.

The problems of widespread implementation of the domestication of agroforestry trees will be exacerbated by the fact that the philosophy to diversify farming systems and economies in the ways described here runs counter to the philosophy promulgated by many agencies that biotechnology is the solution to the issues of poverty, malnutrition, and rural development. Nevertheless, there is a growing body of evidence that supports the agroforestry solution to many of these problems, and the evident congruence between agroforestry outputs and Millenium Development Goals offers hope for the future (Garrity, 2004). Agroforestry centered on indigenous trees is very compatible with the aims of the Ecoagriculture initiative ratified at the World Summit on Sustainable Development in Johannesburg in 2002 (McNeely and Scherr, 2003). In this regard, tree domestication provides a powerful incentive for subsistence farmers to diversify their farms with indigenous trees that provide economic returns and environmental services, including biodiversity conservation (see also McNeely, 2004).

In addition, the future success of AFTP domestication and commercialization will depend on the benefits remaining with the small-scale farmers and their local industries and markets (Clement et al., 2004). This will inevitably depend on finding ways to satisfactorily protect the intellectual property rights (IPRs) of the farmers and communities investing in the process. Currently it is unclear how individual farmers and communities in economically and socially disadvantaged countries can avail themselves of the "rights" conferred by the Convention on Biological Diversity. In this connection, the attempts of the African Union to develop a model law for the protection of the rights of poor farmers is encouraging.

The "chicken-and-egg" linkages between domestication and commercialization will be difficult to resolve, making it crucial that agroforesters find ways to work closely with commercial and industrial partners so that the products match the needs of both the farmers, the processing industries and the consumers. It may not be a beneficial situation for everyone and clear understanding of the winners and losers of greater cultivation and commercialization will be needed.

Finally, however, the strength of the approach lies in its ability to empower the millions of farmers desperate to improve their lot in the world. What they need most is information about what is possible and the simple skills needed to do the job. The public in the developed world can help by buying the newly improved products, and by demanding that policymakers recognize the importance of simultaneously resolving the poverty and environmental crises facing the developing world.

Chapter 13

Agroforestry Tree Products (AFTPs): Targeting Poverty Reduction and Enhanced Livelihoods

This chapter was previously published in Leakey, R.R.B., Tchoundjeu, Z., Schreckenberg, K., Shackleton, S., Shackleton, C., 2005. International Journal of Agricultural Sustainability, 3, 1–23, with permission of Taylor & Francis

SUMMARY

Agroforestry tree domestication emerged as a farmer-driven, market-led process in the early 1990s and became an international initiative. A participatory approach now supplements the more traditional aspects of tree improvement, and is seen as an important strategy towards the Millennium Development Goals of eradicating poverty and hunger, promoting social equity and environmental sustainability. Considerable progress has been made towards the domestication of indigenous fruits and nuts in many villages in Cameroon and Nigeria. Vegetatively-propagated cultivars based on a sound knowledge of 'ideotypes' derived from an understanding of the tree-to-tree variation in many commercially important traits are being developed by farmers. These are being integrated into polycultural farming systems, especially the cocoa agroforests. Markets for Agroforestry Tree Products (AFTPs) are crucial for the adoption of agroforestry on a scale to have meaningful economic, social and environmental impacts. Important lessons have been learned in southern Africa from detailed studies of the commercialisation of AFTPs. These provide support for the wider acceptance of the role of domesticating indigenous trees in the promotion of enhanced livelihoods for poor farmers in the tropics. Policy guidelines have been developed in support of this sustainable rural development as an alternative strategy to those proposed in many other major development and conservation fora.

INTRODUCTION

Agroforestry is now being seen as an alternative paradigm for rural development worldwide, centered on species-rich, low-input agricultural techniques including a diverse array of new indigenous tree crops, rather than on high-input monocultures with only a small set of staple food crops (Leakey, 2001a,b). This alternative paradigm addresses many of the global challenges highlighted by the UN Millennium Development Goals and environmental conventions (Garrity, 2004). These challenges are associated with deforestation, land degradation, unsustainable cropping practices, loss of biodiversity, increased risks of climate change, and rising hunger, poverty and malnutrition (Fig. 13.1). In the last 10–15 years, agroforestry tree domestication strategies, approaches, and techniques, together with the commercialization and marketing of agroforestry tree products (AFTPs), have become one of the "pillars" of this new paradigm (ICRAF, 1997; Leakey and Simons, 1998; Simons and Leakey, 2004). Agroforestry tree domestication is aimed at promoting the cultivation of indigenous trees with economic potential as new cash crops.

© 2017 Elsevier Inc. All rights reserved.

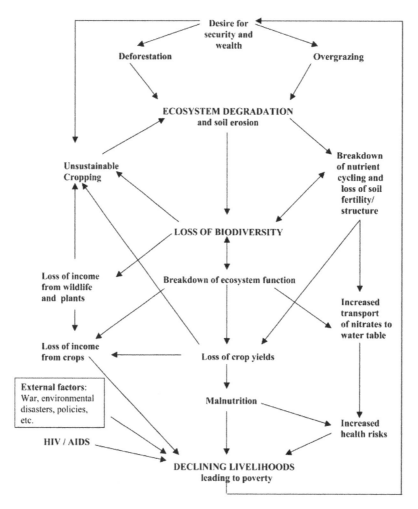

FIGURE 13.1   The cycle of biophysical and socioeconomic processes causing ecosystem degradation, biodiversity loss, and the breakdown of ecosystem function in agricultural land in many tropical countries.

TREES

The Origins of Tropical Tree Domestication

The domestication of many species for food and other products has been carried out for thousands of years in almost every part of the world, often arising from extractive uses by indigenous people (Homma, 1994). Current agroforestry tree domestication initiatives build on the efforts of smallholder farmers who, over a number of generations, have improved the size of some of the more important indigenous fruits, so providing the first steps toward domestication (Leakey et al., 2004). Research programs to domesticate agroforestry trees, particularly for the production of nontimber forest products (NTFPs), were initiated in the 1980s (Leakey et al., 1982b; Okafor, 1980) and emerged as a global program in the 1990s (Leakey and Newton, 1994b,c; Leakey and Simons, 1998). In 1992, the term "Cinderella species" was coined for all the tree species that have provided poor people with their everyday needs for food, medicinal, and other forest products and which have been overlooked by science and the Green Revolution (Leakey and Newton, 1994a). At this time agroforestry tree domestication was defined to encompass the socioeconomic and biophysical processes involved in the identification and characterization of germplasm resources; the capture, selection, and management of genetic resources; and the regeneration and sustainable cultivation of the species in managed ecosystems (Leakey and Newton, 1994b). In agroforestry, domestication is not restricted to tree species, and a range of new herbaceous crops is also being developed (Smartt and Haq, 1997). Many indigenous vegetables are candidates for domestication (Guarino, 1997; Schippers, 2000; Schippers and Budd, 1997; Sunderland et al., 1999) and can be components of multistrata systems, where there is a need for new shade tolerant crops. Agroforestry tree domestication has now been refined and expanded with emphasis on it being a farmer-driven and market-led process (Leakey and Simons, 1998; Simons, 1996; Simons and Leakey, 2004) and with emphasis on a participatory approach to the involvement of local communities (Leakey et al., 2003; Tchoundjeu et al., 1998). Additionally, it has been recognized that not just species but also whole landscapes are domesticated when brought into cultivation (Wiersum, 1996).

In 1995, the International Center for Research in Agroforestry (ICRAF, now called the World Agroforestry Center) established a Tree Domestication Program with projects in six ecoregions of the tropics, each of which developed their own species priorities (Jaenicke et al., 1995, 2000; Maghembe et al., 1998; Roshetko and Evans, 1999; Weber et al., 2001). While the focus of this paper is on the development of marketable AFTPs from agroforestry trees, interest in tree domestication encompasses trees for other purposes such as soil amelioration, fodder, fuelwood, timber, and boundary demarcation. The implementation of sound domestication strategies for all species include the activities of: assessing the demand; evaluating the status of the resource; defining the purpose, objectives and strategy; germplasm collection, conservation and dissemination; reproductive biology; propagation techniques; tree improvement; tree breeding, etc. (Simons, 1996; Simons and Leakey, 2004).

The current concept of domestication of indigenous trees for AFTPs goes hand-in-hand with the commercialization of the products and together, through agroforestry, they provide an incentive for subsistence farmers to plant trees in ways that achieve the Millennium Development Goals (Garrity, 2004), especially the reduction of poverty, and enhancement of food and nutritional security, human health and environmental sustainability (i.e., Goals 1, 3, 4, 5, 6, 7, and 8: see www.un.org/millenniumgoals/).

THE PARTICIPATORY TREE DOMESTICATION APPROACH—THE CASE OF WEST AND CENTRAL AFRICA

Participatory approaches to the domestication of agroforestry trees started with the involvement of many stakeholders in the germplasm collections of *Sesbania sesban* for improved fallow technologies (Ndungu and Boland, 1994) and the principles extended to collections of *Sclerocarya birrea* and *Uapaca kirkiana* for fruit production (Leakey and Simons, 1998; Ndungu et al., 1995). This led to the development of guidelines for species priority-setting in West and Central Africa (Franzel et al., 1996), where the top priority species were identified as *Irvingia gabonensis* (Ladipo et al., 1996; Okafor and Lamb, 1994) and *Dacryodes edulis* (Kengue, 2002; Okafor, 1983; Tchoundjeu et al., 2002a). Since then, it has expanded to a range of other species, including *Prunus africana* (Simons et al., 2000b), *Pausinystalia johimbe* (Ngo Mpeck et al., 2003a; Tchoundjeu et al., 2004), *Ricinodendron heudelottii* (Ngo Mpeck et al., 2003b), *Garcinia kola* (unpublished), etc. The approach has utilized, disseminated and refined a simple low-technology system for the vegetative propagation of tropical trees that is appropriate for use in small, low-cost village nurseries (Leakey et al., 1990; Mbile et al., 2004; Shiembo et al., 1996a,b, Shiembo, 1997), so that cultivars can be produced and multiplied by villagers. The trees for propagation have been identified by quantitatively examining the tree-to-tree variation in a range of fruit and nut traits to determine the potential for highly productive and qualitatively superior cultivars with a high Harvest Index (e.g., Anegbeh et al., 2003, 2005; Atangana et al., 2001, 2002; Leakey et al., 2002, 2005a; Ngo Mpeck et al., 2003b; Waruhiu et al., 2004).

Together these developments result in a model participatory domestication strategy that is now being scaled up to a regional level (Tchoundjeu et al., 1998, 2006), and encompasses 35 villages in southern Cameroon (about 2500 farmers), 11 villages in Nigeria (about 2000 farmers), three villages in Gabon (about 800 farmers), and two villages in Equatorial Guinea (about 500 farmers). The program fulfils the criteria outlined earlier for agroforestry approaches to meet the Millennium Development Goals, and conforms to the Convention on Biological Diversity (Leakey et al., 2003; Simons and Leakey, 2004; Tchoundjeu et al., 1998), by recognizing the rights of local people to their indigenous knowledge and traditional use of native plant species. However, ways have to be found to ensure that the investments made by community members in time and effort in developing cultivars can be protected through recognition of their rights in existing systems of intellectual property rights, and through the development of alternative *sui generis* ("of its own kind") rights systems for access and benefit sharing. Alternatively, ways may be found for indigenous communities and smallholder farmers to register "plant breeders' rights." If protection of these rights is not achieved, there is the danger that other people will reap the benefits of the pioneering work by villagers. Protection of this kind would then ensure that participatory domestication by local farmers could be recognized as a good model of biodiscovery, an alternative to biopiracy by expatriate or local entrepreneurs. There is, however, a need for guidelines on how local communities can establish rights over cultivars, to ensure that they acquire royalties.

IDENTIFICATION, CAPTURE, RETENTION, AND PROTECTION OF GENETIC DIVERSITY

Domestication has been defined as human-induced change in the genetics of the species to conform to human desires and agroecosystems (Harlan, 1975). It is not surprising, therefore, that much of the work to domesticate agroforestry trees has focused on the identification of intraspecific genetic variability of the priority species, and the vegetative propagation techniques to capture the combinations of genetic traits found in superior individual trees. One of the key findings of the characterization studies done in a number of species (*D. edulis, I. gabonensis, S. birrea*) is that each trait shows continuous variation, but that high values of one trait are not necessarily associated with high values of another trait: thus large fruits are not necessarily sweet fruits, and do not necessarily contain large nuts or kernels. This means that the variability between trees is increased the more traits there are that are of interest. This multitrait variation, coupled with the extent of the variability of each individual trait, results in very considerable opportunity for selection of trees with good combinations of traits. Obviously, the more traits that are simultaneously screened the more unlikely it is that a tree with good values for all traits will be found. Thus, large numbers of trees have to be screened to find the rare combinations of traits. This rapidly becomes a major task and very expensive. Consequently, the practical approach is to seek trees that have particular, market-oriented trait combinations—such as big, sweet fruits for the fresh fruit market (a fruit ideotype); big, easily extracted kernels for the kernel market (kernel ideotype), etc. (Leakey and Page, 2006). Following this logic, the kernel ideotype can then be subdivided into those meeting the demands of different markets, such as for food-thickening agents (Leakey et al., 2005d), or other products, such as pectins or oils for food or cosmetic industries (Kalenda et al., 2002; Kapseu et al., 2002; Leakey et al., 2005b,c,d). In the three species studied to date, the screening of only 100–300 trees has successfully identified a number of trees of interest for each ideotype. This approach to genetic selection can result in substantial improvement in crop quality and productivity for a relatively small level of investment, especially when implemented in participatory mode with farmers, rather than through research institutions.

Building on the ideotype concept, little has been done to look at the genetic variability in nutritive value or sensory analysis in any of the new AFTP-producing crops. In Cameroon, a start has been made to examine variability in flavor, taste and aroma in samples of *D. edulis* (Kengni et al., 2001), demonstrating that organoleptic evaluation on a tree-to-tree basis should be possible. Studies of this sort also need to be linked to any processing to extend shelf life and/or create new markets. The limited evidence available indicates that nutritive value of indigenous fruits is as variable as the other characteristics. In *S. birrea*, for example, the tree-to-tree variation in the skin and pulp of fruits from just 15 trees was found to be 30.1–112.6 g per fruit for protein content, and 1.3–3.2 mg per fruit for vitamin C content (Thiong'o et al., 2002). The oil content of kernels was not so variable (44.7–72.3%), but oil yield per nut was even more variable (8–53 g per nut) because of the variation in kernel mass per nut (Leakey et al., 2005c). Characterization studies with another indigenous fruit of southern Africa, *Strychnos cocculoides*, are underway and again finding significant differences in fruit traits, e.g., fruit mass, between sites (Mkonda et al., 2003).

In *I. gabonensis*, a study of the physical properties of the polysaccharide food-thickening agent from kernels again found very extensive tree-to-tree variability in two independent traits of considerable potential market importance: viscosity (34.0–124.1 SNU[1]) and drawability (0.26–3.65 SNU). *Irvingia gabonensis* kernels also varied in oil content (37.5–75.5%) and in protein content (Leakey et al., 2005d). All of this suggests that more detailed study of these characteristics affecting the food value and acceptability of ATFPs from different cultivars of different species is a high priority for future research. Much of this variability in nutritional quality is also likely to affect the potential for food processing and different markets, so there is an urgent need for agroforesters to work closely with the food and other industries to optimize the domestication/commercialization partnerships (Leakey, 1999a). Some studies have been initiated in *D. edulis*, for example, to formulate nutritious biscuits (Mbofung et al., 2002). Research on the nutritional value of AFTPs has important implications for the alleviation of nutritional insecurity and health (Millennium Development Goals 4, 5, and 6). One aspect of the potential health benefits of agroforestry is the fortification of the immune systems of HIV/AIDS sufferers through the selection of especially nutritious cultivars of indigenous fruits and nuts (Barany et al., 2001, 2003), something that requires further investigation as an output from agroforestry (Villarreal et al., 2006). In this connection it is interesting to note that the tree-to-tree variation of B-sitosterol, the major sterol component of the medicinal product from the bark of *P. africana*, used to treat BPH, ranges from 50 to 191 μg/g (Simons and Leakey, 2004).

Another aspect of domestication needing further work is selection for production traits, such as yield, seasonality and regularity of production, reproductive biology, and reduction of susceptibility to pests and diseases which can reduce productivity or quality (Kengue et al., 2002). Interestingly, evidence from South Africa indicates that the yield

---

1. 1 SNU512 mPa.s.

of "marula" (*S. birrea*) in South Africa is increased 5- to 15-fold by cultivation in homestead plots and fields (Shackleton et al., 2003a). Mean fruit size is also greater from trees in these farms, again with some evidence for domestication by farmers (Leakey, 2005; Leakey et al., 2005b,c). High yield is obviously a desirable trait in any cultivar but, within reason, may not be as important in the early stages of domestication as the quality attributes. Many wild fruit trees have phenological variability, resulting in a fruiting season of 2 to 4 months, with individual trees differing in their period of ripening and fruit drop. Within this general pattern, there can be a few trees which fruit outside the normal season, or which fruit more than once per year. Cultivars derived from these trees can extend the productive season for farmers. Seedlessness is another very desirable trait in cultivars of fresh fruit-producing species and trees with a high pulp:kernel ratio have been identified in *D. edulis* (Anegbeh et al., 2005; Kengue, 2002; Waruhiu et al., 2004).

Having identified the superior trees with the desired traits, the capture of tree-to-tree variation using techniques of vegetative propagation is relatively simple and well understood (Leakey, 1985, 2004; Leakey et al., 1990, 1994; Mudge and Brennan, 1999), although the numbers of people with the appropriate skills may be a constraint to its widespread application in the future (Simons and Leakey, 2004). Typically, the techniques of grafting, budding, and air-layering (marcotting) are used to capture superior fruit trees and to multiply them as cultivars. This is because mature tissues with the capacity to flower and fruit can only be propagated by cuttings with great difficulty (low multiplication rates). It is unclear to what extent this is a function of their ontogenetic age (state of reproductive maturity) or their physiological age (Dick and Leakey, 2006). However, propagation by juvenile leafy cuttings is very easy, with high multiplication rates, for almost all tree species (Leakey et al., 1990). Because of the ease of propagating juvenile tissues by cuttings, this is currently the preferred option for participatory domestication in village nurseries (Mbile et al., 2004; Mialoundama et al., 2002; Shiembo et al., 1996a; Tchoundjeu et al., 2002b).

Having indicated previously the great opportunities arising from the identification and capture of intraspecific genetic diversity, it is important then to consider the retention and protection of this genetic diversity (Leakey, 1991). Domestication is generally considered to reduce the genetic diversity of the species that has been domesticated. This is probably true in situations where the domesticated plant replaces or dominates the wild origin, but is probably not the case at the current level of domestication of agroforestry trees. For example, the range of fruit sizes in on-farm populations of *D. edulis* and *I. gabonensis* has been increased by the early stages of domestication (Leakey et al., 2004). Nevertheless, it is essential that domestication activities are undertaken with the realization that it is important to retain and maintain as much diversity as possible. Modern molecular techniques are useful in the development of a wise strategy for the maintenance of genetic diversity. Within the geographic range of a particular species they can be used to identify the "hot-spots" of intraspecific diversity (e.g., Lowe et al., 1998, 2000), places which should if possible be protected for in situ genetic conservation, or be the source of germplasm collections if ex situ conservation is required. In addition, when developing cultivars, it is highly desirable that they originate from unrelated populations with very different genetic structure. It is foreseen that this is one of the benefits of the participatory domestication model being developed in Cameroon, as each village is developing a different set of cultivars, which should retain a wide cross-section of the existing diversity. This raises the question as to how many cultivars/clones each village should produce to achieve their production objectives and, importantly, ensure that there is still a high level of genetic diversity in the cultivated population to minimize the risk of pest and disease epidemics. In the first instance each village may produce 10−15 cultivars for each species, and these will probably have been selected for a number of different attributes. While this number will be reduced as the cultivars are tested, it is hoped that new cultivars will be developed as a result of further screening of wild trees. In due course, seeds perhaps derived from cross-cultivar pollinations will create a new generation of trees for screening and selection. These approaches conform to published strategies for clonal selection and deployment and the wise use of tree improvement techniques (Foster and Bertolucci, 1994; Leakey, 1991; Libby, 1982).

CULTIVATION AND THE GROWTH OF CULTIVARS

The final stage of the domestication process is the integration of selected plants into the farming system in ways that make effective use of natural resources (light, water, nutrients), and have positive socioeconomic and environmental benefits (Leakey and Newton, 1994b,c). In African farmland, a wide range of densities and configurations are grown (Gockowski and Dury, 1999; Kindt, 2002). A study of the fruit tree component in villages with varying mean farm size (0.7−6.0 ha) in Cameroon and Nigeria found that fruit tree density was inversely related to area, with small farms having the greatest tree densities (Degrande et al., 2006). Of these trees about 50% were indigenous species for AFTPs, and, in Cameroon, 21−57% of these indigenous fruit trees are *D. edulis* (Schreckenberg et al., 2002b). Agroforestry is expected to provide positive environmental benefits on climate change and biodiversity (Millennium Development Goal 7).

Evidence is also growing that there are biodiversity benefits arising from the introduction of indigenous trees producing a wide range of marketable products into smallholder farming systems (Bignell et al., 2005; Schroth et al., 2004). However, research is needed to determine the impacts of such diversity on agroecosystem function (Gliessman, 1998; Leakey, 1999b; Mbile et al., 2003); carbon sequestration (Gockowski et al., 2001) and trace gas fluxes; and on the sustainability of production and household livelihoods. However, the domestication of agroforestry trees is only just reaching the point where the density and configuration of cultivars in the farm is becoming a research topic. This information will be fundamental to understanding the processes determining positive impacts of agricultural diversification on biodiversity, land degradation, livelihoods, and income (Simons et al., 2000a). Evidence is also required to determine whether the domestication and commercialization of AFTPs provides incentives for farmers to diversify, as suggested by Leakey (2001a,b). The importance of agricultural biodiversity and traditional food crops is increasingly being recognized internationally for their value in human nutrition and income generation (Frison et al., 2004).

MARKETS

The linkage of tree domestication with product commercialization was the focus of a conference at ICRAF in 1996 (Leakey and Izac, 1996; Leakey et al., 1996). To be effective this linkage requires the involvement of the food, pharmaceutical, and other industries in the identification of the characteristics that will determine market acceptability (Leakey, 1999a).

The term agroforestry tree products (AFTPs) is of very recent origin (Simons and Leakey, 2004) and refers to timber and nontimber forest products that are sourced from trees cultivated outside of forests. This distinction from the term nontimber forest products (NTFPs) for nontimber extractive resources from natural systems is to distinguish between extractive resources from forests and cultivated trees in farming systems, and hopefully will avoid some of the confusion in the current literature (Belcher, 2003). Nevertheless, some products will be marketed as both NTFPs and AFTPs (depending on their origin) during the period of transition from wild resources to newly domesticated crops. Consequently, both terms are used in the following sections.

ECONOMIC AND SOCIAL BENEFITS FROM TRADING AFTPS

In western and central Africa, a number of indigenous fruits and nuts, mostly gathered from farm trees, contribute to regional trade (Ndoye et al., 1997). In Cameroon, the annual trade of the products of five key species has been valued at US$7.5 million, of which exports generate US$2.5 million (Awono et al., 2002). Perhaps because of this trade, evidence is accumulating that AFTPs do contribute significantly to household income (Awono et al., 2002; Gockowski et al., 1997) and to household welfare (Degrande et al., 2006; Schreckenberg et al., 2002b). For example, farm level production of three indigenous fruit and nut species in southern Cameroon has been reported to be worth US$355 (Ayuk et al., 1999a,b,c), from an average farm size of 1.7 ha, and against an average annual expenditure of US$244 (Gockowski et al., 1998). In Cameroon, farmers from four widely dispersed villages indicated that indigenous fruits represent 12.5% of their primary income, and 17% of their secondary income, while the equivalent income from exotic fruits was 6.8 and 3.5%, respectively (Degrande et al., 2006). In Nigeria, the equivalent proportions of income from indigenous fruits were 15% as primary income and 37.5% as secondary income, while exotic fruits had no value as primary income and only 2.5% as secondary income (Degrande et al., 2006). A crop of *D. edulis* fruits can be worth between US$20 and $150 per tree, depending on the quality of the fruits and the yield (Leakey et al., unpublished). Thus, taking the number of *D. edulis* trees per household (Schreckenberg et al., 2002b), a low estimate of the value of their fruits per tree (US$20) gives another estimate of annual income per household of US$380–$2000. This result concurs with an economic analysis of farms in Cameroon with an average size of 1.4 ha, which found that when indigenous fruits are grown with cocoa they have a net present value/ha (over a 30-year period with a 10% discount rate) of about US$500 (Gockowski et al., 1997; Gockowski and Dury, 1999). It seems that similar situations occur outside West Africa. For example, in South Africa, although the absolute income from *S. birrea* fruits, kernels, and beer was not as great as that from *D. edulis* in Cameroon, it was nevertheless in excess of the local wage rate (Shackleton et al., 2003b). In Guyana, subsistence households were able to generate value-added equivalent to US$288 per capita per annum, from utilization of Andiroba oil and other forest resources (Sullivan, 2003).

What would be the impact of domestication on household income? It is anticipated that improved quality and market appeal from the domestication of these fruits would result in farmers getting higher prices, so long as supply does not exceed demand. At present there is high demand, but it is known that although the retail traders recognize the higher value of superior fruits, the wholesale traders do not (Leakey et al., 2002). This is probably because a loaded truck of fruits from the current crop comes from a wide variety of trees of seedling origin and therefore includes the full spectrum of quality from very poor to superior. This would not be the case once farmers are planting recognized cultivars, as it would be possible for the wholesaler to obtain a loaded truck of superior fruits.

Different indigenous fruit species can vary in their seasonality, so income opportunities can be spread across the year. Thus the overall household benefits from several different AFTPs, even without domestication, are almost certainly greater than the preceding examples suggest. To these benefits can also be added those that are derived from AFTP products used in domestic consumption, which represent a saving on expenditure. Evaluation of the economic benefits are further complicated by the fact that cash earned from AFTPs can potentially be invested in fertilizers, or in adding value to products, etc., so increasing the overall income derived from the sale of AFTPs.

Women are often the beneficiaries of this trade and they have especially indicated their interest in marketing *D. edulis* fruits because the fruiting season coincides with the time to pay school fees and to buy school uniforms (Schreckenberg et al., 2002b). It is also the women who are the main retailers of NTFPs (Awono et al., 2002), with men being the wholesalers, and interestingly, it is the retail trade that recognizes the market value of the tree-to-tree variation in size, color, flavor, etc. (Leakey et al., 2002). Evidence has also shown that some communities domesticate valuable species for the purposes of intergenerational security (Sullivan, 2003). Clearly these are social impacts of importance both to sustainable development in general, and in particular to the empowerment of women (Millennium Development Goal 3). Further work is, however, required to get a much better understanding of the market dynamics and potential for expansion. In the case of *D. edulis*, extending the season with early and late fruiting cultivars would be important (e.g., the Nöel cultivar, which fruits at Christmas), as would methods to extend their shelf life through simple fruit storage (bottling, canning, drying, freezing, etc.) and processing into paste, biscuits (Mbofung et al., 2002), etc. Similar trends are emerging in southern Africa, where indigenous fruits have relatively new local and international markets (Brigham et al., 1996; Shackleton et al., 2000, 2002, 2003b).

The production and trading of AFTPs are based on traditional lifestyles, with many products used both for domestic consumption and/or sale depending on the household's cash and nutritional requirements. The ability to use household labor for harvesting/processing, combined with the low requirement for skills, capital and external inputs, makes it relatively easy for poor producers to adopt this approach to intensifying production and enhancing household livelihoods.

With HIV/AIDS now reaching up to 20−30% of the population in the worst-hit countries (Villarreal et al., 2006), one social benefit of special interest is potential health benefits which may accrue from a diet including more indigenous fruits and vegetables, many of which are rich in protein, oils, minerals, and vitamins (Leakey, 1999a; Leakey et al., 2005a). The domestication of species producing these nutritious AFTPs is seen as a way to further enhance nutritional security and health, by strengthening the immune system of HIV/AIDS sufferers. This is seen as a critical component of an integrated natural resource management approach to improving the lives of poor people worldwide, especially in southern Africa (Barany et al., 2003). Worldwide, medicinal products represent an annual international trade valued in excess of US$1 billion (Rao et al., 2004). Many of these are herbs, which can be grown in the shade of agroforestry trees.

THE LINKAGES BETWEEN THE DOMESTICATION AND COMMERCIALIZATION OF AFTPS.

The success of domesticating agroforestry trees is very dependent on having an adequate market for the products (AFTPs). In some instances, species currently being domesticated, such as *D. edulis*, *I. gabonensis*, and *Gnetum africanum*, have local and regional markets, including exports to neighboring countries (Awono et al., 2002). In other cases, such as *P. africana* and *P. johimbe*, there are already established international markets in Europe and the United States, in addition to local ones (Cunningham et al., 2002). As already indicated, market-oriented domestication has the greatest likelihood of being adopted on a scale to have impact on the economic, social and environmental problems afflicting many tropical countries. This requires that agroforesters work closely with the companies processing and marketing the products (Leakey, 1999a). However, in doing this it is important to remember that smallholder farmers are the client of the research and development work and that there needs to be a functional production-to-consumption chain. This principle was apparently overlooked during recent domestication of peach palm in Amazonia (Clement et al., 2004), resulting in the underperformance of the market. This failure has been attributed to a lack of understanding about the consumers' needs, and incorrect identification of the research client (i.e., the smallholder and not the entrepreneur).

In many cases the successful commercialization of AFTPs is dependent on domestication, as frequently initiatives to develop markets for new products collapse (or do not expand to their potential) when supply does not meet the demand. This is especially problematic if the product has a seasonal production pattern, and the product is derived from many small growers, with minimal quality control. Another constraint to commercialization can be the intraspecific variability that is so beneficial to the domestication process. This variability is a major problem when uniformity of quality (taste, size, and purity) is important in the marketplace. Quality control is doubly important if there is any local level processing for value-adding, to extend shelf life or to reduce the costs of bulk transport. Domestication is one way to increase the supply of high-quality product, and through cultivar development can also greatly improve the uniformity of the product. Domestication can also lead to an extended season of production, making it easier to supply industries throughout the year. Good examples of coordinated domestication and commercialization are kiwi fruit (*Actinidia chinensis*) and macadamia nuts (*Macadamia integrifolia*). Kiwi fruits were first grown commercially in New Zealand in the 1930s. By the 1950s there were a number of commercially grown cultivars and fruits were first exported in 1952. The macadamia selection program started in 1934, with considerable market interest.

In many domesticated crops, the market demand for the product has promoted large-scale monocultural production systems that frequently have been the cause of environmental degradation through deforestation, soil erosion, nutrient mining, and loss of biodiversity. Typically, these systems of farming have also resulted in social inequity and the "poverty trap" for small-scale producers who are unable to compete in international trade with large or multinational companies. Concerns about this have rightly raised many questions about the wisdom of domesticating and commercializing agroforestry trees. The key question that agroforesters have to address is whether or not agroforestry can prevent these negative impacts. In theory, agroforestry is beneficial to the environment and beneficial to the poor farmer. At the level of the individual farm there are many examples of these benefits being achieved. For example, extensive intercropping with trees and shrubs provides subsistence households in Amerindian communities with a significant degree of food security (Sullivan, 2003). The problem and the complexity of this issue is exacerbated by the need for agroforestry to be scaled up to the point when it reduces poverty and has environmental benefits at national, regional, and global scales. So, what will happen if the domestication of AFTPs is so successful that the market demand for one of them reaches the point when a company sees the opportunity to develop monocultural plantations as a cash-crop, either in the country of origin or in some overseas location with a similar climate and better access to markets? Will this undermine the whole purpose of developing new crops? The answer has to be a qualified "yes." Having said that, what reservations or limitations can mitigate the problem? These issues were the subject of an ICRAF conference in 1996 (Leakey et al., 1996) and it was concluded that, recognizing the traditional role of nontimber forest products in food security, health, and income generation, the potential benefits from domestication outweighed the risks. Nevertheless, many areas of market, social science, and policy research were recommended by Working Groups. Important areas for more study are the complex issues surrounding commercialization of genetic resources and benefit sharing (ten Kate and Laird, 1999) and traditional knowledge (Laird, 2002). Without markets there will not be the opportunity for subsistence households to increase their standard of living, while expanded market opportunities could lead to their exploitation by businessmen. Thus it is clear that commercialization is both necessary and potentially harmful to small-scale farmers practicing agroforestry (Leakey and Izac, 1996), and that as advocated by Dewees and Scherr (1996), policy scientists need to "stretch their conceptual framework ... and to consider more carefully the links between markets, the environment, household production and household welfare." One risk-alleviating strategy is to support the domestication of a wide range of tree species producing AFTPs, especially those with local and regional market potential. In this way, coupled with strong indigenous rights, it is very unlikely that the market demand will attract major companies and, even if products of a few species do become international commodities, there will be others that will remain only of local and regional importance.

In recent years there have been some very positive outcomes from the involvement of international companies in agroforestry. For example, Daimler-Benz has taken a smallholder, multistrata agroforestry approach to producing raw materials for their C-Class Mercedes-Benz cars in Brazil, and in partnership with International Finance Corporation have been developing this as a new paradigm for Public/Private Sector Partnerships in Development (Mitschein and Miranda, 1998; Panik, 1998) Another example is the leadership being taken by Masterfoods within the chocolate industry, in support of sustainable livelihoods for smallholder cocoa farmers in Africa and Asia, through the diversification of cocoa farms into cocoa agroforests by the promotion of AFTP-producing trees. This development is building on the actions of the smallholder farmers themselves, who have integrated fruit trees (often indigenous species) into the cocoa farm so that the shade trees are also companion crops (Leakey and Tchoundjeu, 2001). This has been done as a risk-aversion strategy to provide new sources of income, in response to fluctuating market prices. Interestingly, cocoa is not the only former plantation cash-crop to now be an important agroforestry species. Rubber is perhaps the best example, especially in Southeast Asia (Tomich et al., 2001), while tea and coffee are moving in the same direction. Taking these developments together, therefore, there are some good reasons for being positive about the potential impacts of commercialization of agroforestry trees. A somewhat different but interesting example of AFTP commercialization is the case of marula (*S. birrea*), a tree of dry Africa which is starting to be commercialized by subsistence farmers for traditional beer and for industrial processing as an internationally marketed liqueur Amarula by Distell Corporation. Marula kernel oil is also breaking into international cosmetics markets. This species thus provides an opportunity to examine the impacts of different commercialization strategies on the livelihoods of the producers, the sustainability of the resource and the economic and social institutions. In other words, who or what are the winners and losers arising from the commercialization of indigenous fruits and nuts? This question has been the focus of the following study to investigate the impacts of commercializing both traditional and new products from emerging agroforestry tree crops.

WINNERS AND LOSERS: IMPACTS ON LIVELIHOODS

The importance of nontimber forest products for the livelihoods of poor forest dwellers has been recognized for some time (Peters et al., 1989; Sunderland and Ndoye, 2004; Vedeld et al., 2004). A multidisciplinary, multiinstitutional study of the impacts of different commercialization strategies for a number of different NTFP products from *S. birrea* within farming systems and in structurally and ethnically different communities has provided very interesting insights as to who are the "winners and losers" (Sullivan and O'Regan, 2003). The study was structured to examine the impacts of commercialization on the five forms of Livelihood Capital (Human, Social, Financial, Natural, and Physical). This review cannot do justice to this comprehensive study, but in brief it was concluded that to improve the livelihood benefits from commercializing NTFPs it is important to improve:

1. The quality and yield of the products through:
   a. domestication and the dissemination of germplasm;
   b. enhancing the efficiency of postharvest technology (extraction, processing, storage, etc.).
2. The marketing and commercialization processes by:
   a. diversifying markets for existing and new products;
   b. investing in marketing initiatives and campaigns;
   c. promoting supply contracts with equitable distribution of benefits, opportunities in national and international cuisine that build on indigenous knowledge and cultural heritage, improved sensory characteristics (taste, aroma, etc.), market chain investments, trading partnerships in local businesses with plans for sustainability (including exit strategies), commercialization pathways that recognize the role of women, health, and nutritional benefits.

The analysis from this study identified the factors that determine who/what are the winners and the losers under different circumstances (Table 13.1). The following lessons were learned for NTFP commercialization from the study of *S. birrea* (abridged from Shackleton et al., 2003b), and they apply equally to AFTPs:

- NTFPs are most important for poor and marginalized people, and make up income shortfalls but do not significantly alleviate poverty. How domestication may change this still needs to be determined.
- Engagement in NTFP commercialization and the extent of benefits is variable even among the poorest households. Households are far from homogenous in their levels of engagement. Entrepreneurship, labor availability, personal drive, and choice play a pivotal role in determining whether or not households take up opportunities. So too do the levels of organization within a community, the availability and quality of information about markets, access to transport, and the extent to which a producer is "networked." Benefits of NTFP commercialization must be weighed against the negative social and cultural costs of commercialization: there are trade-offs, which need to be recognized, between the preservation of traditions, cultures, and social norms, and the benefits derived from increased income.
- Land and usufruct rights must be clear, government intervention pitched at the appropriate level, and political support for the NTFP industry secured: insecure land tenure and resource rights can have a range of negative economic, social, and ecological outcomes, and can severely jeopardize efforts to successfully commercialize NTFPs. The commercialization of marula illustrates the central role that can be played by customary law in NTFP management. The findings argue for greater integration of customary and local law in places where traditional systems have eroded, and minimal governmental intervention in areas where customary law is adequate to deal with the pressures of commercialization.
- NTFP commercialization can lead to improved management and conservation of the resource in certain circumstances. This depends on the particular product, the species, the presence of a conducive policy environment, the feasibility/desirability of cultivation, and the potential for participatory domestication by interested communities.
- An abundance of "winner" qualities (see Table 13.1) need to be in place or developed among the participants in the trade, and across the resource and markets. NTFP cultivation needs to be community-owned and driven: communities harvesting products and domesticating the species need assistance and support to guarantee their ownership of germplasm and knowledge, and to ensure they are the beneficiaries of future commercialization initiatives.

TABLE 13.1 The characteristics of winners and losers among: (1) people and enterprises, (2) marketing, and (3) natural resources, that are due to the commercialization of nontimber forest products.

Winner qualities	Loser qualities
(1) *People and enterprises*	
• Individuals organized as a group	• Poorly organized group structure
• Well informed about markets	• Poorly informed of markets
• Good access to transport	• Poor access to transport
• Coordinated production	• Uncoordinated production
• Small "Input cost: Revenue received" ratio	• Large "Input cost: Revenue received" ratio ratio
• Consistently good quality products	• Variable quality products
• Skilled in bargaining	• Unskilled in bargaining
• Well networked with good partnerships	• Poorly networked
• Easy and equitable access to resource	• Uncertain and restricted access to resource
• Fits with other livelihood strategies and sociocultural norms	• Competes with other livelihood strategies and sociocultural norms
(2) *Marketing*	
• Commercial opportunities	• Undeveloped/poor market interest
• Diversity of end markets	• Limited markets
• Positive marketing image	• No or negative marketing image
• Unique characteristics of product	• Many other substitutes
• Raw product quality well matched to market	• Raw product requires processing
• Many buyers of raw materials and products	• A monopsony—only one buyer of raw materials
• Many sellers of raw materials and products	• A monopsony—only one seller
• Buyers aware of product or brand	• Buyers ignorant of product or brand
(3) *Natural resources*	
• Abundant resource	• Rare resource
• Plant part used is readily renewable	• Slow replacement of harvested product
• Harvesting does not destroy the plant	• Destructive and demaging harvesting
• Easily propagated	• Difficult to propagate
• Genetically diverse with potential for domestication	• Genetically uniform or little potential for selection
• Multiple uses for products	• Narrow use options
• High yield of high-quality product	• Low-yielding and/or poor-quality product
• Valuable product	• Low-value product
• Consistent and reliable yield from year to year	• Inconsistent and unpredictable production
• Already cultivated within farming system	• Wild resource which is difficult to cultivate
• Already being domesticated by local farmers	• Totally wild resource
• Fast growing	• Slow growing
• Short time to production of product	• Long time to production
• Compatible with agroforestry land uses	• Compatible with crops; labor intensive, etc.
• Hardy	• Sensitive to adverse environmental conditions
• Widely distributed	• Only locally distributed

After Shackleton, S.E., Wynberg, R.P., Sullivan, C.A., Shackleton, C.M., Leakey, R.R.B., Mander, M., et al., 2003. Marula commercialization for sustainable and equitable livelihoods: synthesis of a southern African case study. In: Winners and Losers in Forest Product Commercialization, Final Technical Report to DFID, FRP R7795, Appendix 3.5. CEH, Wallingford.

- Benefits can be accrued at the local level: value-adding increases the returns to labor, but does not have to be large-scale or aimed at external markets, e.g., marula beer, in which traders can earn greater income per hour than the suppliers of fruit and kernels to other markets.
- Communities are generally poorly placed to benefit from intellectual property rights: IPRs can play both a potentially positive and negative role in protecting the interests of primary producers, but to realize positive effects communities need substantial finance and support. There is an urgent need for IPR systems that promote poverty alleviation, food security, and sustainable agriculture as NTFPs make the transition to AFTPs.
- Models of commercialization based on partnerships between producer communities, NGOs, and the private sector are most likely to succeed: partnerships between different players can allow for mutually profitable arrangements (e.g., CRIAA SA-DC marula oil model). Most important is the retention of ownership and control of the enterprise at producer level.

- Diversified production and reduced dependence on a single product: The diversification of species used, products produced, markets traded, and players involved, is an extremely important strategy to minimize the risks of NTFP commercialization for rural communities. Often it is best to build on what exists at the local level rather than aiming for new high-value, specialized markets.
- Scaling up and introducing new technologies can shift benefits away from women and the most marginalized producers: the increased commercialization of NTFPs inevitably entails a shift from small-scale to large-scale, male-dominated activities.
- NTFPs form only part of a far broader ecological, economic, social, and political landscape: NTFPs are harvested and used within the context of broader development and land use pressures. For example, continued land clearance, the need for biomass energy, and woodcarving can be a greater threat than the commercialization of a fruit product.
- NTFP trade and industries are dynamic in space and time: there are seldom permanent winners and losers in the NTFP trade and producers' relationships with the resource base, other role players, the industry and the markets will be constantly changing and adapting in response to a range of internal and external drivers and processes and policy contexts.

The conclusion from this on-farm study was that NTFP/AFTP commercialization can create both winners and losers, but positive outcomes can be maximized if the importance of community involvement is appreciated by external players and if the communities themselves work together and use their own strengths to manage and use their resources effectively. This provides encouragement and some endorsement that the approach being developed for the participatory domestication of agroforestry trees is appropriate. This is supported by the findings of a study investigating the role of tree domestication in poverty alleviation (Poulton and Poole, 2001). Nevertheless, to ensure the farmers engaged in participatory domestication are winners, there is the need to resolve the current difficulties facing farmers wishing to protect their rights to the cultivars that they are producing.

FEATURES OF THIS AGROFORESTRY APPROACH TO RURAL DEVELOPMENT

This paper has focused on the role of agroforestry trees to reduce poverty and to enhance smallholder livelihoods. However, in rural development, the problems of poverty, land degradation, loss of biodiversity, social deprivation, malnutrition and hunger, poor health, and declining livelihoods are all inextricably linked and cyclical (Fig. 13.1). Consequently any attempts to alleviate the problems have to target a number of different points within the cycle. Agroforestry is advocated as a means of meeting these global challenges (Fig. 13.2). Leakey and Sanchez (1997) estimated that 1.8 billion people in the world (many of them in urban centers) make some use of agroforestry products and services. Thus potentially the domestication of agroforestry trees and the commercialization of their products should be

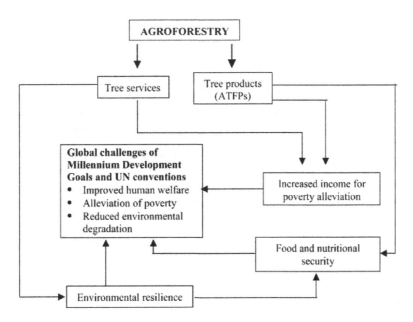

FIGURE 13.2 The relationship between the two functions of agroforestry trees and their potential to mitigate global problems arising from unsustainable land use. *After Leakey and Tomich, 1999.*

able to have positive impacts on the lives of many of the 50% of the world population (currently 6.4 billion) who are living on less than US$2 per day.

Regarding these objectives, the features of this AFTP approach to agroforestry and rural development are that it is based on:

- Traditional knowledge and culture (Cunningham, 2001).
- Participatory techniques to ensure relevance to local people (Franzel et al., 1996). This empowers subsistence farmers to control their own destiny and decentralize business opportunities to the villagers and create employment in processing and marketing.
- Doing research and development directly with communities (Leakey et al., 2003) rather than a "top-down" approach with national or international institutions. This promotes early adoption and impact.
- Indigenous perennial crops and foods (Leakey, 1999a) which:
  - have a reduced requirement for soil tillage;
  - create ecological niches and habitat for wild species above- and belowground so enhancing biodiversity and agroecosystem processes;
  - enhance food security, so reducing hunger;
  - promote nutritional security and dietary quality, so enhancing health and reducing malnutrition and diseases. This addresses food insecurity and micronutrient deficiencies through natural products and systems rather than biofortification;
  - create opportunity for income generation, reducing poverty and improving rural and urban livelihoods;
  - can be developed as new agricultural commodities to diversify the market economy and buffer commodity price fluctuations;
  - can substantially increase the numbers of crop plants available to farmers, adding those of local importance traditionally, culturally, and ecologically.
- Integrated natural resources management and sustainable land use (Leakey, 1999b), based on diversifying the farming system at the local and landscape scale, enhancing agro-ecological function and watershed services. This enhances international public goods and services by reducing trace gas emissions that impact on climate change and by minimizing the loss of biodiversity.
- Low-input polycultures rather than high-input monocultures (Leakey, 2001a,b); i.e., working with nature rather than against it.
- Knowledge of the natural resource (Shackleton et al., 2003a).
- Local germplasm and appropriate technology (Simons and Leakey, 2004).
- Market specialization for a number of niche products with local and regional acceptability, rather than on globalization and exposure to commodity price fluctuations in world trade.

The potential of this approach of course comes with some risks (Fig. 13.3), of which the three most important are the dangers of:

- Reducing intraspecific genetic diversity and losing resistance to pests and diseases; losing genes that may have importance for future developments, etc.
- Losing the traditional and cultural values associated with indigenous species.
- Losing the sustainability of production systems by overemphasis of high-input, monocultural practices (Leakey and Izac, 1996), which would also undermine the markets supporting small-scale agroforestry producers.

There is one negative aspect of the domestication of AFTPs. It is likely to result in a reduction in the market share of wild-harvested NTFPs. This would probably disadvantage landless rural people. However, the number of people benefiting from this domestication probably greatly outweighs those who would be disadvantaged.

There is also one problem that needs to be overcome to implement this major paradigm shift in rural development. Extension services in many countries are virtually moribund, although to some extent they have been replaced by NGOs and CBOs. There is therefore a need for policies to create an extensive new network to transfer agroforestry technologies to farmers. These would need to extend into remote and marginal areas.

POLICY GUIDELINES

The features of this agroforestry paradigm for rural development will require some policy changes at national and international levels, especially to ensure the scaling up to the levels required to achieve the Millennium Development Goals.

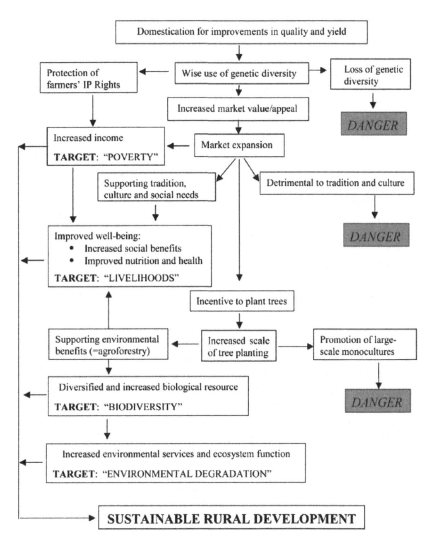

FIGURE 13.3   Potential impacts on sustainability of domesticating agroforestry trees.

An earlier review of the policy issues surrounding the domestication and commercialization of trees producing AFTPs raised many questions (Leakey and Tomich, 1999), some of which have been elaborated previously. Inevitably, however, in a new research area such as this, with only a short history, many questions remain unanswered, and indeed cannot be answered until the techniques and strategies outlined previously have been in use for longer periods and on larger scales. Nevertheless, there seems to be growing confidence on the part of institutions like ICRAF and their donors that this approach to agroforestry and the alleviation of poverty has merit (e.g., Poulton and Poole, 2001). This is emphasized by suggestions that these concepts have a role to play in the achievement of several of the Millennium Development Goals (Garrity, 2004; Sullivan and O'Regan, 2003).

For agroforestry to assist with the achievement of the Millennium Development Goals, there is a need for a major international initiative to create the level of upscaling that is required to bring domesticated AFTP-producing trees into millions of households every year before 2010. Garrity (2004) has, consequently, indicated that many agricultural R&D institutions around the world must be helped to develop new skills in the domestication of indigenous species and the processing/storage of their products, in market analysis and in developing market linkages. This level of upscaling will also require high level policy support to ensure a coordinated and coherent approach to domestication and commercialization across government departments ranging from agriculture and forestry, education and training, infrastructure and transport, to rural development and trade.

The scaling up of participatory domestication to tens of millions of new households every year across the developing world is probably the biggest challenge. Currently participatory domestication is driven by the clients (the subsistence farmers) meeting existing local market demands which are more focused on quantity than quality, but as the process proceeds and supply meets these market demands, it will become increasingly important that new market opportunities are identified, with attention to product processing, adding value, and storage. These new markets are likely to be more interested in quality and thus markets are likely to start to drive the selection processes of domestication. This will require refinements in the identification of ideotypes that focus on the specific needs of a particular market: for nutritious food, medicinal products, cosmetics, etc. Matching these market demands with the domestication process and the supply chain will become a challenge, dependent on much better market information than is currently available.

The upscaling of participatory domestication will be a challenge both in terms of the logistics of training and supervision, and in its adaptation to new species, environments, and markets. In this connection, there is an urgent need to rapidly expand the pool of expertise in core techniques like vegetative propagation. Fortunately, the Commonwealth Science Council has published an excellent manual (Longman, 1993), with supporting videos (Edinburgh Center for Tropical Forests, 1993) for "training the trainers," which are appropriate for use by NGOs, etc. Associated with the development of vegetatively propagated cultivars, an urgent issue to be resolved will be the acquisition and protection of "plant breeder's rights" on the cultivars created by communities practicing participatory domestication. The failure to achieve this will be a severe disincentive for villagers to invest their time, effort and limited resources to a venture that could be taken away from them. This outcome would destroy the potential of agroforestry to enhance rural livelihoods and human welfare, and to meet the Millennium Development Goals. In this connection, policymakers should realize that IPR-protected participatory domestication represents a new and acceptable approach to biodiscovery—the antithesis of biopiracy.

As domestication becomes more widely implemented and the objectives more sophisticated, it will become increasingly important to avoid the potential pitfalls of domestication (Fig. 13.3), like the loss of genetic diversity. Thus the trainers and mentor organizations need to ensure that the communities understand the need to deliberately retain intraspecific variation for pest and disease resistance, environmental stress, etc., through germplasm conservation, and a rolling program of genetic selection (Leakey, 1991). If attention is paid to genetic diversity issues, agroforestry with domesticated indigenous plants can serve a useful *circa situ* conservation function. In addition, as commercial interests increase, it will be important to maintain a focus on diversified agroforestry production that should promote integrated pest management (Leakey, 1999b), rather than shifting to monocultures.

This risk of large-scale monocultural production is one that can develop in-country, or indeed overseas, the latter having particularly serious implications; and of course there are many precedents among the cash-crop commodities. However, in this regard, it is encouraging that rubber, cocoa, coffee, and tea are increasingly becoming smallholder crops, as the profitability of large plantations declines. The growing recognition of the suitability of smallholder production by large companies (Chrysler-Benz, Masterfoods, Bodyshop) is also encouraging. It is, however, clear from the preceding that sound policy interventions will probably be needed to ensure that smallholder farmers are the beneficiaries of the domestication of AFTPs. As mentioned earlier, the desirability of starting to domesticate a wide range of new tree crops with local markets in each region should keep the options open for farmers and so minimize the risks of monopolistic production companies. Policymakers tend not to think much about the differences between a monocultural approach to growing a new crop versus an agroforestry approach. This is illustrated by the recognition that 20 million trees can be grown by four farmers, or by a million farmers each growing only 20 trees. Clearly that latter approach has the greatest potential impact in terms of the Millennium Development Goals.

One clear policy message from many sources is the need to recognize the "chicken and egg" relationship between domestication and commercialization (Leakey and Izac, 1996)—and the folly of doing one without the other. However, it is clear that the relationship between domestication and commercialization is delicately balanced, with both the lack of a market and the excessive growth of a market posing a threat. The latter is undesirable from several points of view: firstly an excessive market demand can result in low quality and nonuniform produce being placed on the market; secondly high market prices can encourage poor farmers to sell produce that should be used domestically to provide food and nutritional security, and thirdly, high market prices could encourage businessmen to embark on large-scale, monocultural plantations, which could undercut marginalized smallholder farmers practicing agroforestry, and so defeat the object of meeting the Millennium Development Goals. However, the complexity of marketing means that creating demand can also have positive effects. For example, the rise of large, multinational "supermarket" consumer companies (Reardon et al., 2003) may perhaps be an asset for the marketing of new food products. These companies operate on a scale that may allow them to take a risk and to offer new products to consumers increasingly interested in the cuisine of other countries and cultures.

The following specific policy interventions are some of those recommended by recent studies on the domestication and commercialization of agroforestry tree products. For example, following a study of the benefits and constraints of domestication of indigenous fruits in Cameroon and Nigeria, Ndoye et al. (2005) and Tchoundjeu et al. (2005) have recommended that governments and international agencies need to:

- Promote the participatory domestication of tree species fitting a variety of on-farm niches. Unlike institutional domestication programs, participatory domestication empowers local communities, and maintains their rights over indigenous knowledge and germplasm, as proposed by the Convention on Biological Diversity.
- Focus domestication activities on the capture and use of intraspecific variation existing in wild/semidomesticated populations and utilize the relatively quick economic and social returns from participatory domestication. These result from early fruiting and rapid improvements in productivity and product quality.
- Promote the local level processing and marketing of indigenous fruits, nuts and other tree products in parallel with domestication to maximize adoption of diversified, sedentary farming (agroforestry).
- Recognize the very considerable training and extension needs of rural communities that are required to achieve the scaling up necessary to meet the Millennium Development Goals.

In a policy brief developed from a study of the commercialization and potential domestication of marula fruits in South Africa and Namibia, Wynberg et al. (2003) have recommended 12 policy interventions, including the following:

- Governments should clarify land and usufruct rights to facilitate the successful and effective commercial development of AFTPs, recognizing that Western approaches to titling may not be appropriate for indigenous resource tenure systems.
- Urgent efforts should be made to develop and implement systems to protect community-based cultivars (through participatory domestication) . . . as part of legislative reforms for biodiversity management, indigenous knowledge protection, and plant genetic resource conservation and use.
- Through effective natural resource management, governments, traditional authorities, and communities should ensure the continued use of a wide range of NTFPs, to support rural livelihoods. Commercial enterprises should promote the development of a wide range of products and markets.
- To ensure that local people capture a greater share of the benefits from commercialization, basic management, financial, and institutional capacities must be in place.

Together these policy guidelines provide some direction on specific interventions to improve the likelihood that agroforestry will contribute substantially to the achievement of the Millennium Development Goals.

DEVELOPMENT ISSUES FOR THE FUTURE

In the 9 years that agroforestry tree domestication has been in progress, great advances have been made. This review has focused on progress in the humid zone of West and Central Africa and in southern Africa, but similar programs are underway in the Sahel, in East Africa, in Amazonia and in Southeast Asia, as well as in programs outside the ICRAF. Hopefully, the experiences reported here for agroforestry based on locally relevant tree species and markets will be of great benefit to other areas of the world embarking on similar people-centered concepts for rural development.

After 25 years of agroforestry research, it has been argued that there are already many examples of modern approaches to agroforestry, achieving the objectives of the Millennium Development Goals at the household and community level. Some of these examples include the enhanced production and marketing of AFTPs to raise incomes above US$2 per day. The challenge identified by these authors for the Millennium Development Initiative was how to scale up agroforestry between now and 2015 to reach the millions of poor rural families (60 million in the humid lowlands of West and Central Africa alone). Scaling up agroforestry is really more a matter of extension and community training than one of developing new technologies, although as is evident from this review, there is very considerable need to expand participatory tree domestication research, with locally relevant species, in very large numbers of communities throughout the tropics.

The upscaling of improved short-term fallows (Buresh and Cooper, 1999), especially in the maize belt of southern Africa, has been one good example of how to promote the adoption of agroforestry. Using this example, it can be argued that the benefits of short-term fallows on maize yields can be the catalyst for further advances into agroforestry based on indigenous fruit trees, which in turn can allow smallholder farmers to make the transition from subsistence into a cash economy. In this, as in other examples, the small amounts of cash generated by selling AFTPs can allow farmers to purchase agricultural inputs, to achieve higher yields from their staple foods, and so create an opportunity for further advances into cash cropping, and increase the returns from the investment in the Green Revolution (Leakey, 2001b; Leakey and Tomich, 1999).

The realization of this vision would be a "Really Green Revolution" (Leakey, 2001b; Leakey and Newton, 1994a). It is an alternative to some of the other approaches being advocated for rural development (e.g., McCalla and Brown, 1999; Serageldin and Persley, 2000), but has synergies with others (e.g., Inter Academy Council Report on African Agriculture, 2004—www.interacademycouncil.net), the World Summit on Sustainable Development's Water, Energy, Health, Agriculture and Biodiversity initiative (http://esl.jrc.it/dc/wehab/WEHAB_indicators.htm) and the proposals for Ecoagriculture (McNeely and Scherr, 2003).

From this review, it is suggested that the domestication of new tree crops provides an incentive for farmers to implement agroforestry practices that target intervention points in the cycle of agroecosystem degradation (Fig. 13.1) and that by so doing, it is possible to reduce land degradation, hunger, malnutrition, disease and poverty and so move toward the achievement of the ambitious targets set by the Millennium Development Goals. This approach to rural development in the tropics will, however, require fundamental changes in the attitudes of many national government and international development agencies and in their policies. Such changes are beyond the scope of this paper.

Note
This paper is adapted from a contribution to the 25th Anniversary Conference of ICRAF (World Agroforestry Center) in Nairobi, Kenya.

Chapter 14

Tree Domestication in Agroforestry: Progress in the Second Decade (2003–2012)

This chapter was previously published in Leakey, R.R.B., Weber, J.C., Page, T., Cornelius, J.P., Akinnifesi, F.K., Roshetko, J.M., Tchoundjeu, Z., Jamnadass, R., 2012. In: Nair, P.K., Garrity, D. (Eds.), Agroforestry — The Future of Global Land Use. Springer, USA, pp. 145–173, with permission of Springer

SUMMARY

More than 420 research papers, involving more than 50 tree species, form the literature on agroforestry tree domestication since the 1992 conference that initiated the global programme. In the fi rst decade, the global effort was strongly led by scientists working in humid West Africa; it was then expanded to the rest of Africa in the second decade, with additional growth in Latin America, Asia (mostly SE Asia) and Oceania. While the assessment of species potential and the development and dissemination of techniques for improved germplasm production were the principal activities in the fi rst decade, the second decade was characterized by a growing research agenda that included characterization of genetic variation using morphological and molecular techniques, product commercialization, adoption and impact and protection of farmers' rights. In parallel with this expanding research agenda, there was also an increasing use of laboratory techniques to quantify genetic variation of the chemical and physical composition of marketable products (e.g. essential oils, food-thickening agents, pharmaceutical and nutriceutical compounds, fuelwood). Looking to the third decade, suggestions are made for further development and expansion of both the science to underpin agroforestry tree domestication and applied research in support of development programmes to enhance the livelihoods of poor smallholder farmers worldwide.

INTRODUCTION

The "International Year of Forests" (2011) is an appropriate time to reflect on progress since the 1992 Conference in Edinburgh, UK, on *Tropical Trees: The Potential for Domestication and the Rebuilding of Forest Resources* (Leakey and Newton, 1994a,b). That international conference was the first to specifically discuss the potential of tree domestication to improve the livelihoods of poor smallholders in the tropics by rebuilding the resource of tree species on which hunter-gatherers had relied. The concept of domesticating specific tropical tree species had been around for a few years before this time (e.g., Clement, 1989; Holtzhausen et al., 1990). However, it was the Edinburgh conference that enunciated the vision of how the improvement and cultivation of these overlooked and underutilized "Cinderella" species could play a critical role in rural development. That was the beginning of what has become a global multidisciplinary research initiative to use agroforestry for the alleviation of malnutrition and poverty in the tropics. This has now been seen as the start of a second wave of domestication to address the needs of societies in the developing world (Leakey, 2012b,c).

The early concepts of tree domestication for agroforestry were rooted in traditional knowledge about the utility of forest species (e.g., Abbiw, 1990) and in ethnobotany (e.g., Cunningham, 2001), especially with regard to the nutritional

© 2017 Elsevier Inc. All rights reserved.

value of indigenous fruits. From an initial focus on about six traditionally important tree species, the international literature of more than 420 research papers has grown to include more than 50 species. This information has been collated (Table 14.1) to illustrate the growth and evolution of agroforestry tree domestication. In this chapter we demonstrate how tree domestication has evolved temporally and spatially over the last two decades to become an important global program. We then highlight some recent developments that enhance the capacity of agroforestry tree domestication to have meaningful impacts on the livelihoods of smallholder farmers around the tropics.

THE FIRST DECADE (1992–2002)

The early history of agroforestry tree domestication has been reviewed in detail elsewhere (Leakey et al., 2005a; Akinnifesi et al., 2008) and is only summarized here. Agroforestry tree domestication research started in the humid zone of West and Central Africa on several fronts; however, the dominant areas of work were the assessment of species potential, the propagation techniques and the variation in fruit and nut morphology (Figs. 14.1, 14.2). This set the pattern which was later followed in other regions, with or without the ICRAF. ICRAF's tree domestication program began with a participatory species priority-setting exercise with rural households (Franzel et al., 1996) which resulted in a subsequent initial focus on the indigenous fruit trees *Irvingia gabonensis* Baillon and *Dacryodes edulis* (G. Don) H.J. Lam in Cameroon and Nigeria. Parallel studies in the Congo, outside the ICRAF program, examined the potential of postharvest product processing (Mbofung et al., 2002; Kapseu et al., 2002). From the start, the interest of poor smallholder farmers in wild fruits and nuts directed the implementation of the program. This led to the emergence of a tree domestication strategy that recognized the capacity of vegetative propagation to capture phenotypic variation among individual fruit and nut trees (Simons, 1996: later refined by Leakey and Akinnifesi, 2008) and the use of simple low-technology polythene propagators (Leakey et al., 1990). These propagators are particularly appropriate for use in remote locations because they do not require running water or electricity. Based on this strategy, priority-setting exercises were subsequently implemented in southern Africa and the Sahel (Franzel et al., 2008; Faye et al., 2011) and Amazonia (Weber et al., 2001). In these regions rural communities expressed interest in species for timber, fodder, medicines, and fuelwood, in addition to local fruits and nuts. Much later, this model was also implemented in the Solomon Islands (Pauku et al., 2010).

In West and Central Africa, much of the work in the first decade (Table 14.1) was associated with the development of village nurseries (Tchoundjeu et al., 1998), the collection and dissemination of germplasm and the refinement of vegetative propagation techniques developed for tropical timber trees. These techniques then had to be augmented with better methods of marcotting so that sexually mature tissues with the existing capacity to flower and fruit could be propagated. The mature material creates cultivars which will start to yield within 2–3 years, while they are still small trees. This makes the cultivation of fruit trees much more attractive to farmers who want quick results from their investment of time and effort.

Before using vegetative propagation to develop cultivars, it is necessary to have some understanding of the extent and patterns of phenotypic variation in wild tree populations. Therefore, detailed studies were made of the tree-to-tree variation in morphological traits (fruit size, shape, color, etc.) within and between villages (e.g., Atangana et al., 2001). This confirmed that the phenotypic variation in all the species studied was very extensive (three- to ten-fold) and continuous—i.e., not clustered into groups that could be considered to be genetic varieties. Importantly, most of this variability was found within individual villages. These results confirmed the appropriateness of village level tree domestication, both from the point of view of giving individual farmers access to the full set of useful variation and that of minimizing the loss of genetic diversity often attributed to domestication activities.

Socioeconomic studies of village communities found that farmers were taking an increasing interest in the cultivation of a mixture of indigenous and exotic fruit tree species (Schreckenberg et al., 2002b), and that indigenous fruits were important at the household level for domestic consumption, as well as being a source of income based on local marketing. Parallel work in the Congo continued to provide a better understanding of product development, particularly nutritive value, oil extraction, postharvest processing, and the properties of *D. edulis* oil (Kapseu et al., 2002; Mbofung et al., 2002).

It also became clear from this early research that market price was not determined by fruit/nut size and morphology alone, but rather that the flavor and chemical composition of fruits and nuts contributed to consumer preference for the fruits of certain trees. This was confirmed by organoleptic studies (Kengni et al., 2001) and physicochemical analyses (Leakey et al., 2005d). However, while market stallholders (retailers) recognized consumer preferences for the products of certain trees, wholesalers did not. Thus, farmers selling a wide range of unselected fruits in mixed batches were not the beneficiaries of consumers' willingness to pay higher prices for desirable fruits. This lack of discrimination by

TABLE 14.1 Summary of research topics published in the first two decades (I = 1992–2002 and 2012) of agroforestry tree domestication worldwide.

Research topic	Global		Humid Africa		West Africa		Sahel		Southern Africa		East Africa		Latin America		Asia		Oceania		Total		
	1	2	1	2	1	2	1	2	1	2	1	2	1	2	1	2	1	2	1	2	1 + 2
Domestication concept	7	4	–	–	–	1	1	2	1	1	1	2	–	1	1	2	1	2	10	9	19
Domestication strategy	4	5	–	–	2	3	–	–	–	5	2	–	6	7	1	5	–	1	16	26	42
Propagation and germplasm	4	2	–	–	7	13	4	7	4	11	–	1	–	4	3	7	1	0.5	23	45.5	68.5
Species potential	–	–	–	2	14	5	6	7	8	1	2	3	3	5	1	–	3	9	37	32	69
Genetic characterization	–	1	–	–	8	7	3	14	4	5	–	–	–	13	–	–	1	4.5	16	44.5	60.5
Morphological molecular	–	1	–	4	2	2	–	8	–	3	1	2	3	3	–	–	–	–	6	23	29
Reproductive biology	–	–	–	–	3	1	–	2	1	–	–	2	–	–	1	–	–	–	4	5	9
Nutritional benefits	1	–	–	2	4	2	1	5	2	–	1	–	2	2	–	–	–	1	11	12	23
Product evaluation and development postharvest																					
Participatory implementation on-farm	–	–	–	–	1	3	–	–	–	1	–	1	1	–	–	–	–	–	2	4	6
Agroforestry enrichment	2	2	–	–	–	–	1	–	1	3	–	8	–	4	1	–	–	–	5	17	22
Socioeconomic issues	–	–	–	–	5	3	1	3	2	4	–	–	–	1	1	–	–	–	9	11	20
Commercial issues	3	–	–	1	2	2	2	–	2	4	–	–	2	1	1	2	–	1	12	10	22
Ecology	–	–	–	–	–	5	–	–	5	4	–	–	–	–	–	–	–	–	5	9	14
Adoption and impact	1	2	–	–	–	2	–	–	–	–	–	–	–	–	–	–	–	–	1	4	5
Policy	3	–	–	–	–	–	–	–	–	5	–	–	–	–	–	–	–	–	3	5	8
Total	25	17	–	9	63	44	19	53	29	50	6	17	17	40	10	14	6	17	172	252	424
Total per region	42		9		107		72		79		23		57		24		23		424		

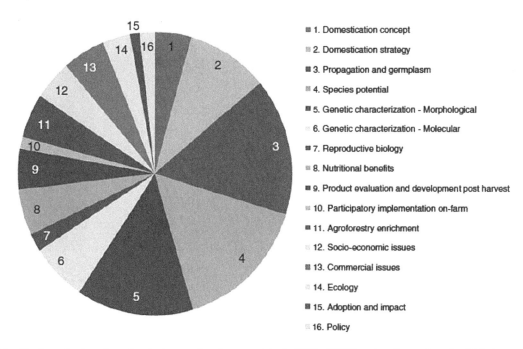

- 1. Domestication concept
- 2. Domestication strategy
- 3. Propagation and germplasm
- 4. Species potential
- 5. Genetic characterization - Morphological
- 6. Genetic characterization - Molecular
- 7. Reproductive biology
- 8. Nutritional benefits
- 9. Product evaluation and development post harvest
- 10. Participatory implementation on-farm
- 11. Agroforestry enrichment
- 12. Socio-economic issues
- 13. Commercial issues
- 14. Ecology
- 15. Adoption and impact
- 16. Policy

FIGURE 14.1   The domestication of agroforestry tree species—by research topic (1992−2012) based on the number of published research papers.

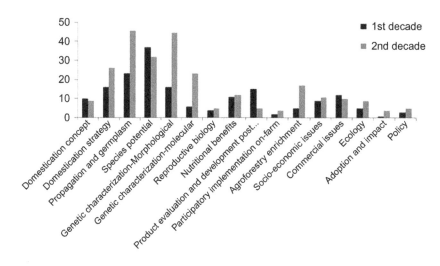

FIGURE 14.2   The domestication of agroforestry tree species by research topic—comparison of 1992−2002 (first decade) with 2003−2112 (second decade)—based on the number of published research papers.

traders emphasizes the potential benefit for farmers to produce and market specific varieties based on selected domesticated cultivars.

Work in humid West Africa in the first decade set the pattern that was subsequently adapted for species in other regions, including the Peruvian Amazon, southern Africa and the Sahel. By contrast, the participatory process in Southeast Asia identified priority topics to advance smallholder tree domestication research as well as a long list of priority species for the region's various biophysical, socioeconomic and farming conditions. Priority topics for smallholder tree domestication were access to tree germplasm through its multiplication and dissemination; development of tree propagation, nursery techniques and silvicultural practices; expansion of species diversity and improved management in agroforestry systems; market integration; and improved agroforestry information and training (Roshetko and Evans, 1999). Subsequent research gave special attention to *Gliricidia sepium* (Jacq.) Kunth ex Walp. (Roshetko et al., 1999; Mangaoang and Roshetko, 1999) and *Eucalyptus* species (Bertomu and Sungkit, 1999).

THE SECOND DECADE (2003–12)

The basic concepts, techniques and strategies developed in the first decade have been endorsed in the second decade and used for a wider range of species, environments and sites (Table 14.1, Figs. 14.3, 14.4). Additionally, they have been modified as required by local biophysical, ecological, and social conditions and applied in the Sahel, the woody savannah of southern Africa, Amazonia, and in some small Pacific islands of Oceania.

The Humid Lowlands of West and Central Africa

Indigenous fruits are important at the household level, as well as being an important source of income (Schreckenberg et al., 2006). In the humid tropics, indigenous trees have many potential on-farm niches, but the importance of shade for the cocoa (*Theobroma cacao* L.) and coffee (*Coffea* spp.) crops creates a great opportunity to increase the profitability of these cash cropping systems by using indigenous trees that produce marketable products as the shade trees. Through domestication of these trees, this multistory system becomes a productive agroforest. The cultivation of domesticated agroforestry trees converts these indigenous trees into new crops, and consequently, their marketable products become farm produce instead of being common property forest resources. To signify this important distinction, the description of these products as nontimber forest products (NTFPs) was changed to agroforestry tree products—AFTPs (Simons and Leakey, 2004).

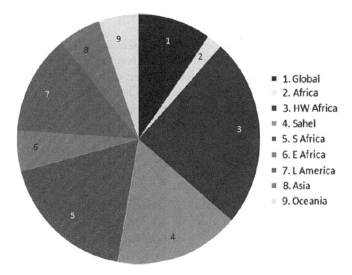

FIGURE 14.3 The domestication of agroforestry tree species by region (1992–2012)—based on the number of published research papers.

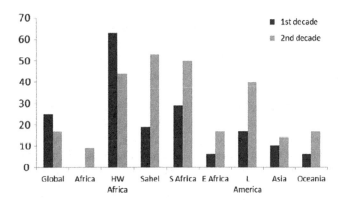

FIGURE 14.4 The domestication of agroforestry tree species by region—comparison of 1992–2002 (first decade) with 2003–2012 (second decade)—based on the number of published research papers.

As part of the domestication process, the tree-to-tree variation in traits affecting yield and quality of fruits and nuts was quantified. From this it became clear that a large fruit is not necessarily a tasty fruit and may not have a large or useful nut. Thus, to select trees for cultivar development by vegetative propagation, the concept of an ideotype was modified so that the desirable combination of different traits to produce a cultivar targeting a particular market opportunity could be visualized (Leakey and Page, 2006). Building on this ideotype, a range of different tree species have now been characterized for traits such as food-thickening agents (drawability and viscosity in *I. gabonensis*, Leakey et al., 2005d) and fatty acid profiles (stearic and oleic acids in *Allanblackia* spp., Atangana et al., 2011).

Another outcome of the morphological characterization was a technique based on the frequency distribution of the data for any particular trait, which quantifies the stage of domestication that has been reached by the farmers' own selections for the most desirable trees. This revealed that, in some Cameroonian villages, out of the five stages of domestication (Leakey et al., 2004) *D. edulis* is at stage 2, while the same is true for *I. gabonensis* in Nigeria. It is therefore clear that farmers are interested in the domestication of their indigenous food species but lacked the knowledge to achieve this other than by the slow route of sexual recombination. Consequently, when ICRAF researchers and their local national research partners approached farmers about the initiation of a program of participatory tree domestication, the farmers were enthusiastic (Tchoundjeu et al., 2006, 2010). This program has now expanded from a few farmers in two pilot villages to over 10,000 farmers in more than 200 villages (Asaah et al., 2011).

Village level participatory domestication is dependent on simple and robust techniques of vegetative propagation. The development and refinement of these techniques have been ongoing processes involving an increasing number of species. This process is helped by the formulation of some basic principles which apply to most, if not all, species (Leakey, 2004). These principles are particularly useful when domestication activities are centered on difficult to ltotaoatpropagate species like those in the genus *Allanblackia*.

The initial focus on fruit and nut trees in this region has been expanded to include overexploited medicinal species, especially *Prunus africana* (Hook. f) Kalkman, *Pausinystalia johimbe* (Schumann) Beille, and *Annickia chlorantha* Oliv., whose barks are used to treat prostate enlargement, cardiac disease, and malaria, respectively. Johimbe (*P. johimbe*) is also an aphrodisiac. The cultivation of these species as herbal medicines for local use is relatively simple, but their domestication for the production of internationally marketed drugs needs to involve industrial partners. This is further complicated by competition from synthetic drugs.

The Drylands of the Sahel

Rural communities in Burkina Faso, Mali, Niger, and Senegal value more than 115 indigenous tree species for the livelihood benefits of their products and services (Faye et al., 2011). The "parkland" is the most common agroforestry system in these countries and combines crops, trees, and livestock. Farmers maintain several indigenous tree species in the parklands for food (e.g., *Adansonia digitata* L., *Parkia biglobosa* (Jacq.) Benth., *Vitellaria paradoxa* C.F. Gaertn., *Ziziphus mauritiana* Lam.); dry season fodder (e.g., *Balanites aegyptiaca* (L.) Del., *Faidherbia albida* (Del.) A. Chev., *Pterocarpus* spp.); wood for fuel, construction, household, and farm implements (e.g., *B. aegyptiaca*, *Combretum glutinosum* Perrott. ex DC., *Guiera senegalensis* J.F. Gmel., *Prosopis africana* (Guill. & Perr.) Taub.); medicines; and environmental services such as shade, soil fertility improvement and soil/water and conservation. The sale of these products contributes 25−75% of annual household revenue in Mali (Faye et al., 2010), with some having international markets.

The provision of human and animal food is particularly important during the peak of the long dry season. Consequently, rural communities in the driest areas of the Sahel use significantly more species than those in wetter areas because this maximizes the chance that at least one species will provide products/services even in a dry year. Therefore, tree domestication programs are focusing on the specific priorities of different regions and diversifying the number of species for each product and service.

To enhance dry season fodder production, fodder banks of exotic (e.g., *G. sepium* from Central America) and indigenous (e.g., *Pterocarpus erinaceus* Poiret) species were developed within thorny hedges for protection from livestock. These fodder banks have considerable economic importance with small bundles of shoots fetching good prices in local markets. The fodder trees have been propagated by both seed and vegetative propagation (Tchoundjeu, 1996; Tchoundjeu et al., 1997). Likewise, both approaches have been used for fruit trees, especially those that are difficult to propagate by cuttings (e.g., *V. paradoxa*).

With international markets for indigenous fruits and nuts such as shea butter (*V. paradoxa*) and *B. aegyptiaca*, oil quality is important. In order to improve quality by genetic selection, studies have been made of phenotypic variation in fruit/seed traits across the Sahel. These have found significant variation both amongst and within provenances (Abasse et al., 2011; Ræbild et al., 2011). For example, fruit and/or kernel size is greater in more humid sites for *A. digitata* and *V. paradoxa* but in drier sites for *B. aegyptiaca*. This variation offers great potential for future selection and domestication. For example, fruits of baobab (*A. digitata*) are very rich in vitamin C, calcium and magnesium, while its leaves contain vitamins C and A. In addition, characterization studies of morphological variation in fruit and seed traits of *A. digitata* in Mali have found considerable potential for selection of trees with superior pulp mass and also with high pulp:seed ratios (De Smedt et al., 2011). Baobabs occur throughout dry Africa, so evidence that trees from Mali and Malawi differ in pulp percentages, seed size and shape illustrates even greater potential for selection in different countries across the continent (Sanchez et al., 2011). The potential gains from this selection will become apparent from ongoing provenance tests (Kalinganire et al., 2008). To further explore the extent of genetic variation in shea nut (*V. paradoxa*) and baobab (*A. digitata*), molecular techniques have been used (Jamnadass et al., 2009). In the latter, superior morphotypes were not genetically related varieties, suggesting that the development and use of many clonal cultivars could maintain considerable genetic diversity.

As there are strong latitudinal and longitudinal gradients in mean annual rainfall in the Sahel, provenance/progeny tests are being used to compare the performance of germplasm collected from sites across these gradients. Results from tests of *B. aegyptiaca* and *Pro. africana* indicated that provenances from drier sites had significantly better aboveground growth than provenances from more humid sites when tested at a relatively dry site (Weber et al., 2008; Weber and Sotelo Montes, 2010). In addition, wood density of *Pro. africana* and calorific value of the wood of both species also varied along the rainfall gradients (Sotelo Montes and Weber, 2009; Sotelo Montes et al., 2011). Based on these tests, it is recommended that germplasm should be collected in the drier sites for future plantings in parklands, especially as this germplasm appears to be better adapted to dry conditions. As part of a participatory tree domestication program, rural communities in Burkina Faso, Mali and Niger are establishing provenance/progeny tests of several species for fruit, wood, fodder and medicines (e.g., *A. digitata*, *F. albida*, *G. senegalensis*, *P. biglobosa*, *Pro. africana*, *V. paradoxa*) in their parklands to compare the performance of their local germplasm with germplasm collected in drier sites (J. C. Weber, personal communication, 2011). The tests will provide basic information about drought adaptation and variation in commercially important traits under farmer-managed conditions. In addition, it is expected that the introduced genes from the drier sites will increase the drought adaptation of the natural regeneration in the parkland species.

Studies are underway to determine if fuelwood properties of trees in natural populations vary with rainfall gradients. In Mali, for example, fuelwood properties were better for *B. aegyptiaca*, *C. glutinosum*, and *Piliostigma reticulatum* (DC.) Hochst. in drier regions, worst for *Z. mauritiana* in the drier regions and good for *G. senegalensis* in both humid and dry regions in Mali (C. Sotelo Montes, personal communication, 2011). Since the climate is becoming hotter and drier in the Sahel than before, these studies could be used to identify the best regions, species and germplasm for fuelwood production in parklands as part of climate change adaptation planning (Nair, 2012).

Woody Savannah of Southern Africa

The Miombo woodlands are rich in edible indigenous fruit trees, for example, *Sclerocarya birrea* (A. Rich.) Hochst., *Strychnos cocculoides* Baker, *Uapac kirkiana* Muell. Arg., *Vangueria infausta* Burch., *Parinari curatellifolia* Planchon ex Benth., *Z. mauritiana*, and *A. digitata*, many of which are traded in the region. However, land clearance for maize (*Z. mays* L.) and other staples has severely reduced their availability in Malawi, Zambia, and Zimbabwe. In contrast, some of these fruit trees like *S. birrea* are commonly found as much appreciated scattered trees in parklands, as in northern Namibia. Due to the local knowledge about these traditionally and culturally important species, the domestication strategy that has been adopted is based on the premise that farmers have adequate knowledge of the natural variability in fruit and kernel traits to be able to locate and identify superior trees in the wild for themselves. Thus, farmers have been trained in techniques of germplasm collection, nursery management, propagation, tree cultivation, and postharvest processing. As the seeds of many of these species have short viability, their collection and germination have to be rapid.

Market research has indicated that traders want a consistent and regular supply of uniform fruits of good quality. Wild fruits do not meet these criteria. Domestication is the best way to achieve uniformity and superior quality. By selecting the best trees in wild populations and then multiplying them vegetatively, large numbers of genetically identical trees (cultivars) can be produced for cultivation in farming systems. As the use of cuttings has been found difficult in Miombo trees, the other options are grafting, budding, and marcotting. These techniques are especially appropriate for fruit trees as they allow already mature trees to be propagated. Experiments have found that while marcotts are very effective, they seem to suffer from fruit bud abortion (Akinnifesi et al., 2009), and so experience suggests that grafting is the most appropriate option in *A. digitata, U. kirkiana, S. birrea*, and *V. infausta*, while budding has been found to be best for *Z. mauritiana*. Boththese techniques require a large supply of seedlings which are then used as the rootstock to which the desired mature scion or bud is attached. This necessitates seedlings with stems the same diameter as the mature scions. To achieve this, the seedlings of *U. kirkiana*, for example, normally have to be at least 2 years old, but this has been reduced to 10 months with intensive nursery management (Mhango et al., 2008). Experience and experiment have found that successful grafting is affected by the experience and skill of the grafter.

The selection of the mother trees can be based on farmer experience or on scientific assessment of the tree-to-tree variation within wild populations. The former is appropriate within village-based participatory domestication programs, while the latter is beneficial to develop understanding of the range of traits that can be selected to maximize market appeal. This latter approach was implemented in South Africa and Namibia with *S. birrea*, which has numerous potential market opportunities for fruits, nuts, kernel oil, and several alcoholic beverages (Leakey, 2005). Other studies have been done with *U. kirkiana* and *St. cocculoides* in Malawi, Zambia, and Zimbabwe (Akinnifesi et al., 2006). Enhanced acceptability and market demand are expected to serve as an incentive for farmers to domesticate their indigenous tree species.

The second way to boost and maintain the market demand is a focus on postharvest quality and shelf life. Detailed studies have been made of the effects of fruit-handling procedures (blanching, drying, handling, and storage) on fruit color, bruising, durability, etc. in *U. kirkiana* (reviewed by Saka et al., 2008). These studies included assessments of nutritional quality of fruits, product processing, certification, and on-farm economics.

In the case of *S. birrea*, a large multidisciplinary project has examined the potential winners and losers from the domestication and commercialization of this fruit tree in Namibia and South Africa. Commercialization activities in the region are both top-down for Amarula liqueur and bottom-up by the ladies of the mineworkers union for fruit juice and kernel oil sold to commercial companies. While potentially the top-down approach might be expected to have negative impacts on the livelihoods of local people, this was not found to be the case, as Distillers Corporation buys fruits directly from community vendors. This finding has important policy implications regarding the development of appropriate models for production, marketing, protection of farmers' rights and traditional knowledge and rural development (Wynberg et al., 2003). In Namibia, other interesting arrangements in the form of trade agreements between community producers and commercial companies both in the region and overseas have provided the prospect of protecting local communities from commercial exploitation (Lombard and Leakey, 2010) and greatly expanding markets for products from indigenous species grown by smallholders.

The demand for seed of tree legumes used in soil fertility restoration and as fodder trees in Malawi has led to the development of Community Agroforestry Tree Seed Banks to produce and distribute improved germplasm to farmers (Nyoka et al., 2011). During 2006–2011, these have distributed nearly 50 t of tree seeds, so overcoming one of the biggest constraints to farmer adoption of agroforestry.

East Africa

There is an active tree planting culture among small-scale farmers in the East African highlands where a range of exotic and indigenous species are cultivated for fodder, poles, fuelwood, timber, and fruits. However, most of the fruit trees are exotics like mango (*Mangifera indica* L.) and avocado (*Persea americana* Miller). This reflects the paucity of indigenous species with potential in the highlands—with the notable exception of *Pru. africana*, an important medicinal tree, which is restricted to Afromontane islands above 1500 m throughout sub-Saharan Africa. This tree is heavily exploited for its bark, resulting in unsustainable harvesting practices. As the active ingredient in the bark is not known, domestication activities have been limited to improving the seed supply and the assessment of genetic diversity. Another species is *Warburgia ugandensis* Sprague, a multipurpose tree found in the lowland rainforest and upland dry evergreen forest of eastern Africa between 1000 and 3000 m. It is widely used by the local communities to cure diseases like measles and malaria. Stem and root barks are harvested for herbal remedies. Overharvesting of the bark and illegal felling of trees in protected natural forests, as well as encroachment of their natural habitat for farming and human settlements, threaten the species survival and conservation. Molecular analyses of both these species have determined that populations to the east and west of the Rift Valley are genetically distinct (Muchugi et al., 2006, 2008). This geographic variation has implications for strategies of germplasm collection and use and perhaps even for genetic characterization and selection with regard to the levels and quality of the medicinal compounds. It may also be relevant to drought responses arising from climate change.

Despite the large number of indigenous species grown by smallholders in the highlands of East Africa, few have been nominated for intensive domestication; consequently, domestication activities have been focused on the provision of seed to farmers. This addresses the major problem that tree seed currently used by farmers is characterized by widespread distribution of inferior seed with an almost complete absence of concern for genetic quality and adaptability of planting material. Prime examples of programs to reverse this trend are those of *Calliandra calothyrsus* Meissner for fodder and *Sesbania sesban* (L.) Merrill for soil fertility improvement. The development of an informal network for delivering *Calliandra* seed has allowed widespread adoption of dairy cattle and goat fodder production by East African smallholders and is now recognized as one of the best models of seed dissemination (Wambugu et al., 2011). Despite much success, seed/seedling production and distribution systems for good quality germplasm only reach a small proportion of smallholders. Efforts are being made to overcome the disconnection between seed sources, tree nurseries and farmers.

In contrast to the East African highlands, the semiarid lowlands have a number of important indigenous species. Some such as *Acacia senegal* (L.) Willd., *Boswellia* spp. and *Commiphora* spp. produce high-value gums, while others (e.g., *A. digitata* and *Z. mauritiana*) are foods for local people and livestock (Simitu et al., 2008). Many of these dry zone species are also native to southern Africa and the Sahel.

Latin America

Some indigenous fruit trees of Latin America, such as *Bactris gasipaes* Kunth and *Chrysophyllum cainito* L., are recognized as being semidomesticated (Parker et al., 2010). ICRAF's domestication research in the Peruvian Amazon focused on studies of genetic variation in four priority species (Weber et al., 2001): *B. gasipaes* for fruit and heart-of-palm, *Calycophyllum spruceanum* (Benth.) Hook. f. ex K. Schum. For construction wood and fuelwood, *Guazuma crinita* (Mart.) for construction wood and *Inga edulis* Mart. for fuelwood and food (fleshy aril of the fruit). Research methods included on-farm tests and molecular genetic approaches.

As a result of selection, domesticated populations typically have lower genetic variation in the selected traits (Cornelius et al., 2006) and perhaps in selectively neutral molecular markers (Hollingsworth et al., 2005; Dawson et al., 2008). Farmers commonly collect germplasm from only a few trees, especially fruit trees, when planting trees on-farm (Weber et al., 1997), and this can lead to serious inbreeding problems in subsequent generations (O'Neill et al., 2001). A provenance/progeny test demonstrated that a low-intensity selection strategy can significantly increase tree growth without significantly reducing genetic variation in growth traits in the subsequent generation (Weber et al., 2009). It was recommended, therefore, that farmers select a larger proportion of trees for future planting, even though this will result in less genetic improvement compared with more intensive selection. In addition, since exchange of fruits/seeds among farmers from different watersheds can counteract the reduction in genetic diversity due to selection and genetic drift on farms (Adin et al., 2004), domestication programs should incorporate germplasm exchange pathways within and among watersheds.

Understanding variation amongst provenances and gene flow patterns is important for tree domestication and conservation programs. For example, provenance tests of *C. spruceanum* and *G. crinita* on farms demonstrated that the provenance from the local watershed generally grew better than most nonlocal provenances when tested in the local watershed (Weber and Sotelo Montes, 2005, 2008). Therefore, it was recommended that farmers use the local provenance for on-farm planting unless there was evidence that nonlocal provenances were significantly better. Some replicates of these on-farm tests were later transformed into seed orchards for production and sale of improved, source-identified germplasm by rural communities. This created a new business opportunity for rural communities as producers of high-quality tree seed for reforestation programs. In addition, if fruits/seeds are dispersed by rivers, as is the case for *C. spruceanum*, genetic diversity may be greater in populations below the confluence of major tributaries (Russell et al., 1999). For species like this, downstream populations therefore could be targeted for *in* or *circa situ* conservation.

Improving tree growth and wood properties depends on the magnitude of genetic variation in the traits, the heritability of each trait and the correlation between traits. Results from provenance and provenance/progeny tests of *C. spruceanum* and *G. crinita* indicate that (1) there is considerable genetic variation in tree growth and wood properties (density, strength, stiffness, shrinkage, color); (2) wood traits have higher heritability than growth traits, especially in sites where trees grow rapidly; and (3) correlations differ among test sites and provenances (e.g., Sotelo Montes et al., 2006, 2008; Weber and Sotelo Montes, 2008; Weber et al., 2011). Therefore, tree domestication programs can simultaneously improve growth and wood properties of these species by selecting trees within provenances and test environments where the heritability of traits is high and the correlations between desirable traits are positive.

Asia

Most smallholder agroforestry systems in Southeast Asia are characterized by limited proactive management and planning, with species composition and genetic material most often a result of chance or opportunity. The quantity and quality of products are often below the systems' potential (Roshetko et al., 2007). Aptly, the prime focus of agroforestry tree domestication in Southeast Asia has focused on the development of germplasm for smallholder and community organizations. These had been shown to play an important role in tree seed collection and dissemination but, like the local seed dealers, were not familiar with proper seed collection guidelines (Koffa and Roshetko, 1999; Roshetko et al., 2008). Through farmer training and field tests, technically sound farmer-appropriate tree seed collection and farmer seed orchard guidelines were developed. This led to the establishment of farmer and community tree seed enterprises (Carandang et al., 2006; Catacutan et al., 2008). Capacity building activities in smallholder nursery management and vegetative propagation skills resulted in the establishment of hundreds of local nurseries and a set of farmer manuals. Through a series of participatory on-farm trials, guidelines for farmer demonstration trials were validated (Roshetko et al., 2004b).

Research to improve smallholder timber production has centered in the Philippines, with some activities in Indonesia. Exotic species like *Gmelina arborea* Roxb. are widely planted (Roshetko et al., 2004a); however, the choice of species is often determined by access to germplasm, knowledge/experience of the operator, market demand and the priorities of donors and government agencies (Carandang et al., 2006). In the Philippines, smallholder farmers have become major timber producers, with trees planted and grown on farms an important source of raw materials and income for themselves and the local timber industry. Government statistics show that since 1999 between 50% and 70% of domestic log production came from smallholder on-farm sources. The two most important factors driving this enterprise are a paucity of forests/trees and the existence of market demand for timber. However, poor management practices led to an oversupply of low quality timber and declining prices for farm-grown timber. Consequently, on-farm research has focused on identifying silviculture regimes that are adoptable by smallholder farmers (Bertomeu et al., 2011).

Recently, there has been increased interest in indigenous timber species. Among the indigenous species, dipterocarps are important for both timber and nontimber products such as dammar resins and are grown by smallholder farmers in Indonesia, the Philippines and other countries in the region often in complex agroforests in association with cinnamon (*Cinnamomum* spp.), rubber (*Hevea brasiliensis* Muell. Arg.) and many local fruit and nut tree species. However, seed supply, due to irregular flowering (masting) and short seed viability, poses a serious constraint to large-scale planting. This can, however, be circumvented by the use of vegetative propagation. So far, however, the opportunity to use these techniques to develop cultivars of these local trees has not been taken.

Oceania

A formal tree domestication program in this region has been led by James Cook University (Agroforestry and Novel Crops Unit) with partners in the Solomon Islands, Papua New Guinea and Vanuatu since 2002. This is a region with many traditionally important nuts, such as *Canarium indicum* L., *Barringtonia procera* (Miers) Knuth, *Inocarpus fagifer* (Parkinson ex Zollinger) Fosberg, and *Terminalia kaernbachii* Warburg, which have had great cultural and social significance for millennia. The region also had very significant resources of sandalwood (*Santalum* species), valued internationally for its scented heartwood, and other valuable export timbers such as *Endospermum medullosum* L. S. Smith, *Instia bijuga* (Colebr.) O. Kuntze, *Pterocarpus indicus* Willd., and *Terminalia catappa* L. Historic overexploitation of these sandalwood and timber resources has severely reduced the livelihood benefits derived from them, and, therefore, they are important candidates for domestication and genetic restoration.

In Oceania, the approach to the domestication of indigenous nuts has been strongly based on the experience of the team in Cameroon. Thus, feasibility (producer and consumer surveys, Nevenimo et al., 2008) and priority-setting exercises (Pauku et al., 2010) were carried out as the first steps, prior to work to characterize the fruits and nuts morphologically. The characterization also included proximal and chemical analyses, demonstrating tree-to-tree variation in oil and protein content and yield as well as in antioxidant activity (mg ascorbate equivalents per gram), vitamin E (tocopherol content—α, β, γ, δ isomers) and antinutrients such as phenolic content (mg catechin equivalents per gram). Most interesting perhaps was the very considerable variation in the antiinflammatory activity (prostaglandin E_2 assay) of kernels (Leakey et al., 2008), demonstrating the possibility of selecting trees for their medicinal properties.

B. procera and *I. fagifer* were easy to propagate by cuttings, but *C. indicum* was very difficult. However, when the stockplants were grown under the shade of a *Gliricidia* canopy, fertilized and well managed, the rooting percentage was greatly improved (from 10% to 80%). Mature shoots of *B. procera* were also easily propagated by marcotting (Pauku et al., 2010). Mature cuttings were also rooted, with success being enhanced when the harvested shoots were taken from marcotted branches, both before and after severance of the marcotts (Pauku, 2005).

The industrial exploitation of sandalwood has depleted the wild resource of *Santalum austrocaledonicum* Vieillard across the region. An expedition to measure the remaining trees in Vanuatu located small remnant populations across seven islands. When solvent extractions of heartwood samples were analyzed for their content of four essential oils (α-santalol, β-santalol, (Z)-β-curcumen-12-ol and *cis*-nuciferol), significant tree-to-tree variation was found for each. Contrary to expectation, some trees exceeded the content of α- and β-santalol as prescribed in the International Standard for Sandalwood Oil conferring acceptability to the perfume industry (Page et al., 2010a). Interestingly, this variation was unrelated to heartwood color, thereby breaking long-held beliefs by some in the industry. Near infrared spectrometry technologies have been found to accurately predict α−santalol content of heartwood. As sandalwood is a hemiparasite, it was not known to what extent the host species would influence oil quality or yield. However, no host: parasite relationships were found. Individual trees with elevated santalol levels were selected and secured as a grafted seed orchard. This orchard has served as a source of both seeds for establishing new agroforestry plantings and scion material for replicating the seed orchard on other islands. These developments offer smallholder producers an economic opportunity to replenish the natural resource and contribute to the industry in Vanuatu (Page et al., 2010b).

E. medullosum is a valuable timber species (whitewood/basswood) found in Vanuatu, Solomon Islands, and Papua New Guinea. As with many valuable timber species throughout the world, the natural resources of *E. medullosum* have been depleted over long periods of commercial exploitation. Significant variation in growth and form characteristics was found within a provenance/progeny trial established in Vanuatu (Vutilolo et al., 2005). Continuing selection in this progeny trial and further efforts to develop both clonal cultivars and clonal seed orchards throughout the islands will give smallholder farmers greater access to this improved planting material. This in turn will increase productivity of smallholder plantings and relieve harvesting pressure on already depleted wild stands of the species.

Studies outside the main domestication program have examined the diversity of existing cultivars of breadfruit (*Artocarpus altilis* (Parkinson) Fosberg), which were developed primarily by selection and vegetative propagation over generations in Oceania (Ragone, 1997; Zerega et al., 2004). Initial diversity evaluations have been used to develop strategies for extending the breadfruit season through development and maintenance of a diverse range of cultivars with complementary fruiting seasons (Jones et al., 2010). In Vanuatu, germplasm was assessed for morphological diversity, and an ex situ strategy for conserving the germplasm was implemented with the view of increasing food security within its agriculturally dependent islands (Navarro et al., 2007). Indigenous methods for drying and preserving the carbohydrate-rich fruits are also being examined for their potential application in processing fruit for export.

It is evident from the preceding that the six regions of ICRAF have not implemented "farmer-driven, market led" agroforestry tree domestication in the same way (Table 14.1). This is partly due to variation in the experience and skills of staff in the different regions, partly determined by the priority of different donors and partly because a participatory priority-setting process was used, and the farmers themselves had different priorities for wood products versus food and medicinal products. In the latter case, the nature of the products selected and the species that produce them required different tree domestication strategies.

RECENT DEVELOPMENTS IN AGROFORESTRY TREE IMPROVEMENT

Molecular Genetics

Modern molecular techniques have been used in 13 agroforestry tree species (*Allanblackia floribunda* Oliver, *A. digitata*, *B. gasipaes*, *B. procera*, *C. spruceanum*, *I. edulis*, *I. gabonensis*, *Pru. africana*, *S. birrea*, *Spondias purpurea* L., *W. ugandensis*, *V. paradoxa*, *Vitex fischeri* Gürke) to determine the structure of genetic variation in natural, managed and cultivated tree stands and to devise appropriate management strategies that benefit users (Jamnadass et al., 2009). The resulting knowledge is used in three ways:

- To determine whether cultivated stands are of local or introduced origin and, if so, assess whether planted material comes from single or multiple sources. This historical information is important for genetic conservation and to derive appropriate management strategies (e.g., sexual reproduction vs clonal multiplication) to ensure that domesticated populations are both diverse and based on the most appropriate resources for future genetic improvement. The use of unrelated individuals is particularly important when developing clonal cultivars.
- To ensure that domesticated populations have sufficient genetic diversity to avoid future problems from inbreeding. Inbreeding results in depressed growth and/or poor reproductive success, both of which have important yield

implications. The use of molecular markers assists the determination of effective population sizes, breeding systems, and gene flow.

- To determine the proportion of a species genetic variation that is available at a local geographic scale. If this is high, then a decentralized approach to domestication is appropriate. On the other hand, if it is low, a more centralized approach with germplasm infusions from outside may be required. To date, most agroforestry trees appear to contain high levels of variation in local populations and to partition most of their total genetic diversity within rather than among stands—which permits the use of a decentralized participatory domestication strategy like that implemented in humid West Africa.

Some tree species have separate male and female trees that are indistinguishable until they are sexually mature and start flowering. This creates a problem in the clonal domestication of fruit trees as it is the females that are productive. Likewise, breeding programs need to include plants of both sexes in an optimal sex ratio. The identification and use of sex-specific molecular markers suggest that the sex of young plants of *U. kirkiana* can be differentiated, and that the relevant genes are autosomal (Mwase et al., 2010). This result has important implications for tree domestication of diecious species in the future. As the understanding of genetic variation based on these genomic studies increases, there are likely to be rapid advances in tree domestication, especially in the areas of nutritional quality, seasonality of production and resistance to pests and diseases and to abiotic stresses like drought, salinity, and extreme temperatures.

The Use of New Technologies

The expanding research agenda and number of species being domesticated have led to the increasing use of sophisticated laboratory techniques to quantify genetic variation in the chemical and physical composition of marketable products and commercial partnerships (Leakey, 1999a). These techniques include the assessment of polysaccharide food-thickening agents (Leakey et al., 2005d), proximate analysis (protein, carbohydrate, oils, fiber, vitamins and minerals, etc.), assessment of nutritional and medicinal factors (Leakey et al., 2008), isolation of essential oils (Page et al., 2010a) and fatty acids (Atangana et al., 2011) and determination of wood density, strength, shrinkage, color, calorific value (Sotelo Montes and Weber, 2009: Sotelo Montes et al., 2011) and other important wood properties correlated with tree growth. This is a good example of how agroforestry is increasingly engaging with modern scientific technologies as it matures.

Community Engagement in Germplasm Production

Studies in Latin America (Cornelius et al., 2010), Asia (Carandang et al., 2006; He et al., 2011), and Africa (Dawson et al., 2009) have sought to determine the best forms of management and dissemination of genetic resources, using local and community infrastructure. There is clear potential for improved commercial community engagement in germplasm production. Across Asia, successful national tree seedling supply systems integrate local, institutional (private sector and NGOs) and government nurseries. The latter two types generally provide better access to technology, germplasm and finance, while local nurseries effectively supply a wide variety of species and facilitate tree planting. They also play an important role in developing appropriate technology, providing feedback on farmers' technical needs and knowledge of indigenous species. Unfortunately, central control over a national supply system can constrain the development of local germplasm enterprises (He et al., 2011; Roshetko et al., 2008). Additionally, such enterprises may have an overreliance on external support, a paucity of leadership and limited business capacity (Catacutan et al., 2008). Helping these enterprises gain institutional and market capacity is relevant for both research and government agencies.

In the case of seed, the generalized current practice of selling seed per unit weight rather than based on reproductive potential (e.g., per 1000 plants) in effect discriminates against small seed, prices of which are often orders of magnitude less than prices of large seed (i.e., when expressed per unit of reproductive potential) (Cornelius et al., 2010). Where pricing practices cannot be modified (e.g., in cases where the concept of reproductive potential finds market acceptance), the potential for commercial smallholder production will lie in large-seeded species and also in value-adding through seedling or clone production, ideally allied with development of new cultivars.

Recognition of the Rights of Small-Scale Producers

As already mentioned, the purpose of engaging directly with communities in participatory domestication is to empower them to help themselves. One crucial element is to ensure that the farmers who produce new cultivars are protected from unscrupulous entrepreneurs. Unfortunately, the international negotiations to develop new legal instruments to ensure this have not made adequate progress. To go some way toward proving protection, Lombard and Leakey (2010) have suggested three activities: developing a register of named varieties developed through participatory domestication together with clear ownership and genetic "fingerprints," defining species descriptors based on published data for the purpose of identifying distinctiveness and establishing comparative field trials of selected cultivars and unselected clones to be protected in a small number of safe locations for purposes of quantifying and confirming yield and quality traits.

NEGOTIATION OF ACCESS TO MARKETS

Expanding farmers' market linkages is critical to the success of tree domestication innovations. In many cases, developing linkages and negotiating favorable access to markets—local, domestic, or international—will depend on farmers adapting management regimes that yield reliable quantities of quality products (fruit, vegetables, timber) that meet market specifications. Improving their product quality will likewise strengthen their bargaining position, enabling farmers to move up the value chain increasing their margin. Part of this progression could be collaborating with traders to assume postharvest processing to assure products of the desired quality (Holding-Anyonge and Roshetko, 2003; Tukan et al., 2006). To help communities to secure long-term access to formal markets, PhytoTrade Africa has been involved in setting up these trade associations on behalf of local communities in southern Africa (Lombard and Leakey, 2010).

Adoption and Impact: Toward Enhanced Farmer Livelihoods and Global Environmental Benefits

To date, one tree domestication project has been outstanding in its achievements. Interestingly, the Food for Progress program in west and northwest regions of Cameroon has placed agroforestry tree domestication at the heart of an integrated rural development project, which simultaneously reduces poverty, malnutrition, hunger, and environmental degradation. This has been the catalyst for farmer adoption, and the socioeconomic impacts have been impressive in only 12 years (Asaah et al., 2011). Success has in effect been the outcome of enthusiastic adoption of participatory tree domestication and the dissemination of knowledge and skills to neighboring communities via rural resource centers (Fig. 14.5).

To rebuild the forest resources of useful indigenous trees and their associated traditional knowledge, this program has taken an innovative three-step approach to promoting adoption and impact (Asaah et al., 2011):

FIGURE 14.5 A satellite village nursery in Batibo, Cameroon.

- To mitigate environmental degradation that constrains food production through the use of nitrogen-fixing trees to restore soil fertility and raise crop yields
- To create income generation opportunities through the establishment of village tree nurseries and then through the production of indigenous fruits and nuts in agroforestry systems for local and regional trade
- To encourage local processing and marketing of food crops and tree products in order to create employment and entrepreneurial opportunities for community members.

This project therefore addresses the key socioeconomic and biophysical problems facing smallholder farmers in Cameroon. Its success can be attributed to the relevance of its work to the farmers' needs and interests and the fact that the program builds on traditional knowledge, local culture, local species, and local markets. This initiative has hit the right set of buttons to appeal to farmers and rural communities. Impressively, this process also "snowballs" as each community draws in neighboring communities in a continuous progression of adoption and knowledge dissemination.

Some 30 life-changing positive impacts have been recorded. These range from income generation and better nutrition to the decision of young men to stay in the community rather than to migrate to town, because they can now see a future in the village (Asaah et al., 2011). Overall, therefore, this agroforestry program is creating a pathway to rural development for the alleviation of hunger, malnutrition, and poverty by delivering multifunctional agriculture (Leakey, 2010, 2012b,e). The challenge is to scale this project up from ten thousand farmers to hundreds of millions of rural people, many of whom will have found employment and business opportunities in the rural economy outside farming.

PUBLIC/PRIVATE PARTNERSHIPS: LOCALIZATION AND THE CASE OF *ALLANBLACKIA* SPP.

One very encouraging aspect of the agroforestry initiative for a multifunctional agriculture approach to Third World development is the recent involvement of a small number of multinational companies in the commercial development of AFTPs—especially their recognition of smallholder agroforestry as a better alternative than large-scale plantation monoculture. Some are also engaging in in-country processing rather than exporting raw materials to industrialized countries for product development. One relevant example of this public–private partnership in agroforestry crops is Unilever's initiative to develop a new margarine from the edible oils of *Allanblackia* trees in Tanzania, Ghana, Nigeria, and Cameroon. The kernels of the large fruits of *Allanblackia* trees contain up to 50 nuts that are very rich (70–100%) in stearic and oleic fatty acids (Atangana et al., 2011). The company has committed to developing this new edible oil industry with smallholder communities in Africa (Jamnadass et al., 2010).

TOWARD THE THIRD DECADE

It is clear from the literature review that tree domestication activities are dynamic and expanding both geographically and in species number. The research agenda is also making increasing use of laboratory techniques to improve product quality (Figs. 14.1, 14.2, 14.3 and 14.4). It is also clear that there is an emerging sequence of steps (Fig. 14.2) at present dominated by direct genetic selection and propagation but leading to marketing, commerce and impacts from social and economic reform, steps which will become more dominant as the process gathers momentum.

Looking forward to the next decade, further progress in agroforestry tree development research will probably come from:

- Improving the capture of ontogenetically mature phenotypes by identifying the principles for success in grafting and marcotting.
- Chemical analyses of a wide range of useful ingredients, including essential nutrients, medicinal compounds, perfumes, flavors and other sensory characteristics, found in AFTPs and their selection as traits for cultivar development to meet the needs of new markets.
- Postharvest processing, storage and packaging of AFTPs to expand local, regional and global trade opportunities.
- Controlled pollination between cultivars with good morphological characteristics and those with high nutritive value and/or with out-of-season fruiting. This will require a more centralized approach to domestication research, but wherever possible this should be done in conjunction with rural resource centers working directly with farmers.
- Upscaling tree domestication, especially in Africa, focusing on species with impact on income generation and nutrition.
- Quantification of impact against baseline data based on well-defined criteria and indicators.

- Better understanding of the integration of domesticated agroforestry trees in different cropping systems for improved livelihoods and greater environmental benefits.
- Development of producer-trader linkages and agreements that expand farmer opportunities, promote transparency and reduce inefficiency for mutual benefit.
- Greater involvement of the private sector through public—private partnerships in the local processing and wider trading of AFTPs. The ideotype approach should be used to formulate trait combinations that meet a wider range of commercial markets.
- Enhanced recognition of the importance of AFTPs in agriculture by national and international policymakers and the adoption of appropriate policies.
- Formulation of intellectual property rights that protect the innovative activities of poor farmers and local communities in developing countries.

In conclusion, great progress has been made in the first two decades of agroforestry tree domestication since its conception. It is expected that the third decade will see further expansion in the size of the overall research effort, in the number of species, sites and research topics, as well as in the depth of the studies.

Chapter 15

Trees: Capturing Useful Traits in Elite Cultivars: An Update

R.R.B. Leakey

As recently as 1982, the idea that deforestation could be addressed by promoting the use and cultivation of the wild tree resources that had been traditionally gathered and used/consumed by local people was in its infancy. Indeed, agroforestry as we now know it was a new concept and primarily focused on soil fertility amelioration. At this time very little was known about these species; indeed the scientific literature on their biology and genetic potential was virtually nonexistent. Only Okafor (1978, 1980) had made a concerted attempt to initiate any domestication research. Additionally, the propagation techniques now being widely used for fruit and nut trees had primarily been developed for the clonal propagation of West African timber trees with seed problems (Chapter 11 [Leakey et al., 1982]). Action to domesticate trees producing nontimber products was triggered by the 1992 conference in Edinburgh (Leakey and Newton, 1994b), leading to ICRAF developing a "farmer-driven" and "market-led" tree domestication program (Chapter 5 [Leakey and Simons, 1998]). This recognized the requirement to address the needs of a very wide range of potential candidate species; the needs of informal, small-scale local markets for these products; and especially the needs of millions of potential stakeholders—the farmers. It was clear from this that an alternative to the conventional approach of a centralized research laboratory engaged in on-station tree improvement (based on plant breeding, as used by other CGIAR International Research Centers) was essential. Thus, from the start, a horticultural approach to cultivar/clonal development based on vegetative propagation was adopted. This rapidly built on the low-technology techniques that had been developed and found to be useful for tropical timber trees (Chapters 17, 18 and 19 [Leakey et al., 1990; Leakey 1983; Leakey and Storeton-West, 1992].). This low-technology approach was particularly appropriate for agroforestry as the stakeholders are widely distributed across remote locations with poor infrastructure (often lacking electricity and piped water), with poor communications and transport. Furthermore, the stakeholders lacked formal education and skills, but conversely they were blessed with centuries of traditional knowledge about their local species. This traditional knowledge was vital as it provided evidence of the important uses of the species.

The domestication initiative was initially hindered by the fact that there was a serious lack of information about the diversity of the genetic resources and the availability of germplasm of the candidate species, making it difficult to determine the opportunity for their genetic selection, improvement and domestication. However, as just mentioned, the absence of scientific knowledge was offset by the availability of traditional knowledge. This unique combination further emphasized the need for a novel decentralized and appropriate farmer-driven approach to agroforestry tree domestication. This is described further in Chapters 27 and 28 (Leakey et al., 2003; Leakey, 2014).

A key challenge for the domestication of agroforestry trees for fruits and nuts centers around the longevity of trees and their outbreeding reproductive system (Leakey, 1991; Leakey and Simons, 2000). Typically trees do not become mature and capable of sexual reproduction until 5—15 years old; consequently this is the minimum time between different generations. This makes tree breeding a slow process. This problem is then exacerbated by the fact that in outbreeding species, the heritability of individual traits is low because genetic material from both parents is segregated during meiosis, so the progeny are genetically heterogeneous and do not necessarily carry the desired trait. As a result, genetic gains are generally small. However, through the selection of elite parents, and the use of controlled pollination, gains from breeding can be improved, but only to a limited extent (Fig. 15.1). In contrast, by vegetative propagation (Chapter 17 [Leakey et al., 1990]), the new plants form a clone or clonal cultivar, and are an exact copy of the selected plant with identical traits (Fig. 15.1) and can capture the sexual maturity of the tree crown (Fig. 28.1). As a result, the level

Multifunctional Agriculture. DOI: http://dx.doi.org/10.1016/B978-0-12-805356-0.00015-5
© 2017 Elsevier Inc. All rights reserved.

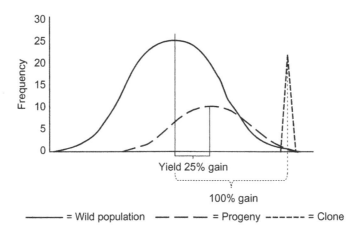

FIGURE 15.1 Diagrammatic representation of genetic gain from sexual reproduction and vegetative propagation. *Modified from Leakey, R.R.B., 1991. Clonal forestry: towards a strategy. Some guidelines based on experience with tropical trees. In: Jackson, J.E., (Eds.), Tree Breeding and Improvement. Royal Forestry Society of England, Wales and Northern Ireland, Tring, England, pp. 27–42.* dashes = sexual reproduction; dots = vegetative propagation.

of gain depends on the identification of rare but elite individuals for asexual propagation. In this way, elite plants can be replicated in large numbers. While this is a huge advantage for producers, the misuse of vegetative propagation can be a threat as the cultivation of only a very few clones would result in a population with an extremely narrow genetic base. In Chapters 16 and 26 (Leakey and Akinnifesi, 2008; Leakey 2017e), we will see how a wise domestication strategy can avoid this potential problem, as well as learning when the use of vegetative propagation is appropriate (see also Leakey and Simons, 2000). This falls within the context of how to manage tree genetic resources for sustainable impacts on production, as well as for positive impacts on various economic, social and environmental outcomes and trade-offs (Dawson et al., 2014). It is clear, however, that in the short-term participatory domestication is starting to deliver tangible and life-transforming social and economic benefits on a small scale (500 villages) in less than 10 years (Chapters 27–30 [Leakey et al., 2003; Leakey, 2014e; Lombard and Leakey, 2010 and Leakey and van Damme, 2014]).

An alternative to this approach has gained popularity in recent years due to advances in genetic science to understand inheritance and the structure of the genome. These have made genetic manipulation possible, opening up opportunities for novel trait combinations to meet a wide range of different needs (Clark and Pazdernik, 2016). These biotechnological advances are really, however, the domain of large multinational companies and universities as they require highly trained specialist scientists and capital-intensive laboratory equipment. Later (Chapter 34 [Leakey, 2013]) we will see that the main problems facing subsistence smallholders in Africa are related to social and environmental issues that prevent existing staple cereal/root crop varieties from performing to their full potential. These nonbiological issues can be simply and cheaply addressed by agroforestry, but not by the use of genetic manipulation.

Modern molecular techniques do, however, have an important role to play in agroforestry and are being used to characterize and explain the genetic structure of variation in a number of traditionally important indigenous food trees as reported previously (Chapter 14 [Leakey et al 2004]; Jamnadass et al., 2009; Fig. 26.1). This is important information for scientists leading both participatory domestication and conventional tree breeding programs. One very important finding from the use of molecular marker technologies on extracted DNA has been that local populations of the wild species being domesticated for the production of agroforestry tree products are genetically very diverse. It is especially interesting to find that in *Barringtonia procera* an indigenous nut of the South Pacific, about 80% of the morphological variation is found in every village population (Pauku et al., 2010). This endorses the strategic benefit of implementing a decentralized program of participatory domestication as a way to maintain intraspecific genetic diversity within a domesticated population (Fig. 26.1). This confirms the findings of studies of tree-to-tree fruit morphological variation in different villages (Fig. 26.2)—such as in *Adansonia digitata* (de Smedt et al., 2011) and many other species (Chapters 22–24 [Leakey et al., 2005a,b; Leakey, 2005]).

Regarding genomics, the latest equipment and techniques are being introduced and employed in agroforestry by The African Orphan Crops Consortium (AOCC) with the objective of genetically sequencing, assembling and annotating the genomes of 100 traditional African food crops. This initiative is aimed at improving the nutritional content, productivity and climatic adaptability of these traditionally important tree species. However, it is not yet clear how the "costs and benefits" from this centralized, high-technology approach will compare in the long-term with the low-technology, decentralized techniques and strategies used in participatory domestication (examined in Chapters 17, 21, 28, and 32 [Leakey et al., 1990; 2000; Leakey, 2014, 2017f]).

So, what is now emerging in agroforestry is an integrated approach to the use of forest genetic resources for the reduction of deforestation/conservation and rebuilding of forests, which works along the full spectrum of genetic resource management. As we have seen, at one end it has a highly effective and tested, decentralized approach to indigenous tree domestication for the relief of food and nutritional insecurity, poverty, social injustice, environmental degradation, and climate change. At the other end of the spectrum, we have studies of genomics in the same range of indigenous species, aimed again at improving the lot of poor smallholder farmers. The focus of the latter approach on nutritional content of the products, could be extended to medicinal qualities, and probably other traits that are very difficult, or even impossible, to improve without laboratory work. Thus these are traits which are much more easily assessed and characterized in a centralized, high capital cost, high operating cost laboratory. Such outputs would be implemented in marker-assisted tree breeding programs. Tree breeding in long-lived perennials, almost by definition, is very slow because trees don't reach sexual maturity for many years. Given the high inherent nutritional value of the traditional food species, it is questionable whether complex, long-term and expensive research to enhance these properties is a high priority in the effort to mitigate malnutrition.

It is already clear that the spread of tree domestication research in terms of new species, new locations and new topics in the second decade (Chapter 14 [Leakey et al., 2012]) has continued into the third decade, heralding a "new wave" of crop domestication (Leakey, 2012c). Previous "waves" have been focused on production. This one, as acknowledged by Gepts (2014), also has social, economic, and environmental benefits as it is based on indigenous, locally adapted, perennial plants that can mitigate environmental degradation, create income and employment opportunities from traditionally and culturally important species. The diversification of domesticated species is also important to ensuring that agriculture satisfies the needs of humanity globally. Interestingly, the domestication of these tree species also has a key role in the "carbonization" of agriculture and hence in the mitigation of climate change (Gepts, 2014)—a role enhanced by the finding that vegetatively propagated mature fruit trees of *Dacryodes edulis* sequester more carbon than seedlings (Asaah et al., 2012).

Domestication alone seldom delivers substantial livelihood impacts, especially the social and economic impacts which are so badly needed to improve the lives of both rural and urban poor in Africa, which depend on the products being widely traded. Likewise, the commercialization of useful and desirable products is also not very effective on its own as markets require quality, uniformity and reliability of supply. Thus, it is important that domestication and commercialization go hand-in-hand, as we saw in Chapter 7 (Leakey, 1999a) with domestication scientists involving industrial partners to help to shape the genetic selection program. We will return to this theme in Chapter 25 (Leakey and Page, 2006). This synergy between domestication and commercialization is further examined in Chapter 30 (Leakey and van Damme, 2014) with regard to the need to understand and regulate a "value-chain" (Fig. 15.2) if the benefits of either process are to be maximized. This maximization of benefits is also important to create the incentive to adopt the technologies, as well as to drive policy change vis-à-vis the mitigation of food and nutritional insecurity, poverty or climate change (Chapters 35, 38 and 39 [Leakey, 2017g,h,i; Leakey and Prabhu, 2017]).

Evidence from Cameroon has shown that the tree-to-tree variation in traits such as size, color, and flavor of wild *D. edulis* fruits are recognized by small-scale retail market traders when setting the price for small samples of fruits from individual trees (Range = from 3 fruits to 1010 fruits for 250 Central African francs [FCFA]; Figs. 5.2 and 28.9). In contrast, wholesalers who collect and sell fruits in truck loads, made up of a mixture of fruits from many different trees, do not recognize this variation paying farm-gate prices equivalent to about 25 fruits for 250 FCA for bulk samples (Leakey et al., 2002). Consequently, it is expected that farmers able to sell bulk samples of a named superior cultivar should receive much higher farm-gate prices. Interestingly, the prices of fruits from the rare trees that fruit outside the normal season are up to 10-fold greater. So, the development of cultivars that extend the fruiting season is important. However, while these traits can be captured and harnessed as vegetatively propagated cultivars (see Chapters 15 and 26 [Leakey, 2017d,e]), it is perhaps unlikely that these traits will coexist with other desirable morphological, organoleptic/sensory or biochemical traits. In this event, it may be necessary to undertake a laboratory-based breeding program between clones selected for one or other trait as a second phase of domestication, in the hope that some individuals in the progeny will contain the desired combinations.

To enhance the efforts to domesticate and market agroforestry tree species aimed at economic and social benefits for poor smallholder farmers in the tropics and subtropics, Leakey and Tomich (1999) posed six research questions regarding likely impacts *vis à vis* international policies, market expansion, land use issues and technical constraints. As we will see later, some progress toward finding answers has been made.

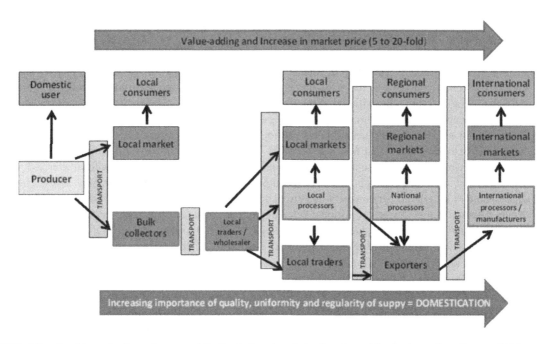

FIGURE 15.2 The role of domestication and commercialization in the value chain of products. *After Leakey and van Damme, 2014.*

Gepts (2014) has also pointed out that the conscious neodomestication of wild species also raises two questions: (1) why have so few species been domesticated so far? and (2) how many more species could be usefully domesticated to complete the transition from hunter-gathering to sustainable agriculture throughout the world? We explore the latter of these two questions in Chapters 6, 10, 15, 26, 32, 35 and 38-40 (Leakey, 2017b,c,d,e,f,g,i; Leakey and Prabhu, 2017).

Part III

Research Methods

Section 5

Strategy and Techniques

This section provides essential information for the implementation of tree domestication. It starts with the presentation of wise strategies for the management of the genetic resource of a species being domesticated by clonal or tree breeding. This includes strategies for the management of the wild resource and gene banks for germplasm conservation; strategies for the use of vegetative propagation; and strategies for clonal selection. These then lead to the wise management of the selection population and the production population.

This is followed by detailed information on the techniques of vegetative propagation using leafy single-node stem cuttings from juvenile stockplants in low-technology, nonmist propagators that don't require electricity or running water, as well as by the marcotting (air layering) of shoots from the crowns of mature trees. Of special importance for maximization of rooting success is the preseverance management of the stockplant's environment, particularly the interactions between light (quantity and quality) and nutrition.

Finally, studies to characterize, quantify, and understand the extent of the intraspecific (tree-to-tree) variation in numerous different morphological and other traits at the village scale are presented. This information facilitates the selection of elite trees for cultivar development. It also informs the selection process about the opportunities to meet the specific needs of different markets and potential new industries in the form of an "ideotype" that presents the ideal combination of multiple traits.

These techniques are now being used in increasingly diverse sets of species for food and nonfood products with potential as new crops and new industries.

Section 5.1

Strategy

Chapter 16

Towards a Domestication Strategy for Indigenous Fruit Trees in the Tropics

This chapter was previously published in Leakey, R.R.B., Akinnifesi, F.K., 2008. In: Akinnifesi, F.K., Leakey, R.R.B., Ajayi, O.C., Sileshi, G., Tchoundjeu, Z., Matakala, P., Kwesiga, F. (Eds.) Indigenous Fruit Trees in the Tropics: Domestication, Utilization and Commercialization. CAB International, Wallingford, UK, pp. 28–49, with permission from CABI

INTRODUCTION

Increasingly, agroforestry trees are being improved in quality and productivity through the processes of market-driven domestication (Simons, 1996; Simons and Leakey, 2004; Leakey et al., 2006), based on strategies that consider: (1) the needs of the farmers, their priorities for domestication (Maghembe et al., 1998; Franzel et al., 2008) and an inventory of the natural resource (Shackleton et al., 2003a); (2) the sustainable production of agroforestry tree products, including fruits, nuts, medicinals and nutriceuticals, timber, etc.; (3) the restoration of degraded land and reduction of deforestation; and (4) the wise use and conservation of genetic resources. These approaches to tree domestication are being implemented in southern Africa (Akinnifesi et al., 2006).

There are two main pathways within a domestication strategy (Fig. 16.1). Domestication can be implemented on-farm by the farmers (Phase 1), who bring the trees into cultivation themselves (Leakey et al., 2004), or through programs of genetic improvement on research stations (Leakey and Simons, 1998). In recent years, however, scientific approaches have also been introduced into on-farm domestication through the application of participatory approaches to tree improvement (Phase 2). In this approach, researchers typically act as mentors, helping and advising the farmers, and sometimes jointly implementing on-farm research. Participatory approaches have numerous advantages (Leakey et al., 2003), building on tradition and culture and promoting rapid adoption by growers to enhance livelihood and environmental benefits (Simons and Leakey, 2004). Both these pathways to domestication should be targeted at meeting market opportunities, which should examine traditional as well as emerging markets (Shackleton et al., 2003a).

In practice, agroforestry plantings are often constrained by the lack of genetically superior seed sources, the traditional source of planting stock (Simons, 1996). Consequently, one of the first decisions in developing a domestication strategy for a particular species has to be whether to use seed and reproductive processes or vegetative propagation to achieve genetic improvements. Foresters have generally adopted seed-based tree breeding approaches, while horticulturalists have adopted clonal vegetative propagation and the development of cultivars. The following economic and biological situations have been identified as favoring a clonal approach (Leakey and Simons, 2000; Akinnifesi et al., 2006):

- The occurrence of individual trees in a wild population, which have a rare combination of traits such as large fruit size, sweetness, precocity, early fruiting, delayed or extended fruiting season, and desirable kernel characteristics.
- The need to combine many desirable traits for simultaneous selection and improvement.
- A requirement for high product uniformity to ensure profitability and to meet market specifications. This contrasts with the genetic heterogeneity which is a characteristic of seedling progenies of outbreeding trees.
- The products are highly valuable and can thus justify the extra expense and care required to ensure quality and productivity, especially when the risks of market saturation are minimal.

Multifunctional Agriculture. DOI: http://dx.doi.org/10.1016/B978-0-12-805356-0.00016-7
© 2017 Elsevier Inc. All rights reserved.

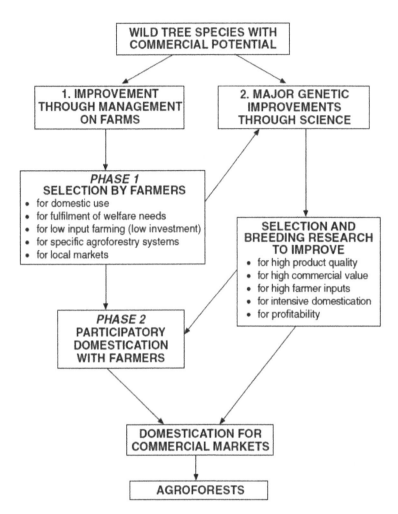

FIGURE 16.1   Two pathways for the domestication of agroforestry tree products. *Modified from Leakey et al., 1996.*

- The species to be propagated is a shy seeder (i.e., does not flower and fruit every year or produces only a very small seed crop).
- The propagation material is limited, as for example in: (1) the progeny of a specific controlled pollination made in a breeding program; (2) the products of a biotechnological manipulation such as the transmission of genetic material; or (3) the result of hybridization where segregation would occur in the F2 generation and beyond (hybrid progenies are also often sterile, so further propagation has to be done vegetatively).
- The timescale in which results are required is insufficient to allow progress through the slower process of breeding. This is particularly relevant in fruit trees with a long juvenile period prior to the attainment of identifiable superiority and sexual maturity, as vegetative propagation can be applied to shoots from trees that have already shown their superiority and that have already acquired sexual maturity. Such shoots will retain both this superiority and their maturity when propagated vegetatively.
- The seeds of the chosen species have a short period of viability (i.e., they are recalcitrant) or very low viability, and hence cannot be stored for later use.
- Knowledge of proven traits is acquired through either the indigenous knowledge of farmers or a long-term experiment. This situation is plagued by the problems of propagating mature tissues vegetatively. Currently, the usual procedure is to use grafting/budding if the mature traits are required (i.e., fruiting ability) and to coppice the tree if rejuvenation is required.
- A participatory tree domestication program is planned. This is because farmers do not generally have the time or genetic knowledge to implement a breeding program.

The antithesis of these situations is that sexual propagation is preferable when the requirements are for large quantities of genetically diverse, low-value plants with unlimited seed supplies.

DEVELOPING A STRATEGY FOR CREATING NEW "CULTIVARS" VEGETATIVELY

The increased interest in vegetative propagation has arisen from the desire to rapidly acquire higher yields, early fruiting, and better quality fruit products in agroforestry trees. The development of cultivars through cloning also results in the uniformity of the products, as all the trees of a given clone are genetically identical. This is beneficial in meeting the market demand for uniform products. A range of vegetative propagation techniques can be utilized to achieve this (Leakey, 1985; Hartmann et al., 2002), including grafting, stem cuttings, hardwood cuttings, marcotting (air layering), suckering, and in vitro techniques, such as meristem proliferation, organogenesis, and somatic embryogenesis (Mng'omba et al., 2007a, 2007b).

The decision to pursue the clonal opportunities offered by vegetative propagation necessitates the formulation of a strategy, as there are a number of factors associated with the process of cloning that need to be considered. These factors relate to: (1) the methods of propagation; (2) the level of technology that is appropriate; and (3) the effects of using juvenile or mature tissues (Leakey, 1991; Leakey and Simons, 2000).

Methods of Propagation and the Cloning Process

The principal reason for cloning is to take advantage of its ability to capture and fix desirable traits, or combinations of traits, found in individual trees. By taking a cutting or grafting a scion onto a rootstock, the new plant that is formed has an exact copy of the genetic code of the plant from which the tissue was taken. In contrast, following sexual reproduction seedlings are genetically heterogeneous, each seed having inherited different parts of the genetic codes of its parent trees, with segregation of genes among the progeny. Vegetative propagation is thus both a means of capturing and utilizing genetic variation and of producing cultivars to increase productivity and quality (Mudge and Brennan, 1999; Leakey, 2004). Vegetative propagation results in the formation of clones (or cultivars), each of which retains the genetic traits of the original tree from which cuttings or scions were collected. Both on-farm and on-station approaches to tree domestication can involve vegetative propagation and clonal selection, but with a growing interest in the participatory domestication of agroforestry trees (Leakey et al., 2003; Tchoundjeu et al., 2006). There is now great interest in vegetative propagation using stem cuttings (Leakey et al., 1990; Shiembo et al., 1996a, 1996b, 1997; Mialoundama et al., 2002; Tchoundjeu et al., 2002) and grafting for *Uapaca kirkiana* and *Strychnos cocculoides* (Akinnifesi et al., 2006), and *Sclerocarya birrea* (Holtzhausen et al., 1990; Taylor et al., 1996). Vegetative propagation gives the tree improver the ability to multiply, test, select from, and utilize the large genetic diversity present in most tree species. It should be noted that, contrary to some misguided opinion, vegetative propagation does not in itself generate genetically improved material. Only when some form of genetic selection is employed in tandem with propagation will improvement result.

To capture the first asexual propagule from proven and mature field trees that have already expressed their genetic traits, it is necessary to use either grafting/budding or air layering techniques. Alternatively, coppicing can be used to produce juvenile material. The latter is preferable for clonal timber production, whereas the former is more suitable for fruit trees (Leakey, 1991). Grafting produces many more individuals with less effort than air layering, although many more individuals can be produced from stem cuttings if the tree resprouts copiously after coppicing. Once clonal stocks have been obtained, the resulting shoots can either be used to provide rooted cuttings for transplanting to the field, additional material for clonal seed orchards, or scions for grafting onto suitable rootstocks. Evidence from simple tests suggests that the majority (probably more than 90%) of tropical trees are amenable to propagation by juvenile stem cuttings (Leakey et al., 1990). In southern Africa, some high-priority indigenous fruit trees are propagated by grafting (e.g., *U. kirkiana, St. cocculoides, Sc. birrea, Adansonia digitata*, and *Vitex mombassae*, while others (e.g., *Sc. birrea*) are easily propagated by large, leafless, hardwood cuttings, or stakes/truncheons. *Parinari curatellifolia* is easily propagated by root cuttings.

Appropriate Technology

Low-cost nonmist propagators have been developed for the rooting of leafy stem cuttings. These do not require electricity or running water and are extremely effective, meeting the needs of most tree improvement projects in developing countries in both the moist and dry tropics (Leakey et al., 1990). In technologically advanced countries, mist or fogging systems are available for the rooting of cuttings. Alternatively, where laboratory facilities are available in vitro culture techniques can be used, but these require highly trained staff and regular power supplies, and are capital-intensive. In many cases, in vitro techniques have resulted from empirical testing of different media and plant growth regulators. As a result, inadequate knowledge about the long-term effects of treatments on the field performance has caused some major failures of field performance. In the same way, the vegetative propagation of mature tissues by marcotting, or air layering, requires a lower level of skill than grafting and budding. The low-technology options are especially appropriate if the participatory approach to domestication is the preferred strategy. In this situation, farmers vegetatively propagate their best trees to create selected cultivars.

Choice Between Mature and Juvenile Tissues

An advantage often ascribed to propagation from mature tissues is that by the time the tree is mature, it has demonstrated whether or not it has superior qualities. However, it is not always easy to take advantage of this proven superiority, as propagation by mature stem cuttings is notoriously difficult. This is in contrast with the rooting of juvenile tissues, which is typically easy. Another important advantage of propagating mature tissues is that they are already capable of reproductive processes, and so will flower and fruit within a few years, reducing the time before economic returns start to flow. Plants propagated from mature tissues will also have a lower stature. On the other hand, timber production requires the vigor and form associated with juvenile trees, making propagation by cuttings attractive and appropriate. For timber trees, propagation from mature trees is generally limited to the establishment of clonal seed orchards within breeding programs.

The Use of Juvenile Tissues

Seedlings, coppice shoots, and root suckers are the sources of juvenile tissues. For tree domestication purposes, coppice shoots from the stumps of felled trees have the advantage that it is possible to propagate trees that have already proven to be superior, as it is possible to determine the phenotypic quality of the tree prior to felling. This is highly beneficial in the domestication of trees producing all kinds of agroforestry tree products, but has the added advantage in diecious fruit-tree species that it allows cultivars to be restricted to trees. Nevertheless, there are three reasons why the use of seedlings may still be preferred over coppicing from trees of known phenotype (Leakey and Simons, 2000):

- The population of mature timber trees may be dysgenic because the elite specimens may have been removed by loggers, which means that seedling populations have a better array of genetic variation.
- The felling of large numbers of mature trees for the purpose of generating cultivars may not be acceptable to the owners. In addition, felling the mature trees may be environmentally damaging.
- The use of seedlings allows the screening of far larger populations, with much more diverse origins, maintaining genetic diversity among the cultivars.

Whether using seedling or coppice stumps as stockplants, it is important to ensure that they are managed for sustained, cost-effective, and easy rooting. The way in which stockplants are managed is probably one of the most important determinants of the long-term success of a cloning program. Good rooting ability is maintained by encouraging vigorous orthotropic growth of shoots from regularly pruned stockplants. This requires a much greater level of knowledge than is available for most, if not all, tree species. Good progress has been made in starting to unravel the sources of variation in rooting ability (Leakey, 2004). For example, it has become clear that cuttings taken from different parts of the same shoot differ in their capacity to form roots (Leakey, 1983; Leakey and Mohammed, 1985; Leakey and Coutts, 1989) and that this is influenced by cutting length (or volume). In addition, there are influences on rooting ability that originate from factors between different shoots on the same plant (Leakey, 1983). These factors are also affected by shading, which determines both the amount and the quality of light received by lower shoots. Both the quality and the quantity of light independently affect the physiology, morphology, and rooting ability of cuttings from differently illuminated shoots (Leakey and Storeton-West, 1992; Hoad and Leakey, 1996). To further complicate this situation, the nutrient status of the stockplant interacts with the effects of light and shading (Leakey, 1983; Leakey and Storeton-West, 1992), with shading and a high level of nutrients combining to enhance rooting. Light, especially light quality, also affects the relative size and dominance of different shoots on managed stockplants (Hoad and Leakey, 1994). There are also interactions between stockplant factors and the propagator environment.

The complexity of the stockplant factors affecting rooting ability means that for the production of large numbers of cuttings from stockplants it is important to develop a good understanding of these factors. This level of knowledge is not common for tropical and subtropical tree species. However, from work on a few tropical tree species it does appear that some generalizations are possible (Leakey et al., 1994; Leakey, 2004) and that through a modeling approach (Dick and Dewar, 1992) it is possible to predict the best management options for new species. It is, however, usually necessary to start out by propagating a larger number of clones than is needed, as some will be lost while going through the various rooting and multiplication cycles.

From this discussion, it is clear that, in addition to economic considerations, in formulating a strategy for clonal forestry it is advisable to consider which forms of propagation have the lowest risk of failure. It seems that the low-tech system of rooting stem cuttings is the most robust.

The Use of Mature Tissues

As trees grow they develop a gradient toward maturity (ontogenetic aging) and after a time reach a threshold above which the newly developing shoots have the capacity to fruit and flower, while those below the threshold are still juvenile. The transition from the juvenile to mature state is called a "phase change," and the coppicing of mature trees is generally regarded as the best way to return to the juvenile state. Because of the difficulty in rooting cuttings from mature tissues, the most commonly used vegetative propagation methods for mature trees of horticultural and cash crops are grafting and budding techniques (Hartmann et al., 2002). Using these techniques requires skill, as the close juxtaposition of the cambium in the scion and rootstock is necessary if callus growth is to heal the wound and reconnect the vascular tissues. Failures also result from dehydration of the tissues.

These techniques can, however, result in severe and often delayed problems because of incompatibility between the tissues of the rootstock and scion, in which graft unions are rejected and broken, sometimes after 5–10 years of growth. This is a form of tissue rejection, and it is less common between closely related tissues (Jeffree and Yeoman, 1983). Mng'omba et al. (2007a) have recently shown that graft incompatibility is caused by the presence of p-coumaric acids, and that greater incompatibility can be expected for heterospecific than for homospecific scion/stock combinations. Another common problem with grafting and budding is that shoots can develop and grow from the rootstock. If not carefully managed, growth from the rootstock dominates that of the scion, which dies, resulting in the replacement of the selected mature cultivar with an unselected juvenile plant.

Once mature tissues are successfully established as rooted propagules, be they marcots, cuttings or grafts, they can be used as stockplants for subsequently harvested cuttings. With good stockplant management, good rooting treatments and an appropriate rooting environment, mature cuttings from these stockplants can usually be rooted easily. Nevertheless, it is clear that the need to propagate from mature tissues does pose a severe constraint and challenge to domestication strategies, especially of fruit trees.

In vitro culture techniques hold some promise of circumventing the problems of maturation, as some rejuvenation of in vitro cultures has been reported (e.g., Amin and Jaiswal, 1993), but the mechanism remains unclear. In vitro micrografting has also been used to rejuvenate shoots (Ewald and Kretzschmar, 1996).

DEVELOPING A STRATEGY FOR CLONAL SELECTION

Cultivar development relies on three processes: selection, testing, and breeding. Selection identifies certain genotypes for cultivar development; testing exposes the new cultivars to appropriate environments; and breeding creates new genetic variability. Within the overall domestication strategy, three interlinked populations are conceptualized: the gene resource population, the selection population, and the production population (Fig. 16.2). The gene resource population is often the wild or unimproved population from which new selections can be derived. The selection population is the somewhat improved population of genotypes that are being tested and that are used in subsequent breeding programs to create the next generation of potential cultivars. A wide range of genotypes may be kept in this population as long as each has at least one characteristic of possible future interest. The production population consists of the highly selected genotypes, which are used for planting.

As mentioned earlier, there are two basic approaches to the genetic improvement of trees: the seed-based breeding approach typical of forestry and the clonal approach typical of horticulture. The seed approach typically involves the selection of populations (provenance testing) and/or families (progeny testing) (Zobel and Talbert, 1984; Leakey, 1991). While this approach could be taken for indigenous fruit trees, it is very likely that an examination of the ten situations outlined in the Introduction of this chapter would indicate that a clonal approach is more appropriate. There are basically two ways to select the best individual trees for cloning from broad and diverse wild populations: (1) selection from a pool of seedlings of virtually unknown quality in a nursery or field trial (although it may be known that the

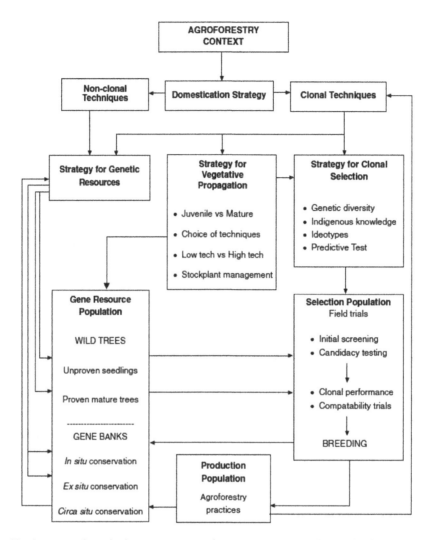

FIGURE 16.2 Relationships between a domestication strategy, a genetic resources strategy and strategies for vegetative propagation and clonal selection.

pool originates from a good provenance or progeny) and (2) selection of proven mature trees in wild or planted populations (Fig. 16.2). In scenario (1), genetic improvement in yield per hectare will undoubtedly require a series of tests, each spanning many years. Typically, there are four levels of testing (Foster and Bertolucci, 1994).

1. Initial screening with large numbers, preferably tens of thousands, of seedlings or, if seedlings have already been cloned, a few ramets per clone.
2. Candidacy testing with large numbers of cloned genotypes fewer than with initial screening (preferably hundreds or thousands) and two to six ramets per clone.
3. Clonal performance trials with moderate numbers of clones (e.g., fewer than 200) and large numbers (e.g., 0.1 ha plots) of ramets per clone.
4. Compatibility trials with small numbers of clones (e.g., 20−50) with very large plot sizes.

It is important to recognize that there is a trade-off between the accuracy of genetic value estimation and the intensity of selection (i.e., greater accuracy is at the expense of numbers of families, individuals per family, or clones). For cost-effective clonal tree improvement programs with limited or fixed resources, it has been found that the best strategy is to plant as many clones as possible with relatively few ramets per clone.

To short-circuit the lengthy process of field trials, Ladipo et al. (1991a,b) developed a predictive test for timber tree seedlings in which the initial screening is done on young seedlings in the nursery; it is then possible to jump straight into clonal or compatibility trials with some confidence. To date there is no similar opportunity for fruit trees.

Like the predictive test, scenario (1) is an alternative and much quicker option. In this case, mature trees, which have already expressed their genetic potential at a particular site over many years of growth, are selected and propagated vegetatively and the propagules are planted either in clonal performance trials or directly into compatibility trials. This raises the question of how the superior mature trees should be identified, especially if it is desirable to select for multiple traits. This can create a problem, as many traits may be weakly or negatively correlated (e.g., fruit size and kernel size in *S. birrea*; Leakey, 2005c). Consequently, as the number of desirable traits increases, the number of genotypes superior for all traits diminishes rapidly. Thus, the selection intensity (and also the number of trees screened) must be substantially increased, or the expected genetic gain will rapidly decline. For this reason, only the few most economically important traits (e.g., fruit flesh or nut mass, taste) should be concentrated upon in the early phases of selection.

Two techniques can be used to assist in the identification of superior mature trees (sometimes called "elite" or "plus" trees) producing indigenous fruits and nuts. The first is to involve indigenous people in the domestication process and to seek their local knowledge about which trees produce the best products. Local people usually have good knowledge about the whereabouts of elite trees, and this knowledge often extends to superiority in a number of different traits, such as size, flavor, and seasonality of production. However, access to this knowledge has to be earned by the development of trust between the holder of the knowledge and the potential recipient. Ideally, the recipient should enter into an agreement that the intellectual property rights of the holder will be formally (and legally) recognized if a cultivar is developed from the selected tree. Unfortunately, at present the process of legally recognizing such cultivars is not well developed and requires considerable improvement.

The second technique for identifying mature elite fruit and nut trees has recently been extended to a study on marula (Leakey, 2005) following its development in Cameroon and Nigeria (Atangana et al., 2002; Leakey et al., 2002, 2005b,c). This technique involves the quantitative characterization of many traits of fruits and kernels, which are associated with size, flavor, nutritional value, etc. This characterization also determines the extent of the tree-to-tree variation, which is typically three- to sevenfold, as found in marula (Leakey et al., 2005b,c), as well as the frequency distribution, which is typically normal in wild populations but tends to become skewed in populations subjected to some selection. The characterization data can then be used to identify the best combination of traits (the "ideotype") to meet a particular market opportunity (for an example see Fig. 16.3). The development of single-purpose ideotypes (Leakey and Page, 2006) provides a tool for the development of cultivars with different levels of market focus and sophistication

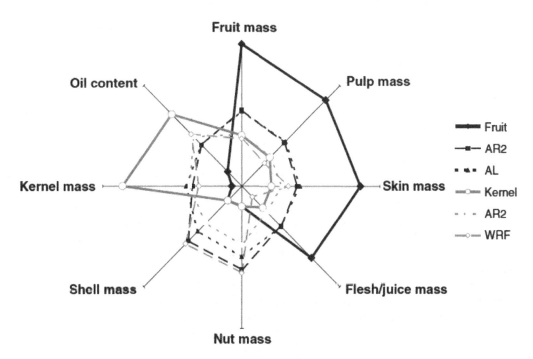

FIGURE 16.3 Fruit and kernel ideotypes for marula (*Sclerocarya birrea*) in South Africa, with the best-fit trees. *After Leakey, R.R.B., 2005. Domestication potential of marula (Sclerocarya birrea subsp. caffra) in South Africa and Namibia. 3. Multi-trait selection. Agrofor. Syst. 64, 51–59.*

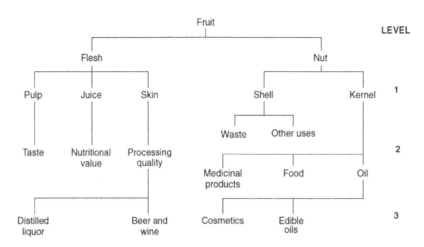

FIGURE 16.4 The development of single-purpose ideotypes provides a tool for the development of cultivars with different levels of market focus and sophistication.

(Fig. 16.4). In addition to advancing the selection process for multiple traits, the ideotypes also provide information about opportunities to select for better partitioning of dry matter between desirable and undesirable traits. For example, in marula 90% of the dry matter in nuts is typically found in the shell and only 10% in the valuable kernel, although the range of variation is 3–16%. The shell:kernel ratio is therefore a trait that could usefully be included in the ideotype. However, the inclusion of each additional trait in a multitrait selection process greatly increases the number of trees that need to be screened, especially if the traits are independent or only weakly related.

It is clear from this discussion that in the early phases of domestication there is sufficient genetic variation in most tropical tree populations to allow considerable progress in the development of cultivars, but a strategy for clonal agroforestry should not forgo any opportunity for creating new variation.

The selection of clones is not a once-and-for-all event. Domestication is a continuous process, which in wheat, rice, maize, oranges, and apples, for example, started thousands of years ago and continues today. Thus, a series of clonal selection trials should be established to seek the best individuals from new accessions of genetically diverse populations or progenies (Fig. 16.2). It is also important to discard old clones as they are superseded, although some of these should be retained in the gene resource population. In the first instance, clones may be selected for yield and quality. With time, this can be extended to include nutritional quality, disease/pest resistance, component products (oils, flavorings, thickening agents, etc.). This continued turnover of the selected clonal population will further ensure the diversity of the commercially planted clones and prevent excessive narrowing of the genetic base. Indeed, it can be argued that in this way clonal plantations of 30–50 superior but unrelated clones can be more diverse than seedlings. This is because seed-lots typically originate from a number of related mother trees and share some genetic material.

Because domestication is a continuous process, commercial plantings have to be made with whatever material is best at a given time, knowing that they will be superseded later. Having a succession of increasingly good planting stock is one of the ways in which the diversity of the genetic base can be maintained, although of course this has to be rigorously enforced as one of the objectives of a breeding and selection program. For species with existing provenance selection and breeding programs, clones should be derived from seed collections sampling a wide range of the known variation, as it is not uncommon for a few elite trees to be found with poor provenances.

As the selection process intensifies with time, new traits will be introduced into the program (e.g., seasonality of production, early fruiting, disease and/or pest resistance, and drought tolerance). Capturing variation in the seasonality of production and expanding the harvesting season are likely to be among the best ways of supporting market growth. As the price of end-of-season fruits is likely to be better than the midseason price, it is also a good way to enhance the benefits of producing households that need sources of income throughout the year. The further the domestication process proceeds, the more important it becomes that the combination of traits being selected is targeted at a particular market (Fig. 16.2). This again is where the ideotype concept can be useful, and ideally advice should be sought from industrial partners who are aware of which characteristics are important in the marketplace or in product processing (Leakey, 1999a).

Using the example of *S. birrea*, fruit-producing cultivars could be developed which have large fruit flesh/juice mass (Leakey et al., 2005b) and which are nutritious (Thiong'o et al., 2002) as raw fruits, or good for traditional beer- or wine-making, or which meet the needs of the distilling industry. Likewise, other cultivars could be developed for the size and quality of the kernel, with a low shell:kernel ratio and with either nutritious or medicinal qualities for eating, or with oil yield and quality traits of importance in the cosmetics industry (Leakey, 2005; Leakey et al., 2005c). Similarly, superior phenotypes of *U. kirkiana* with heavy fruit loads, large fruits, and high pulp content have been identified by communities during participatory selection in Malawi, Zambia, and Zimbabwe (Akinnifesi et al., 2006). Thus, it is clear that in the domestication of multipurpose tree species, highly productive, single-purpose clones, or cultivars are probably the best option. To maximize the market recognition of, and farmer interest in, cultivars, it is a good idea to name them. In Namibia every marula tree already has a name, so it is easy to give a name to a cultivar derived from any particular tree. The name can also be used to recognize the person or community holding the rights to the cultivar.

OPPORTUNITIES FOR INTRODUCING NEW VARIATION

A clonal selection program seeks to utilize as much of the existing variation as possible within wild populations, or within progenies from breeding programs. However, after the initial phases of selection, the opportunity will arise for controlled pollinations between proven elite clones. In this way it may be possible to take advantage of specific combining ability in unrelated superior clones and to produce progeny with heterosis in desirable traits that exceed what is found in the wild populations. The vegetative propagation of new genotypes can become a second clonal generation. The philosophy of adding new variation in clonal populations in the future can also be extended to the possibility of using genetically manipulated materials arising from biotechnology programs.

Ideally, to make rapid progress in a second-generation breeding program, means of inducing early flowering in superior clones are needed in order to shorten the generation time (e.g., Longman et al., 1990).

THE WISE USE OF GENETIC VARIABILITY

The genetic resource of a species is the foundation of its future as a wild plant and as a source of products for human use. It is therefore crucial to protect and use this resource wisely. Within the domestication process, whether clonal or not, one of the first requirements of an appropriate strategy is to conserve a substantial proportion of the genetic variability for future use in selection programs and, subsequently, through breeding to broaden the genetic base of the cultivars in the production population. This also serves as a risk-aversion strategy should it be necessary to breed for resistance to pests and diseases in the future. There are three actions which each contribute to genetic conservation: establishing a gene bank; the wise utilization of the genetic resource in cultivation; and protecting some wild populations. These will be discussed in the following sections.

Establishing a Gene Bank (*Ex situ* Conservation)

The strategy for building up the base population will, to some extent, be influenced by whether or not there has been a tree improvement program for the species in question, and how far it has gone toward identifying genetically superior stock. If, from previous tree improvement work, the population has been subdivided into provenances and progenies, it is easy to ensure that a wide range of unrelated seed sources and clones are set aside for the establishment of a number of living gene banks established at different sites to minimize the risk of loss if a site fails or is destroyed by fire or some other event. If possible, the material conserved in the gene bank should encompass the full geographical range of the species and be obtained from sites with differing soils, rainfall, altitude, etc. The latter is desirable for the subsequent selection of clones that are appropriate for different sites.

In the same way, for species without an existing tree improvement program, clones for field testing should originate from seed collections spanning the natural range of the species, particularly including populations on the edge of the range and any isolated subpopulations that may be important in a later breeding program between selected clones. Ideally, each collection area should be represented by identifiable half-sib progenies of individual mother trees. The first round of germplasm collections of this sort were made for *Sc. birrea* and *U. kirkiana* in eight SADC (Southern Africa Development Community) countries in 1996 (Kwesiga et al., 2000) and established in multilocational provenance trials (Akinnifesi et al., 2004, 2006). In both of these instances, the individual trees to be included in the gene bank should be selected at random from unrelated origins and vegetatively propagated to form as large a clonal population as possible. The unused seedlings should be established at different sites as living gene banks for future use, whether directly by subsequent coppicing and cloning or indirectly through their progeny. The identity of plants within gene banks of this sort should be maintained and different seed-lots should be planted such that cross-pollination between plants of different origins is likely. In this way the gene pool will be maintained with maximum diversity.

Ex situ conservation can also be achieved in species producing seeds, which retain their viability in storage, by creating seed banks. Species differ in their amenability to seed storage at different temperatures and moisture contents, including cryopreservation. The Genetic Resources Unit of the World Agroforestry Center (ICRAF, formerly called the International Council for Research in Agroforestry) contains material of several fruit-tree species. In southern Africa, clonal orchards have been established for the production of superior *U. kirkiana* clones. Numerous studies have also been made on the germination of indigenous fruit-tree species (Maghembe et al., 1994; Phofeutsile et al., 2002; Mkonda et al., 2003; Akinnifesi et al., 2006). It needs to be appreciated that new material should be brought into the gene resource (gene bank), selection and production populations whenever possible (see Fig. 16.2), so that genetic diversity is being continually enhanced.

The Wise Utilization of Genetic Resources in Cultivation (*Circa situ* Conservation)

A wise genetic resource strategy will ensure that the production population is based on a diverse set of cultivars, and that this set of cultivars is continually being supplemented with new and better cultivars from the ongoing screening of wild populations, provenances, progenies, a program of intensive tree breeding, and the regular replacement of cultivars that are no longer the best. In this way selected, highly productive but unrelated clones can be used commercially for agroforestry. Indeed, ten well-selected and unrelated clones may contain as much, or more, genetic variation as a narrowly based sexually reproducing population.

This strategy will also ensure that the cultivars being planted are well adapted to their environment—and thus fully expressing their genetic potential—while at the same time minimizing the risks associated with intensive cultivation. These risks are perhaps greater when indigenous trees, rather than exotics, are planted in the tropics in areas where the complexity of forest ecosystems has been disturbed: for example, by shifting agriculture. The minimization of risk therefore requires that the trees be planted in situations that ensure a minimum of damage to the nutrient and hydrological cycles, food webs and life cycles of the intact ecosystem.

In clonal tree domestication programs, final yield is strongly influenced by the adaptation of the trees, individually and collectively, to the site. This adaptation has two components: clonal development, and clonal deployment or the stand establishment process (Foster and Bertolucci, 1994). Clonal development approaches, including breeding, testing, and selection, largely affect the genetic quality of the resulting clonal population available for planting. Clones must be developed that are highly selected for growth and productivity traits, but that display substantial homeostasis and so can adapt to their changing environment. Clonal deployment, on the other hand, must strike a balance between the need for efficiency of management for the economic production of products and the need to deploy populations that are genetically buffered against environmental changes, including pests.

Clonal development relies on three processes: breeding, testing, and selection. However, despite its overwhelming importance, little research has been conducted to investigate the effects of different deployment strategies on the health, growth and yield of forest stands. This type of research requires a large amount of resources and a long period of assessment. Foster and Bertolucci (1994) identified the following major questions:

- How many clones should be in the production population, and how many should be deployed to a single site?
- Should the clones be deployed at a single site as a mixture or as mosaics of monoclonal stands? This is important both for operational reasons such as planting, and to maximize yield through the optimization of inter- or intraclonal competition.
- What are the key attributes of the clones themselves that cause them to be used either as mixtures or in monoclonal plots?

The question of the number of clones to be deployed has two aspects: the number of clones within the production population, and the number of clones planted per site. Based on probability theory, 7–30 clones provide a reasonable probability of achieving an acceptable final harvest. There is continued debate on the wisdom of releasing a few rather than many clones to optimize gains and minimize risk. The arguments about susceptibility to environmental disasters when only a few clones are deployed may apply more rigorously to agroforestry trees, given the risk-averseness of small-scale farmers and their possible desire to maximize stability of production rather than production per se. Consensus among tree breeders and forest geneticists indicates that production population sizes of 100 genotypes or fewer are acceptable. Actually, the absolute number of clones is less important than the range of genetic diversity among the clones. Ten clones that share the same alleles for a particular trait would express no genetic diversity, in contrast with five clones each of which has different alleles for the same trait. Hence, by sourcing material from diverse origins, the tree breeder must emphasize genetic diversity for traits associated with survival and adaptation while exercising strong selection pressure on production traits. Following rigorous testing, new clones should be added to the production population each year. In the case of diecious species such as *Sc. birrea*, it is important to include male trees in the production population in order to ensure adequate pollination. This is especially important for kernel production as there is some evidence that the number of kernels per nut may be constrained by inadequate pollination (Leakey et al., 2005c).

There is perhaps one situation in which an exception to the preceding strategy to broaden the genetic base of the production population may be acceptable. This situation arises when there are good market reasons to preserve regional variations in the quality of the product. For example, in the wine industry, regional attributes of the wine ("appellation") are a result of gene combinations specific to different regions. Nevertheless, within each region processes to maintain a broad genetic base are still important.

Protecting Some Wild Populations (*In situ* Conservation)

A conservation program for any species would be incomplete without a strategy for the protection of wild populations, which represent "hotspots" of genetic diversity. One advantage of this approach is that a species is conserved together with its symbionts (e.g., mycorrhizal fungi), pollinators and other associated species, something that is not so easy in ex situ conservation and which is important for *circa situ* conservation. Molecular techniques now provide a powerful tool to identify these hotspots rapidly (e.g., Lowe et al., 2000), although of course the data can only be as good as the sampling strategy will allow. In southern Africa, molecular studies have been completed for *U. kirkiana* and *Sc. birrea* (Agufa et al., 2006; Mwase et al., 2006a,b).

SOCIOECONOMIC AND ENVIRONMENTAL CONTEXT FOR THIS STRATEGY

In the introduction to this chapter it was stated that a domestication strategy for clonal forestry/agroforestry in the tropics should take into account not only the commercial production of agroforestry tree products but also the need to provide the domestic needs of rural people, encourage sustainable systems of production, and encourage the restoration of degraded land. There is debate in southern Africa about the impacts of poverty on the natural resources of very many African countries, especially those in southern Africa. The domestication of indigenous plants as new, improved crop species for complex agroforestry land uses offers the opportunity to return to more sustainable polycultural systems, to build on traditional and cultural uses of local plants, and to enhance the income of subsistence farmers through the sale of indigenous fruits, medicines, oils, gums, fibers, etc. (Leakey, 2003). Indeed, it has been argued that the biological and economic constraints on the wider use of indigenous trees in agroforestry can be overcome by cloning techniques and that the economic incentives should promote cultivation with the ecologically more important indigenous species. This approach to agroforestry has been recommended as a sound policy for land use in Africa (Leakey, 2001a,b). Potentially, through enhanced food and nutritional security, the domestication of indigenous fruits may even have positive impacts on HIV/AIDS (Swallow et al., 2007), by boosting the immune systems of sufferers (Barany et al., 2003). Consequently, the application of the strategies developed in this chapter is important if the people of Africa are to benefit from the domestication of indigenous trees.

CONCLUSIONS

This chapter presents three interacting, multifaceted strategies for the development of clonal fruit trees in southern Africa. These strategies are the foundation of a sustainable domestication strategy for indigenous fruit trees based on the establishment of three interlinked populations: a gene resource population for genetic conservation; a selection population for the achievement of genetic improvement; and a production population of trees for farmers to grow. The practice of domesticating a species using these strategies is cyclical and therefore continuous.

Vegetative propagation is a powerful means of capturing existing genetic traits and fixing them so that they can be used as the basis of a clonal cultivar, or in a different role as a research variable. The desirability of using clonal cultivars in preference to genetically diverse seedling populations varies depending on the situation and the type of trees to be propagated. However, the advantages of clonal propagules outweigh those of seedlings when the products are valuable, when the tree has a long generation time, and when the seeds are scarce or difficult to keep in storage.

There are many opportunities to enhance agroforestry practices through the wise application of vegetative propagation and clonal selection. These techniques in turn offer many ways of creating new and greatly improved crop plants. The potential for increased profits from clonal techniques arises from their capacity to capture and utilize genetic variation. The consequent uniformity in the crop is advantageous in terms of maximizing quality, meeting market specifications and increasing productivity, but it may also increase the risks of pest and disease problems; consequently, risk avoidance through the diversification of the clonal production population is a crucial component of the strategies presented. The application of these strategies should lead to benefits that encompass many of the rural development goals of development agencies, as specified in the Millennium Development Goals of the United Nations. Achieving these benefits will, however, require the large-scale adoption of the techniques and strategies presented here in ways that will meet the needs both of farmers and also those of new and emerging markets. This makes it important to ensure that policy makers get the message about good domestication strategies (e.g., Wynberg et al., 2003).

Section 5.2

Techniques: Vegetative Propagation

Chapter 17

Low-Technology Techniques for the Vegetative Propagation of Tropical Trees

This chapter was previously published in Leakey, R.R.B., Mesén, J.F., Tchoundjeu, Z., Longman, K.A., Dick, J. McP., Newton, A.C., Matin, A., Grace, J., Munro, R.C. and Muthoka, P.N., 1990. Commonwealth Forestry Review, 69, 247−257, with permission from Commonwealth Forestry Association

SUMMARY

Stem cuttings of five tree species from dry and semi-arid woodlands (*Acacia tortilis, Prosopis juliflora, Terminalia spinosa, Terminalia brownii* and *Albizia guachapele*) and seven species from moist tropical forests (*Cordia alliodora, Vochysia hondurensis, Nauclea diderrichii, Ricinodendron heudelotii, Lovoa trichiliodes, Gmelina arborea, Eucalyptus deglupta*) have been easily rooted in improved lowtechnology, high humidity polythene propagators in Kenya, Cameroon, Costa Rica and Britain. These propagators, which are cheap to construct, are very effective and have no essential requirements for either piped water or an electricity supply. Experiments have tested different rooting media, auxin applications and compared mist versus non-mist propagation. Assessments of the physical and gaseous environment of the propagators has indicated ways of improving the rooting environment through an understanding of the sensitivity of the relative humidity to radiant energy and to opening the propagator for short periods (eg 2−3 minutes).

INTRODUCTION

It is now widely realized that vegetative propagation and clonal selection offer a means to greatly enhance the yield and quality of forest products from commercial plantings in the tropics (Leakey, 1987). However, there is a need to simplify the technology so that vegetative propagation can be achieved in the absence of mains electricity and a piped water supply. In addition, in many tropical countries, the high capital and running costs of currently available mist propagation systems makes them inappropriate, except for research or large-scale commercial projects.

The environmental requirements for root initiation in leafy stem cuttings are those that minimize physiological stress in the cutting. In general terms this means using shade to lower the air temperature and, by providing a high humidity, to reduce transpiration losses. By the latter, the vapor pressure of the atmosphere surrounding the cutting is maintained close to that in the intercellular spaces of its leaf.

There are numerous propagation systems used in commercial horticulture. These are usually based either on spraying mist, fogging or enclosing the cuttings in polythene. The advantages of polythene systems have been known for many years (Loach, 1977) and they have been used to propagate tropical hardwoods with good success, particularly at the Forest Research Institute of Nigeria, Ibadan (Howland, 1975).

Recent work by the Institute of Terrestrial Ecology (ITE) and its overseas collaborators has applied and improved the design of nonmist propagators for use with a wide range of timber and multipurpose tree species from both tropical moist forests and semi arid areas (Leakey and Longman, 1988). Recent studies with *Triplochiton scleroxylon* cuttings under intermittent mist have indicated that rooting ability is related to the production of reflux-soluble carbohydrates, apparently derived from current photosynthesis while the cuttings are in the propagation unit (Leakey and Storeton-West, 1992).

Multifunctional Agriculture. DOI: http://dx.doi.org/10.1016/B978-0-12-805356-0.00017-9
© 2017 Elsevier Inc. All rights reserved.

Furthermore, it seems that the ability to produce these carbohydrates is related to the preseverance light environment and nutrient status of the cuttings while on the stockplants. Both the total irradiance and the light quality (red:far-red ratio) are important components of the preseverance light environment, and these factors interact with nutrient availability to influence the rates of net photosynthesis and rooting. These variables to a large extent account for the variation in the rooting ability of cuttings of *T. scleroxylon* taken from different shoots of variously treated stockplants (Leakey, 1983).

MATERIALS AND METHODS

General

Juvenile shoots of 12 tree species (Table 17.1) have been used as leafy stem cuttings. The studies presented here were done in either glasshouses in the UK or under nursery conditions in Costa Rica, Kenya, and Cameroon. In all instances, however, the propagator temperature was between 22 and 27°C, and the cuttings were prepared as described in the following subsection and set in randomized blocks. The numbers of replicate cuttings per treatment were between 24 and 117. Standard errors for percentage rooting were calculated using the procedures of Bailey (1959) for data with binomial distributions.

Preparation of Cuttings

Cuttings were harvested from seedlings, managed juvenile stockplants or coppice shoots. Depending on the species, 1- to 4-node cuttings were used. These were usually about 50−60 mm long and with a leaf area of about 50 cm^2 (Leakey, 1985). In large-leaved species, leaf areas were reduced by trimming prior to severance. The basal ends of cuttings were dipped briefly in indole-3yl-butyric acid solutions (0.2−0.4% IBA in industrial methylated spirit) to a depth of about 2−5 mm, and the alcohol then rapidly evaporated off in a stream of cold air from a fan (Leakey et al., 1982a, Leakey, 1989). To minimize stress, the cuttings were inserted in the propagator as soon as they were dry. Alternatively, commercial auxin-based rooting powders "Strike" and "Seradix 2" (May & Baker Ltd) were used.

THE NONMIST PROPAGATOR

The propagator design currently in use is based on that of Howland (1975), modified by Leakey and Longman (1988) and now further modified so that it does not require daily watering (Fig. 17.1). Basically, a wooden or metal frame is

TABLE 17.1 Tropical tree species vegetatively propagated using simple, low-tech propagators in Costa Rica, Cameroon, Kenya, and Great Britain.

Species	Family	Range	Uses
Gmelina arborea Roxb.	Verbenaceae	Indo-Burma region, and S.E. Asia, a Pan-tropical exotic	Timber
Eucalyptus deglupta Bl.	Myrtaceae	Tropical Australasia, a pan-tropical exotic	Timber
Nauclea diderrichii (DeWild & Th Dur.) Merr.	Rubiceae	W. and C. Africa	Timber
Lovoa trichilioides Harms	Meliaceae	W. and C. Africa	Timber
Ricinodendron heudelotii (Baill.) Pierre ex Pax	Euphorbiaceae	W. and C. Africa	Fruit
Cordia alliodora (Ruiz & Pav.) Oken	Ehretiaceae	C. America	Timber
Vochysia hondurensis Sprague	Vochysiaceae	C. America	Timber
Albizia guachapele (Kunth) Dug.	Mimosaceae	C. America	Timber
Prosopis juliflora (Swartz) D.C.	Mimosaceae	C. America	Multipurpose
Acacia tortilis (Forsk.) Hayne	Mimosaceae	W. and E. Africa	Multipurpose
Terminalia spinosa Engl.	Combretaceae	E. Africa	Multipurpose
Terminalia brownii Fresen	Combretaceae	E. Africa	Multipurpose

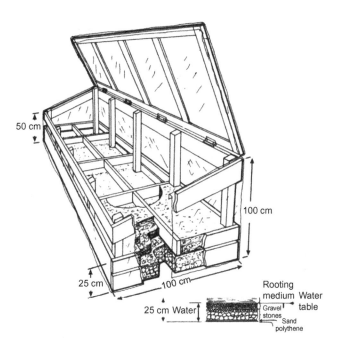

FIGURE 17.1 The design of ITE's improved nonmist propagator.

enclosed in clear polythene so that the base is watertight (Leakey, 1989). The frame also provides support for the enclosed volume of water. The polythene base of the propagator is covered in a thin layer of sand to prevent the polythene from being punctured by the large stones (6–10 cm) which are placed on it to a depth of 10–15 cm. These stones are then covered by successive layers of small stones (3–6 cm) and gravel (0.5–1.0 cm) to a total depth of 20 cm. The gravel provides support for the rooting medium which is the uppermost layer, while the spaces between the stones are filled with water. A length of hollow bamboo provides an open cylinder inserted into the medium and stones, which is used both to observe the water level and to add water if necessary. The rest of the frame is covered tightly with a single piece of clear polythene, and a closely fitting lid is attached. Internal supports to the frame at the level of rooting medium also provide subdivisions allowing the independent use of different rooting media (Fig. 17.1). As a result of the studies reported here, further refinements to the design of nonmist propagators are discussed later. A similar frame to that of the nonmist propagator, with roll-up polythene sides, can be used, as in Costa Rica, as a weaning area.

RESULTS

The Propagator Environment

In tests run in ITE glasshouses, in which air temperatures were maintained at about 20°C, temperatures in nonmist propagators rose to a midday peak of about 34°C during bright, sunny, midsummer weather (e.g., 28th July). This rise in temperature was associated with a decrease in relative humidity from about 95% to about 75% (Fig. 17.2).

This represents a substantial increase in the saturation vapor pressure deficit (SVPD) of the air from 0.02 kPa to 1.37 kPa. An important decrease in relative humidity also occurred when the propagator was opened for 5 minutes at midday (Fig. 17.3). In this instance, relative humidity decreased by about 40–50% to glasshouse ambient within 2 minutes, representing an increase in evaporation rate of approximately ×4.5 (SVPD = 0.45–2.08 kPa). Relative humidity increased rapidly again following closure of the lid. Decreases in air temperature to ambient were also associated with this period of opening. Subsequent gains in the temperature resulted from closing the propagator, but the response time for temperature was considerably slower than for relative humidity

When the easy-to-root species *Nauclea diderrichii* (Leakey, 1990) was used for physiological studies in a nonmist propagator at Edinburgh University (Matin, 1989), it was found that cuttings had maximum rates of photosynthesis that were typical of intact plants, up to $6\,\mu$ mol CO_2 m^{-2} s^{-1} at an irradiance of $1000\,\mu$ mol m^{-2} s^{-1}. However, the photosynthetic capacity of these cuttings was influenced by changes in the CO_2 concentration inside the propagator. In the middle of the day, the CO_2 concentration fell to $150\,\mu$ mol mol^{-1}, while by midnight it rose to $550\,\mu$ mol mol^{-1}, reflecting daytime assimilation and nighttime respiration, respectively.

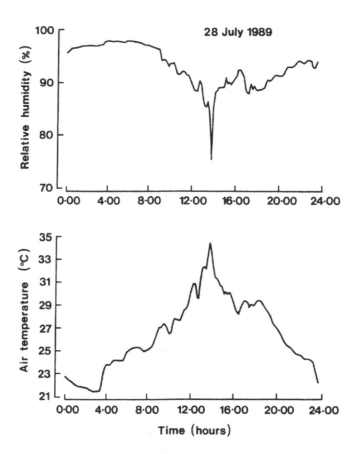

FIGURE 17.2   Effects of a rise in air temperature on the relative humidity inside a nonmist propagator.

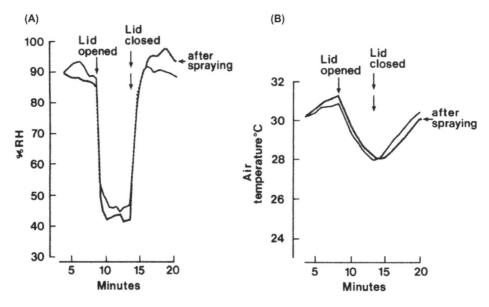

FIGURE 17.3   Effects of opening the lid of a nonmist propagator on its: (A) relative humidity and (B) air temperature.

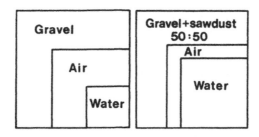

FIGURE 17.4 Relative composition by volume of a gravel rooting medium with and without sawdust.

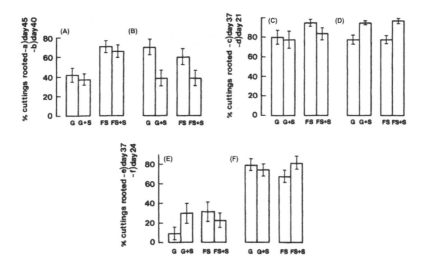

FIGURE 17.5 Effects of rooting medium (*G*, gravel; *FS*, fine sand; and *S*, sawdust) on the rooting of leafy stem cuttings of (A) *Cordia alliodora*, (B) *Vochysia hondurensis*, (C) *Gmelina arborea* (juvenile), (D) *Eucalyptus deglupta*, (E) *Gmelina arborea* (mature), and (F) *Albizia guachapele*.

As regards rooting media, the water-holding capacity of a fine-gravel (2−3 mm diameter) medium was considerably increased, at the expense of the volume of the air spaces (Fig. 17.4), by the addition of rotted sawdust (50% by volume).

Rooting Tests

In Costa Rica, studies using five tree species investigated the effects of four different rooting media: (1) gravel, (2) 50:50 gravel with sawdust, (3) fine sand, and (4) 50:50 fine sand and sawdust. Each medium was tested with cuttings dipped in a range of IBA concentrations (0, 0.05, 0.1, 0.2, 0.4, and 0.8%). There were, however, substantial differences between species with regard to rooting success on the different media. Single-node juvenile cuttings of all five species rooted well (70−95%) on their best medium (Fig. 17.5). *Cordia alliodora* rooted best in fine sand, with or without sawdust, while rooting of *Vochysia hondurensis* cuttings was detrimentally affected by the incorporation of sawdust into both gravel and fine sand. On the other hand, sawdust enhanced the rooting of *Eucalyptus deglupta* cuttings in both gravel and sand, while *Gmelina arborea* and *Albizia guachapele* rooted well in all media. Unlike these juvenile cuttings, mature cuttings from vigorous shoots in a heavily pruned crown of *G. arborea* rooted much less well (Fig. 17.5E) especially in pure gravel. In juvenile *G. arborea*, a comparison between cuttings set in fine sand under an intermittent mist propagator and the nonmist propagator, showed better rooting in the nonmist propagator.

Cordia alliodora, *A. guachapele*, and *V. hondurensis* differed in their responses to the range of IBA concentrations (0−0.8%). Optimal concentrations would appear to be 0.4, 0.1, and 0.2% IBA respectively for the three species (Fig. 17.6). In addition *C. alliodora* did not root at all without applied auxins, while rooting in untreated *V. hondurensis* cuttings exceeded 40%. Cuttings of *Al. guachapele* were very responsive to all auxin treatments.

Studies using nonmist propagators in Edinburgh and in Kenya, with tree species from semi arid areas, investigated the relative merits of two commercially available rooting powders: (1) "Strike" = 0.25% NAA, and (2) "Seradix

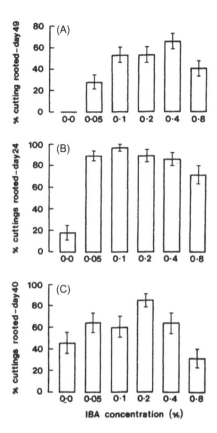

FIGURE 17.6 Effects of IBA concentrations on the rooting of leafy stem cuttings of (A) *Cordia alliodora*, (B) *Albizia guachapele*, and (C) *Vochysia hondurensis* in a non-mist propagator.

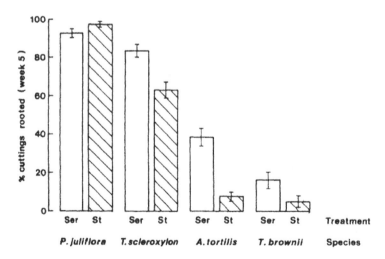

FIGURE 17.7 Percentage of cuttings of *Prosopis juliflora*, *Terminalia spinosa*, *Acacia tortilis*, and *Terminalia brownii* rooted in a nonmist propagator when treated with commercial rooting powders (Ser = "Seradix 2," St = "Strike").

2" = 0.8% IBA. Seradix-treated cuttings of *Acacia tortilis*, *Terminalia spinosa,* and *Terminalia brownii* rooted better than those treated with Strike, while for *Prosopis juliflora*, rooting percentages were above 90% with both treatments (Fig. 17.7).

Two-node cuttings of *P. juliflora* also rooted relatively easily without applied auxins, although IBA solutions (0.4−0.8%) did hasten rooting and increase the numbers of roots formed. Long cuttings tended to root better than short ones, although there was no relationship between cutting length and its position of origin on the stockplant. Cuttings

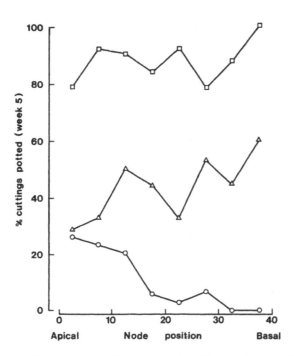

FIGURE 17.8 Effects of node position on the percentage of 2-node *Prosopis juliflora* cuttings rooted under three propagation environments (□ = non mist, ○ = enclosed mist, Δ = open mist).

from lower on a stem generally rooted better than apical ones. An experiment comparing rooting under mist, with rooting in a nonmist propagator, clearly demonstrated the advantages of conditions without mist (Fig. 17.8). A third treatment in which a mist propagator was enclosed in polythene, resulted in even less rooting and higher mortalities due to rotting. When nonmist propagators have been used in Kenya to produce clonal material of *P. juliflora* for field experiments, the success rate was greater than 75%.

Like those of *P. juliflora*, cuttings of *A. tortilis* and *T. spinosa* are also much more easily rooted in the high humidity conditions of a nonmist propagator. There is, however, very considerable clonal variation in rooting ability in these species, and cuttings of *T. brownii* have so far only been rooted with a low rate of success.

In Cameroon, the percentage rooting of cuttings of *Lovoa trichiloides* was relatively poor (c. 40−50%), and, in addition, the results were frequently unexpected. For example, no beneficial effects on rooting were found following the application of NAA or a range of IBA concentrations (0−200 μg/cutting), either in terms of the percentage of cuttings rooted, or the number of roots per cutting. Furthermore, the optimum leaf area seemed to be very high at about 200 cm^2 per cutting (Tchoundjeu, 1989b).

When the rooting of cuttings of *Ricinodendron heudelotii* was tested both under mist and in nonmist propagators with Seradix 2, rooting by day 21 was best without mist (75% vs 50% under mist).

DISCUSSION

While there are reports of cuttings of *Prosopis, Gmelina*, and *Eucalyptus* species being rooted (Felker and Clark, 1981; Sim and Jones, 1985; Delwaulle, 1983; Delwaulle et al., 1983), there are apparently none on the other species tested here. *Terminalia spinosa, T. brownii, V. hondurensis, A. guachapele*, and *R. heudelotii* may not have been previously tested, while for example, a number of previous attempts to root *C. alliodora* have failed (Dyson, 1981). *Acacia* spp. of semi arid/arid areas of Africa also have the reputation of being difficult-to-root (Roche et al., 1989). The low-tech, nonmist propagators described here therefore seem to provide a very practical solution to the problem of how a wide range of tropical tree species can be propagated vegetatively. The advantage of these propagators seems to be particularly great for the dry-zone species, which can be very susceptible to rotting under mist. While it is clear that the environment within the nonmist propagator fluctuates considerably over the day in response to variations in ambient temperature and incident radiation, it is also clear that, by being enclosed, and continuously moist, the cuttings are not subjected to the extremes of saturation and water stress that can occur when misting frequency is poorly matched to changes in the weather.

An improvement that might further stabilize relative humidity in the nonmist propagator would be to construct the lid in several sections. Then the whole propagator would not have to be opened to access the cuttings. Ideally, the propagators are opened as little as possible and especially not during the heat of the day. When propagators are opened the cuttings should be sprayed frequently with a fine spray of water from a hand-held or knapsack sprayer. During bright sunny weather, shading is essential and spraying is highly desirable to prevent low relative humidity developing at midday (see Fig. 17.2). Attention to these points of detail are likely to be particularly important when propagators are used in hot, sunny, and dry areas.

The reasons why the rooting of cuttings of different species have slightly different requirements with respect to propagation media and auxin concentration is unknown. Studies are in progress at the Institute of Terrestrial Ecology and are aimed at the identification of fundamental principles determining rooting ability in a range of tropical tree species. In this regard it appears that stockplant light/nutrient interactions prior to severance are important (Leakey and Coutts, 1989; Leakey and Storeton-West, 1992) affecting the subsequent capacity of unrooted cuttings to photosynthesize. This photosynthesis could also be limited to low daytime CO_2 concentrations in the propagator. Thus it may be necessary in future to enhance CO_2 diffusion into the propagators.

Chapter 18

Stockplant Factors Affecting Root Initiation in Cuttings of *Triplochiton scleroxylon* K. Schum., an Indigenous Hardwood of West Africa

This chapter was previously published in Leakey, R.R.B., 1983. Journal of Horticultural Science, 58, 277−290, with permission from Journal of Horticultural Science and Biotechnology

SUMMARY

The ability of auxin-treated *Triplochiton scleroxylon* cuttings to root was affected by the prior management of potted stockplants. In undecapitated single-stem stockplants more cuttings from upper rather than lower mainstem nodes rooted; a difference paralleled by leaf water potential immediately after severance, although there was also a positive relationship with internode length. The rooting percentage of mainstem cuttings from unpruned stock plants ranged from 15% to 43%, whereas that of cuttings from the lateral shoots of pruned stockplants ranged from 40% to 83%. Considerably more cuttings rooted from stockplants which were severely pruned than from those where decapitation removed only the top node; there seemed to be an inverse relationship with the number of shoots per plant and the carbohydrate:nitrogen ratio. However, in tall pruned stockplants, more cuttings from lower lateral (basal) than from upper (apical) shoots rooted, although the differences between cuttings from basal and apical lateral shoots were less when the stockplants' mainstems were orientated at 45 degrees or kept horizontal, instead of vertically. Adding NPK 16 weeks before harvesting cuttings from 10-node vertical stockplants increased the rooting ability of cuttings from basal shoots without affecting the rooting of those from apical shoots. More lateral shoot cuttings rooted when two, instead of one or four, lateral shoots were allowed to develop per stockplant, this being associated with less cutting mortality than occurred in pruned stockplants. In stockplants with two shoots, cuttings from basal lateral shoots rooted better than those from apical shoots, although without competition from basal shoots. The rooting of apical shoots was enhanced by application of a complete fertilizer. The presence of basal shoots reduced the rooting ability of apical shoots even with the fertilizer application. Many of the effects of lateral branch position on rooting may be related to light intensity, for greater rooting percentages occurred among cuttings from lower, more shaded than from upper, less shaded branches. This positional effect was eliminated when branches were uniformly illuminated.

INTRODUCTION

Unless somatic mutations occur, plants of the same clone (i.e., vegetatively propagated from stockplants originating from a single seedling) have identical gene complements. Foresters, like horticulturists, can utilize this genetic uniformity to advantage by selecting and planting superior clones (Longman, 1976a; Heybroek, 1978; Leakey et al., 1982b). To exploit the advantages of vegetative propagation and overcome seed shortages, attempts are being made with *Triplochiton scleroxylon* K. Schum, a commercially important native hardwood (obeche) of West African high forests, to: (1) develop techniques for rooting stem cuttings (Howland, 1975; Leakey et al., 1975, 1982a), and (2) establish

Multifunctional Agriculture. DOI: http://dx.doi.org/10.1016/B978-0-12-805356-0.00018-0
© 2017 Elsevier Inc. All rights reserved.

experimental clonal plantations in Nigeria, which also serve the purpose of a gene bank (Bowen et al., 1977; Longman et al., 1978; Leakey et al., 1980). Within a short time marked clonal variations in growth and form were apparent and studies in progress are identifying selection criteria, particularly the role of branching habit in the determination of yield (Ladipo, 1981; Ladipo et al., 1980).

In the long-term the commercial utilization of clones will depend on the ability to sustain continued multiplication of selected clones without the loss of easy rooting and the incidence of persistent within-clone variations, such as plagiotropism. In woody plants both these phenomena are generally considered to be associated with the loss of juvenility and the attainment of the mature phase, although some workers consider that they may be independent of this process (Borchert, 1976).

Although easy rooting, and perhaps juvenility, can be prolonged by repeatedly pruning stockplants (as with stoolbeds and the practice of hedging), cuttings from pruned stockplants still vary considerably in their rooting ability, even within a batch of cuttings from the same plant. Such differences may presumably be attributable to the position of a cutting on a shoot and the position of that shoot on a plant. Variations of this type have been reported in *Pisum, Malus,* and *Hedera helix* to be related to internode length and light intensities (Veierskov, 1978; Christensen et al., 1980; Poulsen and Andersen, 1980). Because little was known of the extent or causes of positional variation in rooting in trees, and particularly in *T. scleroxylon*, a series of detailed experiments was initiated hoping that they would lead to improved methods of stockplant management so maximizing uniformity and minimizing the "carry-over" effects reported by Libby and Jund (1962).

MATERIALS AND METHODS

Cuttings were taken from vegetatively propagated stockplants originating from seeds of *T. scleroxylon* (Fig. 18.1) collected in different parts of Nigeria: (1) Ilugun, Oyo State (clone numbers: 8019, 8032, 8034, 8035, 8036, 8038), (2) lgbo Ora, Ogun State (8046, 8047, 8049, 8054), (3) mile 19 on the Iwo to lbadan road, Oyo State (8045, 8053), and (4) Ezillo, Anambra State (8074, 8075). The stockplants were grown in pots of John Innes compost in automatically controlled tropicalized glasshouses at the Institute of Terrestrial Ecology, Bush Estate, near Edinburgh. The glasshouses were kept at 25–30°C, with a constant daylength throughout the year of 19.5 hours (Leakey et al., 1982a). Plants were watered daily and, except where otherwise stated, a liquid fertilizer (1% "Solufeed") with 23% N: 19.5% P: 16% K was added once a week.

FIGURE 18.1 A three-month-old, undecapitated plant of *Triplochiton scleroxylon*.

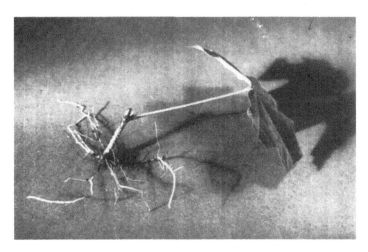

FIGURE 18.2   A rooted single-node, leafy cutting of *Triplochiton scleroxylon* with a trimmed leaf of about 50 cm².

The leaf lamina of each single-node cutting was trimmed with scissors to c. 50 cm² by removing the fingers of the palmate leaf and, except where otherwise stated, the basal cut- surfaces of their stems were treated with a 10 μL d roplet of methanol containing 20 μg of an equal mixture of two auxins [α-napthalene acetic acid (NAA) and indole-3-butyric acid (IBA)]. After drying in a stream of air the cuttings were set in randomized blocks in coarse sand/gravel on mist propagation beds heated to $30 \pm 1.5°C$ at the base of the cuttings. Air temperature of the propagating area was $20 \pm 2°C$, except during very hot weather. Cuttings were lifted and examined at weekly intervals, those with one or more roots (Fig. 18.2) were scored as rooted and, if the roots were more than 1 cm long, were potted in compost and removed from the experiment.

Unrooted cuttings, or those not ready for potting, were replaced in the propagation bed. Potting compost (7:3:1:: peat:sand:loam by vol.) was enriched with "Enmag" (8.2 g kg^{-1} containing 5.5% N: 20.5% P: 9% K: 8% Mg) and John Innes Base (52 g kg^{-1} containing 4.7 N: 7.7% P: 10% K), both slow-release fertilizers. Potted cuttings were "weaned" for 2 to 3 weeks before being transferred to standard glasshouse conditions.

The amounts of total N and carbohydrate in dried (95°C), ground-up samples of lateral shoots were determined using the Kjeldahl and Anthrone methods, respectively.

Standard errors for percentage rooting were transformed and analyzed using the procedures of Bailey (1959) for data with binomial distribution.

EXPERIMENTAL AND RESULTS

Effects of Node Position

Fifteen plants of each of three clones (8038, 8049, and 8053) were grouped in three height classes (21, 26, and 31 nodes) and arranged in five randomized blocks in the glasshouse. The shorter plants were raised on blocks so that tops of all plants were at the same height above ground. After 16 weeks, eight single-node mainstem cuttings were taken per plant, excluding the most apical node. Each cutting, whose position on the parent plant was noted, was treated with 40 μg IBA and then set in five randomized blocks on the propagation bed. The mean lengths of the stem sections of these cuttings below the node decreased sequentially from 54.7 cm to 24.4 cm, the longest originating from the top of the shoot.

Final rooting percentages after 10 weeks usually decreased sequentially from c. 70% for cuttings taken from the youngest apical nodes to c. 10% from the oldest basal nodes, with mortalities (associated with leaf shedding) being inversely related (Fig. 18.3). However, in clone 8049 cuttings from node 3 rooted more readily than those from nodes 1 and 2. The heights of the stockplants did not markedly affect the percentage rooting.

In a supporting series of observations on six replicate plants of clone 8038, c. 1 m tall, it was found that leaf water potential, as determined by a pressure bomb (Waring and Cleary, 1967), was least in leaf 2 and increased steadily at successive nodes down the stem (Fig. 18.4). The water tension of leaf 1, that at the uppermost node which was discarded in the propagation experiment, was similar to that at node 11.

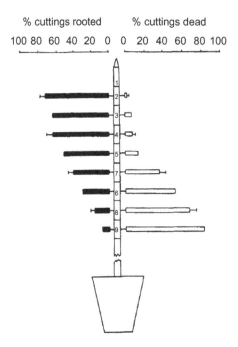

FIGURE 18.3 Mean effects, after 10 weeks, of node positions on rooting and death of single-node leafy mainstem cuttings from undecapitated stockplants of three *Triplochiton scleroxylon* clones. Bar = +SE.

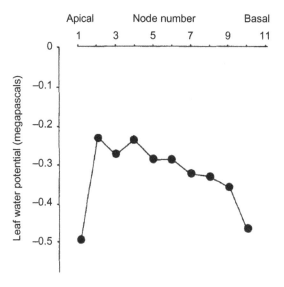

FIGURE 18.4 Effects of node positions on leaf water potential in undecapitated, unbranched plants of *Triplochiton scleroxylon*.

Effects of Stockplant Orientation

Twenty-one plants of each of two clones (8034 and 8036) were decapitated 1 m above ground and seven of each clone were then planted and staked: (1) vertically, (2) angled at 45 degrees, or (3) horizontally, always using 19 cm pots. To minimize shading, plants at different angles were randomized in separate batches. Over a period of 16 weeks lateral shoots developed at nodes along much of the length of these decapitated plants. To examine positional effects 36 cuttings were taken from shoots near the apical and basal ends of the stems at all three orientations. In the event four types of lateral shoot were sampled:

1. Very *vigorous orthotropic shoots* from the apical nodes of vertical plants. The phyllotaxis of these shoots changed fairly rapidly from distichous to radially symmetrical.
2. *Suppressed plagiotropic shoots* with distichous phyllotaxis from *basal* nodes of vertical plants.

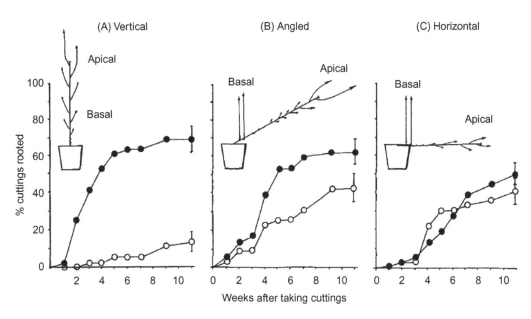

FIGURE 18.5   Effects of stockplant orientation on the rooting of leafy single-node cuttings from apical (○) and basal (●) lateral shoots of decapitated plants of *Triplochiton scleroxylon*. Bar = ± SE.

3. Fairly *vigorous plagiotropic shoots* with distichous phyllotaxis from *apical* nodes of angled and horizontal plants: shoots from angled plants were more vigorous than those from horizontal plants.
4. Very *vigorous orthotropic shoots* with radial symmetry from basal nodes of angled and horizontal plants.

The cuttings were arranged in six randomized blocks during propagation. Although overall numbers of cuttings rooted were unaffected by stockplant orientation, cuttings from basal lateral shoots generally rooted more rapidly and prolifically than cuttings from apical shoots. This contrast was greatest from vertical and least from horizontal stockplants, differences not being significant in the latter (Fig. 18.5). Similarly basal cuttings from vertical stockplants produced more roots (2.0 ± 0.22) than apical cuttings (1.25 ± 0.22) whereas those from horizontal plants produced similar numbers of roots (1.3 ± 0.18 and 1.4 ± 0.19 respectively). When rooted, cuttings from vigorous plagiotropic shoots frequently retained a persistent plagiotropic growth habit with the associated distichous phyllotaxis as described by Longman (1976b).

Effects of Different Pruning Regimes

Plants, 30 nodes in height, of two clones (8034 and 8038) were either decapitated at one of four positions or left intact. There were 10 replicates of each treatment. Decapitation removed either 1, 5, 10, or 20 nodes to leave plants with 29, 25, 20, or 10 nodes respectively. In all plants, irrespective of these treatments, leaves were retained at only the two uppermost nodes of the remaining section of stem. Within-treatment plants were randomized, but the different treatments were arranged in ascending order of height, to minimize shading.

Irrespective of treatment, 21−24% of all remaining buds sprouted. Inevitably, therefore, the absolute numbers of lateral shoots were directly related to plant height (Table 18.1). Twelve weeks after decapitation seven cuttings were taken from the uppermost shoot of each plant and set in 10 randomized blocks on a propagating bed. Cuttings from stockplants with 10 mainstem nodes rooted sooner and to a greater extent than cuttings from stockplants with 20 nodes, and each successive increase in the size of stockplants decreased the rooting ability of their cuttings (Fig. 18.6). However, the numbers of roots per rooted cutting were unaltered by stockplant height. About 50% of mainstem cuttings from unbranched control (undecapitated) plants rooted, a proportion exceeded by cuttings from lateral shoots of all but the tallest decapitated stockplants. Interestingly, although more of the mainstem cuttings of clone 8038 rooted than of 8034 (54.0 and 32.0 respectively), the rooting abilities of cuttings from decapitated plants of these clones were very similar.

In another, slightly different experiment, the mainstem apices (terminal bud and first node) of plants of different heights (10, 15, and 20 nodes) were removed by decapitation. There were 20 replicates of each treatment, and leaves were retained at only the two uppermost nodes of each plant. After 13 weeks growth all lateral shoots were sampled, oven-dried and analyzed for total N and carbohydrates. Amounts of N and carbohydrates were inversely related, the

TABLE 18.1 Effects on lateral shoot growth of different decapitation treatments in 30-node plants of *Triplochiton scleroxylon*. Mean of clones 8034, 8038.

Decapitated plants with:	Nodes removed	1	5	10	20
	Nodes remaining	29	25	20	10
Number of shoots per plant		6.9	5.3	4.7	2.4
% of lateral buds forming shoots		23.8	21.2	23.5	24.0

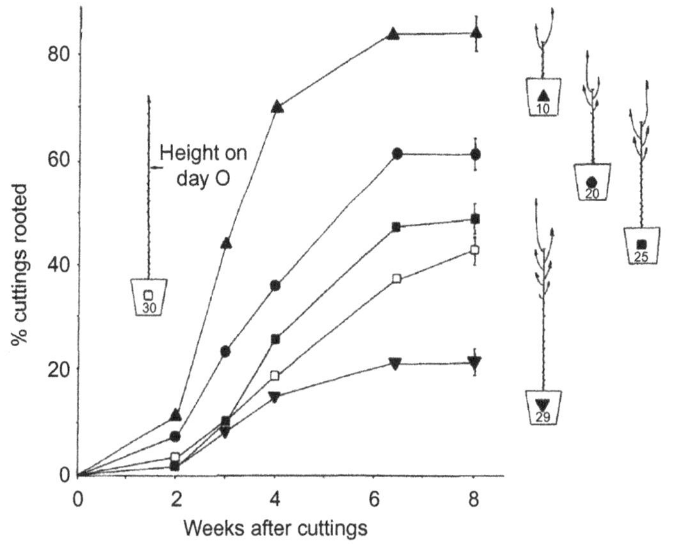

FIGURE 18.6 Effects of stockplant height (10 (▲), 20 (●), 25 (■), 29 (▼) nodes) on the rooting ability of single-node leafy cuttings from the uppermost lateral shoots of decapitated, but previously uniform, 30-node plants of *Triplochiton scleroxylon* (undecapitated control (□)). Bars = ± SE.

former being in greater quantities in shoots from short rather than tall stockplants. As a result, the carbohydrate:nitrogen ratio was directly proportional to plant height (Table 18.1). Within a batch of plants the carbohydrate:nitrogen ratio was relatively constant between successive lateral shoots, although there was a tendency for basal shoots to have slightly smaller carbohydrate:nitrogen ratios than apical shoots.

Effects of Applying Nutrients

Forty plants of clone 8035 were decapitated at node 10, ten replicates being allocated to each of the combinations in the factorial arrangement of: (1) two concentrations of liquid fertilizer (50 ml of 0.4% and 100 ml of 4.0% "Solufeed") and (2) two sizes of pot (13 and 23 cm). Sixteen weeks later 40 cuttings were taken from each of the apical and basal groups of lateral shoots, and set in 10 randomized blocks in a propagating bed. The leaves of apical shoots on plants with 0.4% "Solufeed" were already yellowing. Increasing the concentration of "Solufeed" from 0.4% to 4.0% accelerated and increased the proportion of cuttings that rooted (Fig. 18.7A). In particular, it significantly increased the rooting ability of cuttings from basal shoots (Fig. 18.7B). At the time of potting, more than twice the numbers of roots per rooted cutting were produced by cuttings originating from plants in small pots with higher concentration of fertilizers, than in the other three treatments (4.2 ± 0.67 roots/cutting against 1.8–1.9 ± 0.32).

Effects of the Numbers and Positions of Shoots on Stockplants

Eight plants of each of four clones (8019, 8035, 8045, and 8046) were decapitated above node 12, and two of each clone were left undecapitated. Before reducing the numbers of leaves on all plants to four, six of the eight decapitated plants were further treated by the removal of all lateral buds except either the 1, 2, or 4 nearest to the decapitated apex. After 8 weeks' growth 40–50 cuttings/treatment were set in 10 randomized blocks in a propagating bed.

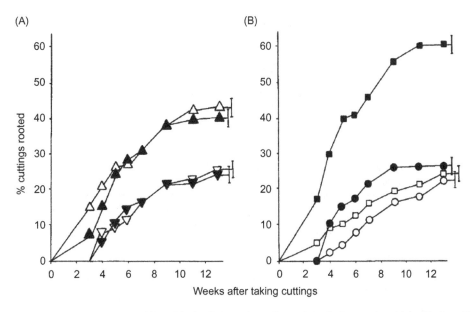

FIGURE 18.7 Effects of fertilizers on rooting ability of leafy single-node cuttings *of Triplochiton scleroxylon*. (A) Overall effects of pot size (△ = 13 cm: ▲, ▼ = 23 cm) and nutrients (△,▲ = with 100 ml of 4.0% liquid fertilizers: ▼ = with 50 ml of 0.4% liquid fertilizer). (B) Partitioning of nutrient effects to basal (● = with 50 ml of 0.4% liquid fertilizer, ■ = with 100 ml of 4.0% liquid fertilizer) and apical lateral shoots (○ = with 50 ml of 0.4% liquid fertilizer: □ = with 100 ml of 4.0% liquid fertilizer). Bars = ± SE.

The percentage rooting of cuttings from the lateral shoots of decapitated plants was significantly greater than that of mainstem cuttings of undecapitated plants (Fig. 18.8). More cuttings from the uppermost (first) laterals rooted than from the second, but none of the cuttings of the third or fourth shoot rooted. In addition, the ability to root of cuttings from the first lateral was generally better from plants with two shoots than where one or four were allowed to develop (Fig. 18.8). In a repeat experiment with seven clones it was confirmed that, on average, cuttings from twin-shooted plants rooted better than those from single-shooted stockplants.

In a third experiment 25 plants of each of two clones (8034 and 8038) were decapitated above the tenth node, allocated to five treatments and randomized within five blocks. Treatments were the factorial combination of two concentrations of "Solufeed" fertilizer (50 ml of 0.4 or 4.0%) applied twice weekly and the removal of all axillary buds from plants except those at two positions on the mainstem (nodes 3−4 and 9−10 from the ground). The fifth treatment, done with the larger amount of fertilizer, retained buds at nodes 3, 4, 9, and 10, the nodes at which mainstem leaves were also retained on plants of all treatments. Between 50 and 70 cuttings were taken after 17 weeks' growth, treated with 20 μg IBA each, and set in 10 randomized blocks in a propagating bed.

Comparable sets of cuttings from plants with 4.0% "Solufeed" rooted better than those with 0.4% (*cf* 2 A vs 4 A and 3B vs 5B in Fig. 18.9), but at both concentrations cuttings from basal shoots rooted significantly more rapidly and more successfully than cuttings from apical shoots (*cf* 1B vs 1 A; 3B vs 2 A; 5B vs 4A in Fig. 18.9). With the larger amounts of fertilizer, "competition" from apical shoots did not significantly decrease the rooting of cuttings from basal shoots (*cf* 1B vs 3B in Fig. 18.9), but basal shoots considerably decreased the rooting ability of apical cuttings (*cf* 1 A vs 2 A in Fig. 18.9). Numbers of roots per rooted cutting from apical shoots (1.56 ± 0.29) was unaffected by either fertilizers or competition from basal shoots. Basally originating cuttings produced 3.6 ± 0.5 roots in response to the prior application of large amounts of fertilizers to their parent plants and the removal of competing apical shoots, as against 2.48 ± 0.37 if amounts of fertilizer were small or if competition from nodes 9 and 10 was allowed to persist.

Effects of Light Environment

To eliminate the effects of shading, which might affect the relative performance of cuttings from apical and basal laterals, plants with shoots developing at different nodes were differentially raised on pots so that their lateral shoots were more or less at the same level above ground (Fig. 18.10). Plants of three clones (8035, 8046, and 8049) were decapitated and then debudded so that lateral shoots could only develop at nodes 3−4, 5−6, 7−8, or 9−10.

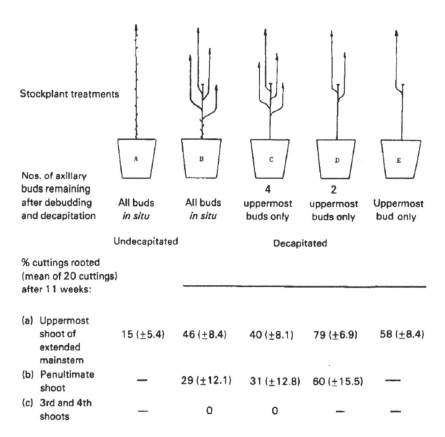

FIGURE 18.8 Effects of the number of lateral shoots per stockplant on % rooting (±SE) of leafy, single-node cuttings of *Triplochiton scleroxylon.*

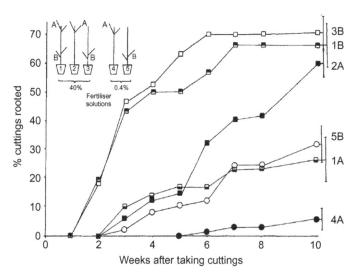

FIGURE 18.9 Effects of node position and competition between shoots on rooting ability of leafy single-node cuttings of *Triplochiton scleroxylon.* Bars = ±SE.

Leaves were removed from all nodes except those whose buds were retained. Plants were arranged in one fully randomized block. After 17 weeks, six cuttings were taken per plant, usually from the uppermost of the two shoots (the lower one sometimes being too small). They were each treated with 20 μg IBA and set in 10 randomized blocks on a propagating bed. Positional differences between the rooting abilities of cuttings from apical and basal shoots were eliminated by these treatments (Fig. 18.10), a result confirmed when the experiment was repeated. However, there was

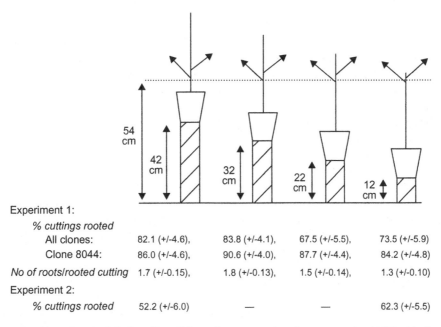

Experiment 1:			
% cuttings rooted			
All clones:	82.1 (+/-4.6),	83.8 (+/-4.1),	67.5 (+/-5.5),	73.5 (+/-5.9)
Clone 8044:	86.0 (+/-4.6),	90.6 (+/-4.0),	87.7 (+/-4.4),	84.2 (+/-4.8)
No of roots/rooted cutting	1.7 (+/-0.15),	1.8 (+/-0.13),	1.5 (+/-0.14),	1.3 (+/-0.10)
Experiment 2:			
% cuttings rooted	52.2 (+/-6.0)	—	—	62.3 (+/-5.5)

FIGURE 18.10 Effects, after 12 weeks, of relatively uniform light environment on the subsequent rooting (±SE) of leafy, single-node cuttings of *Triplochiton scleroxylon* from lateral shoots at different positions on the mainstem.

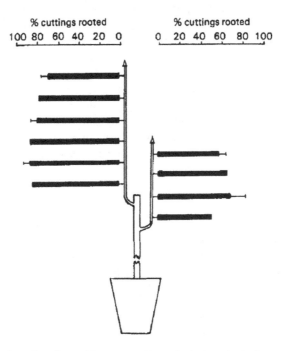

FIGURE 18.11 Effects, after 12 weeks, of node position on rooting of single-node leafy cuttings from decapitated stockplants of *Triplochiton scleroxylon*, with two lateral shoots. Bar = +SE.

still an effect on numbers of roots per rooted cutting: 1.7, 1.8, 1.5, and 1.3 (±0.13) on cuttings from laterals 3−4, 5−6, 7−8, 9−10 respectively. In addition to the positions of shoots the positions of cuttings within shoots were recorded, and in contrast to an earlier experiment with undecapitated stockplants, it was found that, in addition to generally better and less variable rooting ability, cuttings taken from nodes 4 and 5 of the uppermost lateral shoot and at nodes 2 and 3 on the lower shoot rooted best (Fig. 18.11).

DISCUSSION

The intention to propagate forest tree species vegetatively has often been thwarted by: (1) inadequate knowledge of the conditions necessary for root initiation, (2) the use of inappropriate shoots which have low rooting abilities and/or undesirable nonerect growth habits after rooting. In recent years techniques for rooting leafy cuttings and maximizing the development of erect plants of *T. scleroxylon*, and many other tropical trees, have been developed (Leakey et al., 1982a,b). Now the factors limiting the multiplication of selected clones for commercial forestry are mainly those controlling the regular supply of easily rooted, and thus cheaply produced, cuttings.

Stockplant management procedures aimed at the continuous and prolific production of cuttings are being introduced increasingly for fruit and ornamental trees. In *Salix* and *Populus* stooling has been found effective, while in *Camellia sinensis, Malus* spp., and *Pinus radiata* the production of hedges has provided suitable material for propagation. In small plants (<1 m) of *T. scleroxylon* removing mainstem apices results in a short-term flush of lateral shoots which are subsequently rapidly suppressed when the uppermost one becomes dominant. This reestablishment of dominance is contrary to the propagator's desire for numerous shoots as a supply of cuttings. However, prior to the present study, little was known about the extent of variation in rooting ability between cuttings taken from different shoots. Interestingly, it was found that the rooting of cuttings from small intensively managed stockplants of "easy rooting" clones varied as much as, if not more than, that normally found between clones (Leakey et al., 1982a).

When stockplants were subjected to a considerable range of manipulative treatments it was found that differences in rooting ability can be attributed to many contributory factors. The simplest form of variation is that occurring between cuttings taken sequentially down a stem. In young unpruned trees, which are usually unbranched, the mean rooting ability of the eight uppermost greenwood cuttings (excluding the soft expanding apical node) was low (Fig. 18.8). However, this mean obscures systematic differences in rooting attributable to position. Leafless cuttings are known to root rarely (Leakey et al., 1982a) and the greater chance of cutting mortality at lower nodes (Fig. 18.3) probably results from leaf shedding while on the propagating bench. This may reflect the greater internal water tensions and perhaps lower photosynthetic efficiency of the inevitably older leaves at lower nodes. Additionally, cuttings taken from the upper parts of the stem usually had longer internodes and, as in peas (Veierskov, 1978) and ivy (Poulsen and Andersen, 1980) in which sequential differences in internode length were related to the number of roots formed, this was strongly related to their rooting ability (Fig. 18.12).

The extent of variation attributable to node position was considerably less among cuttings from lateral shoots of small pruned stockplants which had been previously decapitated by removing the uppermost sections of the mainstem. Overall, rooting was enhanced in this second crop of cuttings, as those from lower nodes retained their leaves and rooted well (Fig. 18.11). This advantage of pruned stockplants was, however, restricted to small plants, for when stockplants were cut to different heights, up to 1 m, the rooting ability of cuttings from uppermost lateral shoots declined rapidly (Fig. 18.6). This decrease in rooting ability with increasing height of the stockplant appeared to be related to

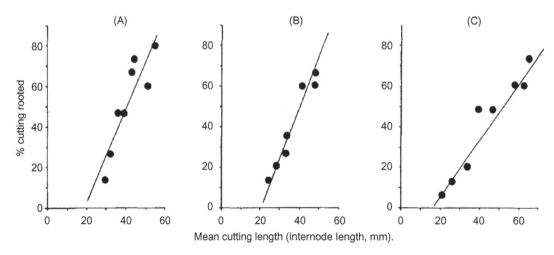

FIGURE 18.12 Relationships between the percent of cuttings rooted (week 10) and the mean length of single-node cuttings from undecapitated, unbranched plants of *Triplochiton scleroxylon* with: (A) 21 nodes ($y = 42.3 + 2.28x$. $r = +0.88$); (B) 26 nodes ($y = 53.8 + 2.52x$. $r = +0.97$); (C) 31 nodes ($y = 21.8 + 1.43x$. $r = +0.97$).

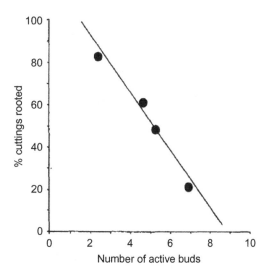

FIGURE 18.13 Relationship between the percent of *Triplochiton scleroxylon* cuttings rooted (week 8) from the top lateral shoot of 30-node stockplants cut to different heights, and the numbers of shoots sprouting from inactive buds following decapitation. ($y = 119.5 - 13.64x$. $r = -0.98$).

TABLE 18.2 Effects of plant height (10−20 nodes) on the total nitrogen, total carbohydrate, and carbohydrate/nitrogen ratio of lateral shoots 11 weeks after the decapitation of *Triplochiton scleroxylon* plants.

	No. of nodes per plant		
	10	**15**	**20**
Total carbohydrate content (% of dm)	7.7 (±0.2)	9.7 (±0.3)	9.5 (±0.3)
Total nitrogen content (% of dm)	3.2 (±0.03)	2.9 (±0.02)	2.7 (±0.04)
Carbohydrate:nitrogen ratio	2.4	3.0	3.5

competition between the shoots, for there is a good correlation between rooting and the number of shoots per stockplant (Fig. 18.13). Interestingly, however, cuttings from the lower lateral shoots of the taller stockplants, unlike those of short plants (Fig. 18.8), rooted well (Fig. 18.5A), suggesting that the lower shoots well away from the mainstem apex had a competitive advantage. In apples *(Malus)* comparable variation has been attributed to lesser amounts of phloem sclerenchyma (Howard and Blasco, 1979). In *T. scleroxylon* orientation of the mainstem away from the vertical eliminated these differences in rooting ability between lateral shoots of tall plants (Fig. 18.5) without affecting the mean numbers of cuttings that rooted per stockplant.

Despite having little apparent effect on the growth of basal shoots, the application of complete fertilizer (NPK) to pruned stockplants considerably enhanced the rooting ability of these lower laterals. In contrast, fertilizer enhanced the growth of apical lateral shoots but did not affect the rooting of cuttings from them (Fig. 18.7), except when competition from basal shoots was removed (Fig. 18.9). The dependence of root initiation on supplies of N and carbohydrates is well known (Hess, 1969; Haissig, 1974), but there are few reports of the role of fertilizers in stockplant management; most investigations have been concerned with internal concentrations of N and carbohydrate, although the relevance of carbohydrate/nitrogen ratios to root initiation is unknown (Haissig, 1974). With *T. scleroxylon* this ratio was affected by stockplant height, the contents of total N being larger in short than in tall plants, with the converse relationship for amounts of carbohydrate (Table 18.2). It seems, therefore, that *T. scleroxylon* shoots containing large amounts of N relative to carbohydrate are likely to yield cuttings that root readily.

Surgically manipulating stockplants confirmed the importance of competition between shoots (Figs. 18.8, 18.9), fertilizers (Fig. 18.9) and lateral shoot position (Fig. 18.9) on rooting ability, although lateral shoot position appeared to a very considerable extent to be a result of differing light intensities within the stockplant (Fig. 18.10). In practice, therefore, the high rooting ability of basal shoots (Figs. 18.5A, 18.9) may be attributable to shading, as low stockplant irradiation is known to enhance rooting (Hansen et al., 1978; Eliasson and Brunes, 1980; Poulsen and Andersen, 1980), particularly following auxin application (Christensen et al., 1980).

To conclude, the production of easily rooted cuttings from pruned stockplants is a complex process with many interacting factors influencing rooting ability. For the successful management of potted stockplants of *T. scleroxylon* it is important to maintain small decapitated plants (about 10 nodes high) from which only about six cuttings are taken from the two upper shoots. These plants should be given regular supplies of fertilizers and kept under shade. As yet the effects of repeated harvests of cuttings on the complex relationships between shoots are unknown.

ACKNOWLEDGEMENTS

My thanks are due to the U.K. Overseas Development Administration for its financial support, Mrs. R. Wilkinson, Mrs. M. Gardner, and Mrs. M. Ferguson for their technical assistance. Dr. R. Manurung carried out the measurements of water potential, and I.T.E. glasshouse staff, Messrs Ottley and Harvey, are thanked for their care and attention of the plants; as is Professor F. T. Last for his support and interest in the project.

Chapter 19

The Rooting Ability of *Triplochiton scleroxylon* K. Schum. Cuttings: The Interactions Between Stockplant Irradiance, Light Quality and Nutrients

This chapter was previously published in Leakey, R.R.B., Storeton-West, R., 1992. Forest Ecology and Management, 49, 133–150, with permission of Elsevier

SUMMARY

Stockplants of *Triplochiton scleroxylon* were grown in controlled-environment cabinets at the Institute of Terrestrial Ecology, Edinburgh, to test the effects of stockplant llumination on the rooting ability of leafy stem cuttings. The environmental variables were: (1) irradiance (PAR = 106, 202 and 246 μmol m^{-2} s^{-1}) with a uniform light quality (red: far red ratio = 1.75); (2) light quality (R: FR = 1.6 and 6.3) with a uniform irradiance (PAR = 294 μmol m^{-2} s^{-1}), and (3) irradiance (PAR = 250 and 650 μmol m^{-2} s^{-1}) and nutrients (with and without 0.2% solution of 1 : 1 : 1, N:P:K) at a uniform light quality (R: FR = 6.3). In all experiments, measurements were made of shoot length and leaf size and in the third experiment, the net photosynthetic rates of each leaf were determined prior to taking cuttings. Leaf area and leaf and stem dry weights were measured, as were their starch and refluxsoluble carbohydrate contents. The resuits showed that decreasing R: FR and irradiance independently increased both shoot growth and rooting ability. Strong positive relationships between photosynthesis and rooting were found when stockplants were grown at low irradiance (250 μmol m^{-2} s^{-1}) with and without fertilizers. A similar relationship was found, at high irradiance (650 μmol m^{-2} s^{-1}) only when nutrients were added. A strong negative relationship between the same parameters occurred without fertilizers at high irradiance. In addition, a weak negative relationship was found between rates of photosynthesis and the starch content of cuttings. It is concluded that end-product inhibition prevented the rooting of cuttings from stockplants grown without fertilizers at high irradiance with an R: FR ratio of 6.3.

INTRODUCTION

Vegetative propagation of *Triplochiton scleroxylon* K. Schum., an indigenous hardwood of West Africa, has been the basis of a genetic improvement program for this species over the last 17 years (Leakey, 1987). To ensure the long-term, sustainable rooting of cuttings from managed stockplants, studies have investigated the sources of variation in rooting ability. These have ranged from the determination of the best rooting environment and optimal auxin requirements (Howland, 1975; Leakey et al., 1975, 1982a) to the role of the leaf on the physiological status of the cutting (Leakey et al., 1982a; Leakey and Coutts, 1989). Additional studies have examined both within and between shoot variables attributable to cutting position within the shoot, shoot position within the plant and to stockplant management and pre-severance treatments (Leakey, 1985, 1992; Leakey and Mohammed, 1985).

During a number of the preceding studies it has become apparent that the amount of light received by individual leaves prior to severance may, to a considerable extent, determine the cutting's subsequent rooting ability. The level of

Multifunctional Agriculture. DOI: http://dx.doi.org/10.1016/B978-0-12-805356-0.00019-2
© 2017 Elsevier Inc. All rights reserved.

irradiance received by stockplants of conifers, deciduous trees, and herbaceous plants is known to affect the subsequent rooting ability of cuttings taken from them (Hansen and Eriksen, 1974; Hansen et al., 1978; Eliasson and Brunes, 1980; Moe and Andersen, 1988). It is, as yet, unclear whether the reduced rooting ability of shoots grown at high irradiance is due to excessive carbohydrate accumulation, water stress, or other factors.

Within a stockplant canopy, shading results in changes in both the quantity and quality of light. The present study investigates whether irradiance or spectral composition are likely to be the main factors affecting rooting ability.

MATERIALS AND METHODS

Fourteen clones of *T. scleroxylon,* originating from controlled cross-pollinations made at the Institute of Terrestrial Ecology (ITE) between clones 8057 and either 8001 or 8002 (Leakey et al., 1981), were grown as managed two-shoot stockplants in glasshouses at ITE, near Edinburgh. Except where otherwise stated, all plants received weekly applications of Sangral SS20 liquid fertilizer (L and K Fertilizers Ltd., Lincoln, UK) with a N:P:K ratio of 1:1:1 (166 ppm). This fertilizer was applied using a Keylutor MKIII in place of the daily watering. Glasshouse air temperatures were maintained at $30 \pm 2°C$ and daylight was supplemented by mercury vapor lighting to provide a year-long daylength of 19.5 hours and a minimum photon flux density (PAR) 150 μmol m^{-2} s^{-1} at plant level. Maximum irradiance in the glasshouse on bright summer days was about 1400 μmol m^{-2} s^{-1}.

Experimental plants were vegetatively propagated using the technique and mist propagators described by Leakey et al. (1982a). Assessments of rooting were made weekly from 2 to 8−9 weeks after severance. Those cuttings which had rooted were potted up and removed from the experiment. The rooted cuttings were potted and later repotted into 7 and 13 cm pots using a compost (7:3:1 mixture of peat:sand:loam) with 4.2 g kg^{-1} Enmag, 2.6 g kg^{-1} John Innes base and 0.3 g kg^{-1} trace elements. Controlled-environment growth cabinets/chambers were used to provide a range of photon flux density and spectral compositions (R:FR ratios). Experiments 1 and 2 utilized modified Fisons (140-G2 MK III) cabinets and Experiment 3 larger growth chambers (Environment and Air Conditioning Ltd., Blackpool, UK). Cuttings were harvested from these plants, their leaves trimmed to 50 cm^2 and 40 μg indole-3-butyric acid (IBA) applied per cutting, in alcohol, as described by Leakey et al. (1982a). Standard errors for percentage rooting were calculated using the procedure of Bailey (1959) for data with binomial distributions. Correlation coefficients were calculated on relations between: (1) the percentage of cuttings rooted and net photosynthesis per leaf and (2) the starch content (rag) in the stem portion of cuttings and net photosynthesis per unit dry weight of leaf. Differences are described as significant where $p = \leq 0.05$, unless otherwise stated.

Six Fisons growth cabinets were used, two for each of three light regimes. Each cabinet contained 12 two-shoot stockplants which had been cut back to 15 cm stumps 1 week prior to insertion in the cabinets. The 12 plants were one of each of clones 8001, 8100, 8102, 8104, 8105, 8110, 8111, 8112, 8114, 8116 and two plants of 8101. The different light regimes were established by using a mixture of fluorescent warm-white tubes and incandescent bulbs with a wattage ratio of about 1.0 (i.e., watts incandescent: watts fluorescent of 110:120, 255:240, and 355:360). The respective mean red:far-red ratios determined at 660 and 730 nm with a Skye red:far-red meter (SKR 100) were 1.77 ± 0.09, 1.69 ± 0.08, and 1.78 ± 0.03. The mean irradiances (PAR) as determined 27 cm below the light source by a Skye PAR meter (SKP 200) with a quantum sensor, were 106, 202, 246 μmol m^{-2} s^{-1}, with mean total radiation as measured by a Kipp solarimeter at the same point, of 96, 179, and 215 W m^{-2}. All light readings were measured at five positions in each cabinet. The plants were maintained with their top leaf 27 cm below the light source by lowering the shelves and adjusting individual blocks under each pot. The day and night temperature was $30 \pm 1.5°C$.

Effects of Two Light Qualities With Uniform Irradiance

Four Fisons growth cabinets were used, two for each light regime. Each cabinet contained 12 single-shoot stockplants, eight of clone 8081 and two each of clones 8085 and 8095.

The light regimes were established by different combinations of fluorescent cold-white, warm-white tubes, and incandescent bulbs. A red:far-red ratio (R:FR) of 1.6 resulted from the combination of four 40 W warm-white tubes with 2×40 W and 2×25 W incandescent bulbs, while four 40 W cold-white tubes gave a R:FR ratio of 6.3. The action spectra (400−750 nm) of these regimes are presented in Fig. 19.1. The mean levels of irradiance (PAR) measured 27 cm below the light source, at ten positions in each cabinet were 288 ± 5.9 and 299 ± 6.1 μmol m^{-2} s^{-1}. As in the previous experiment the light received by the uppermost leaf was maintained constant by lowering the shelving in the cabinets. Both the day and night temperatures were $30 \pm 1.5°C$.

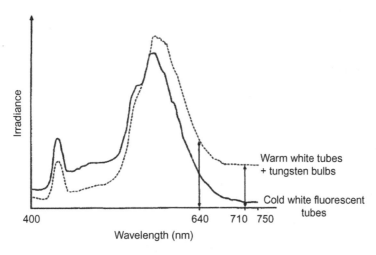

FIGURE 19.1 The action spectra of two artificial light regimes (- - - -,1.6 R:FR; -----, 6.3 R:FR) with a constant irradiance (PAR) of 294 μmol m^{-2} s^{-1} used for the growth of *Triplochiton scleroxylon* stockplants prior to harvesting cuttings.

Effects of Irradiance and Nutrients at a R:FR Ratio of 6.3

Using two large growth chambers, 48 plants of four *T. scleroxylon* clones (8084, 8085, 8094, and 8095) were grown in 19 cm diameter pots at each of two levels of irradiance, 250 and 650 μmol m^{-2} s^{-1} (or 82.5 and 164 Wm^{-2}). Temperatures were maintained at 28−30°C and relative humidity at 80%. The daylength was 19.5 hours. Half the plants were grown with added fertilizer (250 mL of 2% Sangral SS20 liquid), N:P:K, 1:1:1 (2.54 g l^{-1}) and half without. Twelve of the plants were used to assess leaf areas, rates of net photosynthesis, and subsequently, for dry weight and chemical analysis, while the remaining 36 plants per chamber were harvested for cuttings.

The plants were cut back to 15 cm tall stumps and were presprouted in the glasshouse for 2 weeks prior to the start of treatment. After 25 days, growth assessments of photosynthesis started using a differential infrared gas analyzer. All the leaves (4−7) of two to four plants were measured each day, the complete job (24 plants) taking 8 days, starting first with the plants grown at 150 μmol m^{-2} s^{-1} and alternating between those with and without fertilizer. Prior to placing the large leaves of *T. scleroxylon* in a specially made hexagonal aluminum and plexiglass assimilation chamber (230 mm × 230 mm × 50 mm) with internal ventilation (Wilson, 1978) the leaf area was measured by outlining the leaves on paper. The leaf was then trimmed to fit the cuvette, the area of the trimmed-off portion assessed and retained. Gas exchange measurements were made on leaves from the top six node positions (Nl − N6), each at its own level within the stockplant canopy. Flow rates of the IRGA and cuvette were 1000 cm^3 min^{-1} and 5000 cm^3 min^{-1}, respectively. Cuvette temperature, as determined by a copper constantan thermocouple, was 28−29°C. Ambient CO_2 and atmospheric pressures were 365−405 μmol mol^{-1} and 98−103 KPa, respectively. An estimate of boundary layer resistance was determined by Ladipo (1981), using a brass leaf model, as described by Grace et al. (1980). After completing gas exchange measurements, the plant was cut up into single-node pieces and the leaf (with its respective trimmed portion) and stem portions were taken for dry weight and chemical analysis. Carbohydrates were extracted by boiling in 0.5% ammonium oxalate solution for 2 hours under reflux (Deriaz, 1961) and the reflux-extracted soluble carbohydrate (RSC) content determined by the anthrone reaction. Starch content was assessed by hydrolyzing the residue and similarly assessing its glucose content. The cuttings were set under intermittent mist in nine blocks. Five cuttings of each of two plants per treatment were randomized in each block.

RESULTS

Effects of Three Levels of Irradiance With Constant Light Quality

The major growth effects of irradiance were on the length of the top shoot and the length of the uppermost fully expanded leaf on the top shoot. Top shoot lengths were significantly different between each treatment but longest at the medium level of irradiance, and shortest at the highest irradiance (Table 19.1). The length of the second shoot was longest at the highest irradiance. The consequences of these different lengths were differences: (l) in mean cutting length,

TABLE 19.1 Effects of irradiance on the growth of shoots and leaves from two-shoot stockplants of *Triplochiton scleroxylon* at uniform light quality (R:FR ratio = 1.75).

Growth parameters of shoots and leaves	Level of irradiance (μmol m^{-2} s^{-1})		
	106	202	246
Length of top shoot (mm)	228 ± 14.4	250 ± 15.1	191 ± 9.2
Length of second shoot (mm)	67.4 ± 12.7	67.3 ± 9.8	75.0 ± 7.5
Dominance ratio	3.4	3.7	2.6
No. of nodes of top shoot	8.52 ± 0. 19	9.33 ± 0.26	9.08 ± 0.25
No. of nodes of second shoot	5.00 ± 0.31	5.88 ± 0.30	5.21 ± 0.22
Leaf length of top shoot (mm)[a]	185 ± 5.5	177 ± 4.0	163 ± 4.2
Leaf length of second shoot (mm)[1]	116 ± 7.9	116 ± 5.1	116 ± 4.4

[a]*Length of uppermost expanded leaf.*

TABLE 19.2 Effects of stockplant irradiance on the mean length (ram) of cuttings (i.e., internode length) taken from the top shoot of *Triplochiton scleroxylon* stockplants grown at uniform light quality (R:FR ratio = 1.75).

Irradiance (μmol m^{-2} s^{-1}) at the top of the plant	Nodes				
	Apical				Basal
	1	2	3	4	5
106	44.2 ± 3.2	37.5 ± 2.6	30.6 ± 2.2	24.3 +2.5	19.9 ± 2.4
202	54.3 ± 3.3	47.5 ± 2.9	36.3 ± 2.9	30.0 ± 2.6	21.2 ± 1.9
246	35.8 ± 3.3	32.4 ± 1.7	27.7 ± 1.5	22.3 ± 1.9	17.0 ± 1.8

which was longest within each treatment at the uppermost node (Table 19.2); and (2) in the dominance ratio (length of topshoot: length of the second shoot), with the strongest dominance at the medium level of irradiance (Table 19.1). Leaf length decreased significantly between the medium and highest levels of irradiance.

Five cuttings were harvested from the top shoot of each plant after 4−5 weeks shoot growth, and were each treated with 40 μg IBA in l0 μl of industrial methylated spirit (IMS). The cuttings of two plants per treatment were allocated at random to each of 12 blocks and set under intermittent mist. After 5 weeks the percentage of cuttings rooted was greater the lower the level of irradiance (Fig. 19.2). Three weeks later, there was no significant difference between rooting from the two lower levels of irradiance, which continued to be better than the highest level of irradiance. The number of roots rooted cutting was greatest at the intermediate level of irradiance (Table 19.3), especially in weeks 3−5.

Effects of Two Light Qualities With Uniform Irradiance

Light quality affected shoot elongation but not leaf growth. Mean shoot length increments in l month under the cabinet regimes were 227.8 ± 13.0 mm and 103.0 ± 3.7 mm under the 1.6 and 6.3 R:FR light qualities, respectively. This affected mean cutting length (Table 19.4). In contrast, mean leaf lengths per plant were very similar in both treatments (176.2 ± 4.5 mm and 178.9 ± 3.5 mm for 1.6 and 6.3 R:FR ratios, respectively). The subsequent rooting ability of cuttings taken from plants grown under R:FR ratios of 1.6 was greater than those from R:FR of 6.3 (Fig. 19.3).

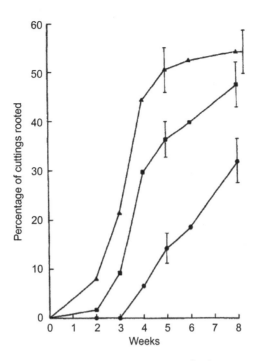

FIGURE 19.2 Effects of stockplant irradiance (▲ = 106, ■ = 202, ● = 246 μmol m^{-2} s^{-1}) on the rooting ability of *Triplochiton scleroxylon* cuttings from stockplants grown at uniform light quality (R:FR ratio = 1.75).

TABLE 19.3 Effects of stockplant irradiance on the mean accumulated number of roots per rooted cutting of *Triplochiton scleroxylon* grown at uniform light quality (R:FR ratio = 1.75).

Irradiance (μmol m^{-2} s^{-1}) at the top of the plant	Weeks after taking cuttings					
	2	3	4	5	6	8
106	6.8	5.6	4.6	4.3	4.2	4.2
202	1.5	7.6	5.1	4.7	4.5	4.4
246	–	–	3.6	2.9	2.6	2.7

TABLE 19.4 Effects of stockplant light quality on mean length (mm) of single-node cuttings of *Triplochiton scleroxylon* grown under uniform irradiance (294 μmol m^{-2} s^{-1}).

R:FR	Nodes				
	Apical				Basal
	1	2	3	4	5
1.6	70.0 ± 4.2	55.5 ± 4.2	35.7 ± 2.7	27.5 ± 1.6	25.0 ± 1.6
6.3	32.5 ± 1.2	24.0 ± 1.1	18.7 ± 0.9	16.5 ± 0.8	20.5 ± 1.3

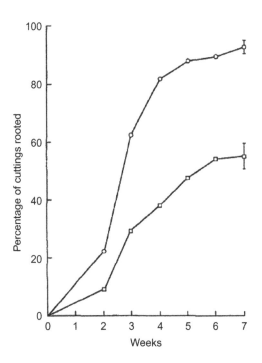

FIGURE 19.3 Effects of light quality (○ = 1.6, □ = 6.3 R:FR ratios) as a stockplant pretreatment on rooting ability of single-node *Triplochiton scleroxylon* cuttings grown on uniform irradiance (294 μmol m^{-2} s^{-1}).

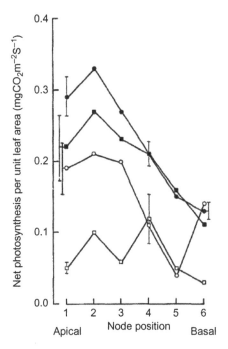

FIGURE 19.4 Effects of stockplant irradiance (●, ■ = 250; ○, □ = 650 μmol m^{-2} s^{-1}) and fertilizer (○, ● with; and □, ■ without 2% SS20 liquid feed) on rates of net photosynthesis of *Triplochiton scleroxylon* stockplants grown at a uniform light quality (R:FR ratio of 6.3).

Effects of Irradiance and Nutrients: at an R:FR Ratio of 6.3

Rates of net photosynthesis per unit dry weight were on average more than double in stockplants grown at 250 μmol m^{-2} s^{-1} compared with those of stockplants grown at 650 μmol m^{-2} s^{-1} (Fig. 19.4). At the lower irradiance, the rates were greatest at the apical nodes, whereas at the higher irradiance they were higher at central nodes. There was no

TABLE 19.5 Effects of irradiance and fertilizer (with and without 2% N:P:K 1:l:l liquid feed) on rates of net photosynthesis per unit of leaf dry weight and per leaf, in *Triplochiton scleroxylon* stockplants grown at a uniform light quality (R:FR ratio = 6.3).

	Nodes					
	Apical					Basal
	1	2	3	4	5	6
Net photosynthesis per unit dry wt. (mg CO_2 g^{-1} s^{-1} $\times 10^{-2}$)						
650 μmol m^{-2} s^{-1}						
Without fertilizer	0.11	0.25	0.13	0.34	0.11	0.12
With fertilizer	0.29	0.42	0.51	0.25	0.08	0.09
250 μmol m^{-2} s^{-1}						
Without fertilizer	1.03	1.04	1.07	0.61	0.45	0.35
With fertilizer	0.91	1.19	0.85	0.71	0.57	0.38
Net photosynthesis per leaf (mg CO_2 g^{-1} s^{-1} $\times 10^{-2}$)						
650 μmol m^{-2} s^{-1}						
Without fertilizer	0.07	0.19	0.09	0.22	0.06	0.08
With fertilizer	0.39	0.46	0.02	0.09	0.07	0.09
250 μmol m^{-2} s^{-1}						
Without fertilizer	0.46	0.66	0.61	0.43	0.22	0.35
With fertilizer	0.66	0.91	0.64	0.48	0.30	0.14

effect of fertilizers at the lower irradiance, while at the higher irradiance, rates of net photosynthesis were very low in plants without fertilizer. Similar patterns of response were shown when net photosynthesis was expressed per unit leaf area and per cutting (Table 19.5).

Chemical analysis of single unreplicated samples of leaf and stem portions from the shoot at the end of the growing period, and prior to propagation, indicated that without fertilizers both the leaves and stems of shoots grown at 650 μmol m^{-2} s^{-1} had a high starch concentration especially at basal nodes (Table 19.6). The same pattern occurred when the data were expressed as weight per cutting. The leaves of nonfertilized stockplants also had a higher starch concentration than fertilized plants, those at 650 μmol m^{-2} s^{-1} being greatest at basal nodes, while those at 250 μmol m^{-2} s^{-1} were greatest at apical nodes (Table 19.6). In leaves, reflux-soluble carbohydrate concentrations were similar among treatments and nodes, while in stems concentrations were greater at apical nodes.

Cuttings from plants receiving 650 μmol m^{-2} s^{-1} without fertilizers had particularly high soluble carbohydrate concentrations (Table 19.6). Both stem and leaf dry weights were greatest in plants from 650 μmol m^{-2} s^{-1} without fertilizers. Stem dry weights were lowest, however, from plants at 650 μmol m^{-2} s^{-1} without fertilizers, while leaf weights were lowest from plants at 250 μmol m^{-2} s^{-1} without fertilizers (Table 19.6). The rooting of cuttings from stockplants grown at high irradiance without fertilizers was very poor (9.3 $\pm$ 3.1%). The addition of fertilizers improved rooting (28.7 $\pm$ 4.8%), while at the lower irradiance, rooting without fertilizers was better (33.3 $\pm$ 5.0%) than with fertilizers (22.1 $\pm$ 4.5%). Generally cuttings from apical nodes rooted better than those from more basal cuttings (Fig. 19.5).

The Light Environment of a Stockplant Canopy

Using a quantum spectroradiometer (Tectum QSM 2500), the spectral composition of light at the top, middle, and bottom of *T. scleroxylon* stockplants was measured at midday under standard glasshouse conditions. The red:far-red ratios were found to be 0.92, 0.37, and 0.28, respectively. At the midpoint of the canopy light interception was about 80%, rising to about 95% at the bottom of the canopy.

TABLE 19.6 Effects of irradiance (low = 250 μmol m^{-2} s^{-1}; high = 650 μmol m^{-2} s^{-1}) and fertilizer (with = 2% N:P:K 1:1:1 liquid feed, without = 0) on the content of starch and reflux-soluble carbohydrate (RSC) contents (% dm) and dry weights of the stem (Nodes 1–6) and leaves of *Triplochiton scleroxylon* stockplants grown at a uniform light quality (R:FR ratio of 6.3).

Treatments	Node no.	% Starch		% RSC		Dry wt. (g)	
		Stem	Leaf	Stem	Leaf	Stem+S.E.	Leaf+S.E.
High light with fertilizer	NI (apical)	0.10	0.19	6.7	2.6	0.11 ± 0.01	1.02 ± 0.05
	N2	0.23	0.20	5.7	4.4	0.20 ± 0.10	1.06 ± 0.07
	N3	0.23	0.19	5.7	3.3	0.15 ± 0.05	0.92 ± 0.05
	N4	0.82	0.11	7.5	1.7	0.32 ± 0.14	0.89 ± 0.14
	N5	0.49	0.10	4.8	3.9	0.20 ± 0.07	0.60 ± 0.10
	N6 (basal)	0.48	0.17	4.4	5.3	0.23 ± 0.08	0.49 ± 0.05
High light without fertilizer	NI (apical)	0.22	0.58	9.7	4.3	0.06 ± 0.02	0.66 ± 0.08
	N2	0.22	0.38	9.7	5.0	0.06 ± 0.01	0.69 ± 0.06
	N3	0.70	0.31	8.2	4.2	0.09 ± 0.02	0.67 ± 0.06
	N4	0.70	0.73	8.2	4.9	0.11 ± 0.03	0.80 ± 0.06
	N5	1.40	2.40	7.9	6.1	0.14 ± 0.03	0.59 ± 0.11
	N6 (basal)	1.40	2.00	7.3	8.3	0.12 ± 0.01	0.41 ± 0.19
Low light with fertilizer	NI (apical)	0.10	0.96	7.5	6.8	0.07 ± 0.01	0.72 ± 0.05
	N2	0.21	0.21	5.8	4.1	0.09 ± 0.01	0.75 ± 0.06
	N3	0.21	0.19	5.8	5.0	0.11 ± 0.02	0.79 ± 0.04
	N4	0.20	0.70	5.0	3.3	0.22 ± 0.11	0.67 ± 0.06
	N5	0.15	0.15	5.1	4.4	0.16 ± 0.04	0.58 ± 0.07
	N6 (basal)	0.28	0.19	5.3	4.5	0.23 ± 0.11	0.39 ± 0.03
Low light without fertilizer	NI (apical)	0.10	1.90	6.6	5.9	0.06 ± 0.01	0.71 ± 0.08
	N2	0.10	1.30	6.6	5.1	0.06 ± 0.01	0.63 ± 0.08
	N3	0.16	0.69	6.4	5.3	0.13 ± 0.04	0.65 ± 0.10
	N4	0.31	0.43	6.6	5.6	0.13 ± 0.02	0.72 ± 0.08
	N5	0.28	0.34	5.2	4.8	0.17 ± 0.04	0.53 ± 0.07
	N6 (basal)	0.35	0.19	5.2	4.5	0.11 ± 0.01	0.27 ± 0.04

DISCUSSION

Cuttings of *T. scleroxylon* rooted well from stockplants grown at low irradiance and at qualities with a high far-red content, illustrating that these factors can independently affect rooting, whereas in nature changes in irradiance and light quality usually occur together.

Little is known about the effects of light quality on rooting ability (Moe and Andersen, 1988), but the benefits of low stockplant irradiance on rooting ability have previously been shown in a wide range of plant species (Hansen and Eriksen, 1974; Hansen et al., 1978; Eliasson and Brunes, 1980). However, the ways light affects root initiation are not clear (Baadsmand and Andersen, 1984). The most obvious possibility is through direct effects of photosynthesis on assimilate production and hence on the initial carbohydrate content of the cuttings. However, indirect effects could also

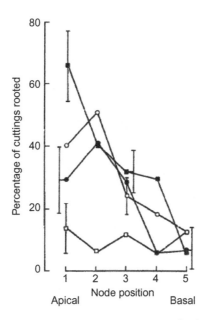

FIGURE 19.5 The effects of stockplant irradiance (●, ■ = 250, ○, □ = 650 μmol m^{-2} s^{-1}) and fertilizers (○, ●, with; and □, ■, without 0.2% N:P:K 1:1:1 liquid feed) on the subsequent rooting of cuttings taken from *Triplochiton scleroxylon* stockplants grown under a uniform light quality (R:FR ratio of 6.3).

occur. For example, from this study, the plants grown under different irradiances were morphologically and probably physiologically different as they had different shoot lengths, internode lengths, and leaf lengths even when grown under uniform light qualities. Furthermore, treatment differences between these characteristics in the top shoot of two-shoot stockplants were not reflected in the growth of the second shoot. Thus, the distribution of assimilates, the correlative relationships between shoots and presumably many other factors were also different in plants at different irradiances. From the work of Kwesiga and Grace (1986) and Kwesiga et al. (1987), it is likely that changes in R:FR ratio may have resulted in differences in specific leaf area, quantum efficiency, and photosynthetic rates. The effects of irradiance and light quality treatments are, therefore, complex. For example, Baadsmand and Andersen (1984) have shown that, in *Pisum sativum,* high levels of stockplant irradiance enhanced the transport of auxin to the base of subsequently collected cuttings. They postulated that this resulted from the higher tissue temperatures at high irradiance. It is suggested that the effect of high irradiance in their study was the formation of fewer roots per cutting because of the earlier establishment of dominance by the first-formed roots. A reduction in root numbers per cutting at high irradiance was also found in *T. scleroxylon* in the present study.

In the present study, the levels of irradiance used (106−650 μmol m^{-2} s^{-1}) were low by comparison with normal conditions in moist deciduous forests of West Africa, where levels in excess of 1200 μmol m^{-2} s^{-1} can occur on bright days. The light response curve for assimilation in *T. scleroxylon,* a light-demanding pioneer species (Hall and Bada, 1979), indicates that photosynthesis reaches saturation at about 800 μmol m^{-2} s^{-1} (Ladipo et al., 1984). Thus it is surprising that maximum shoot growth occurred at about 200 μmol m^{-2} s^{-1}. The reasons for this are likely to be the unnatural light spectrum generally obtained in growth cabinet experiments. Under artificial lighting conditions, it is difficult to get high irradiances with near natural light quality. In particular, natural red:far-red ratios are hard to obtain, as demonstrated in the present study where values were always in excess of 1.2, the level occurring in full sunlight in West Africa. Care must, therefore, be taken in interpreting responses to growth cabinet experiments and extrapolating the results to field conditions. When *T. scleroxylon* was grown at R:FR ratios of 1.6 and 6.3, there were major effects on shoot growth and particularly internode elongation. This is in accordance with the known effects of phytochrome on shoot growth (Morgan and Smith, 1976) especially of light-demanding trees (Warrington et al., 1988). Since cutting length is known to affect rooting ability in single-node cuttings of *T. scleroxylon* (Leakey, 1983; Leakey and Mohammed, 1985), it would be reasonable to assume that the shorter cuttings formed at R:FR of 6.3 would root less well than the longer ones formed at R:FR of 1.6. Rooting was affected by light quality (Fig. 19.3), but with neither treatment was the relationship between cutting length and rooting ability like that reported for glasshouse-grown cuttings. The cuttings formed at R:FR of 1.6 rooted in excess of 90% regardless of cutting length, while those formed at R:FR of 6.3 rooted better than would be expected for their lengths. In the latter group, however, there was only a very

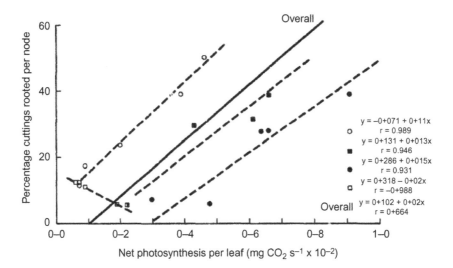

FIGURE 19.6 The relationships between net photosynthesis per leaf in *Triplochiton scleroxylon* stockplants and subsequent rooting ability when stockplants were grown at different levels of irradiance (●, ■ = 250, ○, □ = 650 μmol m^{-2} s^{-1}) and fertilizer (○, ●, with; and □, ■, without 0.2% SS20 liquid feed) (R:FR ratio = 6.3).

weak relationship between length and rooting ability. Similarly, the relationships between these parameters were weak in the experiment with different levels of irradiance at R:FR of 1.75. Leakey and Coutts (1989) have previously reported that the relationship between cutting length and rooting ability was strongest in small leaf cuttings where there seemed to be dependence on stored reserves. The weak relationships found in the present study may suggest that different light qualities and irradiances influence the need for stored reserves.

At the R:FR ratio of 6.3 in the third experiment, rooting was again poor, but cuttings from stockplants grown at low irradiance, with and without nutrients, rooted relatively well, while in cuttings from high irradiance stockplants, there was an interaction, and those without nutrients rooted very poorly. The explanations for these results appear to center on the photosynthetic ability of the shoots. Over all treatments there was a fairly strong positive relationship ($r = 0.66$) between net photosynthesis per leaf and the rooting ability of cuttings (Fig. 19.6). However, within each treatment this relationship was stronger (high light with fertilizers, $r = 0.99$; low irradiance without fertilizers, $r = 0.95$; low irradiance with fertilizers, $r = 0.93$), but with considerable differences between treatments. Thus for the same level of rooting, the difference between high and low irradiance with fertilizers was that the rate of net photosynthesis per leaf was $0.4–0.5$ mg CO_2 s^{-1} × 10^{-2} greater at low irradiance. In the very difficult-to-root cuttings from high irradiance without fertilizers, the relationship between photosynthesis and rooting was very different and strongly negative ($r = -0.99$). In this instance the higher the rate of photosynthesis, the lower the rooting percentage. These cuttings were the ones with the greatest starch content. It thus appears that the very poor level of rooting in these cuttings was the result of suppressed photosynthetic ability, resulting from end-product inhibition. This conclusion is supported by the relationship between net photosynthesis per leaf and the starch content of the cuttings from all the treatments in this experiment (Fig. 19.7).

The conclusion that the rooting ability of cuttings is a function of their preseverance ability to produce assimilates is interesting as cuttings of *T. scleroxylon* were clearly capable of assimilate production after severance (Leakey and Coutts, 1989; Leakey and Coutts, unpublished data (1983)), indicate that the cutting's ability to produce assimilates postseverance is important to the rooting process. This is seen in the lack of a relationship between rooting ability and the content of reflux-soluble carbohydrates on the day of severance (day 0), and the existence of relationships between these parameters after 28 days under intermittent mist (Fig. 19.8). These relationships between rooting ability and preseverance rates of net photosynthesis imply that treatments which limit photosynthesis preseverance may predetermine the level of assimilate production during propagation. The latter can, however, also be manipulated by trimming the leaf to different sizes.

Although performed under unnatural illumination in growth cabinets, the results of the present study would appear to support the conclusion that much of the within-stockplant variation reported by Leakey (1983) can be attributed to differences in levels of irradiance and light quality within-stockplant canopies, including the apparent interactions with nutrients. Experiments are planned to examine the effects on rooting of R:FR ratios lower than 1.0.

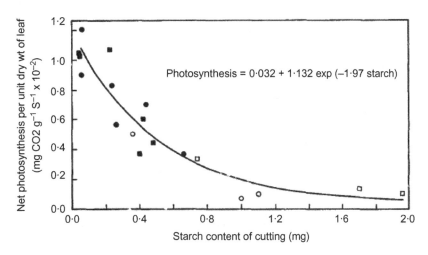

FIGURE 19.7 Relationship in *Triplochiton scleroxylon* stockplants between net photosynthesis per unit dry weight of leaf and the starch content (mg per cutting) of the stem portion of cuttings when stockplants were grown at different levels of irradiance ($\bullet$, $\blacksquare$ = 250, $\circ$, $\square$ = 650 μmol m^{-2} s^{-1}) and fertilizer ($\circ$, $\bullet$, with; and $\square$, $\blacksquare$, without 0.2% SS20 liquid feed) at a uniform light quality (R:FR ratio = 6.3).

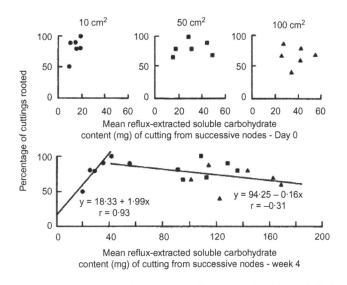

FIGURE 19.8 Relationships between rooting ability (week 6) and mean reflux-extracted soluble carbohydrates at 0 and 28 days after taking *Triplochiton scleroxylon* cuttings with different lamina areas ($\bullet$ = 10, $\blacksquare$ = 50, $\blacktriangle$ = 100 cm^2). Each point represents the mean value for a different node position.

To conclude, light quality and irradiance independently affect rooting ability and provide some evidence that photosynthesis is important in the determination of rooting ability. Practically, this study indicates a relatively easy way of enhancing rooting ability in clonal tree improvement programs. For example, using the results presented here, the rooting success of cuttings in the World Bank-funded tree improvement unit in the National Office for Forest Regeneration in Cameroon has been raised from 21 to 74% by growing plants of *Leucaena leucocephala* as shade in rows on either side of hedged *T. scleroxylon* stockplants (Longman, unpublished data, 1989; Tchatchou, 1989a).

ACKNOWLEDGMENTS

Denise Graham is thanked for her technical assistance and Ray Ottley and Frank Harvey for their care of the plants, both in the glasshouses and in the growth cabinets.

Chapter 20

Plant Cloning: Macro-Propagation

This chapter was previously published in Leakey, R.R.B., 2014. In: van Alfen, N., et al., (Eds.) Encyclopedia of Agriculture and Food Systems, vol. 4. Elsevier, San Diego, pp. 349–359, with permission of Elsevier

SUMMARY

Techniques of macro-propagation have been used for millennia, however it was difficult to identify clear principles applying to all species. Research over the last 20-30 years has provided better understanding of the numerous pre- and post-severance interacting factors – both morphological and physiological – and led to greater clarity. This research has in particular helped to explain what was an apparently contradictory scientific literature based on experiments which did not adequately record details of the stockplant and its environment. The improved understanding has led to the use of macro-propagation in the domestication of new tropical tree crops.

INTRODUCTION

Cloning is a process by which individual organisms are multiplied asexually—a process of vegetative regeneration or reproduction (Longman, 1993). Consequently, the individual plants forming a clone are genetically identical. Cloning can be both a natural and an artificial process. The natural process embraces the lateral spread of creeping plants by their shoots or roots, and the production of new plantlets from dispersed, separated, or fragmented plant parts. Vegetative regeneration is a common characteristic of undesirable and invasive weeds. It is often associated with organs that have evolved as part of a perennial life form involving specialized storage organs (e.g., tubers, bulbs, rhizomes). However, vegetative regeneration also has advantages in agriculture. Firstly some of the specialist storage organs are good sources of carbohydrates for human food and so become crops—potatoes, yams, cassava, onions, etc. In addition, the capacity to regenerate asexually can be used to multiply these crops without the alteration of their genetic characteristics during the segregation phase of sexual reproduction. This also applies to many trees and other plants which can be artificially regenerated by stem cuttings, grafting, budding, or marcotting (air layering).

The development of clonal crops producing specialist storage organs will not be considered further here as the process just involves the selection of the best individuals from natural populations or from the progeny of breeding programs. Instead, we will examine how the artificial process of vegetative regeneration is used by agriculturalists, horticulturalists, and foresters in domestication programs (Gepts, 2014) to capture and multiply individual genotypes and so to produce cultivars and clones of crops which would not normally be multiplied clonally. This is especially important in trees as less progress has been made in tree crop domestication because of their relatively long generation times (approximately 3–20 years), irregularity in flowering and fruiting due to climate and physiological rhythms, predominantly outbreed nature with low heritability in many traits, and high genetic diversity of base populations.

In this Encyclopedia two chapters examine the processes of vegetative regeneration. One involves the relatively new processes of micropropagation (Read and Preece, 2014) which have arisen from the development of modern biotechnology, and the other, this chapter, which involves the more traditional techniques of macropropagation.

Multifunctional Agriculture. DOI: http://dx.doi.org/10.1016/B978-0-12-805356-0.00020-9
© 2017 Elsevier Inc. All rights reserved.

THE USE OF MACROPROPAGATION

The use of cloning is not a new concept associated with the advent of biotechnology. It has been used mainly in perennial crops for thousands of years by agriculturalists and horticulturalists, and used by foresters for at least 800 years (Hartmann et al., 2010; Leakey, 2014c). This is because there are many instances when multiplication by vegetative means is a more appropriate strategy than multiplication by seed (Mudge and Brennan, 1999). It is the capacity to rapidly develop cultivars by clonal selection and propagation that is especially important in the domestication of food and nonfood crops from trees, shrubs, and woody vines (Leakey, 2012b,c). However, the techniques are not exclusively used to multiply woody plants, as many ornamental herbaceous plants are also propagated vegetatively because the techniques are simple, inexpensive and result in high rates of multiplication. This chapter focuses on woody perennials as they are the most challenging subjects for vegetative propagation.

Vegetative propagation has the advantage that it captures nonadditive as well as additive gene effects, but it should be noted that vegetative propagation only replicates existing genetic traits within newly formed plantlets and does not in itself improve the genetic quality of the material being propagated.

There are a number of important practical issues that need to be resolved when making the decision to use vegetative propagation (Table 20.1). These are associated with the use of appropriate technologies, techniques, and plant tissues (Table 20.2) to ensure the efficient and wise use of clones for improved and sustainable production

TABLE 20.1 When to use vegetative propagation as a tool in tree domestication.

When elite trees have a rare combination of a few inherited traits

When there are many desirable traits for simultaneous selection

When high uniformity is needed to ensure profitability and to meet market specifications

When the products have a high value that can justify the extra expense

When the trees to be propagated are "shy" seeders and the material for propagation is scarce

When the timescale required does not allow progress through the slower and less efficient process of breeding

When seeds have a low level or short period of viability

When the knowledge of proven traits is acquired through long-term experiments or the traditional knowledge of local people

Extracted from Leakey, R.R.B., Simons, A.J., 2000. When does vegetative propagation provide a viable alternative to propagation by seed in forestry and agroforestry in the tropics and sub-tropics? In: Wolf, H., Arbrecht, J., (Eds.), Problem of Forestry in Tropical and Sub-tropical Countries – The Procurement of Forestry Seed – The Example of Kenya. Ulmer Verlag, Germany, pp. 67–81.

TABLE 20.2 Strategic opportunities to consider when using vegetative propagation to domesticate trees.

What is the most appropriate level of technology to use?

Which tissues are most appropriate—juvenile or mature?

When using juvenile tissues, which is the best source?

When using mature tissues, what are the best methods to use?

How can an easy, sustainable and cost-effective approach be ensured?

How can the best individuals for propagation be selected from broad and diverse wild populations?

What are the opportunities for introducing new variation?

How can clones be wisely used and deployed?

How can a wide genetic base be maintained in clonal populations?

Extracted from Leakey, R.R.B., Simons, A.J., 2000. When does vegetative propagation provide a viable alternative to propagation by seed in forestry and agroforestry in the tropics and sub-tropics? In: Wolf, H., Arbrecht, J., (Eds.), Problem of Forestry in Tropical and Sub-tropical Countries – The Procurement of Forestry Seed – The Example of Kenya. Ulmer Verlag, Germany, pp. 67–81.

(Longman, 1993; Leakey and Simons, 2000) as part of a wise strategy for tree domestication (Leakey and Akinnifesi, 2008; Leakey, 2012c) for the delivery of agroforestry (Leakey, 2012b; Nair et al., 2014; Chapter 28 (Leakey, 2014a).

TECHNIQUES OF MACROPROPAGATION

All techniques of macropropagation are dependent on the capacity of undifferentiated meristematic cells to divide and differentiate to form new shoots or roots (Hartmann et al., 2010). While it is sometimes possible to develop new shoots from root tissues, it is generally easier to develop new roots from shoot tissues.

GRAFTING AND BUDDING

Grafting and budding techniques are especially important in the vegetative propagation of woody plants that have a long period (3−20 years) of juvenile vegetative growth before becoming sexually mature and capable of flowering and fruiting. This is because mature tissues of trees seem to be very difficult to propagate by the formation of roots at the base of a piece of stem (a stem cutting). Consequently, grafting and budding techniques are used in this situation as they don't involve root formation. Instead they are dependent on the fusion of tissues from two different shoots (Hartmann et al., 2010). In this way grafting and budding form multiple copies of large mature trees on seedling rootstocks. Grafting and budding techniques involve the placement of a severed piece of shoot (a scion), or an axillary bud, from the chosen tree in immediate contact with similar tissues on the stump or rootstock of an unselected tree, such that the tissues grow together, fuse and the buds on the attached scion grow out to form a copy of the chosen tree. A number of well known techniques have been used and practiced for thousands of years—known as cleft grafts, approach grafts, whip/tongue grafts, and side veneer grafts (Hartmann et al., 2010).

The success of all these techniques depends on the juxtaposed cambial cells producing callus to form a functional and strong graft union. The ability to achieve this is thought to be dependent on a combination of environmental, anatomical, physiological, and genetic factors.

Genetically, grafting is most successful when the scion and rootstock are closely related—ideally scions of a mother plant grafted on her own seedling progeny. Successful grafts between species and between genera are less common. Even if a union is formed between these poorly related plants, there is the likelihood that there will be tissue rejection at a later date—even many years later—which results in the union breaking. Often this tissue incompatibility is seen as a differential in the growth rate between the rootstock and scion, with one having a larger diameter than the other.

Graft incompatibility is thought to be biochemically mediated and to depend on recognition events between juxtaposed cells through their plasmodesmatal connections. One suggestion is that it involves differences in peroxidase activity across the union which may regulate lignification processes. A difference in the peroxidase banding patterns in both the scion and rootstock is thought to predict graft incompatibility and hence weak graft unions (Gulen et al., 2002). However, despite considerable recent experimentation the mechanisms remain elusive and general principles are hard to elucidate.

Environmentally, probably the main causes of graft failure are nonoptimal temperature for cell division, loss of cell turgidity, movement at the scion/rootstock interface, and disease (Hartmann et al., 2010). Consequently, grafting should be done during the growing season and with protection from water stress and movement. However, almost nothing is known about the importance of the preseverance irradiance, the light quality, and the nutrient status of the severed shoot. These environmental conditions are important for the successful rooting of cuttings. Likewise, not much is known about the best pregrafting environment for rootstocks.

Physiologically, the need for vigorous growth is widely recognized. However, in temperate trees success is often greatest when dormant scions that have been held in chilled storage are grafted to rootstocks already in active growth. This may reflect the reduced chance of water stress in the scion, although it may be associated with changes in the gibberellin and abscisic acid content of plant tissues.

One problem that can arise with grafting is that the scion dies while the rootstock sprouts. If good observations are not made regularly, it can be difficult to know whether the shoots of a grafted plant are of the selected scion clone, or of the unselected rootstock seedling. Of course, if the latter occurs the grafting exercise has been a waste of time, effort, and money.

MARCOTTING OR AIR LAYERING

Layering, the stimulation of roots on intact stems in contact with the ground is a natural feature of many plants, including some trees. This has been modified as an artificial process of vegetative propagation in two main ways—stooling and air layering (or marcotting). In the former, soil mounds are built up around the shoots emerging from coppiced stumps and then the rooted shoots are severed from the stump and planted. This typically captures the juvenile characteristics of the tree associated with the base of the tree trunk.

Air layering is usually applied to mature (capable of flowering and fruiting) branches within the tree crown, usually a long way from the ground (Tchoundjeu et al., 2010). In this case, a ring of bark is removed to promote the accumulation of photosynthates. Then the exposed cambium is typically treated with an auxin rooting powder to promote rooting. It is then wrapped in black polythene enclosing damp compost, peat, or other rooting medium and left for some weeks or months to form roots (Fig. 20.1). This treated part of the branch should be close to the main stem. Once rooted the branch is severely pruned and detached from the tree and potted in a nursery.

Typically air-layered shoots form roots on the underside of the stem, which means that when subsequently planted out the tree does not have a radially orientating root system and so is prone to fall over as the tree gets bigger. To avoid this it is preferable to air layer vertical shoots, such as those formed after pollarding a tree. Alternatively, if a nonvertical branch is used, it can be potted and managed as a stockplant from which to regularly harvest cuttings for subsequent repropagation.

Rigorous studies are needed to determine the best environmental or physiological conditions for successful air layering, although it is affected by branch diameter/age and by season.

FIGURE 20.1   A marcott set on a vertical shoot from a decapitated branch of *Dacryodes edulis. Courtesy of Roger Leakey.*

STEM CUTTINGS

The vegetative propagation of plants by rooting stem cuttings is relatively easy in annual herbaceous species but becomes progressively more difficult as the subject becomes larger and more long-lived. Consequently, large trees provide the nursery manager with the ultimate test of skill, knowledge, and understanding. For this reason, this section focuses on some principles developed for the propagation of tree species.

A wide range of physical facilities are available for the propagation of stem cuttings. These range from sophisticated, electronically controlled glasshouses with fogging equipment, through mist propagation benches to nonmist, polypropagators which are cheap and simple to build and use. The polypropagators are very effective and appropriate for use in remote locations without access to capital and reliable water or electricity services. All these facilities are designed to reduce the postseverance physiological stress (wilting and leaf abscission) resulting from water loss through transpiration by keeping the cuttings cool, moist, and turgid.

Stem cuttings can come in many forms, but the two major groups are leafy softwood cuttings from relatively unlignified, young shoots which are dependent on current photosynthates for rooting, and leafless hardwood cuttings from older and more lignified shoots which depend on the mobilization of carbohydrate reserves stored within the stem tissues. The former are typically a short piece of stem—perhaps a single-node with its bud and leaf and the internode immediately below it (Fig. 20.2), while the latter is often a longer piece of stem with 2–10 nodes and internodes. It is leafless as it has already shed its leaves due to the onset of winter or a dry season. Typically, these large leafless cuttings are taken toward the end of the dormant season.

Although widely practiced for hundreds of years, much of the literature on the rooting of cuttings has been anecdotal, based on the experience of individuals who have neither used standardized conditions nor adequately described

FIGURE 20.2 A single-node rooted cutting of *Triplochiton scleroxylon. Courtesy of Roger Leakey.*

their protocols. This has left a confused literature with many contradictory statements about what determines success (Leakey, 2004). It is only in the last few decades that research has examined many of the factors influencing the rooting process under standardized conditions with the intention of identifying some key principles that ensure success. Furthermore, although much research has been focused on postseverance factors such as the use of auxin "rooting hormones" and the physical environment of the propagation system, much less, indeed very little, research has examined the effects of the preseverance environment of shoots on their stockplants or the effects of stockplant environment. Likewise, very little was known about the interactions between pre- and postseverance factors.

This rest of this section examines five key sets of factors that play a crucial role in determining whether or not leafy cuttings rapidly form a good root system, and hence determine the success of macropropagation for the development of clonal approaches to agricultural production.

The Propagation Environment

The act of severing a cutting from its stockplant creates a physiological shock. To ensure good rooting, the duration and severity of this shock has to be minimized (Mesén et al., 1997a,b). Thus, the most important aspect of the propagation environment is that it minimizes the physiological stresses arising from: (1) the loss of water from the tissues due to transpiration, and (2) the loss of carbohydrate reserves due to tissue respiration. In addition the propagation environment should encourage photosynthesis in leafy cuttings, and promote meristematic activity (mitosis and cell differentiation) in the stem. These processes are linked to the need to transport assimilates and nutrients from the leaf to the base of the stem, and of water from the base of the stem to the leaf.

To minimize these stresses the basic principles are to keep the cuttings well supplied with water at the cutting base, while also maintaining the leaves in an environment with high humidity (low vapor pressure deficit, VPD). The aerial environment in the propagation area is kept cool by shading, and the leaves themselves are cooled by the evaporation of moisture from the leaf surfaces. It is important that the level of shading does not restrict photosynthesis in the leaves (i.e., typically above 400 μmol m^{-2} s^{-1}).

Comparative studies have identified that one of the major advantages of the environment in nonmist polypropagators (Fig. 20.3) is their greater uniformity in both the moisture content of the rooting medium and the VPD across the beds, resulting in overall lower air and leaf temperatures (Newton and Jones, 1993). Mist systems have lower uniformity as a result of both the temporal and spatial distribution of moisture disbursed by intermittent bursts of pressurized mist from jets typically 0.5−1.0 m apart. In addition, peaks of VPD are associated with peaks in irradiance. Species vary in their sensitivity to VPD. This is a result of differences in leaf morphology which affect stomatal conductance, and is often associated with evolutionary adaptations to the physical environment found in the natural range of the species (e.g., rain forest vs woody savannah).

The predominant use of mist and fogging systems by research teams has resulted in few studies on the relationship between photosynthesis and the rooting process due to the difficulty of measuring gas exchange in cuttings with wet leaves. However, studies of gas exchange during propagation have been made in nonmist polypropagators and these have illustrated the importance of photosynthesis for good rooting success, speed of rooting, and the number of roots produced per cutting (Hoad and Leakey, 1994, 1996).

There is a general assumption that the successful rooting of cuttings is associated with a positive carbon balance (i.e., assimilate production > losses through respiration) and a concentration gradient within a cutting which enhances the transport of assimilates from the leaves toward the cutting base. However, little is known about the relationships between respiration losses and cutting origin, leaf area, stem length/diameter, or the rooting environment.

FIGURE 20.3 Left: Nonmist propagator with its lid open. Right: A cross-section showing the different layers of gravel and medium. The arrow marks the top of the water. *Courtesy of Roger Leakey.*

In addition to the aerial environment of the propagator, the environment of the rooting medium is also important. However, the choice of medium is often based on availability of the materials or personal experience. Nevertheless, studies suggest that the medium should hold the cutting firm with its leaf held above the surface of the medium. In addition to providing moisture at the cutting base the medium should allow respiration from the tissues and prevent anoxia which encourages rotting and cutting mortality. To prevent this, the medium should have an air:water ratio that optimizes the oxygen diffusion rate vis à vis the needs of tissue respiration. In practice this involves the use of various sized particles (sand and gravel) and a water-holding medium (compost, perlite, vermiculite, peat, coir, or other organic products) alone or in various mixtures. As in the relationship between leaf morphology and the aerial environment, there is some evidence that morphological and physiological adaptations to different environments may affect the optimum porespace.

Root development and growth is generally enhanced by the cutting base being warmer (artificially enhanced by providing "bottom-heat") than the leaves. This promotes meristematic activity at the cutting base, while the leaves remain cool and free from stress. The converse differential tends to promote the growth of buds creating competition for assimilates.

Postseverance Treatments

Auxin Applications

The application of auxin is considered to promote rooting by stimulation of cell differentiation, the promotion of starch hydrolysis and the attraction of sugars and nutrients to the cutting base, but a better understanding is needed of the mechanisms regulating auxin concentrations and pathways (Atangana et al., 2011a). Indole-3-butyric acid (IBA) is typically the most effective auxin. Auxins are not, however, a "cure-all" treatment and exogenous applications of auxin do not promote rooting in cuttings which are morphologically or physiologically dysfunctional. Thus for auxins to have their stimulatory effects, the cuttings should have been taken from shoots which are preconditioned by stockplant management to be physiologically active when in the propagator, free from water and respiratory stresses, and with the capacity to mobilize stored or current assimilates.

A point of practical importance is that the stems of some species are hairy, while others are waxy. This affects the retention of applied auxin. In addition, uptake is affected by the cross-sectional area of the cutting base (i.e., stem diameter).

Leaf Area

The rooting of softwood cuttings is typically dependent on the presence of a leaf. Studies using infrared gas analyzers to measure the rates of photosynthesis in nonmist polypropagators during the propagation process found that rooting ability was maximized in photosynthetically active cuttings. However, a large photosynthetically active leaf is also actively losing water by transpiration and can suffer water stress—so closing its stomata or shedding its leaf. This then prevents further photosynthesis. Conversely, small leaved cuttings with inadequate assimilate production rapidly decline in their carbohydrate (sugars and starch) content due to respiration losses and hence cannot support root development.

Successful rooting therefore requires an optimal leaf area which balances the positive effects of photosynthesis and the negative effects of transpiration (Leakey, 2004). The leaf area associated with this balance varies depending on the adaptations in leaf morphology of different species to different environments. Such variation is also determined by leaf age (node position) and the position of a shoot with the stockplant.

In addition to leaf area, the correct balance between photosynthesis and transpiration will also be determined by the photosynthetic efficiency of the leaf due to the light environment of the propagator (level of irradiance) and the cutting's water relations (affected by the aerial environment of the propagator and the air:water ratio of the medium). Studies of these important aspects of propagation have found a relationship between rooting ability and the content of reflux-extracted soluble carbohydrates, which confirm the importance of assimilate production throughout the rooting process.

The determination of the optimum leaf area is therefore a very important aspect of developing a successful propagation protocol, especially in difficult-to-root species.

In general, rooting experiments focus on the factors enhancing rooting success. However, to learn more about the processes affecting rooting, there is much that could be learned from paying more attention to the causes of rooting failure. For example, more information is needed about how and when cutting death can be attributed to: water and heat stress, leaf abscission, photoinhibition, negative carbon balance, etc. In addition, some cuttings neither die nor root.

Other related reasons for the failure of cuttings to root can be the death of the leaf due to microbial infection, anoxia and rotting, necrosis, bleaching, or leaf abscission. These problems can arise due to the use of old shoots with senescent or photosynthetically inactive leaves that are past their compensation point (photosynthetic activity vis à vis senescence); water stressed or starch-filled. The most common symptoms are leaf shedding, leaf rot, and stem rot. Gaining an understanding of these causes of cutting death and hence the failure of the propagation process can be as important as determining how to achieve good rooting.

Despite the importance of photosynthesis for the rooting of leafy softwood cuttings, there is some evidence that some species are also able to mobilize and use stored reserves of carbohydrates to contribute to the rooting process. This may reflect differences in stem anatomy and perhaps also adaptations to seasonally harsh environments.

Cutting Length

The stem length of a cutting is another important variable affecting rooting success (Leakey, 2004). It is important in two ways. Firstly it affects the depth of insertion into the rooting medium, as well as the height of the leaf above the surface of the rooting medium, and secondly it inherently affects the capacity of the cutting to root. This is something we will investigate later. Both the depth of insertion and the height of the leaves above the medium can be important in terms of providing uniform conditions for water uptake and preventing competition between cuttings for light. It is also desirable to prevent the leaves from touching the medium and getting saturated by water droplets.

Cuttings can be cut to a constant length (these may vary in the number of nodes and leaves present), or can be cut to the length determined by a chosen number of nodes. The simplest cutting is a single-node cutting. It has one internode and generally has one leaf and one bud. In this case the internode will generally vary in length depending on its position within the stem (Fig. 20.4). As a general rule, basal internodes are shorter than more apical ones, this reflecting the vigor of growth at the time the node was formed in the terminal bud. We will examine this within stem variability later. For practical purposes it is probably best to use a constant length close to the optimum, although this may not result in the availability of greatest number of cuttings.

Stockplant Factors: Cutting Origin and Environment

There are two major sources of variation attributable to the stockplant: (1) Within-shoot factors, and (2) Between shoot factors. Both of these are subject to the stockplant environment and so can be influenced by stockplant management.

Within-Shoot Factors

We saw earlier that single-node cuttings vary in their internode length (Fig. 20.4). This variation is associated with gradients in numerous other variables which basically run from the bottom to the top of the stem in parallel with chronological age, such as leaf size, leaf water potential, leaf carbon balance, leaf senescence, internode diameter, stem

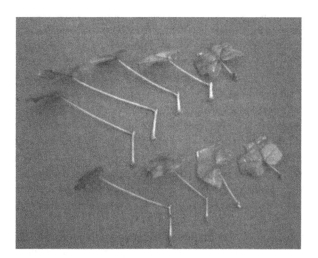

FIGURE 20.4   The node-to-node variation in the length of the internode and the petiole in cuttings from the top two shoots of a *Triplochiton scleroxylon* stockplant (uppermost nodes on the left). *Courtesy of Roger Leakey.*

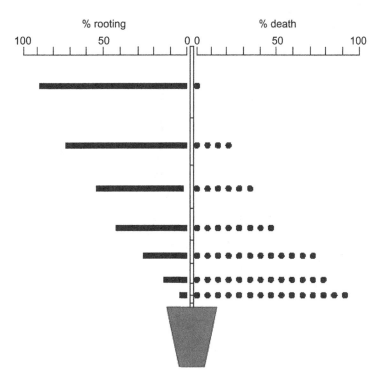

FIGURE 20.5 The gradient of variation in rooting ability within a shoot: the long uppermost cuttings root best, whereas the short basal cuttings die.

lignification, nutrient and stem carbohydrate content, and respiration (Leakey, 2004). Thus no two cuttings are identical and, as a consequence, no two cuttings have the same rooting ability (Fig. 20.5). This is because all these variables affect the physiological processes in the cutting. A number of studies have used this node-by-node variation as an experimental variable to examine some of the key factors affecting rooting ability. For example, a comparison of the normal gradient in cutting length with an inverse gradient showed the importance of cutting length while a comparison with standard length cuttings showed the importance of cutting diameter/volume. It seems that the stem volume of the cutting determines the storage capacity of the cutting for current assimilates. In this connection, there are interactions between stem volume and leaf area. The stockplant environment and/or stockplant management also interact with node position in ways that help to elucidate the effects of different treatments. Studies of this sort have shed light on the relative importance of carbohydrates and nutrients in the rooting process.

Between Shoot Factors

As plants grow and increase in size they branch and become more complex. This creates competition between shoots for assimilates and nutrients, as well as causing mutual shading between leaves. This competition is made more complex by the processes of correlative inhibition and the development of dominance between shoots which are mediated by growth regulators as well as by the plant's environment, especially light and nutrients (Leakey, 2004).

To maintain stockplants in a good condition for easy rooting they need to be pruned back hard to leave a short stump of about 20 cm. So the simplest form of stockplant has just been cut back once and then sprouts from its uppermost remaining buds. A comparison of the rooting ability of these different shoots has found that cuttings from the uppermost one roots best and those from lower shoots root progressively less well. However, if such stockplants are cut back to different heights (say between 20 and 100 cm) then usually the taller stockplants produce more shoots. In this case the rooting ability of cuttings from the top shoot declines with increasing height and a relationship is found between the number of shoots and the percentage of cuttings rooted (Fig. 20.6). This implies that competition between the shoots reduces rooting ability and the removal of the lower shoots increases the rooting ability of cuttings from upper shoots.

Further studies to test the competition hypothesis in plants of the same height have found that other factors also seem to be involved. For example, the lower shaded shoots can root very much better than less shaded shoots, especially under conditions of high soil nitrogen. Additionally, reorienting the stockplant so that its stem is not vertical

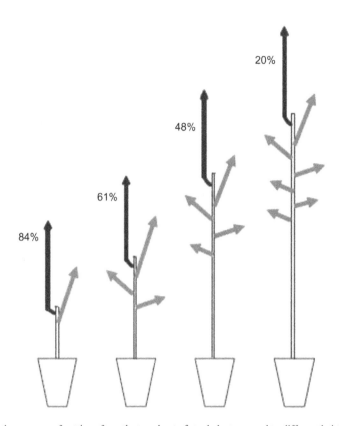

FIGURE 20.6 Effects on the rooting success of cuttings from the top shoot of stockplants pruned to different heights. Cuttings from short stockplants with few shoots root best.

(e.g., 45 degrees or 90 degrees from the vertical) alters the location of vigorous growth and the fast-growing lower basal shoots become the most easily rooted. Thus in addition to competition there are effects of shoot position and of stock-plant environment.

When studies tried to elucidate these interactions between shoot position and intershoot competition in two-shoot stockplants of the same height, it was found that basal shaded shoots had a higher rooting ability than upper shoots. However, if these basal and upper shoots were under conditions with the same light environment (irradiance) the differences between shoot positions were eliminated (Fig. 20.7). This finding led to experiments in controlled-environment growth chambers to investigate the role of light (Hoad and Leakey, 1994, 1996).

STOCKPLANT ENVIRONMENT

As seen previously, both nutrients and shade seem to affect the rooting ability of cuttings from shoots on relatively simple stockplants, but these complex preconditioning processes are poorly understood. It is clear that both the amount of light (irradiance) and the spectral quality of light (red:far-red ratio)—both features of shade—are important and independently affect the rooting ability of cuttings. Shade light is also commonly associated with shoot etiolation and hence with long internodes, and with large thin leaves. These affects are then further complicated by an interaction with soil nutrients which promote shoot growth (Leakey and Storeton-West, 1992). The light/nutrient interaction affects photosynthetic processes. The most dramatic of these interactions seems to be the combination of high irradiance and low nutrients which results in short starch-filled stems, the inhibition of photosynthesis and very poor rooting (Fig. 20.8). At the other extreme, low irradiance, and high nutrients are associated with active photosynthesis, low starch, and good rooting.

These preconditioning effects of irradiance, light quality, and nutrients on morphology are also associated with physiological differences in the stems and leaves of the shoots preserverance (Hoad and Leakey, 1994, 1996). Physiologically, shade reduces the tendency of the top shoot to dominate and reduce the growth of other shoots

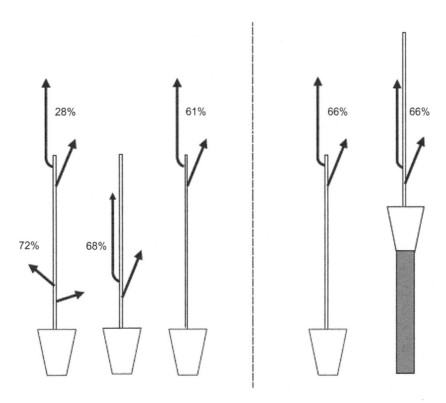

FIGURE 20.7 Effects of shoot position on the rooting success of cuttings from stockplants with apical and basal shoots. The effects of shade were examined by adjusting the height of the pots.

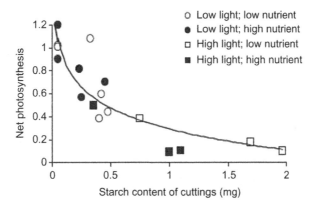

FIGURE 20.8 The effects of growing stockplants at two levels of light (low and high irradiance) and two levels of nutrients (low and high). This affects the rates of photosynthesis and the starch content of cuttings. Starch-filled cuttings with low rates of photosynthesis rooted very poorly. *Modified from Leakey, R.R.B., Storeton-West, R., 1992. The rooting ability of Triplochiton scleroxylon K. Schum. cuttings: the interactions between stockplant irradiance, light quality and nutrients. For. Ecol. Manage., 49, 133—150.*

(i.e., promotes codominance), lowers the rates of preseverance net photosynthesis, lowers leaf chlorophyll concentration, but stimulates higher rates of net photosynthesis per unit of chlorophyll. Preseverance conditioning thus actively promotes postseverance photosynthesis during the rooting process. The shoots are therefore said to be physiological "young," while shoots with low vigor and inhibited photosynthesis are physiologically "old."

A mechanistic model of carbohydrate dynamics during the rooting process can be used to gain further insights into the rooting process (Dick and Dewar, 1992) and to define key principles for the development of robust rooting protocols.

TABLE 20.3 The importance of stockplant management for successful vegetative propagation.

	Standard hedged stockplant	Standard hedged stockplant	Preconditioned (far-red light and fertilizers) hedged stockplant
	No knowledge of stockplant management	Knowledge that rooting is best from upper shoots	Knowledge that preconditioning improves physiological activity and rooting capacity
No. of cuttings harvested	31	16	45
No. of cuttings rooted	16	16	40
Percentage of cuttings rooted	52%	100%	89%
Number of plants established	16	16	40

Note the lack of relationship between rooting percentage and the numbers of plants established—this reflecting the understanding of the importance of stockplant management (which shoots to use and the role of stockplant environment.
Adapted from Leakey, R.R.B., 2004. Physiology of vegetative reproduction. In: Burley, J., Evans, J., Youngquist, J.A., (Eds.), Encyclopedia of Forest Sciences. Academic Press, London, pp. 1655–1668.

Stockplant Management

The major importance of all of the previous morphological and physiological stockplant factors makes it clear that stockplant management (a combination of regular pruning, fertilizer use, and light management) should be a key component of any macropropagation protocol. To retain physiological youth, stockplants are often managed as hedges. Using nitrogen-fixing tree/shrub species to shade these hedges then maximizes the rooting success by provision of desirable light and soil nutrient environments.

One of the issues which has created much of the confusion in the scientific literature of vegetative propagation is the very common (even ubiquitous) use of percentage rooting as the measure of success. Rooting percentage is, however, affected by the standards of stockplant management, use of the best shoots (the shoots with optimal morphology and highest physiological activity, Table 20.3). Contradictory results arise in the literature when authors do not provide the relevant information in the description of their experimental protocols.

Phase Change

The vegetative propagation literature has numerous references to what is called "Phase Change" or ontogenetic aging. This literature basically attributes the loss of rooting ability as perennial plants age and get larger to their gradual transition from the juvenile phase of vegetative growth to a phase of sexual maturity. The propagation of mature tissues is one of the major constraints to many tree improvement programs focusing on cultivar development through macropropagation. This loss of rooting ability with increasing size and structural complexity is perhaps not surprising given the already mentioned effects of stockplant factors in even small and simple stockplants. Nevertheless, the importance of phase change vis à vis rooting ability is an unresolved aspect of vegetative propagation (Leakey, 2004).

The phase change hypothesis assumes that plants (and trees in particular) gradually progress from juvenility (sexually immature and easy-to-root) to maturity (sexually mature and capable of flowering and fruiting associated with low rooting ability) over time. In trees, maturity may not be achieved for 3–20 years. It is generally recognized that the best way to return a tree to the juvenile state is to cut it down and to allow coppice shoots to grow from the stump. These coppice shoots are young and vigorous and cuttings from them can usually be rooted easily. Studies have suggested however that there is a decline in rooting ability with the increasing diameter of the stump. It is not known whether this is due to some aspect of the aging process or to stockplant factors like those mentioned for seedlings above. For example, large stumps generally produce more shoots than smaller stumps. This loss of rooting ability could therefore be due to increased intershoot competition.

There are some inconsistencies between the concept of phase change and observation. Originally the hypothesis was based on work in ivy (*Hedera helix*) which has various leaf shapes. It was assumed that the normal palmate leaves of climbing ivy vines were characteristic of the juvenile easy rooting phase, while the ovate leaves of flowering shoots were assumed to be characteristic of the difficult-to-root mature phase. Observation of large ivy plants suggests a different interpretation as palmate leaves can be seen to be associated with main stems (the climbing vine), while ovate leaves are associated with free-standing branches on which flowers are formed. The association of flowering with branches is common in many woody plants, and it is not unusual for branches and main stems to differ in the arrangement of their leaves and buds. Certainly in ivy it is common to see old and large vines climbing a cliff or dead tree with palmate leaves tens of meters above the ground and flowering branches with ovate leaves all the way up the vine.

A second way in which the hypothesis is not supported is the evidence that ontogenetically mature plants (with the capacity to flower and form fruits) propagated by cuttings, grafts or marcotts can be easily repropagated by cuttings when they are used as small managed stockplants. Although often plagiotropic, such stockplants have the vigor of juvenile seedlings and coppice shoots. This suggests the poor rooting ability of "mature" shoots can be attributed to "physiological aging" rather than to "ontogenetic aging." In other words, the difficulty in rooting crown shoots in large mature trees is due to a severe case of the combination of undesirable morphological and physiological factors resulting from within and between stockplant factors and their interaction with the stockplant environment (Dick and Leakey, 2006). A start has been made to examining this experimentally within the crown of mature trees (Pauku et al., 2010).

Genetic Variation in Rooting Ability

Early in this chapter we saw that there are genetic differences in the ease of propagation between species and even between provenances and clones within species. In severe cases this has led to the conclusion that some species are "impossible to root." We have also seen that species vary in the leaf and stem morphology, and in their adaptations and responses to environmental factors. There is now good evidence that species previously thought to be "impossible to root" can now be preconditioned to be easier to root by good stockplant management based on an understanding of the morphological and physiological factors affecting rooting ability. Thus the apparent genetic differences in rooting ability can instead be attributed to genetic differences in the growth and development of shoots.

The concept that inherent genetic differences in rooting ability are not critically important is also supported by the use of stepwise regression in the analysis of experimental data from rooting studies. Such analysis commonly finds that the factors explaining much of the variance are cutting length, leaf abscission, leaf area, etc., rather than clone or provenance (Dick et al., 1999).

TABLE 20.4 The analogy between athletic ability and rooting ability in the vegetative propagation of tree stem cuttings.

Sporting ability	Rooting ability
Young	Juvenile
Physiologically fit—lean, well-prepared, strong	Physiologically young—high rates of photosynthesis—long internodes, large leaf area, chronologically young node position
Genetic morphology—muscles, body build, etc.	Genetic morphology—stem and leaf anatomy
Preparation in gym—physiologically fit and full of energy	Preconditioning by stockplant management —physiologically active
Highly competitive and ready for action	Preconditioned to contain low starch/high soluble sugars and free from competing shoots
Use of performance-enhancing drugs	Use of auxins as rooting hormone
Stress-free Olympic village	Stress-free propagation environment
Different sports—refine training regime	Different species—refine propagation regime

Adapted from Leakey, R.R.B., 2012b. Living with the Trees of Life – Towards the Transformation of Tropical Agriculture, CAB International, Wallingford, UK, 200 pp.

Contrary to this, some studies to detect quantitative trait loci affecting vegetative propagation have, however, reported that phenotypic variation has a meaningful genetic component.

CONCLUSIONS

There are many facets to developing a robust approach to macropropagating plants, especially by the rooting of stem cuttings. Over the last 15−20 years much progress has been made to gain a good understanding of the numerous interacting factors (stockplant environment × stockplant management × topophytic variables × node position × nursery management × postseverance treatments × propagation environment) by studying them in tree species. This understanding now provides some general principles that can be applied to the propagation of new species, and especially those considered to be difficult to ltotaoatpropagate.

Leakey (2012b) has likened many of the factors that determine the success of propagation by stem cuttings to those that determine the success of an athlete competing in the Olympic Games (Table 20.4). This analogy has been found to help people to understand the key principles.

The greatest challenge remaining is to improve the design and reporting of experiments so that the quality of the literature is improved by researchers adequately describing the material they used, as well as the pre and postseverance environments. This should remove the contradictory results in the literature created by people not using comparable material in terms of the physiology and morphology of the tissues used. Then there is a need for more extensive studies of the importance of stockplant management and the stockplant environment across a wider range of species.

Courses

5 Training videos by Edinburgh Center for Tropical Forests (DVD)

Relevant Websites

ICRAF website − Narrated PowerPoint lectures − www.worldagroforestry.org/Units/training/downloads/tree_domestication

Section 5.3

Techniques: Genetic Characterization

Chapter 21

Quantitative Descriptors of Variation in the Fruits and Seeds of *Irvingia gabonensis*

This chapter was previously published in Leakey, R.R.B., Fondoun, J-.M., Atangana, A., Tchoundjeu, Z., 2000. Agroforestry Systems, 50, 47–58, with permission of Springer

SUMMARY

Methods were developed to quantify variation in the fruit, nut and kernel traits using the fruits from four trees of *Irvingia gabonensis*, an indigenous fruit tree of west and central Africa. The measurement of 18 characteristics of 16–32 fruits per tree, identified significant variation in fruit, nut and kernel size and weight, and flesh depth. Differences were also identified in shell weight and brittleness, fruit taste, fibrosity and flesh colour. Relationships between fruit size and weight, with nut and kernel size and weight were found to be very weak, indicating that it is not possible to accurately predict the traits of the commercially-important kernel from fruit traits. Seven key qualitative traits are recommended for future assessments of the levels of genetic variation in fruits and kernels. These traits describe ideotypes for fresh fruit and kernel production.

INTRODUCTION

Recent work in the humid lowlands of West Africa has established that there are substantial local and regional markets for the nontimber forest products (NTFP) of certain indigenous trees, such as *Irvingia gabonensis* (Aubry-Lecomte ex O'Rorke) (Ndoye et al., 1997). Some NTFPs have even entered the international market and a recent market survey in Europe has emphasized their high value (Tabuna, 1999).

Studies on the biological variability of indigenous fruit-tree species, their propagation using cheap and simple low-technology methods appropriate for rural development projects, and their suitability for domestication have been progressively increasing in West Africa, over the last 20 years (see Leakey et al., 1990; Okafor and Lamb, 1994). In the last 5 years, the World Agroforestry Center (ICRAF, formerly International Center for Research in Agroforestry) initiated a coordinated initiative across the Region, based on the priorities of subsistence farmers and national research scientists (Jaenicke et al., 1995; Franzel et al., 1996). This species prioritization activity involved field activities such as farmer preference surveys (e.g., Adeola et al., unpublished), product ranking (e.g., Aiyelaagbe et al., 1996), and the valuation and ranking of priority species (Franzel et al., 1996). From this process, the top priority species identified for humid West Africa was *I. gabonensis* (bush mango/dika nut). Domestication of indigenous fruits through agroforestry is seen as one of three important issues in the transformation of land use in Africa (Sanchez and Leakey, 1997), through the establishment of a better balance between food security and natural resource utilization. It should also help to alleviate the poverty that drives deforestation (ICRAF, 1997; Leakey and Simons, 1998) and is an important element of the "Woody Plant Revolution" (Leakey and Newton, 1994a). To be successful in these terms, domestication has to be linked to commercialization and market expansion. This has to be done in ways that will provide smallholder farmers with socioeconomic (Leakey and Izac, 1996), policy (Leakey and Tomich, 1999), environmental (Leakey, 1998), and product development benefits (Leakey, 1999a).

Multifunctional Agriculture. DOI: http://dx.doi.org/10.1016/B978-0-12-805356-0.00021-0
© 2017 Elsevier Inc. All rights reserved.

In Nigeria, Okafor (1974) identified two varieties of *I. gabonensis*, one with sweet fruits and the other with bitter fruits. Harris (1996) subsequently revised the taxonomy calling the bitter one *Irvingia wombolu* Vermoesen, which is more important for its kernels than for its fruits. The kernels of both species are, however, used as a food-thickening agent. In Cameroon, the trade in these kernels to Gabon, Nigeria, Equatorial Guinea, and the Central African Republic has been valued at US$260,000 per annum (Ndoye et al., 1997). The kernels are an important source of both a polysaccharide, which forms the glutinaceous thickening agent, and an oil. The oil content of these kernels is known, however, to vary from 51% to 72%, depending on the geographic origin of the fruits (reviewed by Leakey, 1999a). More knowledge of this sort of variability is needed if market opportunities are to be expanded in support of domestication.

The geographic range of *I. gabonensis* is from Nigeria to Congo, while that of *I. wombolu* is much greater: Senegal to Uganda (Harris, 1996). In order to capture much of the genetic variation, germplasm collections were implemented during the fruiting seasons of 1994/95 (Ladipo et al., 1996). These collections, which were made in collaboration with the authorities and farmers in the host countries, were targeted at visible variation and farmers' perceptions of superiority. They were made at 10–30 locations per country in the humid forest belt of Ghana, Nigeria, Cameroon, and Gabon, exchanged between countries and established as living gene banks at three sites (Ibadan and Onne in Nigeria; Mbalmayo in Cameroon) in 1995. Each gene bank contains approximately 60 accessions (single progenies). ICRAF adheres to the FAO Code of Conduct for Germplasm Collection and Transfer and the Convention on Biological Diversity, and through its Policy on Intellectual Property Rights, holds this germplasm in trust for humankind to ensure the availability of the best germplasm to farmers in developing countries.

The germplasm collections are being evaluated by ICRAF and partners for growth and phenological traits and will become part of a program of genetic testing, with selection of superior individuals. Molecular studies of the extent of genetic diversity in these populations are underway to provide a baseline for the domestication program. These studies by Institute of Terrestrial Ecology have determined that the center of diversity for *I. gabonensis* is in southern Cameroon, while that of *I. wombolu* is in southeast Cameroon and western Nigeria (Lowe et al., 1998), areas postulated as forest refugia during the last Ice Age. For the latter species, the center of diversity could extend either further east or west into areas not covered by this study. As would be expected for outbreeding species, the genetic diversity within progeny arrays was high, while that between populations was low. This information provides a fundamentally important basis on which to build a farmer-oriented cultivar selection program.

The second phase of the domestication program is being initiated in Nigeria and Cameroon. In these two areas, field teams are working with farmers to identify, select, and multiply superior trees. To date over 2000 mature plus-trees have been identified and these are being propagated vegetatively by air layering (Tchoundjeu et al., 1998). The first rooted marcotts from these selected trees are currently in nurseries and will become stockplants for further multiplication and, subsequently, these cultivars will be established in on-farm trials. In parallel with more strategic studies, village nurseries are being established so that the participating farmers are the beneficiaries of the program. This participatory approach to domestication is in harmony with the ideals and aspirations of the Convention of Biological Diversity.

Vegetative propagation techniques are being used for the capture and multiplication of superior phenotypes as putative cultivars. Currently, studies are in progress in Nigeria and Cameroon to improve existing methods of air layering, grafting and budding, as these are all means of propagating from the mature crowns of selected trees. However, an attractive alternative would be to propagate by cuttings as this avoids grafting incompatibilities and potentially gives high multiplication rates. Simple, low-technology methods have been developed in Cameroon for juvenile shoots of tropical trees (Leakey et al., 1990), including *I. gabonensis* (Shiembo et al., 1996a), but it is known that mature shoots are more difficult as they become physiologically and ontogenetically old. Research is needed to try to rejuvenate mature crown shoots physiologically, while retaining the benefits of reproductive capacity due to ontogenetic age.

ICRAF's tree domestication program for *I. gabonensis* is currently focusing much of its resources on the vegetative propagation of trees that farmers are identifying as superior. There is at present, however, little knowledge about farmers' perception of genetic variation or their capacity to identify and capture the potential for improvement through the formation of cultivars based on the propagation of elite trees in a wild population. Evidence from farmer interviews suggests that there is substantial variation in certain fruit traits, such as fruit size and the taste of fresh fruit pulp, as well as some variation in the size of the commercially important kernels. Attempts to identify elite trees are, however, subject to both the cooperation of farmers and their willingness to disclose knowledge of local trees. The most commonly available information is about variation in fruit sweetness, but this is usually limited to the differentiation into two classes: sweet and bitter. Farmers indicate that kernel size is important in the market, although there is as yet no indication if this is recognized by price discrimination. It seems to be generally considered that large kernels will be found in large fruits.

The purpose of the present study was to develop a methodology for the quantitative characterization of tree-to-tree variation in *I. gabonensis* fruits in Cameroon and Nigeria. This assessment of genetic variation is aimed at the determination of:

- the level of diversity available to farmers within the area of their communal ownership
- how farmers evaluate superiority between trees
- how farmers criteria for selection matches with recognized horticultural traits of fruit quality, yield, or chemical composition
- farmers' concepts of genetic variation in fruit traits and their "improvement"
- the levels of selection intensity being applied by farmers, and
- the relationships between the variation of fruit/kernel traits and market prices.

METHODS AND MATERIALS

Fruits were collected from four *I. gabonensis* trees in August 1998 from different locations in Cameroon (Table 21.1). Fruits were scarce that year as many trees were not fruiting, or fruiting poorly. Consequently, a sampling procedure was only tested in the case of Tree 2 Ngalan, where the area below the tree crown was divided into four quadrants. Eight ripe, but undamaged, fruits were collected from each quadrant at a point two-thirds of the distance from the trunk to the edge of the crown along transects at 90 degrees from each other. The mean weight of 5, 10, 15, 20, 25, and 30 fruits from this sample was compared with that of a larger sample of 90 fruits in order to determine optimal sample size.

On the day of collection the fruits were weighed using a small portable kitchen scales graduated to 2 g, and measured using calipers graduated to 0.1 mm. Fruit flesh depth was measured by a spike attached to the same calipers. The spike was inserted until it hit the hard nut at the center of the fruit. The external measurements of the fruits were made in three dimensions (length, width, and thickness), and the flesh depth measurements were made on both sides of the fruit in these three dimensions (Fig. 21.1).

The next day, the fruits were depulped to expose the endocarp (nut) and a record made of fruit taste (scored 1 [bitter] to 5 [sweet]), color (yellow or orange) and fibrosity (scored 1 [nonfibrous] to 5 [fibrous]). The nuts were then left to dry for 7 to 10 days until the remnants of flesh had dried. They were then weighed and measured using the calipers in the same three dimensions as the fruits. The nuts were then carefully broken open so that

TABLE 21.1 Locations of *Irvingia gabonensis* trees that provided fruits for this study.

Tree no	Village	Farmer	No of fruits	Latitude	Longitude
1	Ngomedzap	Market	16	3° 16′	11° 11′
1	Nkoevos 2	Moise Nkomo	22	2° 56′	11° 26′
1	Ngalan	Odile Essong	26	2° 54′	11° 12′
2	Ngalan	Odile Essong	32	2° 54′	11° 12′

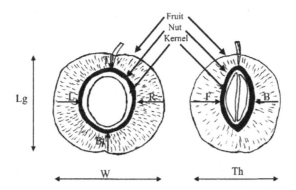

FIGURE 21.1 Diagram of the positions of fruit, nut (endocarp) and seed (kernel) measurement (*Lg*, length: [*T*, Top, *B*, Bottom], *W*, width [*L*, left; *R*,right], *Th*, thickness [*F*, front; *B*, back]).

the seeds (kernels) could be extracted. The fresh kernels were then weighed using a laboratory balance (Mettler Toledo PB 3002) and measured in the same three dimensions using the calipers. The weight of the nutshell was determined by the difference between the nut and kernel weights, and a "shell brittleness" factor derived as 50/shell weight. Statistical analyses of the data were done using Microsoft Excel 97, and skewness was calculated using GENSTAT 5 Version 4.1.

RESULTS

Sample Size

A comparison between the average weight of 32 randomly selected fruits and a larger general fruit collection of 90 fruits from Tree 2 at Ngalan showed that the sample had the same mean value.

A comparison of the standard errors of the mean values of the first 5, 10, 15, 20, 25, and 30 fruits from the 32 fruit sample from Tree 2 at Ngalan found that the mean weight of a sample of 20 or more fruits differed by less than the standard error (Table 21.2). The weights of the 30 individual fruits forming a sample were normally distributed (skewness = -0.02).

Fruit Weight and Size

Mean fruit weight, fruit length, fruit width, and fruit thickness differed significantly between trees (Table 21.3). Fruits from Tree 1 at Ngomedzap were greater in length than they were in width or thickness, while those of Tree 1 at Nkoevos 2 were greater in width than they were in length or thickness (Table 21.3). The overall relationship between fruit weight and fruit length was linear ($r^2 = 0.923$) as it was from Ngalan and Nkoevos 2 (Table 21.4). Similar relationships were found between fruit weight and fruit width or thickness.

Nut Weight and Size

Nut weight was significantly different between trees from the different villages (Table 21.3), with those from Ngomedzap being on average twice as heavy as those from Ngalan. In general, nuts were significantly longer than they were wide, and wider than they were thick (Table 21.3).

Kernel Weight and Size

The patterns of variation in the weight of kernels were different from those of fruits and nuts, as the differences between kernels from Tree 1 Ngalan, Ngomedzap, and Nkoevos 2 were not statistically significant (Table 21.3). The greater weight of kernels from Tree 1 at Ngalan was matched by their greater thickness. Kernels varied most in length and least in thickness (Table 21.3).

TABLE 21.2 Effect of sample size on the mean weight and standard deviation of *Irvingia gabonensis* fruits from Tree 2 at Ngalan.

Sample size	Mean fruit weight (g)	Standard error	Skewness
5 fruits	44.0	± 4.7	−0.76
10 fruits	44.7	± 2.5	−0.81
15 fruits	43.7	± 2.0	−0.50
20 fruits	42.7	± 1.8	−0.29
25 fruits	42.7	± 1.6	−0.44
30 fruits	43.9	± 1.6	−0.02

TABLE 21.3 Tree-to-tree variation in fruit, nut, and kernel weight (g) and size (mm) in *Irvingia gabonensis*.

Village	Ngomedzap		Nkoevos 2		Ngalan: Tree 1		Ngalan: Tree 2	
	Mean	± SE	Mean	± SE	Mean	± SE	Mean	± SE
Fruit weight	103.8	1.40	58.2	2.88	36.3	2.29	43.3	1.62
Fruit length	58.4	0.51	47.0	0.81	42.4	0.90	43.1	0.53
Fruit width	56.6	0.30	55.6	0.99	44.2	1.09	45.4	0.53
Fruit thickness	49.4	0.42	46.8	0.93	36.2	0.79	42.3	0.63
Nut weight	11.9	0.55	7.7	0.42	6.0	0.36	6.0	0.20
Nut length	42.0	0.61	40.2	0.86	31.7	0.71	31.7	0.39
Nut width	33.3	0.74	36.9	0.74	30.0	0.94	27.9	0.38
Nut thickness	22.2	0.35	19.3	0.46	19.0	0.63	19.0	0.28
Kernel weight	3.5	0.14	3.1	0.19	3.3	0.30	1.8	0.12
Kernel length	32.0	0.55	26.5	0.73	25.5	0.59	23.1	0.41
Kernel width	21.4	0.72	22.0	0.68	21.8	1.06	18.7	0.45
Kernel thickness	11.6	0.35	11.3	0.26	12.5	0.43	10.1	0.22
Nut shell weight	8.5	0.19	4.6	0.24	2.6	0.53	4.2	0.14

TABLE 21.4 Relationships between fruit and nut traits in *Irvingia gabonensis* in Cameroon.

Traits	Tree	Equation	r^2
Fruit weight *versus* fruit length	Ngomedzap	$y = 0.13x + 44.80$	0.135
	Nkoevos 2	$y = 0.26x + 31.89$	0.843
	Ngalan—tree 1	$y = 0.36x + 29.25$	0.849
	Ngalan—tree 2	$y = 0.30x + 30.22$	0.807
Fruit thickness *versus* flesh depth	Ngomedzap	$y = 0.27x + 1.84$	0.278
	Nkoevos 2	$y = 0.31x + 0.20$	0.655
	Ngalan—tree 1	$y = 0.22x + 2.24$	0.219
	Ngalan—tree 2	$y = 0.32x + 0.70$	0.612
Fruit length *versus* Nut length	Ngomedzap	$y = -0.48x + 69.72$	0.150
	Nkoevos 2	$y = -0.36x + 57.05$	0.116
	Ngalan—tree 1	$y = 0.21x + 23.02$	0.086
	Ngalan—tree 2	$y = 0.01x + 31.11$	0.0003
Fruit length *versus* Kernel length	Ngomedzap	$y = -0.40x + 55.07$	0.126
	Nkoevos 2	$y = -0.27x + 39.09$	0.090
	Ngalan—tree 1	$y = 0.11x + 21.07$	0.036
	Ngalan—tree 2	$y = 0.14x + 17.25$	0.032
Fruit weight *versus* Kernel weight	Ngomedzap	$y = 0.04x - 0.127$	0.128
	Nkoevos 2	$y = -0.03x + 4.569$	0.162
	Ngalan—tree 1	$y = 0.01x + 2.782$	0.014
	Ngalan—tree 2	$y = 0.02x + 1.010$	0.067

Shell Weight

Shell weight varied greatly between trees, with that of fruits from Ngomedzap being three times that of fruits from Tree 1 at Ngalan (Table 21.3). The shells of the latter had also been found to be the most brittle during depulping, with the shells of five of the 30 fruits of Tree 2 Ngalan breaking during the depulping process.

Fruit Flesh Depth

Flesh depth was not uniform over the whole fruit. It was least in the length dimension of the fruits (i.e., top to bottom), and greatest in the thickness dimension (see Fig. 21.1 for explanation). In all dimensions, except top and bottom, flesh depth was the same on opposite sides of the fruit. Flesh depth at the top of the fruit was less than at the bottom because of the attachment of the peduncle. Fruits from Tree 1 at Ngalan had the least flesh depth.

Relationships between flesh depth and fruit size (length/width/thickness) were found to be weak (Table 21.4) but strongest in the thickness dimension.

Relationships Between Fruit, Nut, and Kernel Traits

Nut and kernel length was shown to be virtually independent of fruit length as was kernel weight of fruit weight (Table 21.4). The relationship between nut length and kernel length was however, stronger, but still relatively weak (Table 21.4).

Fruit Taste, Color, and Fibrosity

There were marked differences in fruit color (yellow for Ngomedzap and Nkoevos 2, to orange for Ngalan trees 1 and 2), fibrosity (score 1 for Ngomedzap and Nkoevos 2 to 5 for Ngalan trees 1 and 2) and taste (score 1 for Ngalan tree 1−4 for Ngomedzap and Nkoevos 2) between trees.

Multitrait Assessment

To visualize the combination of these traits, as a single "variety" the key traits can be expressed as a web diagram (Fig. 21.2A), showing that in this sample of only four trees, there were differences in the relative superiority/inferiority between traits and combinations of traits.

DISCUSSION

Prior to this study, ideas about the traits of interest for domestication of I. gabonensis were based on descriptive accounts of the variation in fruit characteristics between trees (e.g., Ladipo et al., 1996). There is still no good assessment of the range of genetic variation in traits of likely importance, although a study is in progress (Shiembo et al., 1996a,b).

The present study establishes which of a number of possible fruit, nut, and kernel traits of I. gabonensis will be useful for future studies aimed at quantifying the extent of tree-to-tree variation in a larger and more extensive study of the variation in I. gabonensis populations.

It is clear from the present study that there is considerable phenotypic variation in almost every parameter measured and that a few traits are closely related. The lack of a relationship between fruit and kernel weight and size is clearly very important, as it indicates that, contrary to farmers' presumptions, it is not possible to accurately predict the kernel traits from an observation of the size and weight of the fruit.

With respect to the identification of desirable traits for further characterization of I. gabonensis fruits, it is particularly interesting to observe the stronger relationship between flesh depth and fruit thickness, compared with fruit length or width. Future fruit measurements with ongoing studies of genetic variation in I. gabonensis will be restricted to measurements of fruit length and width, flesh depth in the thickness dimension, fruit, nut, and kernel weights. The variation in nut-shell weight found in the present study may be of special significance as kernel extraction is a labor-intensive activity (about 1 kg/person/hour: Acworth, 1992), done by women. Shell brittleness would therefore be a socially important trait. Easily cracked nuts have been reported twice before, once from Gabon and once from Cameroon. However, this trait may be relatively uncommon, as it has not been previously observed by the ICRAF tree domestication team.

The lack of a relationship between fruit and kernel traits also suggests that, as in many other tree crops (Dickmann, 1985), there is the opportunity to identify a small number of key traits that together would form an "ideo-type" that combines a number of highly desirable characteristics of potential commercial value. Currently, fruits of I. gabonensis are most important as a source of kernels for use as a food-thickening agent, and of lesser importance as a fresh fruit. Consequently, it would be useful to identify traits that would contribute to the creation of both "fruit"

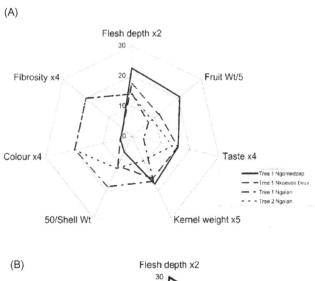

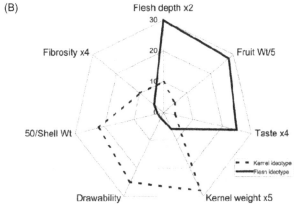

FIGURE 21.2   (A) Tree-to-tree variation webs for four trees of *Irvingia gabonensis* in Cameroon. (B) Fruit and nut ideotypes for *I. gabonensis*.

and "kernel" ideotypes. Use of the previously identified traits in different combinations could achieve this (Fig. 21.2B). In this case, the "fruit ideotype" would be described by large values for flesh depth, fruit weight, good taste, and low fibrosity, while the kernel ideotype would be described by high values for kernel weight, shell brittleness (low shell weight) and drawability (the thickness of the glutinaceous polysaccharide exudate formed by boiling the kernels). Obviously, as information becomes available these ideotypes can be refined. For example, flesh color may become important if large sweet fruits are marketed, while other traits associated with suitability for wine making might also become important (see Leakey, 1999a). Similarly, the kernels are also a good source of vegetable oil and it may be that the kernel ideotype could be subdivided into two: one rich in high quality oil and the other with desirable levels/quality of the polysaccharide.

When the data from the present study are displayed to illustrate their closeness toward these two ideotypes, it is interesting to observe that one (Tree 1 at Ngomedzap) has some of the characteristics of the fruit ideotype, and that another (Tree 1 at Ngalan) has two of the characteristics of the kernel ideotype (Fig. 21.2B). From such a small sample of trees, this result gives much encouragement that ideotype selection is a worthwhile approach to the domestication of this species. Future studies will apply the ideotype concept to much larger tree populations of *I. gabonensis*.

ACKNOWLEDGMENTS

This publication is an output from a research project (R7190 Forestry Research Program) funded by the Department for International Development of the United Kingdom (DFID). The authors are indebted to DFID for funding the project, understanding that DFID can accept no responsibility for any information provided or views expressed. They also wish to thank their collaborators in UK Overseas Development Institute (Dr. Kate Schreckenberg and Miss Charlotte Boyd) and ICRAF/IRAD Agroforestry Research Network in Cameroon and Nigeria (Dr. Tony Simons, Dr. Bahiru Duguma and Mr. Joseph Kengue) for their help and encouragement.

Chapter 22

Domestication Potential of Marula (*Sclerocarya birrea* subsp *caffra*) in South Africa and Namibia: 1. Phenotypic Variation in Fruit Traits

This chapter was previously published in Leakey, R.R.B., Shackleton, S., du Plessis, P., 2005. Agroforestry Systems, 64, 25–35, with permission of Springer

SUMMARY

Studies of tree-to-tree variation in fruit traits are a pre-requisite for cultivar development. Fruits were collected from each of 63 marula (*Sclerocarya birrea*) trees in Bushbuckridge, South Africa and from 55 trees from the North Central Region of Namibia. The South African trees were in farmers' fields, communal land and natural woodland, at three sites: Acornhoek road, Allandale/Green Valley and Andover/Wits Rural Facility. The Namibian trees were all from farmers' fields in three areas: North east, North west and West. The fruits were partitioned into skin and flesh/juice to examine the extent of the variation found in different components of marula fruits from different trees. Namibian fruits were significantly larger than those from South Africa (26.7 *vs* 20.1 g), due to their greater pulp mass (22.2 *vs* 16.2 g), especially the flesh/juice component. In South African fruits, those from farmers' fields were significantly larger in all components (Fruit mass = 23.6 *vs* 19.3 and 18.0 g in natural woodland and communal land respectively). In Namibia, mean fruit mass did not differ significantly across sites (25.5–27.0 g). However, within each sample there was highly significant and continuous variation between trees in the pulp (S Africa = 7.5–31.3 g; Namibia = 8.3–36.0 g) and flesh/juice mass (S Africa = 2.2–7.6 g; Namibia = 3.8–22.6 g), indicating the potential for selection of trees producing superior products. The fruits of the Namibian trees were compared with the fruits from one superior tree ('Namibian Wonder') with a mean fruit mass of 69.9 g. The percentage frequency distribution of fruit mass from trees in farmers' fields in South Africa was skewed, while being bimodal in North east and North west populations from Namibia, suggesting that at these sites farmers are engaged in domestication through truncated selection of the best mother trees. It is concluded that there are trees in on-farm populations that have great potential to be propagated vegetatively as selected cultivars.

INTRODUCTION

Sclerocarya birrea (A. Rich.) Hochst. subsp. *caffra* (Sond.) Kokwaro is one of the traditionally important indigenous fruits of southern Africa, which in recent years has also become commercially important, as its fruits and other products have entered local, regional, and international trade (Shackleton et al., 2002b; Wynberg et al., 2002). Marula fruits have traditionally been used to make a "beer," which is now being marketed locally (Shackleton, 2002). Fruits are also used to make Amarula cream liqueur (Distell Corporation), jams, wine, fruit juice, etc. (Mander et al., 2002). This commercial interest grew from studies on marula juice (von Teichmann 1982, 1983; Pretorius et al., 1985). Consequently, several domestication initiatives have emerged such as that by Holtzhausen et al. (1990) of Pretoria University, who started to develop cultivars from "plus-trees" using grafting techniques. Another initiative was led by Veld Products Research

© 2017 Elsevier Inc. All rights reserved.

in Botswana (Taylor et al., 1996). In addition, an external domestication program was initiated by Ben Gurion University in Israel, using material obtained in southern Africa and planted in the Negev Desert (Nerd and Mizrahi 1993; Mizrahi and Nerd, 1996). More recently, a fourth program was launched in 1995 by the International Center for Research in Agroforestry (ICRAF), now the World Agroforestry Center, with a participatory mandate, in which subsistence farmers are the planned beneficiaries of the domestication activities (Maghembe et al., 1995; 1998). Rangewide germplasm collections of *S. birrea* and *Uapaca kirkiana* have been made as the first step in a domestication strategy (Leakey and Simons, 1998). These collections provide material for conservation and future utilization.

In many countries, nontimber forest products (NTFPs) are an underutilized resource, and it is only in recent years that domestication projects for agroforestry trees have been initiated (Leakey et al., 1996). Recent collections of marula fruits and nuts from individual trees in Makueni district of Kenya have been analyzed for a wide range of nutritional compounds and minerals. The skin and flesh was found to vary considerably in the vitamin C content (85–319 mg/100 g), while the kernels were rich (56–64%) in oils (Thiong'o et al., 2002). The present study, as part of a resource inventory, quantifies the phenotypic variation in fruit traits in marula (*S. birrea* ssp. *caffra*), within the framework of a broader project examining the benefits and opportunities for domesticating and commercializing the fruits and kernel oil of marula in South Africa and Namibia (Sullivan et al., 2003).

METHODS AND MATERIALS

Marula fruits fall from the tree just before they ripen. Fruits were collected in villages in Limpopo Province, South Africa (Bushbuckridge) and Namibia (North Central Region), between January 21 and February 5, 2002 (Table 22.1).

TABLE 22.1 Number of trees sampled in South Africa and Namibia in 2002, by village and land use.

South Africa	Natural woodland	Communal land	Farmers' fields	Total
Bushbuckridge				
Acornhoek road	0	29	6	35
Allandale	0	0	10	10
Green valley	0	0	4	4
Andover	11	0	0	11
Wits rural facility	3	0	0	3
Total	14	29	20	63

Namibia	Farmers' fields	Total
North West		
Omanjoshi	7	7
Omankango	9	9
Okamukwa	4	4
North east		
Onangwe	8	8
Oilyateko	1	1
Elope	1	1
West		
Tsandi	10	10
Eunda	6	6
Ongosi	4	4
Total	5	5

Farms were chosen at random when driving along minor roads in areas known to have marula trees. Fruit samples were collected from all fruiting trees on the selected farms. In South Africa, marula trees were found in farmers' fields, in communal grazing land, and in natural woodland, while in Namibia, fruiting trees were only found in farmers' fields. In general, fallen fruits were usually plentiful beneath the tree crown, and so ripe, unblemished fruits were collected at random, sampling from four quadrants (six fruits per quadrant), following the procedures described by Leakey et al. (2000). Fruits from each tree were separately bagged and labeled. As soon as possible (usually 2–3 days later), the fresh fruits were weighed using a 0.1 g electronic balance. Keeping the fruits in the same order, the skins were then peeled off and weighed, while the nuts, still in the same order, were soaked and scrubbed to remove the flesh before being set in the sun to dry for about 10 hours. When dry, the nuts were weighed and numbered so that their identity was maintained for subsequent cracking and kernel removal (Leakey et al., 2005c). Flesh mass were derived by difference (Fruit - skin - nut = flesh).

In addition, one fruit sample was collected from the Mhala Development Center (MDC), a project of the Mineworker's Development Agency in Bushbuckridge that is processing fruits for fruit juice and for kernel oil. For comparative purposes, a further sample from a superior tree on a different farm was analyzed in the same way. The origin of this tree is not disclosed in order to protect the villagers' rights to this germplasm. The name "Namibian Wonder" is used here to identify it.

SPSS 10.0 for Windows was used for the analysis of variance, Duncan's multiple range tests, and tests for skewness and kurtosis.

RESULTS

Variation Between Sites

Comparison of Mean Values Between South Africa and Namibia

The significantly greater mean fruit mass of Namibian fruits is attributable to the greater mass of pulp, as opposed to nut (Table 22.2). In turn, the greater pulp mass is attributable to a greater mass of fresh fruit flesh and juice, as opposed to skin. Although there were differences in the time between collection and laboratory processing of the fruit samples in South Africa (1 day) and Namibia (2–3 days) which could have resulted in greater ripening of the Namibian samples, and hence the juiciness of the samples (i.e., partitioning of water between flesh and skin), it is unlikely that there could have been any change in the overall fresh weight of the pulp as the fruits were stored in sealed polythene bags.

Comparison of Mean Values Between Sites in South Africa

There were highly significant differences in mean fruit, skin, pulp and flesh/juice mass, between sites in South Africa (Table 22.3). Fruits from Allandale were typically the largest. The trees from Acornhoek road were located in both farmers' fields and communal land, while those from Allandale and Green Valley were only in farmers' fields. Trees from Andover and Wits Rural Facility were in natural woodland.

Comparison of Mean Values Between Land Uses in South Africa

The mean mass of fruits, pulp, flesh/juice, and skin were significantly greater in fruits from farmers' fields than from communal land or natural woodland in South Africa (Table 22.4). Mean fruit, pulp, flesh mass were significantly lower in communal land than in natural woodland.

TABLE 22.2 Comparison of marula (*Sclerocarya birrea*) fruit traits between South Africa and Namibia in 2002.

Trait	South Africa	Namibia	Probability
Mean fruit mass (g)	20.11	26.68	$p \leq 0.001$
Mean skin mass (g)	8.91	9.60	$p \leq 0.001$
Mean flesh mass (g)	7.24	13.37	$p \leq 0.001$
Mean pulp mass (g)	16.15	22.23	$p \leq 0.001$

TABLE 22.3 Comparison of marula (*Sclerocarya birrea*) fruit traits across South African sites in 2002.

Trait	Site				Probability
	Acornhoek road	Andover and Wits Rural Facility	Allandale	Green Valley	
Mean fruit mass (g)	18.8 c	19.3 c	24.8 a	22.1 b	$p \leq 0.001$
Mean skin mass (g)	8.6 b	8.6 b	10.1 a	9.8 a	$p \leq 0.001$
Mean flesh mass (g)	6.5 c	6.5 c	10.4 a	8.7 b	$p \leq 0.001$
Mean pulp mass (g)	15.0 c	15.1 c	20.5 a	18.5 b	$p \leq 0.001$

Means in a row followed by the same letter are not significantly different ($p = 0.001$) according to Duncan's Multiple Range test.

TABLE 22.4 Comparison of marula (*Sclerocarya birrea*) fruit traits across land use systems in South Africa in 2002.

Trait	Site			Probability
	Farmers' fields	Communal land	Natural woodland	
Mean fruit mass (g)	23.60 a	18.03 c	19.34 b	$p \leq 0.001$
Mean skin mass (g)	9.92 a	8.33 b	8.65 b	$p \leq 0.001$
Mean flesh mass (g)	9.50 a	6.02 c	6.48 b	$p \leq 0.001$
Mean pulp mass (g)	19.42 a	14.35 c	15.13 b	$p \leq 0.001$

Means in a row followed by the same letter are not significantly different ($p = 0.001$) according to Duncan's Multiple Range test.

TABLE 22.5 Comparison of marula (*Sclerocarya birrea*) fruit traits across Namibian sites in 2002.

Trait	Site			Probability
	Northwest	Northeast	West	
Mean fruit mass (g)	26.89 a	25.49 a	26.96 a	$p = 0.137$
Mean skin mass (g)	9.52 b	10.28 a	9.40 b	$p = 0.003$
Mean flesh mass (g)	13.26 a	11.19 b	14.29 a	$p \leq 0.001$
Mean pulp mass (g)	22.78 a	21.47 b	22.10 ab	$p = 0.094$

Means in a row followed by the same letter are not significantly different ($p = 0.001$) according to Duncan's Multiple Range test.

Comparison of Mean Values Between Sites in Namibia

There were significant differences in mean skin, pulp, flesh/juice mass, between areas in Namibia, but not in fruit mass (Table 22.5).

Variation Within Sites

Within all sites, in both South Africa and Namibia, there was highly significant variation between individual trees in all the morphological traits of fruits that were measured.

Fruit Mass

There was highly significant and continuous variation in fruit mass within each site in both South Africa and Namibia (Fig. 22.1A and B), which was most evident in samples in excess of five trees. The fruits of "Namibian Wonder" were, however, very much heavier (69.9 g) than those of any other tree assessed (largest was N38 at 41.7 g). The fruits of

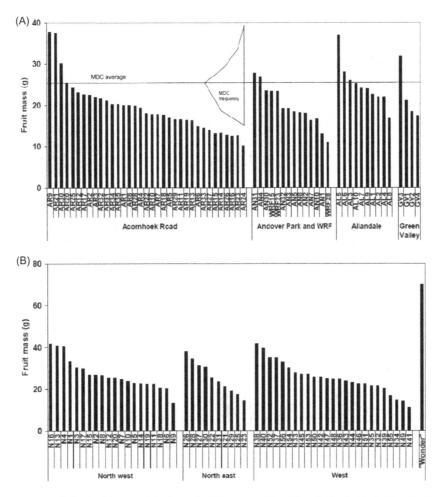

FIGURE 22.1   (A) Tree-to-tree variation in fruit mass (g) of marula (*Sclerocarya birrea*) by site in Bushbuckridge, South Africa; (B) Tree-to-tree variation in fruit mass (g) of marula (*S. birrea*) by site in North Central Region, Namibia.

"Namibian Wonder," which ranged in mass from 57–79 g, were heavier than any reported to date, exceeding those developed as cultivars by Holtzhausen et al. (1990). Assessment of the fruits passing through the MDC fruit juice processing unit indicated that the fruits brought to this unit are comparable in mass with the best of the fruits sampled in this study, indicating that the women discarded the smaller fruits.

When the data from South African trees was aggregated by land use, the mean mass of fruits from farmers' fields was significantly greater than that from communal land or natural woodland, there also being a significant difference between the populations outside farmers' fields (Fig. 22.2). This comparison cannot be made in Namibia, as all the trees sampled were in farmers' fields.

Pulp Mass

Pulp and nut mass are the two major constituents of fruits. Pulp mass showed highly significant and continuous variation (7.5–36.0 g) within site samples, in a sequence similar to that for fruit mass. The pulp component of marula fruits is approximately 50% skin and 50% flesh and juice. In the South African fruits, skin was the larger proportion, while in Namibia, flesh and juice made the larger proportion (Fig. 22.3A and B); this may reflect differences in ripeness, due to the longer time before Namibian fruits were weighed and processed. Alternatively it may reflect the greater proximity to the watertable, in the Cuvelai drainage system of the Owambo Basin of Namibia, which drains into the Etosha Pan.

The fruits of "Namibian Wonder" had very much greater pulp mass (more than twice the population mean) than the fruits of most other trees.

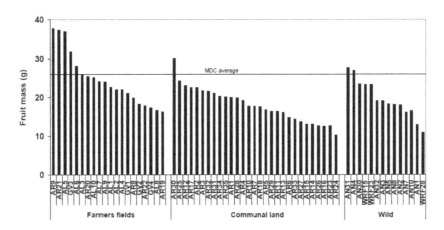

FIGURE 22.2 Tree-to-tree variation in fruit mass (g) of marula (*Sclerocarya birrea*) by land use in Bushbuckridge, South Africa.

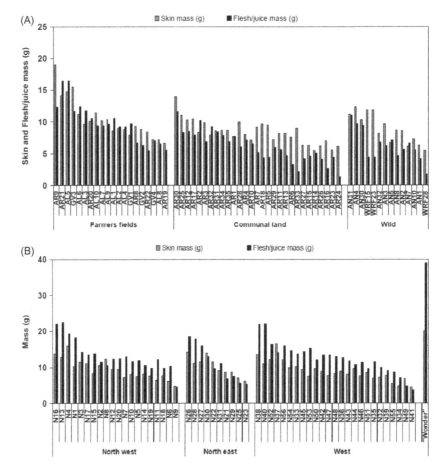

FIGURE 22.3 (A) Tree-to-tree variation in skin and flesh/juice mass of marula (*Sclerocarya birrea*) in Bushbuckridge, South Africa (in order of fruit mass); (B) Tree-to-tree variation in skin and flesh/juice mass of marula (*S. birrea*) in North Central Region, Namibia (in order of fruit mass).

Frequency Distribution

The data for fruit mass of each tree in each site displayed considerable variance about its mean (Fig. 22.4A and B). The frequency distribution of this data, and of the overall site mean were close to a normal distribution (Table 22.6). When the data for South African trees were aggregated by land use, the frequency distribution for natural woodland was also close to normality (skewness = −0.03 to 0.83), but that for communal lands and farmers' fields had a greater tendency

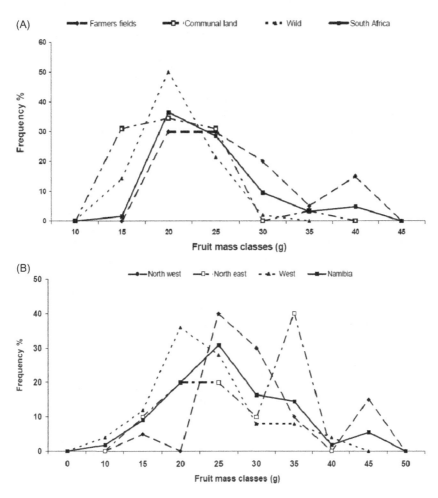

FIGURE 22.4   (A) Frequency distribution (%) of fruit mass of marula (*Sclerocarya birrea*) by land use in Bushbuckridge, South Africa; (B) Frequency distribution (%) of fruit mass of marula (*S. birrea*) by site in North Central Region, Namibia.

toward being positively skewed (Table 22.6), the latter also having some bimodality. In a similar way, the distribution pattern of the same traits in Namibia was close to normal, but with some degree of bimodality in fruit mass from the populations from the northeast and northwest (Table 22.7).

DISCUSSION

The very considerable tree-to-tree variation in fruit characteristics in marula is consistent with results from other indigenous fruit trees, such as *Irvingia gabonensis* (Atangana et al., 2001) and *Dacryodes edulis* (Waruhiu, 1999; Leakey et al., 2002) in West and Central Africa. As in the case of these moist forest species of West and Central Africa, this variation, coupled with the considerable market opportunities, indicates that there are domestication opportunities for marula, as has already been recognized (Holtzhausen et al., 1990; Taylor et al., 1996; Mizrahi and Nerd, 1996; Maghembe et al., 1998). However, this study quantifies the variation in dry matter partitioning between constituent parts of the fruits for the first time, and provides good fundamental knowledge about the range of variation in several important traits across geographically separated, as well as environmentally and culturally different, sites. This allows an analysis of the relationships between different commercial traits of importance to the development of cultivars that can be used to determine "ideotypes" that maximize the Harvest Index across several different fruit traits, and hence to identify potential cultivars for cultivation that could meet the needs of farmers with proximity to different markets (Leakey, 2005).

It is clear from this study of the phenotypic (tree-to-tree) variation in populations in South Africa and Namibia (Figs. 22.1–22.4) that there is very considerable opportunity to identify individual trees with fruit characteristics well

TABLE 22.6 Frequency distributions of marula (*Sclerocarya birrea*) fruit trait data from different sites and land uses in South Africa in 2002.

	Allandale and Green Valley				Andover and WRF				Acornhoek Rd			
	Mean	SE	Skewness	Kurtosis	Mean	SE	Skewness	Kurtosis	Mean	SE	Skewness	Kurtosis
South Africa by site												
Fruit mass (g)	24.00	0.35	0.78	0.64	18.84	0.23	0.96	1.78	19.34	0.31	0.43	0.07
Skin mass (g)	10.03	0.16	1.15	1.38	8.56	0.09	0.80	0.88	8.65	0.14	0.44	−0.09
Flesh mass (g)	9.93	0.18	0.66	1.16	6.45	0.13	0.72	0.84	6.48	0.18	0.53	0.35
Pulp mass (g)	19.96	0.30	0.76	0.63	15.01	0.20	0.88	1.47	15.13	0.27	0.36	−0.03

	Farmers' fields				Communal land				Natural woodland			
	Mean	SE	Skewness	Kurtosis	Mean	SE	Skewness	Kurtosis	Mean	SE	Skewness	Kurtosis
South Africa by site												
Fruit mass (g)	23.60	0.32	0.97	0.86	18.03	0.23	0.48	0.19	19.34	0.31	0.43	0.07
Skin mass (g)	9.92	0.13	1.07	1.21	8.33	0.10	0.80	0.95	8.65	0.14	0.44	−0.09
Flesh mass (g)	9.50	0.17	0.68	0.60	6.02	0.13	0.31	−0.41	6.48	0.18	0.53	0.35
Pulp mass (g)	19.42	0.28	0.85	0.57	14.35	0.20	0.52	0.43	15.13	0.27	0.36	−0.03

Skewness/kurtosis = 0 = normality (+skewness = a long tail of high values −skewness = a long tail of low values + kurtosis = tails longer than normality −kurtosis = tails shorter than normality)

TABLE 22.7 Frequency distributions of marula (*Sclerocarya birrea*) fruit trait data from different sites and land uses in Namibia and South Africa and Namibia in 2002.

Namibia by site

	Namibia NE				Namibia NW				Namibia W			
	Mean	SE	Skewness	Kurtosis	Mean	SE	Skewness	Kurtosis	Mean	SE	Skewness	Kurtosis
Fruit mass (g)	24.49	0.54	0.29	−0.82	26.89	0.38	0.66	0.33	25.24	0.34	0.53	0.24
Skin mass (g)	10.28	0.21	0.40	−0.24	9.52	0.14	0.63	0.24	8.98	0.13	0.86	0.95
Flesh mass (g)	11.19	0.35	0.22	−0.95	13.26	0.24	0.48	0.29	12.50	0.21	0.55	0.58
Pulp mass (g)	21.47	0.51	0.23	−0.93	22.78	0.35	0.58	0.30	21.42	0.30	0.41	0.20

South Africa and Namibia

	South Africa				Namibia				South Africa and Namibia			
	Mean	SE	Skewness	Kurtosis	Mean	SE	Skewness	Kurtosis	Mean	SE	Skewness	Kurtosis
Fruit mass (g)	20.11	0.17	0.74	1.01	26.68	0.28	1.60	4.92	23.21	0.17	1.58	5.45
Skin mass (g)	8.91	0.07	0.83	1.08	9.60	0.09	0.95	1.17	9.24	0.06	0.98	1.46
Flesh mass (g)	7.24	0.10	0.53	0.45	13.37	0.22	3.28	16.78	10.14	0.13	3.26	20.36
Pulp mass (g)	16.15	0.15	0.68	0.83	22.23	0.22	0.47	0.01	19.03	0.14	0.72	0.48

Skewness/Kurtosis = 0 = normality (+skewness = a long tail of high values −skewness = a long tail of low values + kurtosis = tails longer than normality −kurtosis = tails shorter than normality)

above the average of the species. While this is evident from the range of continuous variation identified in each of the populations studied, populations of only 14−63 trees, the potential is really highlighted by the discovery of Namibian Wonder, a much rarer member of the population. Such extreme examples of phenotypic variation are only found at far greater selection intensities. In the present study, the Namibian population seems to have greater promise of superior trees than that of South Africa, but the geographical range surveyed in South Africa was considerably narrower than in Namibia, although neither survey was comprehensive.

In this study, a sample of the fruits passing through the MDC juicing process was collected and characterized. It was found (Fig. 22.1A) that the women were processing fruits representative of the best 84% of the trees in the area (=they had discarded the worst 16%). Through domestication, they could improve their trees and produce fruits equivalent to the best 5% in the area. This would reduce the labor involved in juicing the fruits, increase the proportion of the pulp extracted (nb. this project has identified that the women are not recovering a high proportion of the juice; Mander et al., 2002), and with appropriate selection could probably improve the flavor of the product. Interestingly, the Amarula factory of Distell Corporation uses both the skin and the pulp in its process, so maximizing the product they obtain from the fruits. Future studies should use already developed techniques for flavor profiling fruits (Schaefer and McGill, 1986) in an assessment of the genetic and nongenetic (consequences of storage and processing) organoleptic (flavor and fragrance) properties of different fruit samples, as these are important commercially in the juice and liqueur industry. It would be important to interact with commercial companies to determine which properties are desirable and which are undesirable (Leakey, 1999a).

Studies of the frequency distribution of fruit and kernel trait data in *D. edulis* and *I. gabonensis* in West Africa, similar to that reported here, have indicated that farmers, by their own procedures of genetic selection (truncated selection), have made a 40−65% gain in fruit mass (Leakey et al., 2004). A similar analysis with marula is not conclusive, but there are certainly some results that suggest that a similar process of farmer domestication is underway in southern Africa. For example, several data sets (e.g., fruit mass from trees in farmers' fields—Tables 22.6, 22.7) are positively skewed (with a tendency to bimodality), with a tail of unusually large fruits. This contrasts with more normally distributed data from trees in natural woodlands. This suggests that farmers, through truncated selection over several generations, have achieved at least the second stage of domestication (Leakey et al., 2004).

Interestingly, other studies within this project (Botelle et al., 2002; Shackleton et al., 2002a) have found large differences in marula tree fruit yield between trees in farmers' fields and in communal land and natural woodland. Some of these trees were the same as in this study and in these trees the mean yield/tree from 13 trees in crop fields was 33,187 fruits, while that from 12 trees in natural woodlands was only 6135. It is not clear to what extent these differences in fruit yield represent genetic selection or cultivation and reduced competition in farmers' fields from other plants, but the evidence from farmer interviews indicates that selective nurturing of superior trees and the planting of seedlings and vegetative truncheons from them clearly supports the previous suggestion that subsistence farmers in southern Africa have initiated the domestication of marula.

It is concluded that there is an opportunity to select cultivars for fruit production to meet the needs of traditional beer/wine markets and new markets for fruit juices, flavorings, liqueurs, etc. Cultivar selection would increase uniformity in the product, increase productivity and provide an incentive for farmers to plant marula trees in their farming systems. This should lead to socioeconomic and environmental benefits from the adoption of agroforestry practices (Leakey, 2001b).

ACKNOWLEDGMENTS

Jenny Botha is gratefully acknowledged for assistance with the field trip in South Africa. This publication is an output from a project partly funded by the United Kingdom Department for International Development (DFID) for the benefits of developing countries. The authors are indebted to DFID for funding this project (Project No R7190 of the Forestry Research Program) and the views expressed here are not necessarily those of DFID.

Chapter 23

Domestication Potential of Marula (*Sclerocarya birrea* subsp. *caffra*) in South Africa and Namibia: 2. Phenotypic Variation in Nut and Kernel Traits

This chapter was previously published in Leakey, R.R.B., Pate, K., Lombard, C., 2005. Agroforestry Systems, 64, 37−49, with permission of Springer

SUMMARY

As part of a wider study characterizing tree-to-tree variation in fruit traits as a pre-requisite for cultivar development, fruits were collected from each of 63 marula (*Sclerocarya birrea*) trees in Bushbuckridge, South Africa and from 55 trees from the North Central Region of Namibia. The nuts were removed from the fruit flesh, and the kernels extracted, counted and weighed individually to determine the patterns of dry matter partitioning among the nut components (shell and kernel) of different trees. Mean nut, shell and kernel mass were not significantly different between the two countries. Between sites in South Africa there were highly significant differences in mean nut mass, shell mass, kernel mass and kernel number. In Namibia, there were highly significant differences between geographic areas in mean shell mass, kernel mass and kernel number, but not in nut mass. These differences had considerable impacts on shell:kernel ratios (8.0 − 15.4). In South Africa, mean kernel mass was significantly greater in fruits from farmers' fields (0.42g) than from communal land (0.30g) or natural woodland (0.32g). Within all sites, in both South Africa and Namibia, there was highly significant and continuous variation between individual trees in nut mass (South Africa = 2.3 − 7.1g; Namibia = 2.7 − 6.4g) and kernel mass (South Africa = 0.09 − 0.55g; Namibia = 0.01 − 0.92g). The small and valuable kernels constitute a small part of the nut (Namibia = 6.1 − 11.1%; South Africa = 7.6 − 10.7%). There can be 4 kernels per nut, but even within the fruits of the same tree, kernel number can vary between 0-4, suggesting variation in pollination success, in addition to genetic variation. The nuts and kernels of the Namibian trees were compared with the fruits from one superior tree ('Namibian Wonder': nuts = 10.9g; kernels = 1.1g). Oil content (%) and oil yield (g/fruit) also differed significantly between trees (44.7 − 72.3% and 8.0 − 53.0 g/fruit). The percentage frequency distribution of kernel mass was skewed from trees in farmers' fields in South Africa and in some sites in Namibia, suggesting a level of anthropogenic selection. It is concluded that there is great potential for the development of cultivars for kernel traits, but there is also a need to determine how to increase the proportion of nuts with four kernels, perhaps through improved pollination success.

INTRODUCTION

Sclerocarya birrea (A. Rich.) Hochst. subsp. *caffra* (Sond.) Kokwaro is one of the traditionally important indigenous fruits of southern Africa and is now gaining commercial importance (Shackleton et al., 2002b; Wynberg et al., 2002). A previous paper in this series (Leakey et al., 2005b) has indicated its growing importance as indicated by a number of domestication initiatives in South Africa, Botswana, and in Israel. More recently, the World Agroforestry Center (ICRAF), has also initiated germplasm collection and field trials, as part of its international agroforestry tree domestication program (Leakey and Simons, 1998; Simons and Leakey, 2004).

Multifunctional Agriculture. DOI: http://dx.doi.org/10.1016/B978-0-12-805356-0.00023-4
© 2017 Elsevier Inc. All rights reserved.

Initiatives in four regions of Africa are developing techniques and strategies for the domestication and commercialization of trees producing agroforestry tree products (AFTPs) for integration into farmland (Simons and Leakey, 2004). This is seen as an approach to poverty alleviation (Leakey and Simons, 1998; Poulton and Poole, 2001) and the environmental rehabilitation of degraded farmland (Leakey, 1999b; Leakey, 2001a,b). The present study, as part of a resource inventory, quantifies the phenotypic variation in fruit, nut and kernel traits in marula, within the framework of a broader project examining the benefits and opportunities for domesticating and commercializing the fruits and kernel oil of marula in South Africa and Namibia (Sullivan et al., 2003). Chemical analysis of fruits and kernels has indicated the potential of marula nutritionally, and as a source of high quality oil, rich in tocopherol (Burger et al., 1987; Leakey, 1999a). Recent collections of marula fruits and nuts from individual trees in Makueni district of Kenya have been analyzed for a wide range of nutritional compounds and minerals, and the kernels were rich (56−64%) in oils (Thiong'o et al., 2002). Kernels are traditionally used extensively in some areas in southern Africa (e.g., Inhambane, Mozambique; Owambo, Namibia; KwaZulu-Natal, South Africa), but little used in other areas (e.g., Kavango, Namibia; Northern Province, South Africa) as a nutritious food, a meat preservative and as a skin moisturizing agent (Shackleton et al., 2002b; Wynberg et al., 2002). The oil is also starting to become important in the cosmetics industry (Wynberg et al., 2002).

METHODS AND MATERIALS

As already reported for a study of marula fruit characterization, ripe and unblemished fruits were collected from beneath the crown of marula trees in villages in Limpopo Province, South Africa (Bushbuckridge) and Namibia (North Central Region), in 2002 (see details in Leakey et al., 2005b). Fruits from each tree were separately bagged and labeled for use in the study of fruit characteristics (Leakey et al., 2005b) and for the present study. As soon as possible (usually 2−3 days later), the nuts were soaked and scrubbed to remove the flesh before being set in the sun to dry for about 10 hours. When dry, the nuts were weighed and numbered still in the same order as for the study of fruit traits (Leakey et al., 2005b), so that their identity was maintained for subsequent cracking and kernel removal. The kernels were then weighed using a laboratory (0.001 g) balance (Mettler Toledo PB 3002) and packaged for later oil extraction. Shell mass was derived by difference (Nut−kernel = shell).

Additional samples were collected (see Leakey et al., 2005b for details) from the Mhala Development Center (MDC), in Bushbuckridge, South Africa, and from a superior tree (identified here as "Namibian Wonder").

SPSS 10.0 for Windows was used for the Analysis of Variance, Duncan's Multiple Range tests, and tests for skewness and kurtosis.

Oil extraction from the South African and Namibian kernels was done by Analytical Laboratory Services in Windhoek using a petroleum ether extract, according to the Deutsche Einheitsmethoden zur Untersuchung von Fetten, Fettprodukten, Tensiden und Verwandten Stoffen (Method code = DGF 8-15 (B7)) method.

RESULTS

Variation Between Sites

Comparison of Mean Values Between South Africa and Namibia

Mean nut, shell, and kernel mass were not significantly different between the two countries (Table 23.1).

TABLE 23.1 Comparison of Marula (*Sclerocarya birrea*) nut traits between South Africa and Namibia.

Trait	South Africa	Namibia	Probability
Nut mass	3.96	4.06	$p = 0.040$
Kernel mass	0.34	0.36	$p = 0.027$
Shell mass	3.62	3.68	$p = 0.212$
No. of kernels	1.54	1.50	$p = 0.286$
Shell:kernel ratio	9.4	9.8	

Comparison of Mean Values Between Sites in South Africa

There were highly significant differences in mean nut, shell, and kernel mass, as well as kernel number, between sites in South Africa (Table 23.2). Fruits from Allandale were found to be the largest. The trees from Acornhoek Road were located in both farmers' fields and communal land, while those from Allandale and Green Valley were only in farmers' fields. Trees from Andover and Wits Rural Facility were in natural woodland. An analysis by land use follows.

Comparison of Mean Values Between Land Uses in South Africa

Mean kernel mass was significantly greater in fruits from farmers' fields than from communal land or natural woodland in South Africa (Table 23.3). The commercially undesirable trait of a large shell mass was also significantly greater in fruits from farmers' fields than in fruits from communal land, although not significantly different from those from natural woodland.

Comparison of Mean Values Between Sites in Namibia

There were significant differences in mean shell and kernel mass between areas in Namibia, but not in nut mass (Table 23.4).

Variation Within Sites

Within all sites, in both South Africa and Namibia, there was highly significant variation between individual trees in the morphological traits of nuts and kernels that were measured.

Nut and Kernel Mass

Tree-to-tree variation in nut mass was statistically significant within all land uses in South Africa (Fig. 23.1A) and within all sites in South Africa and Namibia (Fig. 23.1A and B). Fig. 23.1A and B present nut mass data in the order of

TABLE 23.2 Comparison of Marula (*Sclerocarya birrea*) nut traits across South African sites.

Trait	Site				Probability
	Acornhoek road	Andover and Wits Rural Facility	Allandale	Green Valley	
Nut mass	3.8 b	4.2 a	4.2 a	3.6 b	$p \leq 0.001$
Kernel mass	0.31 c	0.32 c	0.45 a	0.38 b	$p \leq 0.001$
Shell mass	3.5 b	3.9 a	3.7 a	3.2 c	$p \leq 0.001$
No. of kernels	1.4 b	1.4 b	1.8 a	1.8 a	$p \leq 0.001$
Shell:kernel ratio	11.3	12.2	8.2	8.4	

TABLE 23.3 Comparison of Marula (*Sclerocarya birrea*) nut traits across land use systems in South Africa.

Trait	Site			Probability
	Farmers' fields	Communal land	Natural woodland	
Nut mass	4.18 a	3.68 b	4.20 a	$p \leq 0.001$
Kernel mass	0.42 a	0.30 b	0.32 b	$p \leq 0.001$
Shell mass	3.78 a	3.39 b	3.87 a	$p \leq 0.001$
No of kernels	1.81 a	1.40 b	1.44 b	$p \leq 0.001$
Shell:kernel ratio	9.0	11.3	12.1	

TABLE 23.4 Comparison of Marula (*Sclerocarya birrea*) nut traits across Namibian sites.

Trait	Site			Probability
	Northwest	Northeast	West	
Nut mass	4.11 a	4.02 a	4.04 a	$p = 0.666$
Kernel mass	0.25 c	0.32 b	0.45 a	$p \leq 0.001$
Shell mass	3.86 a	3.55 b	3.59 b	$p = 0.002$
No. of kernels	1.21 b	1.68 a	1.68 a	$p \leq 0.001$
Shell:kernel ratio	15.4	11.1	8.0	

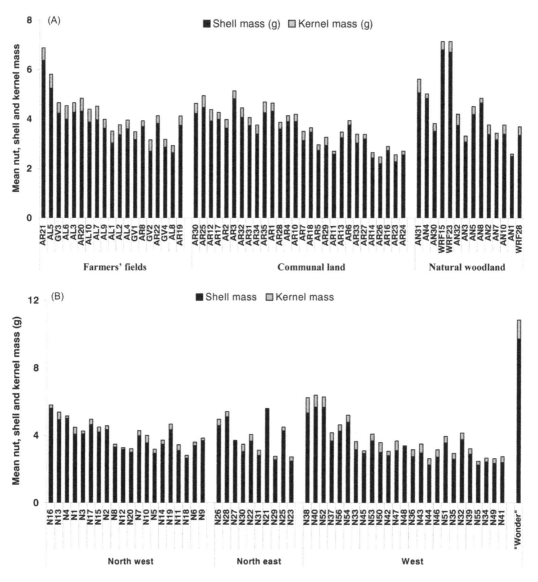

FIGURE 23.1   (A) Tree-to-tree variation in nut (shell + kernel) mass (g) of marula (*Sclerocarya birrea*) by land use in Bushbuckridge, South Africa (in order of increasing fruit mass); (B) Tree-to-tree variation in nut (shell + kernel) mass (g) of marula (*S. birrea*) by site in North Central Region, Namibia (in order of increasing fruit mass).

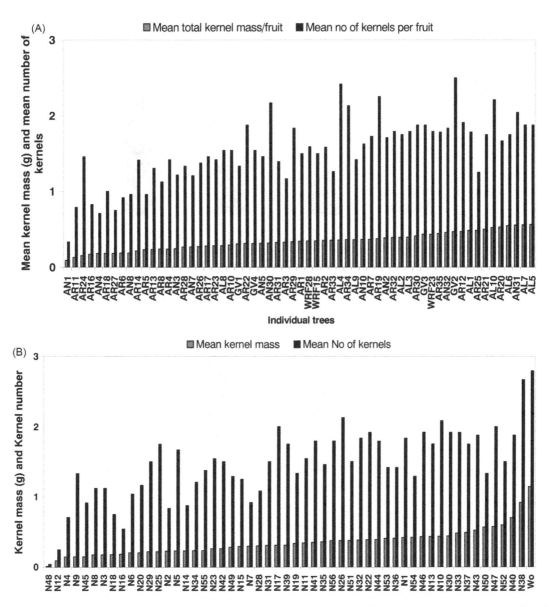

FIGURE 23.2 (A) Tree-to-tree variation in number of kernels per nut in marula (*Sclerocarya birrea*) in Bushbuckridge, South Africa (in order of increasing kernel mass/nut); (B) Tree-to-tree variation in number of kernels per nut in marula (*S. birrea*) in North Central Region, Namibia (in order of increasing kernel mass/nut).

decreasing fruit mass, so indicating a weak relationship with fruit mass. The nuts of marula are mostly composed of shell, with the important kernels making up only 3.5–14.8% of the mass in South Africa and 2.8–16.0% in Namibia. Mean kernel mass per nut is the sum of the mass of all individual kernels in the nut, and while there is continuous variation in mean kernel mass per nut, this is not matched by the number of kernels per nut (Fig. 23.2A and B). The mean kernel mass of "Namibian Wonder" was more than twice that of the mean kernel mass per nut of most other trees. However, the shell mass:kernel mass ratio was similar to that of trees from the West district, but was less than that of trees from the northeast and northwest (Table 23.4). In South Africa, the nut mass:kernel mass ratio was lowest in farmers' fields and greatest in the natural woodland (Table 23.3).

Number of Kernels

There was significant variation between trees in the mean number of kernels per nut (Fig. 23.2A and B). Kernel number per nut also varied within individual tree fruit samples from 0 to 4 kernels per fruit (Fig. 23.3A and B). Nuts with 2–3 kernels were the most common. In both South African and Namibian fruits the mean mass of individual kernels

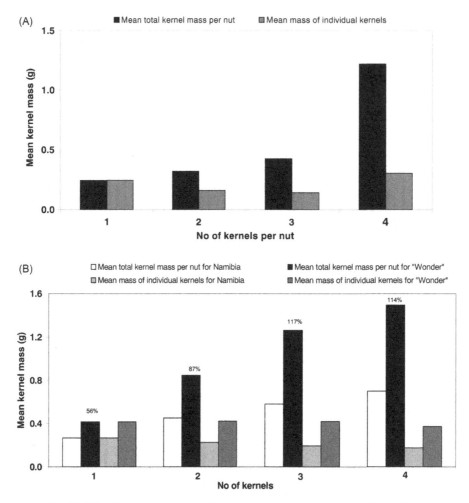

FIGURE 23.3   (A) Mean total and individual kernel mass in nuts of marula (*Sclerocarya birrea*) with 1–4 kernels per nut, in Bushbuckridge, South Africa; (B) Mean total and individual kernel mass in nuts of marula (*S. birrea*) with 1–4 kernels per nut, in North Central Region, Namibia.

declined, the greater the number of kernels per nut (Fig. 23.3A and B). In contrast, the mean mass of quadruple kernels in South African trees was greater than for other groups, but the sample size was $n = 2$, and consequently can be ignored. In "Namibian Wonder," the mean mass of individual kernels was constant, regardless of the number of kernels per nut (Fig. 23.3B).

Oil Content

The percentage oil content of both South African and Namibian kernels was not significantly different between land uses or site (South African range = 44.7–72.3%; Namibian range = 50.2–63.8%). The oil content of "Namibian Wonder" kernels was not dissimilar to that of other trees.

In South Africa, the oil yield per fruit (% oil content × kernel mass) was significantly greater in fruits from farmers' fields (Fig. 23.4A), while in Namibia it was significantly greater in fruits from west district than in those from the northeast (Fig. 23.4B). The oil yield of "Namibian Wonder" was very much greater than from any other tree (Fig. 23.4A and B).

Frequency Distribution

The data for nut and kernel mass of each tree in each site displayed considerable variance about its mean. The frequency distribution of the South African site data sets, and of the overall country mean were close to a normal distribution (Table 23.5). However, when the nut and kernel data for South African trees were aggregated by land use, the frequency distributions for natural woodland remained close to normality (skewness = −0.03 to 0.83), but that for

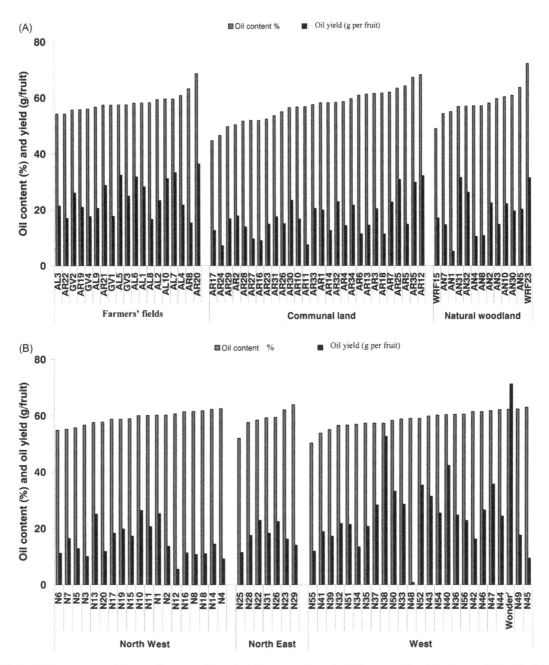

FIGURE 23.4 (A) Tree-to-tree variation in oil content (%) and yield per nut of marula (*Sclerocarya birrea*) by land use in Bushbuckridge, South Africa (in order of increasing oil content); (B) Tree-to-tree variation in oil content (%) and yield per nut of marula (*S. birrea*) by site in North Central Region, Namibia (in order of increasing oil content).

communal lands and farmers' fields had a tendency toward being positively skewed (Fig. 23.5A and B), the latter also having some bimodality (especially kernel mass). In the Namibian data, the distribution pattern of the same traits had similar distributions (Table 23.6), but the kernel mass in populations from the West had a small number of large kernels (Figs. 23.6A and B).

DISCUSSION

This study quantifies the variation in dry matter partitioning between constituent parts of marula nuts for the first time, and provides good fundamental knowledge about the range of variation in several important traits across geographically separated, as well as environmentally and culturally different sites, as well as across different land uses in South Africa.

TABLE 23.5 Frequency distributions of marula (*Sclerocarya birrea*) nut trait data from different sites and land uses in South Africa.

	Allandale and Green Valley				Andover and WRF				Acornhoek Rd			
	Mean	SE	Skewness	Kurtosis	Mean	SE	Skewness	Kurtosis	Mean	SE	Skewness	Kurtosis
South Africa												
Nut mass	4.04	0.06	0.59	0.33	3.82	0.04	0.82	1.78	4.20	0.08	0.83	0.65
Kernel mass	0.43	0.01	0.28	0.21	0.31	0.07	0.55	0.52	0.32	0.01	0.31	−0.38
Shell mass	3.60	0.05	0.28	0.50	3.53	0.04	0.79	1.87	3.88	0.07	0.93	0.79
Kernel No	0.43	0.01	0.28	0.21	1.45	0.03	0.05	−0.49	1.44	0.05	−0.03	−0.79

	Farmers fields				Communal land				Natural woodland			
	Mean	SE	Skewness	Kurtosis	Mean	SE	Skewness	Kurtosis	Mean	SE	Skewness	Kurtosis
South Africa												
Nut mass	4.18	0.05	1.10	2.16	3.68	0.04	0.41	0.37	4.20	0.07	0.83	0.65
Kernel mass	0.42	0.01	0.41	0.26	0.30	0.01	0.42	0.17	0.32	0.01	0.31	−0.38
Shell mass	3.78	0.05	1.00	2.54	3.39	0.04	0.38	0.47	3.87	0.07	0.83	0.89
Kernel No	1.81	0.04	−0.13	−0.64	1.40	0.03	0.06	−0.47	1.44	0.05	−0.03	−0.79

Skewness/Kurtosis = 0 = normality (+skewness = a long tail of high values
+ kurtosis = tails longer than normality)

− skewness = a long tail of low values
− kurtosis = tails shorter than normality)

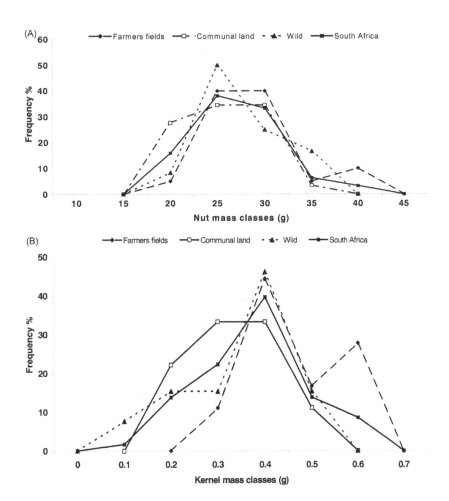

FIGURE 23.5 (A) Frequency distribution (%) of nut mass of marula (*Sclerocarya birrea*) by land use in Bushbuckridge, South Africa; (B) Frequency distribution (%) of kernel mass of marula (*S. birrea*) by land use in Bushbuckridge, South Africa.

Unlike the skin, flesh and juice components of marula fruits (Leakey et al., 2005b), the mean nut, shell, and kernel mass were similar in South Africa and Namibia. However, there were differences in kernel mass between all sites and land uses, and differences in nut and shell mass between sites in South Africa and in just shell mass in Namibia, which may reflect differences in the environment, in anthropogenic activity or evolutionary responses to different survival/regeneration pressures.

Evidence from the frequency distribution data of marula nut and kernel mass suggests that as in West African indigenous fruits (Leakey et al., 2004), subsistence farmers have initiated the domestication process. However, in marula, the process does not seem to be well advanced, although several data sets (e.g., nut and kernel mass from the west of Namibia and from trees in farmers' fields in South Africa) are positively skewed (with a tendency to bimodality). This suggests that farmers, through truncated selection over several generations, have achieved at least the second stage of domestication (Leakey et al., 2005b,c).

The very considerable tree-to-tree variation in nut and kernel characteristics in marula mirror those found in the fruit traits (Leakey et al., 2002) and are consistent with results on nut and kernel traits from other indigenous fruit trees, such as *I. gabonensis* (Atangana et al., 2001; 2002; Anegbeh et al., 2003). There is, however, one major difference in that *I. gabonensis* typically has a single kernel within the nut, while marula has up to four kernels per nut. Nevertheless, there was extensive phenotypic variation in dry matter allocation to kernels in both species that is probably genetic in origin. Interestingly, the mean kernel mass of "Namibian Wonder" was more than twice that of the mean kernel mass per nut of most other trees, indicating the potential for individual tree selection for cultivar development.

In marula, the number of kernels per nut greatly affected the total kernel mass per nut/fruit, with an indication that although individual kernel mass is relatively constant, there is a slight reduction in mean kernel mass as the number of kernels per nut increases. This was particularly clear in the Namibian data set (in the South African case only two nuts

TABLE 23.6 Frequency distributions of marula (*Sclerocarya birrea*) nut trait data from different sites and land uses in Namibia, and in South Africa and Namibia together.

Namibia

	Namibia NE				Namibia NW				Namibia W (excluding Namibian Wonder)			
	Mean	SE	Skewness	Kurtosis	Mean	SE	Skewness	Kurtosis	Mean	SE	Skewness	Kurtosis
Nut mass	4.02	0.08	0.26	−0.68	4.11	0.05	0.41	−0.21	3.77	0.05	1.22	1.56
Kernel mass	0.32	0.01	0.70	0.89	0.25	0.01	0.54	−0.07	0.43	0.01	0.64	0.97
Shell mass	3.55	0.08	0.30	−0.77	3.86	0.05	0.48	−0.17	3.34	0.05	1.25	1.61
Kernel No	1.68	0.05	0.30	−0.59	1.21	0.04	0.26	−0.64	1.63	0.03	−0.17	0.09

Skewness/Kurtosis = 0 = normality (+skewness = a long tail of high values
+ kurtosis = tails longer than normality)

− skewness = a long tail of low values
− kurtosis = tails shorter than normality)

South Africa and Namibia

	South Africa				Namibia				South Africa and Namibia			
	Mean	SE	Skewness	Kurtosis	Mean	SE	Skewness	Kurtosis	Mean	SE	Skewness	Kurtosis
Nut mass	3.96	0.03	0.80	1.33	4.06	0.04	1.96	6.66	4.01	0.03	1.58	5.50
Kernel mass	0.34	0.01	0.06	0.13	0.36	0.01	1.12	2.75	0.35	0.004	0.86	1.95
Shell mass	3.62	0.03	0.80	1.57	3.68	0.04	1.86	5.95	3.65	0.02	1.50	4.96
Kernel No	1.54	0.02	−0.02	−0.61	1.50	0.02	−0.02	−0.26	1.52	0.02	−0.02	−0.46

Skewness/Kurtosis = 0 = normality (+skewness = a long tail of high values
+ kurtosis = tails longer than normality)

− skewness = a long tail of low values
− kurtosis = tails shorter than normality)

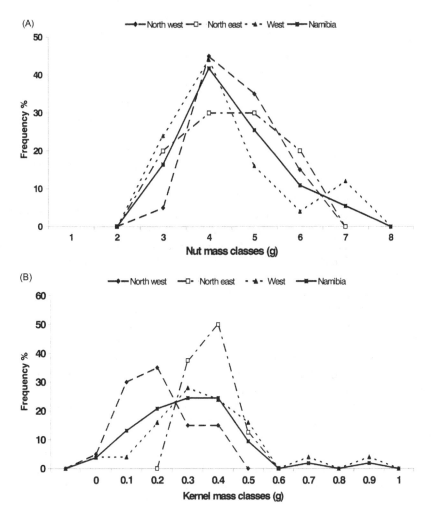

FIGURE 23.6   (A) Frequency distribution (%) of nut mass of marula (*Sclerocarya birrea*) by site in North Central Region, Namibia; (B) Frequency distribution (%) of kernel mass of marula (*S. birrea*) by site in North Central Region, Namibia.

had four kernels and so the mean is not comparable with that for nuts with one to three kernels). Competition between kernels for assimilates is suggested by this declining mean mass of individual kernels as the number of kernels per nut increases (Fig. 23.3A and B). The constant mean mass of individual kernels in "Namibian Wonder," regardless of the number of kernels per nut (Fig. 23.3B), indicates that potentially all kernels can have the same mass and that the partitioning of dry matter does not have to be limited by competition. Clearly this would be a desirable trait in any domesticated cultivar. The finding that nuts with single kernels do not have much larger kernels than nuts with more kernels indicates that selection for nuts with a single kernel is not an appropriate strategy.

The varying number of kernels per nut (0—4 kernels per fruit) within individual tree fruit samples indicates that this trait is affected by some environmental factors and is not only a genetic trait. A possible explanation for this variability in kernel number per nut is that not all ovules were successfully pollinated. This is perhaps the result of poor pollinators due to the elimination of nonproductive male trees (Nghitoolwa et al., 2003), excessive distances between trees, or inappropriate weather for pollinator activity. If poor pollination success is the cause of low kernel numbers, then the siting of beehives in male marula trees, or the grafting of male scions into the crowns of female trees, might be advantageous.

Typically kernel mass is about 10% of nut mass (Wynberg et al., 2002), but this study found a range from 2.8—16% between individual trees. Ideally, in developing marula as a nut crop, selection would seek to increase the partitioning of dry matter from shell to kernel, with the additional benefit of perhaps making the nut easier to crack. However, there was little evidence from this study to suggest that shell mass can be greatly reduced, although the shell:kernel ratio of nuts from the west and of "Namibian Wonder" was considerably lower (8.0) than in nuts from the northwest (15.4), offering some hope that shell mass could be reduced to some extent. It is not clear whether this variability in shell:kernel ratio is the result of anthropogenic or environmental selection. The similarity in nut:kernel ratio between "Namibian Wonder" and many other trees, despite its abnormally large fruit is interesting as it demonstrates that "Namibian Wonder" probably represents the upper end of the continuous variation in fruit nut and kernel traits rather than some abnormality in terms of the relative partitioning of dry matter between different components of the nut.

When the marketable product is only a small proportion of the overall production (i.e., represents a low Harvest Index) it is important to maximize its value, or to derive a number of different products from the crop at the same time. In marula, both the fruit pulp and the kernels are marketable, and while the fruits (pulp, juice, shell, and kernel) are fairly small, they can be numerous when grown in farmers' fields (up to 128,610 fruits have been recorded from a single mature tree; Shackleton et al., 2002a,b). The fruit pulp (skin and juice) is used to make local beer (Shackleton, 2002), and is being marketed as a fruit juice and fermented as an ingredient in Amarula liqueur (Mander et al., 2002). Traditionally kernels are eaten as a highly nutritious food supplement and currently kernel oil is starting to be used in the cosmetics industry. A parallel study (Leakey et al., 2005b) has examined the tree-to-tree variation in fruit traits and found very extensive variation and, thus, the potential for genetic selection. This study has identified equivalent variation in kernel production. Kernels are both a nutritious local food additive, rich in proteins, and a source of oils for a variety of uses from cooking to cosmetics. This study has determined that there is considerable tree-to-tree variation in oil percentage content of kernels (Namibia = 50−64%: South Africa = 45−72%), which when combined with the variation in kernel mass (0.147−1.144 g) results in large differences in oil yield per fruit (0.01−0.71 g). The superiority of "Namibian Wonder" in oil yield indicates the opportunity for genetic selection, an opportunity which could be further improved by increasing the mean number of kernels per nut through better pollination, as described previously. There is great interest in commercial oil processing in the region, but currently, oil entering the cosmetics industry is a bulk sample of unselected quality. However, there could perhaps be a premium for oil quality, so it is now important to determine whether or not there is tree-to-tree variation in the oil composition and quality.

To meet the evident potential of developing cultivars of marula, it is not yet clear to what extent cultivar development in marula needs to be separately focused on selection for fruit and kernel traits, as in *I. gabonensis* from West Africa (Leakey et al., 2000; Atangana et al., 2002), or whether selection could be on the basis of the combination of fruit and kernel traits. This will be examined elsewhere (Leakey, 2005).

It is concluded that in addition to the potential for cultivars selected for fruit characteristics, there is equal opportunity to select cultivars for kernel production to meet the needs of markets for nutritious kernels and various oil industries. Cultivar selection would increase uniformity in the product, increase productivity and provide an incentive for farmers to plant marula trees in their farming systems, with concomitant socioeconomic and environmental benefits (Leakey, 2001b).

ACKNOWLEDGMENTS

Pierre du Plessis and Sheona Shackleton are gratefully acknowledged for assistance with the field trips in Namibia and South Africa, respectively. This publication is an output from a project partly funded by the United Kingdom Department for International Development (DFID) for the benefits of developing countries. The authors are indebted to DFID for funding this project (Project No R7190 of the Forestry Research Program) and the views expressed here are not necessarily those of DFID.

Chapter 24

Domestication Potential of Marula (*Sclerocarya birrea* subsp. *caffra*) in South Africa and Namibia: 3. Multiple Trait Selection

This chapter was previously published in Leakey, R.R.B., 2005. Agroforestry Systems, 64, 51−59, with permission of Springer

SUMMARY

An understanding of the inter-relationships between the traits characterising tree-to-tree variation in fruits and kernels is fundamental to the development of selected cultivars based on multiple trait selection. Using data from previously characterised marula (*Sclerocarya birrea*) trees in Bushbuckridge, South Africa and North Central Region of Namibia, this study examines the relationships between the different traits (fruit pulp, flesh/juice mass, and nut shell and kernel mass) as a means to determine the opportunities to develop cultivars. Strong and highly significant relationships were found between fruit mass and pulp mass in trees from South Africa and Namibia, indicating that size is a good predictor of fruit pulp production. However, fruit size is not a good predictor of nut or kernel production, as there were weak relationships between fruit and nut and/or kernel mass, which varied between sites and landuses. Generally, the relationships between fruit mass and kernel mass were weaker than between fruit mass and nut mass. Relationships between kernel mass and shell mass were generally weak. The lack of strong relationships between fruit and kernel mass does, however, imply that there are opportunities to identify trees with either big fruits/small nuts for pulp production, or trees with large kernels in relatively small fruits for kernel oil production. However, within fruits from the same tree, nuts could contain 0−4 kernels, indicating that even in trees with an inherent propensity for large kernels, improved pollination may be required to maximise kernel mass through an increase in kernel number. Finally, the relationships between percentage kernel oil content and the measured morphological traits were also very weak. The conclusions of these results are that there is merit in identifying different combinations of traits for the selection of trees producing either pulp or kernels. Consequently, fruit and kernel "ideotypes" are presented as guides to the selection of elite trees for cultivar development. These results have important implications for the domestication of the species as a producer of fruits or kernels for food/beverages or cosmetic oils.

INTRODUCTION

Sclerocarya birrea (A. Rich.) Hochst. subsp. *caffra* (Sond.) Kokwaro is one of the traditionally important indigenous fruits of southern Africa, which in recent years has also become commercially important, as its fruits and other products have entered local, regional, and international trade (Shackleton et al., 2002a,b; Wynberg et al., 2002). In many countries, nontimber forest products (NTFPs) are an underutilized resource, and it is only in recent years that the potential to domesticate these products to enhance the livelihoods of poor people has been appreciated (Leakey, 2001a,b; Leakey et al., 2003).

The present study is part of a broader project examining the benefits and opportunities for further commercializing the fruits, kernels, and kernel oil of marula in South Africa and Namibia for the social and economic benefit of local communities, without negative impacts on the resource base, the culture and traditions of the local people

Multifunctional Agriculture. DOI: http://dx.doi.org/10.1016/B978-0-12-805356-0.00024-6
© 2017 Elsevier Inc. All rights reserved.

(Shackleton et al., 2003). A component of this study has quantitatively examined the extent of the phenotypic variation in fruit, nut, and kernel traits in marula in farmland, communal land, and natural woodlands. This study has found very extensive tree-to-tree variation in all measured traits (Leakey et al., 2005b,c), indicating great potential for further domestication through the development of cultivars. This tree domestication strategy is being developed for agroforestry trees in southern Africa (Maghembe et al., 1998).

This paper analyzes the relationships between different traits of importance to the development of cultivars so that it is possible to identify multiple trait combinations that can be simultaneously combined into fruit and nut "ideotypes" to guide the development of cultivars that could meet the needs of farmers with proximity to different markets. This approach builds on experience in West and Central Africa, where the identification of ideotypes (Atangana et al., 2002; Leakey et al., 2002) is a tool within a program of participatory domestication aimed at the development of agroforestry tree products—AFTPs (Leakey et al., 2003; Simons and Leakey, 2004).

METHODS AND MATERIALS

As previously reported, 24 ripe marula fruits were collected from beneath the crown of 63 marula trees in villages in Limpopo Province, South Africa (Bushbuckridge), and of 55 marula trees in Namibia (North Central Region), between January 21 and February 5, 2002, and the tree-to-tree variation in dry matter partitioning between skin, flesh/juice, nut-shell, and kernel determined (Leakey et al., 2005b,c). In South Africa, marula trees were found in farmers' fields, in communal grazing land, and in natural woodland, while in Namibia, fruiting trees were only found in farmers' fields.

Analysis of variance, Duncan's Multiple Range tests, and tests for skewness and kurtosis were done on raw data on the phenotypic variation in a range of fruit and kernel traits using SPSS 10.0 for Windows (Leakey et al., 2005b,c). The present study further analyzes the previously reported data to determine the relationships between different traits using linear and nonlinear regression (SPSS 10.0 for Windows), and examines certain trait:trait ratios. Unless otherwise indicated, all relationships were significant ($p \geq 0.001$). Using this information, the study seeks to identify the combinations of multiple traits that could be used as both fruit and kernel ideotypes for cultivar development.

RESULTS

Ratios

The ratios between one fruit trait and another showed very considerable variation between trees, but little or no overall difference between sites or land uses. This, for example, was the pattern for fruit mass:nut mass ratio (Table 24.1). It was also the case for fruit mass:skin mass and fruit mass:flesh/juice mass, but interestingly, the fruit mass:pulp mass (= skin + flesh/juice) ratio was very consistent between trees.

The fruit mass:kernel mass and nut mass:kernel mass ratios were also very variable between trees but unlike the fruit traits showed marked variation between land uses in South Africa and between sites in Namibia (Table 24.1).

Relationships Between Traits

There were highly significant and strong relationships between fruit mass and pulp mass in South Africa ($r^2 = 0.98$) and Namibia ($r^2 = 0.99$); however, the relationships between fruit mass and nut mass varied between sites and land uses, being weakest in natural woodland and stronger in communal land and farmers' fields (Table 24.2), although the slope of the lines were very similar (Fig. 24.1A). By contrast, in Namibia (Fig. 24.1B), the relationships between fruit mass and nut mass were strong in the west and northwest and weak in the northeast (Table 24.2).

Generally, the relationships between fruit mass and kernel mass were weaker than between fruit mass and nut mass (Table 24.2). In South Africa, the fruit mass versus kernel mass relationship was very weak in natural woodland and weak in communal land and farmers' fields, while in Namibia, it was weak in the west and northeast and very weak in the northwest. The number of kernels per nut was completely unrelated to the pulp mass of individual fruits.

Relationships between kernel mass and shell mass were always weak (South Africa: $r^2 = 0.14-0.34$ and Namibia: $r^2 = 0.08-0.31$). Interestingly, the large mean kernel mass of "Namibia Wonder" was associated with a large mean shell mass.

At the population level, the relationships between mean kernel mass and mean number of kernels per nut (log $r^2 = 0.62$ and 0.65) were not linear (Fig. 24.2A and B) because the individual kernel weight of nuts averaging one to two kernels was greater than in nuts averaging three to four kernels (except in "Namibian Wonder" in which all four

TABLE 24.1 Site variation in fruit:nut, fruit:kernel, and nut:kernel ratios of *Sclerocarya birrea* fruits from South Africa and Namibia.

South Africa		Namibia	
Fruit:Nut ratio			
Farmers' fields	5.7	Northwest	6.7
Communal land	4.9	Northeast	6.6
Natural woodland	4.5	West	6.9
Fruit:Kernel ratio			
Farmers' fields	58.3	Northwest	123.0
Communal land	65.4	Northeast	80.5
Natural woodland	71.1	West	64.3
Nut:Kernel ratio			
Farmers' fields	10.5	Northwest	18.5
Communal land	13.6	Northeast	12.8
Natural woodland	15.8	West	9.5

TABLE 24.2 Relationships between fruit mass and either nut mass or kernel mass in *Sclerocarya birrea* in South Africa and Namibia.

South Africa	Natural woodland	Communal land	Farmers' fields
Fruit mass *vs* nut mass	$r^2 = 0.43, p \geq 0.001$	$r^2 = 0.69\, p \geq 0.001$	$r^2 = 0.82, p \geq 0.001$
	$y = 0.183x + 0.893$	$y = 0149x + 0.994$	$y = 0.139x + 0.934$
Fruit mass *vs* kernel mass	$r^2 = 0.13, p = 0.205$	$r^2 = 0.42, p = 0.003$	$r^2 = 0.42, p = 0.003$
	$y = 0.009x + 0.145$	$y = 0.015x + 0.017$	$y = 0.011x + 0.164$
Namibia	**Northeast**	**Northwest**	**West**
Fruit mass *vs* nut mass	$r^2 = 0.20, p = 0.192$	$r^2 = 0.58, p \geq 0.001$	$r^2 = 0.73, p \geq 0.001$
	$y = 0.060x + 2.478$	$y = 0.084x + 1.852$	$y = 0.132x + 0.438$
Fruit mass *vs* kernel mass	$r^2 = 0.46, p = 0.065$	$r^2 = 0.03, p = 0.486$	$r^2 = 0.41, p = 0.002$
	$y = 0.007x + 0.149$	$y = 0.002x + 0.193$	$y = 0.016x + 0.014$

kernels were of similar mass). In individual trees, however, there were fruits with up to four kernels per nut and the relationships between kernel mass and mean number of kernels per nut was linear (Fig. 24.3A and B). The r^2 values for specific trees were: AR23, $r^2 = 0.79$; AN31, $r^2 = 0.89$; AN32, $r^2 = 0.93$; AL3, $r^2 = 0.86$; AL4, $r^2 = 0.74$; AL5, $r^2 = 0.89$; GV3, $r^2 = 0.71$, N10, $r^2 = 0.80$, N30, $r^2 = 0.46$, N41, $r^2 = 0.51$, N42, $r^2 = 0.75$, N47, $r^2 = 0.03$; Namibian Wonder, $r^2 = 0.76$, although in different trees these relationships could be both strong and weak. Interestingly, however, the slopes of the lines for different trees varied considerably.

Relationships between oil content (%) and all of the morphological traits measured were very weak.

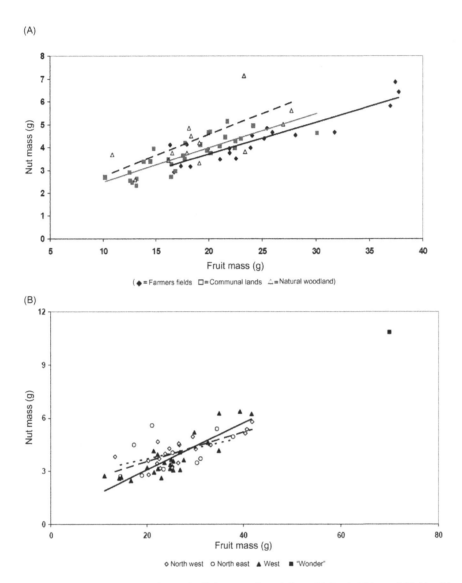

FIGURE 24.1 Relationship between fruit and nut mass in marula (*Sclerocarya birrea*) from (A) South Africa and (B) Namibia.

Development of Ideotypes

Simultaneous examination of all the data from this study illustrates that trees producing large fruits often have high values for flesh/juice mass and pulp mass but only some of these also have high values for kernel mass and oil content. Examination of the interrelationships between the traits indicates some changes in the relative ranking of trees, particularly between kernel mass and oil content, but also to a lesser degree between the components of pulp (skin and flesh/juice). These differences in ranking and the poor relationships between fruit traits and kernel traits suggest that there will be benefits from identifying the multiple trait combinations ("ideotypes") that would be important for tree selections directed toward different marketable products—e.g., fruits and kernels. However, examination of how well the superior trees for each particular trait fit the "ideotype" reveals that few trees conform well (Fig. 24.4A and B), although the data from "Namibian Wonder" conforms fairly well to these ideotypes (Fig. 24.4C).

DISCUSSION

Earlier papers in this study on marula (Leakey et al., 2005b,c) have quantified the tree-to-tree variation in dry matter partitioning between constituent parts of the fruits for the first time, and provided good fundamental knowledge about

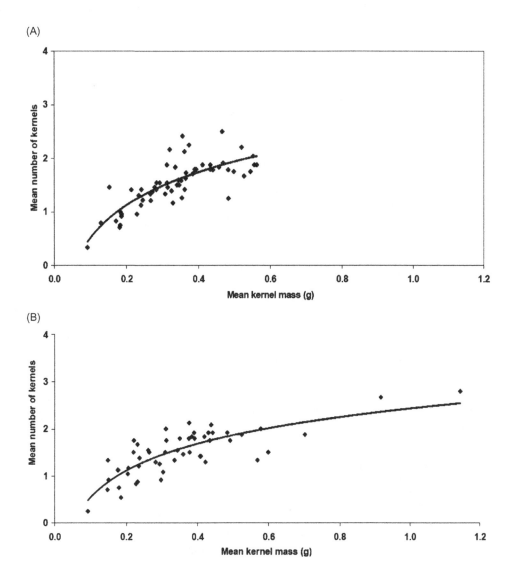

FIGURE 24.2 Relationship between kernel mass and number of kernels per nut in marula (*Sclerocarya birrea*) populations from (B) South Africa and (B) Namibia.

the range of variation in several important traits across geographically separated, as well as environmentally and culturally different, sites.

The strong relationships between fruit mass and pulp mass in marula found in this study indicate that selection for fruit pulp can be based on fruit mass. However, the weak relationships between fruit mass and kernel mass indicate, as in *Irvingia gabonensis* (Atangana et al., 2002), the difficulty of predicting kernel size from fruit size, especially as the shell is a major component of the nut and is of no value. The variability between land uses in South Africa in both the fruit mass versus nut mass relationships and the fruit mass versus kernel mass relationships suggests that the relationship has been strengthened through contact with man, presumably through selection for the larger fruits (Leakey et al., 2005b). If this is the case, then the stronger relationship in the Namibian fruits from the west and northwest implies stronger influences of man in these areas. The somewhat different relationships between fruit mass and kernel mass in the Namibian sites probably reflect the differences between the sites in the nut:kernel ratio, as trees from northwest are characterized by a considerably greater proportion of the nut mass as shell. The lack of strong relationships between fruit and kernel mass indicates the difficulty of predicting kernel mass from fruit size; however, it also implies that there are rare trees in the population with big fruits and small nuts, which would be good for pulp production. Equally, there are opportunities to identify rare trees whose fruits contain large nuts. However, a high proportion of nut mass is shell and this together with the variation in kernel number will make selections for oil production more complicated.

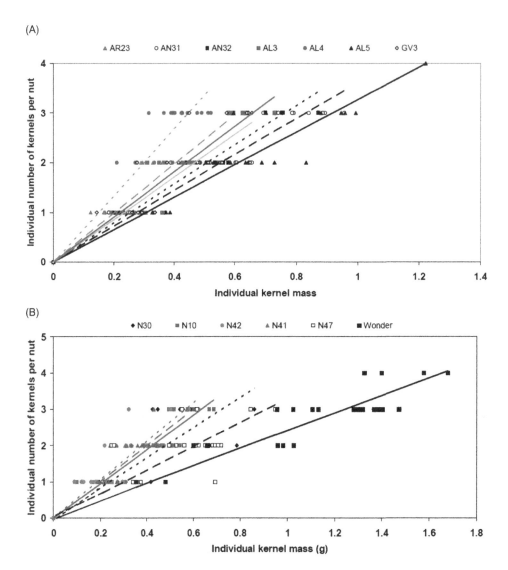

FIGURE 24.3 Relationship between kernel mass and number of kernels per nut from selected individual trees of marula (*Sclerocarya birrea*) from (A) South Africa and (B) Namibia. (*AR*, Acornhoek Road; *AN*, Andover; *AL*, Allandale; *GV*, Green Valley; *N*, Namibia)

The nonlinear relationships between mean kernel mass and the mean number of kernels per nut at the population level indicate that the mean mass of individual kernels in nuts with one to two kernels per nut was greater than in nuts with a mean of two to three kernels (very few had four kernels). However, at the level of individual trees, the different slopes of the linear relationships (Fig. 24.3) between kernel mass and mean number of kernels per nut suggest considerable genetic variation in the potential size of individual kernels from different trees. This implies that trees differ in their capacity to partition dry matter to developing kernels, a trait that would be desirable in cultivars being developed for kernel production. However, these relationships clearly indicate that the different nuts from the same tree vary in the numbers of kernels that they contain. This occurrence of individual fruits with few kernels per nut could result from poor pollination success. If this is the case, the linear relationship between kernel mass and mean number of kernels per nut suggests that increased pollination success would have considerable impact on kernel production (Fig. 24.3), especially in those trees with a genetic propensity to partition similar amounts of assimilate to all four kernels, such as "Namibian Wonder." Thus, improved pollination and genetic selection are required to maximize kernel mass and oil yield. In terms of selection, the lack of any evidence that kernel mass is any greater in nuts with a single kernel suggests that a strategy to select for a single large kernel is not appropriate; therefore the selection process should be focussed on selection of four large kernels per nut.

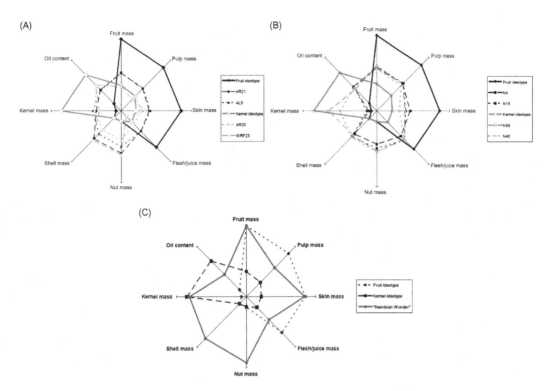

FIGURE 24.4 Fruit and nut ideotypes for marula (*Sclerocarya birrea*) with the "best-fit" trees from (A) South Africa and (B) Namibia, for comparison with (C) Namibian Wonder.

The suggestion that inadequate pollination is having negative impacts on marula kernel production illustrates the importance of maintaining male trees in the population. Evidence from the resource inventory of this project (Shackleton et al., 2003; Botelle, 2002; McHardy, 2002) indicates that there are fewer male trees in the homesteads and fields of the households, compared with communal land and natural woodlands.

In many species the number of seeds (kernels) per fruit can influence pulp mass. However, the lack of a relationship between the number of kernels per nut and the pulp mass of individual marula fruits suggests that this is not the case in this species.

The case of "Namibian Wonder" illustrates that rare individuals that greatly outperform the majority of other trees, and conform well to the ideotype, do exist in the wild population. It is clear from this study that there is very considerable opportunity to identify individual marula trees with fruit and kernel characteristics well above the average of the species; however, it is also clear that much larger populations will need to be examined if these rare trees worthy of development as cultivars are to be found. From the limited study to date, it appears that the Namibian population has greater promise of superior trees than that of South Africa, but the geographical range surveyed in South Africa was considerably narrower than in Namibia, although neither survey was comprehensive. Clearly, if the domestication of marula is to be pursued, more comprehensive surveys will be required, and these should extend into the neighboring countries of Botswana, Zimbabwe, etc. From the studies so far, it is not possible to predict where the most promising trees might be found, and of course, it is possible that trees with desirable fruit traits may be found in different locations from those with desirable kernel or other traits. The ideotype concept should assist this process, making it easier to identify the specific trait combinations required to maximize the commercial production of fruits and kernels for juice/pulp and oil production. To date little is known about the influence of environmental factors on the phenotypic variation of pulp or kernel mass, but the evidence presented here is that it has a major genetic component.

The extent to which all the trees characterized to date are allocating large amounts of dry matter to shell production is clear from the ideotype diagrams. The nut:kernel ratios and weak relationship between kernel mass and shell mass indicate that it may be possible to find trees that partition more dry matter to marketable products and less to the nut-shell, which is the only part of the fruit that does not have a market value (although it is used as a fuel). A lighter, more brittle shell would also make kernel extraction easier.

An opportunity for the future would be to add an assessment of organoleptic (flavor and fragrance) properties of different fruit samples, using techniques of Schaefer and McGill (1986), as these are important commercially in the juice and liqueur industry, remembering that differences in flavor and fragrance can be genetic in origin, as well as the consequences of poor storage and processing. Recent marula data from Kenya (Thiong'o et al., 2002), illustrate that in nutritional terms, traits for nutritional values (protein, carbohydrate, vitamins, and minerals), and for taste/sweetness (carbohydrates, acidity) are traits, which should be added to the ideotypes in due course. The phenotypic evaluation of organoleptic traits is being added to those of morphological traits in the similar studies with the indigenous fruit, *Dacryodes edulis*, in West Africa (Kengni et al., 2001; Leakey et al., 2002). For marula, it may also be possible to add oil quality to the kernel ideotype at a later date. When developing cultivars for any commercial use it is important to interact with commercial companies to determine which properties are desirable and which are undesirable (Leakey, 1999a).

The results of this study with marula have highlighted the extent of tree-to-tree phenotypic variation at the village level in many commercially important traits and indicate the opportunities for capturing various multiple trait combinations appropriate to various commercialization opportunities. The similarities with the results from other indigenous fruits, which are being domesticated in West Africa using a Participatory Domestication approach (Tchoundjeu et al., 1998; Leakey et al., 2003), suggest that a similar participatory strategy to marula domestication should be taken. This would ensure that the local communities are the beneficiaries. The participatory domestication of indigenous fruits has been proposed as an appropriate means to alleviate poverty (Leakey and Simons, 1998; Poulton and Poole, 2001), and could also have positive benefits on the environment. New plantings of marula would help to restore the resource of this important tree and could also provide future ecological niches for a wide range of wildlife, above and below ground (Leakey, 1999b). In addition, the move toward agroforestry may encourage farmers to plant other useful trees, so diversifying the farming system with likely benefits on sustainability, through the creation of an agroecological succession culminating in a mature or climax phase (Leakey, 2001a,b). Hopefully, such developments would also help to sustain some of the traditional values of marula in the culture, for example the "first fruits" ceremony and the traditional role of marula beer in the society.

ACKNOWLEDGMENTS

Sheona Shackleton, Jenny Botha, Pierre du Plessis, Cyril Lombard, and Kris Pate are gratefully acknowledged for assistance with the field-work. This publication is an output from a project partly funded by the United Kingdom Department for International Development (DFID) for the benefits of developing countries. The authors are indebted to DFID for funding this project (Project No R7190 of the Forestry Research Program) and the views expressed here are not necessarily those of DFID.

Techniques: Ideotypes

Chapter 25

The "Ideotype Concept" and its Application to the Selection of "AFTP" Cultivars

This chapter was previously published in Leakey, R.R.B., Page, T., 2006. Forests, Trees and Livelihoods, 16, 5–16.

SUMMARY

The ideotype concept has been developed and modified for a number of different crops, including forest trees, over the last 35 years. In recent years it has been used in the domestication of agroforestry trees producing Agroforestry Tree Products (AFTPs), as an aid to the multiple-trait selection of superior trees for cultivar development. For example in west and central Africa, "fruit" and "kernel" ideotypes have been identified in *Irvingia gabonensis* (Bush Mango), based on quantitative characterisation of a number of fruit, nut and kernel traits. Subsequently, recognising the opportunity to develop different markets for the commercially important kernels, the "kernel ideotype" has been subdivided into "oil" and "food-thickening" ideotypes, with options for further subdivision of the latter into ideotypes with either high viscosity or high drawability characteristics. Similar opportunities for the development of single-purpose cultivars from multi-purpose species have been identified in *Sclerocarya birrea* (Marula), indicating that with increasing information about the variability of AFTP's it is possible to develop a hierarchy of different ideotypes to meet different market opportunities. In *Sc. birrea* and *Dacryodes edulis*, future options exist to extend ideotypes with recognition of the variability in both nutritional and organoleptic qualities. In Australia, a very broad-based ideotype has been developed for the indigenous shrub *Kunzea pomifera* (Muntries) established on both morphological and physiological traits. Similar approaches are identified for timber trees (e.g. *Triplochiton scleroxylon*) and those like *Santalum austrocaledonicum* producing essential oils. It is concluded that ideotypes are a useful tool for visualising and conceptualising how to combine specific rare combinations of visible and invisible traits, aimed at the maximisation of Harvest Index, even when the traits are only weakly related.

INTRODUCTION

The concept of the ideotype—"a form denoting an idea"—was conceived by Donald (1968) as an aid to crop breeding programs, specifically based on physiological factors conferring close to optimum yield in wheat (*Triticum aestivum*) in the first instance. The ideotype is in effect the ideal model phenotype, which can be expected to perform in a predictable way within a defined environment, so providing the strategy and goal for genetic selection.

Since Donald's formulation of the concept, crop ideotypes have been derived for many other species: e.g., barley (*Hordeum vulgare*) (Rasmusson, 1987), chickpea (*Cicer arietinum*) (Siddique and Sedgeley, 1987), including forest trees (Dickmann, 1985), such as spruce and pine (*Picea abies* and *Pinus sylvestris*) (Kärki and Tigerstedt, 1985), as well as fruit trees like mango (*Mangifera indica*) and apple (*Malus* spp.) (Dickmann et al., 1994). With the inclusion of trees, the concept has evolved and Dickmann (1985) describes it as "the first step towards bioengineering an improved plant" in which a combination of characters provides a guide to the selection of potential breeding stock from wild populations. This is important because individuals in a wild population, especially long-lived perennials like trees, have

Multifunctional Agriculture. DOI: http://dx.doi.org/10.1016/B978-0-12-805356-0.00025-8
© 2017 Elsevier Inc. All rights reserved.

evolved to survive and reproduce—a quite different strategy from that introduced by plant breeders to enhance performance under cultivation. Thus the "ideal" combination of traits for a crop plant, derived from trees, may be rare in individual genotypes in a wild population. Dickman (1985) emphasizes that "an ideotype is not a wistful construction born of unproven assumptions or opinions; it is rather a deductive product founded on a detailed understanding of plant morphology and physiology": for example, dry matter partitioning, competition and sink dynamics, phenology, nutrient storage, metabolism of secondary compounds. Given the successful application of ideotype breeding in many domesticated perennial species the method is now being applied in domestication of perennial tree and shrub species; some examples are reported in this paper.

DOMESTICATION OF AGROFORESTRY TREES

In recent years the domestication of trees for agroforestry has focused on the potential of a wide range of indigenous trees in all the main ecoregions of the world to become a new suite of crop plants for the diversification of farming systems, in ways that also create alternatives to shifting cultivation (Leakey and Simons, 1998; Simons and Leakey, 2004). In West and Central Africa, this domestication is being done using a participatory approach, which empowers the participating farmers to create their own cultivars (Leakey et al., 2003). This strategy is seen as a model for biodiscovery that is in accordance with the Convention on Biological Diversity, and an alternative to other approaches that do not recognize the rights of farmers and rural communities to benefit from their traditional knowledge and intellectual property (Leakey et al., 2006). The success of this domestication process as an incentive for farmers to plant a wide array of tree species within their farm systems is dependent on there being markets for the products. Typically these markets are local and based on the demand for traditional products, but they may also have larger regional and even international potential (e.g., Shackleton et al., 2003). Indeed, over the last decade there has been considerable progress in the identification and capture of some of these opportunities (Leakey et al., 2006). Domestication for traditional domestic markets is relatively straightforward, as the preferred characteristics and properties of the products are well known locally. However, this is not the situation when domesticating trees for new markets in the food, cosmetic, and pharmaceutical industries. Consequently, it is important that there be some interaction between the industries and the domestication team so that the industries can provide some guidance about the desired physical and chemical properties of the product (Leakey, 1999a), i.e., the process is market-driven.

The domestication of indigenous plants to meet the needs of the population is recognized as part of a continuum of processes associated with increasing human land use pressure on natural resources (Homma, 1994), that starts with hunter-gathering and progresses to widespread cultivation, often with extensive forest clearance and the planting of exotic crops. Recognition of the social, cultural, and economic importance of natural products, especially tree products, has led to their definition as nontimber forest products (NTFPs), or nonwood forest products (NWFPs). Until very recently this term has also been applied to these products when they are produced by cultivation. This has resulted in confusion about the relative importance of such products originating from open-access natural resources versus those being brought into cultivation. To recognize the growing importance of new crops, especially new tree crops with potential as cash crops for agroforestry, the term agroforestry tree products (AFTPs) has been defined (Simons and Leakey, 2004).

Early approaches to the domestication of trees producing AFTPs were based on the identification of "varieties" (Okafor, 1983) and often these were characterized on the basis of qualitative descriptions (Ladipo et al., 1996) rather than on quantitative data. One of the techniques now being used to assist the domestication of AFTPs is the application of the ideotype concept based on the systematic characterization of the tree-to-tree variation in a range of fruit and nut traits (Leakey et al., 2000).

The ideotype concept, originally conceived for annual grain crops, will need to be adapted to suit individual perennial agroforestry species. The economic products of agroforestry crops varies between species, i.e., leaf, fiber, fruit, oil, seed, medicine, and therefore the characters that influence yield, particularly for secondary metabolites, are going to be different from those used for annual crops. For many agroforestry species further investigation is required to identify traits that can be used to improve yield. Evaluation of like species in which such relationships have been established can offer direction for research in relatively unstudied species. In *Pic. abies* and *Pin. sylvestris* for instance, Kärki Tigerstedt (1985) found extremely narrow crowned trees produce both high quality timber and a high yield per hectare. The development of an ideotype is therefore not static, as it requires an iterative cycle of formulation, application, evaluation, and reformulation (Dickman et al., 1994). A number of ideotype applications in agroforestry and new horticultural crop species are presented in this paper.

IDEOTYPES IN AGROFORESTRY SPECIES

Fruit/Nut Crops

Subsistence farmers can potentially generate cash income from local, regional, and even international trade through the development and sale of cultivars produced in village nurseries using participatory techniques. These cultivars can be derived from elite individual trees that best-fit an ideotype that meets a specific market demand (Leakey et al., 2005b,c). In this way farmers may cultivate indigenous trees to improve their livelihoods (Leakey et al., 2005a).

In recent years a number of studies have quantified tree-to-tree variation in a range of fruit and kernel traits in indigenous African trees producing marketable AFTPs. This has been aimed at helping the identification and selection of the elite trees, which could be developed as putative cultivars through vegetative propagation. These data have then been used to formulate ideotypes targeting different market opportunities, based on the range of phenotypic variation found in the individual trees from a number of different populations. Thus, multitrait web-diagrams illustrating the tree-to-tree variation in each character arranged on radial lines (Fig. 25.1A–C) have been used widely as a conceptual method of simultaneously presenting variation in a number of traits, as well as illustrating the relationships between these traits. These diagrams then allow the many different phenotypes to be compared with a proposed ideotype (Fig. 25.1). The development of ideotypes in this approach to tree improvement is thus rather different from the long-term iterative approach used by plant breeders improving either annual crops or long-lived tree crops.

Typically, when tree improvement is achieved through breeding, simultaneous changes in several traits are sought, with the aim of combining a number of desirable characters into a new cultivar. During this process, the selection of one trait will inevitably change other traits either by chance or through genetic linkages between them (Namkoong et al., 1988). In this situation, multitrait web-diagrams may be a useful tool for representing multitrait variation for a given generation, but it is also necessary to have knowledge of the heritability of and genetic correlations between characters.

Kunzea pomifera *(Muntries)*

Muntries are small edible berries produced by a small indigenous shrub in southern Australia found on sandy coastal plains, which are highly valued by Australian Aboriginal peoples (Clarke, 1998). The fruits, approximately 10–12 mm in diameter, were produced in such abundance that, in addition to fresh consumption, Australian Aboriginal tribes dried and stored the fruit as large cakes to be eaten in the winter months or traded for items such as stone tools and weapons (Tindale, 1981). Muntries have a very distinct taste, similar to that of the apple, but with a unique spicy or sultana-type flavor. Today the fruits are used fresh to complement desserts and salads, cooked in salsas, jams, and sauces, and also prepared in cakes, pies, and muffins.

Muntries are considered to have commercial potential (Graham and Hart, 1997), and the species is currently cultivated in Victoria and South Australia on a somewhat limited scale for both the fresh, and processed, fruit markets (Hele, 2001). In an examination of morphological and genetic diversity of the species natural populations, Page (2003) found considerable variation, indicating that the species has potential for improvement through breeding. The characters considered important for the ideotype were examined and include a range of morphological and physiological traits associated with plant growth habit, edaphic adaptability, reproductive biology, fruit disposition and quality, yield potential, and yield stability.

Irvingia gabonensis *(Bush mango)*

The fruit of *I. gabonensis* is eaten as fresh fruit but rarely marketed as such, although there is some potential to produce a wine. The fruits contain a large nut, which is difficult to crack, and it is the kernel of this nut which is the most important product. The kernel is used traditionally as a food-thickening agent in soups and stews in Cameroon and Nigeria, producing a mucilaginous polysaccharide when heated. These oil and protein-rich kernels can be dried and stored for long periods. They are extensively marketed nationally and are exported regionally (Ndoye et al., 1997). The quantitative characterization of these products has resulted in the recognition that fruit and kernel traits are weakly correlated and thus that trees that produce large fruits do not necessarily produce large or easily extracted kernels (Atangana et al., 2001; Anegbeh et al., 2003). Consequently, separate fruit and kernel ideotypes (Fig. 25.1A) have been identified to aid the selection of elite trees for domestication (Atangana et al., 2002). Subsequently physicochemical studies of the kernels have identified that the oil and protein contents are not strongly related to the food-thickening properties, and that two independent traits (viscosity and drawability) are involved in different aspects of

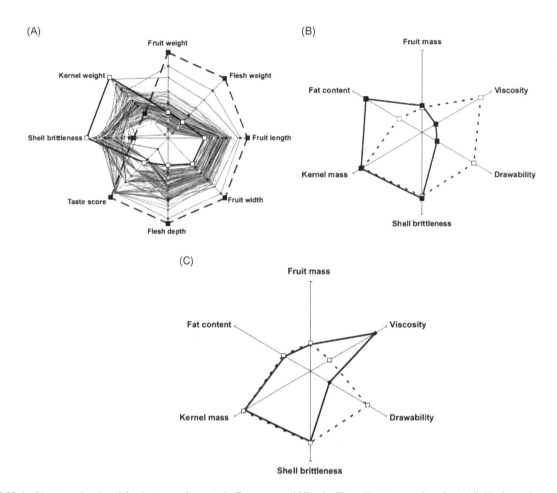

FIGURE 25.1 Ideotypes developed for *Irvingia gabonensis* in Cameroon and Nigeria. These ideotypes are based on individual tree data collected in studies of the range of variation in different traits included in the ideotype, and their interrelatedness, as found in wild populations (Atangana et al., 2001; 2002; Anegbeh et al., 2003; Leakey et al., 2005d). (A) Level 1: Web-diagrams of the "Fruit ideotype" (black dotted line) and "Kernel ideotype" (black solid line). For clarity, the axes are drawn on different scales and the scale omitted, as it is the relative differences between individual trees that are important in defining the ideotype. (B). Level 2: Web-diagrams of the "Food-thickening" (dotted line) and "Edible oil" (solid line) ideotypes. (c) Level 3: Web-diagrams of the "Drawability" (dotted line) and "Viscosity" (solid line) ideotypes for the food-thickening properties.

food-thickening (Leakey et al., 2005d). The desire to make rapid improvements for each character has resulted in the subdivision of the kernel ideotype for either oil production or food-thickening (Fig. 25.1B), the latter being further subdivided into viscosity and drawability[1] ideotypes (Fig. 25.1C).

Sclerocarya birrea *(Marula)*

Traditionally, Marula fruits are used to make a fermented drink of social and cultural significance, and the kernels are important nutritionally (Shackleton et al., 2003). In contrast, the new products are a distilled liqueur made from the fruits for the expatriate and international market and cosmetics made from kernel oil. So, the market and value-adding opportunities for new products are very different from the domestic markets of traditional products. Like the bush mango, different ideotypes have been identified for fruits and kernels (Leakey, 2005), but it is likely that after further study these ideotypes will be subdivided because fruits (pulp and skin) are also known to vary in protein and vitamin C content (Thiong'o et al., 2002). Selection of cultivars with high nutritive value could be of great importance in reducing the incidence of diseases and malnutrition in developing countries. Furthermore, the oil yield per nut was variable (8−53 g per nut), mainly because of the variation in kernel mass per nut (Leakey et al., 2005c).

1. The ability of a mucilaginous extract to be drawn out in a string. This is a desirable trait.

Dacryodes edulis *(African plum/Safou)*

In the case of Safou, the fresh fruits are a highly nutritious staple food (with a short shelf-life), while the kernels are a waste product. Potentially, however, processed or preserved fruits could be a gourmet food and the kernels a source of vegetable oil (Schreckenberg et al., 2002). Once again, quantitative characterization of a number of fruit traits has indicated considerable tree-to-tree variation (Waruhui et al., 2004; Anegbeh et al., 2005), with great potential for selection of superior cultivars. In addition, some studies have been initiated in the variation of organoleptic traits affecting customer preferences (Leakey et al., 2002). This together with future studies in nutritional content, the opportunities for selection for the commercial production of vegetable oil, and the possibility of selecting for ease of processing and storage quality, will lead to the identification of market-oriented ideotypes. Future ideotypes are likely to include nutritive value, pulp:kernel ratio as a few trees seem to have a high incidence of parthenocarpy (Anegbeh et al., 2005).

Timber/Wood Crops

The domestication of timber trees has usually been based on a combination of provenance selection and progeny selection, which can be a very lengthy process particularly for improvement of component characters under quantitative genetic control. To circumvent this problem, Dickman et al. (1994) proposed that an ideotype can be considered a single quantitative trait in which individuals are selected for their degree of fit to the ideotype and captured through clonal propagation. For this approach to be successful the screening of large populations is essential, including both cultured and wild populations, but rapid genetic gains are possible. A few species (e.g., *Populus, Salix* spp., and *Cryptomeria japonica*) have traditionally been grown clonally, and cloning has been extended to a number of other species in the last 20–30 years. This clonal approach to tree improvement is particularly important for characters that are difficult to maintain through sexual reproduction because of their nonadditive inheritance. This clonal approach to tree improvement is assisted by the use of the ideotype concept to conceptualize combinations of often unrelated traits (Dickman, 1985).

Triplochiton scleroxylon *(Obeche/Wawa)*

Detailed assessment of numerous growth parameters in clonal trials of this West and Central African hardwood, the main roundwood export from West Africa, identified that over the first 5 years the main stem volume was greatest in trees with the least number of primary branches per meter of main stem length (Longman et al., 1978: Longman and Leakey, 1995). This species conforms to Rauh's model of branching architecture (Hallé et al., 1978). Low branching frequency was found to result from a single whorl of proleptic branches per growth flush, with a low number of branches per branch whorl, rather than from trees in which sylleptic[2] branching occurred between the proleptic[3] branch whorls. It was hypothesized that this branching pattern was the result of intraspecific variation in the strength of apical dominance (or correlative inhibition) and thus that the desired ideotype was one where low branching frequency resulted from strong expression of apical dominance (Leakey and Ladipo, 1987). It was subsequently found that variation in the strength of apical dominance could be evaluated by decapitation tests in young seedlings conducted under standardized conditions (Leakey and Longman, 1986; Ladipo et al., 1991a, 1992). These studies resulted in a Predictive Test for Branching Habit, which could be used to screen large numbers of seedlings for their likely conformity with the expression of a desired ideotype in forest plantations, which combined variations in the "strength" of apical dominance and dry matter partitioning between main stem and branches in specific branching architectures (Ladipo et al., 1991).

Santalum *spp. (Sandalwood)*

Sandalwood describes a number of small tree species in the genus Santalum, which occur in south and Southeast Asia, Australia and the Pacific. The trees produce an oil, deposited in the heartwood, which when extracted by distillation is used in the international perfumery market. The oil-bearing powdered wood is also the primary ingredient in incense joss-sticks. Little is known about the tree-to-tree variation in morphological and anatomical characteristics of sandalwood and their relationships with essential oil quality or content. Currently this variability in *Santalum austrocaledonicum* in Vanuatu and *Santalum lanceolatum* in Cape York Peninsula in Far North Queensland is being examined in order to identify ideotypes that fulfill the needs of the different sandalwood industries. Initial evaluation of each species' natural

---

2. Sylleptic branches are formed, within the current shoot flush, immediately after the growth of the terminal bud and without a period of bud dormancy (correlative inhibition).
3. Proleptic branches are formed usually in the next phase of terminal shoot flushing, following a period of bud dormancy (correlative inhibition).

populations revealed significant variation in a number of important morphological and biochemical characters. The substantial variation found for oil quality and content has indicated that this commercial character can be readily improved through selection.

CONCLUSION

The application of the ideotype concept to tree and shrub species producing AFTPs builds on the earlier applications to staple cereals and leguminous crops and is found to be a useful means of targeting the domestication process at a range of different market opportunities for fruits, kernels, vegetable oils, timber, essential oils, etc. It is clear from the case studies presented that web-diagrams provide a way to visualize the tree-to-tree variability found in wild or on-farm populations. They also allow the conceptualization of the relationships between easily seen morphological variables and the impossible-to-see variables such as taste, nutritional quality, and physicochemical properties of food-thickening agents. The lack of food and nutritional security in many tropical countries make the nutritional aspects of AFTP domestication very important, especially as these sources of nutrients have been reported to boost the immune systems of HIV/AIDS sufferers (Barany et al., 2001, 2003). With increased understanding of the multitrait variability in products of importance to farmers and to markets, it is possible to develop a hierarchy of market-driven ideotypes (Fig. 16.4), which will maximize Harvest Index through the creation of single-purpose cultivars of multipurpose trees. This is especially important when the desired combinations of weakly related traits are only found as rare individuals in wild populations.

Chapter 26

Trees: Skills and Understanding Essential for Domestication: An Update

R.R.B. Leakey

The domestication strategy and vegetative propagation techniques used for agroforestry trees have been widely adopted by many researchers. Certainly, the low-tech polypropagator has spread across the globe and can be found in remote locations in all corners of Africa, the Amazon, the high Andes, Central America, Southeast Asia, Far East and Oceania (Solomons, Vanuatu, Papua New Guinea), not to mention Australia, Europe, and the United States.

Sadly there is a severe lack of knowledge about the basics of tree biology and vegetative propagation in many areas of agriculture and rural development. This dirth of understanding results from the swing in many university departments from pure botany and zoology to aspects of biotechnology. The result is an inadequate resource of well-founded young biologists with broad knowledge to drive the upscaling of broadly based sciences like agroforestry to meet the challenges of the Sustainable Development Goals (see Chapter 40 [Leakey, 2017i]).

STRATEGY

As reported in Chapters 5 and 11 (Leakey and Simons, 1998; Leakey et al., 1982b) the domestication strategy emerged in the 1990s from work in tropical forestry (Leakey, 1991; Leakey and Newton, 1994b; Simons, 1996; Leakey and Simons, 2000) and it has subsequently been developed (Fig. 16.2), widely approved and adopted. Nevertheless, there are still some who are concerned about the potential issues of narrowing the genetic base of target species through the misuse of vegetative propagation to produce clonal cultivars (Dawson et al., 2009, 2013; Thomas et al., 2014). In Chapter 15 (Leakey, 2017d), evidence is presented from the morphological and molecular characterization of tree-to-tree variation in *Barringtonia procera* that 70 to >80% of genetic variation can be found in individual populations in the Solomon Islands (Pauku et al., 2010) and that many of the individual elite trees selected for vegetative propagation are unrelated (Fig. 26.1). A similar result has been reported in *Adansonia digitata* (Assogbadjo et al., 2006), a species with extensive tree-to-tree variation in fruit morphology across different populations (Fig. 26.2), as also reported for *Sclerocarya birrea* (see Chapters 22 and 23 [Leakey et al., 2005a,b]). It seems clear from this data that the use of a number of selected cultivars, developed through decentralized domestication in different village populations across a region, will not be seriously narrowing the genetic base of these species. Nevertheless, every care must be taken to prevent any such unwise and irresponsible event from happening. Fortunately, in all the cases of agroforestry tree domestication to date, wild populations of the different tree species still exist in local forests and woodlands; and in a few cases randomly selected trees have been planted in gene banks. While the latter is highly desirable as a safety net, it does impose long-term financial obligations on local organizations, and the risks of fire or other forms of destruction are high.

VEGETATIVE PROPAGATION

The use of low-technology propagators (see Chapter 17 [Leakey et al., 1990]: Figs. 17.1, 20.3, Fig. 26.3) has many advantages, both practical and for research (Table 20.1). The physiological measurements within the propagator of gas exchange during photosynthesis and transpiration mentioned in Chapters 19 and 20 (Leakey and Storeton-West, 1992;

Multifunctional Agriculture. DOI: http://dx.doi.org/10.1016/B978-0-12-805356-0.00026-X
© 2017 Elsevier Inc. All rights reserved.

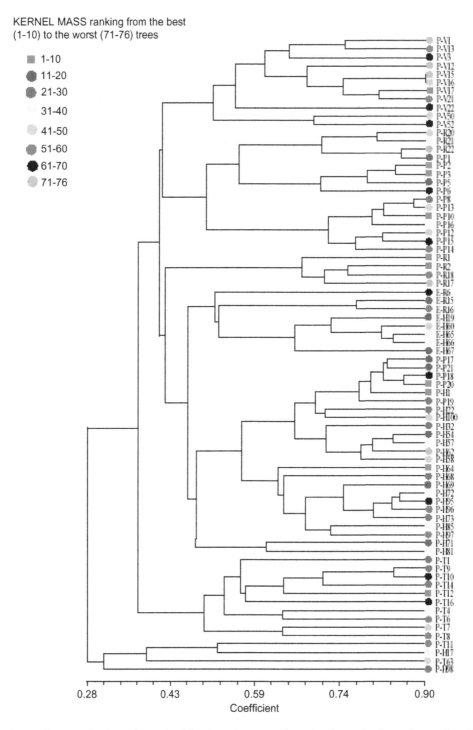

FIGURE 26.1 Dendrogram from a molecular marker study of *Barringtonia procera* illustrating the unrelatedness of trees selected for greatest kernel mass (square shapes) (Pauku et al., 2010).

Leakey, 2014c) could not have been done under a mist system. Thus scientific progress has resulted from this simple but efficient system because the environment is more uniform and less challenging for scientific equipment. Likewise, propagation in practice also benefits from this more uniform environment, as the more sophisticated mist systems tend to have patches that are either too wet or too dry. However, the overriding characteristic from a practical perspective is

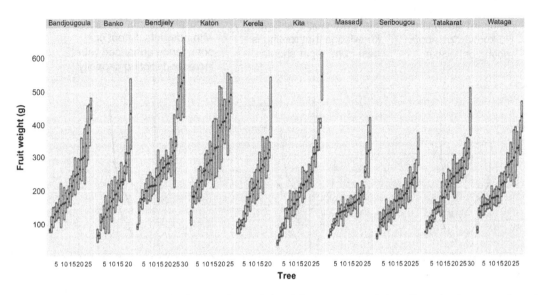

FIGURE 26.2 Data on the tree-to-tree variation of *Adansonia digitata* fruit weight across ten populations, illustrating the opportunity to develop elite cultivars at each site as part of a decentralized approach to tree domestication (de Smedt et al., 2011).

FIGURE 26.3 Examples of the small- and large-scale use and simple and deluxe forms of polythene nonmist propagators.

the low-cost of the propagator and the release from a need for running water and electricity. On the first point, so long as the basic principle of the propagator as a watertight, airtight box is achieved, the construction of the propagator can be very basic and made from waste wood and polythene, through to deluxe made with a metal frame and plastic or fiberglass base enclosed in horticultural grade polythene (Fig. 26.3).

The basics of vegetative propagation have been known for thousands of years. This has been applied by horticulturalists with "green fingers" and over the last 60−70 years much research has advanced the application of modern techniques of plant physiology. However, the result has been a very large body of contradictory data that has led to a lack of clear principles (Leakey, 2004). The studies summarized in Chapter 20 (Leakey, 2014c) have resulted from many years work on the preseverance and postseverance factors affecting rooting success in *Triplochiton scleroxylon* and a number of other species. It seems to be particularly important that cuttings in the propagator be capable of active photosynthesis (see Chapter 19 [Leakey and Storeton-West, 1992]; Newton et al., 1996). These and other studies have identified some important preseverance factors, especially stockplant irradiance and stockplant light quality (red:far-red ratio). These then have important interactions with soil nutrients (especially nitrates) (see Chapter 19 [Leakey and Storeton-West, 1992; Hoad and Leakey, 1994, 1996]) and stockplant morphology (position within-shoot and differences between shoots, see Chapter 20 [Leakey, 2014c]). Typically, none of these preseverance factors are reported by researchers

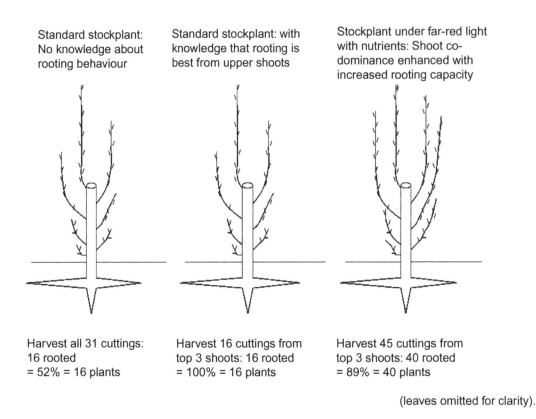

Standard stockplant: No knowledge about rooting behaviour

Standard stockplant: with knowledge that rooting is best from upper shoots

Stockplant under far-red light with nutrients: Shoot co-dominance enhanced with increased rooting capacity

Harvest all 31 cuttings: 16 rooted = 52% = 16 plants

Harvest 16 cuttings from top 3 shoots: 16 rooted = 100% = 16 plants

Harvest 45 cuttings from top 3 shoots: 40 rooted = 89% = 40 plants

(leaves omitted for clarity).

FIGURE 26.4 The effects of different preseverance stockplant management regimes and the location of cuttings harvested on rooting success, expressed as % rooted and number of plants produced (Leakey, 2004).

when publishing their work, so it seems likely that these omissions explain the contradictory data in the literature. This is illustrated in Fig. 26.4.

Following on from the previous studies it now seems to be possible to identify some key principles for successful rooting of leafy stem cuttings from managed stockplants (Leakey, 2004; Chapter 24 [Leakey, 2014e]). To enhance understanding the key points have been described using models (Dick and Dewar, 1992) and "sporting analogies" (Table 20.4). Recognition of the key principles has led to the successful propagation of species considered to be difficult to propagate, like *Cordia allidora* (Mesén et al., 1997a,b), *Shorea leprosula* (Aminah et al., 1995), *Eucalyptus grandis* (Hoad and Leakey, 1994, 1996), as well as commercially important timbers like *Khaya ivorensis* (Tchoundjeu and Leakey, 2000) and *Milicia excelsa* (Ofori et al., 1997). More recently, these principles have also been applied to many indigenous food trees like *Irvingia gabonensis, Ricinodendron heudelotii, Gnetum africanum* (Shiembo et al., 1996a, 1996b, 1997) and medicinal species like *Prunus africana* (Tchoundjeu et al., 2002). Current work with the difficult-to-root species *Allanblackia floribunda* is putting these principles to the test (Tsobeng et al., 2016; Tsobeng et al., in press).

Likewise, these principles have been applied to try to resolve the problems of rooting leafy stem cuttings collected in the crown of mature trees, with some success (Dick and Leakey, 2006; Pauku et al., 2010). This raises many questions about much that has been written about the maturation process in trees *vis à vis* vegetative propagation by cuttings (Leakey, 2004, 2014c). This is an area for future research.

Like the history of propagation by cuttings, the processes of marcotting (Fig. 20.1) and grafting have also been heavily reliant on "green fingers" and practical experience, although there have been some recent studies (Munjuga et al., 2011; Ofori et al., 2015; Asaah et al., 2011). Unfortunately, there have been few studies which have made a dedicated and detailed analysis of all the various factors likely to influence success. Unpublished experience suggests, for example, that young actively growing vertical shoots root the best (Fig. 20.1). Thus, again as we have seen previously for stem cuttings, it seems likely that many of the existing publications have inadequately described the physiological state of the shoots involved. It is also important to determine and report the physical environment

FIGURE 26.5 Setting a marcot on a physiologically active, fast-growing shoot seems to result in more rapid success. Marcotting on vertical sprouts from a cutback branch results in a well-formed radially arranged root system, while marcots on horizontal or oblique branches typically result in a one-sided root system.

of both the stockplant's environment and the microenvironment of the marcot or graft site. Thus, there is a need for much more research on these techniques which are important for the capture of the mature phase of selected genotypes (Fig. 28.3).

GENETIC CHARACTERIZATION

The capacity to propagate trees vegetatively, and hence to capture specific trait combinations as a cultivar, opens doors to the development of new crops. However, it is obviously important to first of all be convinced that there is both a need and a desire to go down this path (Nevenimo et al., 2008). In the early days of this tree domestication work there was a view in the donor community that indigenous trees had little, if any, potential because fruits could be seen left lying under trees to rot. What was not being recognized about this scenario was that the desirable fruits were being gathered at dawn, and that the fruits left to rot were those under trees with undesirable characteristics. Thus it was data on the tree-to-tree variation (see Chapters 22 and 23 [Leakey et al., 2005b,c]) like that which follows in the next paragraph, which brought about a significant change in perceptions about these little-known fruits, nuts and other products. Furthermore, it was also a revelation to find that this variation was recognized in local informal retail markets, but unrecognized by local wholesalers who could not acquire sufficient quality products (Leakey et al., 2002).

Since the development of a simple field-based technique to assess the tree-to-tree variation in the fruit, nut and kernel morphology of 24 fruits from individual trees of *I. gabonensis* within different populations in Cameroon (Leakey et al., 2000), it has been used in more intensive studies in this and a number of other indigenous fruit trees of Africa (*Dacryodes edulis, S. birrea, A. floribunda*). The results have been spectacular with 3- to 10-fold variation between the best and worst trees. To summarize these results, it has become clear that rather than there being "varieties" in these wild populations, as suggested by Okafor (1983), there is continuous variation. In other words, there are a series of small differences from the smallest fruit/nut/kernel, right up to the largest (e.g., see Fig. 22.1), and similar variation is found at different land uses and sites (Fig. 22.2 and Fig. 26.2). The frequency distribution of this variation in a truly wild population follows what statisticians describe as a "normal curve" (Fig. 8.1, Stage 1). This variation is therefore what is expected to result from random breeding between outbreeding individuals of any population. This creates an enormous opportunity for tree domestication, allowing those elite individuals to be selected for vegetative propagation to create a cultivar, or clone—i.e., an almost instantaneous "super" tree with substantial superiority for whatever trait is selected. As mentioned earlier, it is important to minimize the risk that this selection process will result in a narrowing of the genetic variation in a cultivated population (see Chapter 15 [Leakey, 2017d]).

Of course when it comes to capturing this variation and implementing a tree domestication program, it is common to want to select for superiority in more than one trait. For example, in a fruit to be consumed fresh, it is probably desirable to have a large quantity of pulp or juice, an attractive skin, a good taste (typically the combination of several different sensory traits), and a high nutrient content. Likewise in kernels, a large size may be combined with oil content, a desirable fatty-acid profile, good taste, and an easily cracked nut for ease of kernel extraction. Thus it becomes important to study and understand the relationships between different traits (e.g., Figs. 24.1–24.3). In such studies, it often becomes clear that while some desirable traits are closely related, others are not (See Fig. 25.1a). This situation has led to the recognition of different "ideotypes" or ideal trait combinations for different markets or industries. In some species (e.g., *I. gabonensis*) it also becomes clear that there are opportunities to identify both fruit and kernel ideotypes within a single species (Figs. 21.2B and 25.1A), or indeed perhaps even two or more kernel ideotypes (Figs. 25.1B/C) for different industries (e.g., edible oils and oils for cosmetic products or medicinals) forming a hierarchy of opportunities (Fig. 16.4).

Since 1990, in addition to studies on the tree-to-tree variation in morphological traits there have been similar studies with similar outputs for kernel oil content and yield (see Chapter 23 [Leakey et al., 2005c]), physicochemical traits of food-thickening products (Leakey et al., 2005d), nutrient/micronutrient contents and antinutrients (Thiong'o et al., 2002; Leakey et al., 2005d; Leakey et al., 2008; Kimondo, et al., 2012; van den Bilcke et al., 2014; Fungo et al., 2015); fatty acids profiles (Atangana et al., 2011b) sensory traits (Kengni et al., 2001, Pauku, 2005, van den Bilcke et al., 2014); medicinal properties (Leakey et al., 2008), and perfumed essential oils from heartwood (Page et al., 2010). Likewise, there has been an increase in the number of species assessed. The list now includes: *I. gabonensis* (Atangana et al., 2001, 2002, Anegbeh et al., 2003, Leakey et al., 2005d), *D. edulis* (Waruhiu et al., 2004, Anegbeh et al., 2005), *S. birrea* (see Chapters 22 and 23 [Leakey et al., 2005b,c]), *A. digitata* (Soloviev et al., 2004, de Smedt et al., 2011, Cuni Sanchez et al., 2011, Simbo et al., 2013), *Vitellaria paradoxa* (Maranz and Wiesman, 2003, Sanou et al., 2006, Diarrasouba et al., 2007), *Uapaca kirkiana* (Kadzere et al., 2006, Mwase et al., 2006a,b), *B. procera* (Pauku, 2005), *Strychnos cocculoides* (Mkonda et al., 2003), *Tamarindus indica* (Soloviev et al., 2004, Fandohan et al., 2011, van den Bilcke et al., 2014), *Canarium indicum* (Leakey et al., 2008), *Pentaclethra macrophylla* (Tsobeng et al., 2015) *A. floribunda* (Atangana et al., 2011b), *Carapa procera* (Lankoandé et al., 2015), and *Cola acuminata* (Egbe Enow et al., 2013). The future will probably see this list greatly expanded.

IDEOTYPES

The very great intraspecific variation revealed previously in almost every characteristic of the fruits, nuts and other products of these species provides the motivation, process and catalyst for domestication, as already discussed. Crucially, however, the magnitude of this variation and the opportunity of bringing together different combinations provides the scope for novel processing and marketing opportunities, e.g., for *S. birrea* (Figs. 16.3; 24.4) and *I. gabonensis* (Figures 21.2; 25.1). But in addition to the challenge of identifying specific multitrait combinations that match particular marketing opportunities, there is the need to examine the relationships between traits (Chapter 24 [Leakey 2005]; Figs. 24.1; 24.3 and Fig. 28.5).

Chapter 25 (Leakey and Page, 2006), the 'ideotype concept' and its application to the selection of 'AFTP' cultivars, presented relatively easy to identify trait combinations that meet different market or use opportunities. These may be combinations that maximize "harvest index," or certain quality and/or production traits. Interestingly, also there are often opportunities even within a single species to identify a hierarchical suite of different ideotypes (Fig. 16.4). In this case, some more sophisticated trait may require guidance from industry partners (see Chapter 7 [Leakey, 1999a]). Currently, however, ideotype usage seems to be restricted to the needs of informal local markets. Nevertheless, for the future, increasingly sophisticated selection protocols can lead to evermore innovative and complex marketing strategies and value chains. Ultimately, we can see that each of the hundreds or thousands of tree species that could be domesticated for agroforestry production systems could give rise to cultivars as different as the dog breeds that have been teased out of the wolf genome (Leakey, 2012c). Fortunately, due to the availability of vegetative propagation techniques, this potential to capture a diverse set of ideotypes is much easier and quicker in trees than through breeding, as is necessary in animals like dogs.

As we will see in Chapter 30 (Leakey and van Damme, 2014) value-adding along the value chain, and ultimately marketing and trade, are crucial processes in the commercialization step—the third step—toward the development of the sustainable intensification of tropical and subtropical agriculture, especially, but not exclusively, in Africa. Understanding the needs of producers and consumers therefore has to be as much a part of designing and organizing the processes of domestication (Nevenimo et al., 2008) as is the prediction of how the market for the products of a single species could develop locally, regionally, and internationally (Bunt and Leakey, 2008).

Finally, Gepts (2014) has pointed out that the existence of sets of traits (e.g., ideotypes) that are consistently relevant across different, unrelated, crops are being called "domestication syndromes" (Parker et al., 2010), and that, as with ideotypes, there may be potential for different domestication syndromes within a single species.

Part IV

Towards Delivery

Section 6

A Bottom-Up Approach

A unique feature of this study has been the development of a participatory approach, linked to community capacity building, to empower local smallholder farmers to develop new crops from the forest trees, which prior to deforestation used to produce food and non-food products of day-to-day importance. This self-help process creates an incentive to overcome food and nutritional insecurity and generate income from agroforestry practices. A review of the biological, environmental, social and commercial components of the strategies and techniques has found that the outcomes and expected future impacts indicate that this approach has very successfully improved the livelihoods of poor smallholder farming communities. This success is attributed to the benefit flows obtained by farming households; however, there is still need to ensure that the intellectual property derived from these innovations is not misappropriated by unscrupulous entrepreneurs as, in contrast to commercial companies and academic plant breeders, there is currently no legislation that sufficiently protects the cultivars developed by poor, smallholder farmers. Some progress has, however, been achieved, to address the commercialization of indigenous natural products processed by local communities through the establishment of partnerships and trade agreements between the small-scale producers and the local-to-global cosmetic, food, beverage, herbal medicine and pharmaceutical industries. Interestingly, there is growing interest in increasing the international trade in "green" market products, such as organic, fair trade, for the reduction of deforestation and forest degradation and mitigation of climate change, and environmental goods and services. Thus agroforestry developments are focusing on the improvement of access to "green" business opportunities for poor smallholder farmers in Africa, by maximizing the benefits and minimizing the risks.

Participatory Tree Domestication

Chapter 27

The Participatory Domestication of West African Indigenous Fruits

This chapter was previously published in Leakey, R.R.B., Schreckenberg, K., Tchoundjeu, Z., 2003. International Forestry Review, 5, 338–347, with permission from Commonwealth Forestry Association

SUMMARY

This study obtained quantitative data on fruit and nut traits in two indigenous fruit trees from West Africa (*Irvingia gabonensis* and *Dacryodes edulis*), which have led to the identification of trees meeting ideotypes based on multiple morphological, quality and food property traits desirable in putative cultivars. The same data also indicate changes in population structure that provide pointers to the level of domestication already achieved by subsistence farmers. *Dacryodes edulis* represents 21–57% of all fruit trees in farmers' fields and play an important part in the economy of rural communities. An investigation of the socio-economic and biophysical constraints to indigenous tree cultivation found that indigenous fruits could play an even greater role in the rural economy of west and central Africa. The opportunity to build on this through further domestication of these species is considerable, especially as retailers recognise customer preferences for certain *D. edulis* fruit traits, although at present the wholesale market does not. This project was linked to a larger participatory tree domestication programme within ICRAF's wider agroforestry programme with traditionally valuable indigenous trees. Together these projects provided insights into the value of domesticating indigenous fruit trees, which are of strategic importance to poverty alleviation and sustainable development worldwide.

INTRODUCTION

Throughout the tropics there are indigenous tree species that produce locally important fruits and other non-timber forest products, and that have the potential to be domesticated to provide economic and livelihood benefits to subsistence farmers (Leakey and Simons, 1998). Many of these species are valuable sources of nutrition (Leakey, 1999a) with important health benefits against malnutrition and possible nutritional benefits conferring enhanced resilience to epidemics such as AIDS/HIV (Barany et al., 2001). The integration of these species as novel crops within existing farming systems can also provide environmental benefits (Leakey and Tchoundjeu, 2001). The need for greater emphasis on the cultivation and domestication of these overlooked "Cinderella" species in "development" programs poses important policy questions which need to be addressed (Leakey and Tomich, 1999).

The purpose of this paper is to draw attention to a participatory approach to agroforestry tree domestication, which has been developed in West Africa and that may have application in other tropical areas. This is done in the knowledge that there is considerable current interest in tree domestication in Latin America (Clement and Villachica, 1994; Prance, 1994; Sotelo Montes and Weber, 1997; Jaenicke et al., 2000; Weber et al., 2001), southern Africa (Maghembe et al.,

Multifunctional Agriculture. DOI: http://dx.doi.org/10.1016/B978-0-12-805356-0.00027-1
© 2017 Elsevier Inc. All rights reserved.

1998), East Africa (Simons, 1996) and Southeast Asia (Roshetko and Evans, 1999) and that at the Regional Preparatory Conference of Latin America and the Caribbean for the World Summit on Sustainable Development, in Rio de Janeiro (23–24 October 2001) there were recommendations that:

1. International cooperation should be strengthened in order to address the issues of extreme poverty, underdevelopment, unsustainable production and consumption patterns, environmental degradation, and inequities in wealth distribution.
2. Programs should be promoted for the conservation and sustainable use of biodiversity, which also ensure equitable access to the benefits afforded by the use of genetic resources.

Although domestication does not necessarily assure conservation and sustainable use of biodiversity, this study suggests that at least while there is a substantial wild resource, as in the case of most indigenous fruits, domestication can increase intraspecific diversity (Leakey et al., 2004). Consequently, one way to address these resolutions would be to initiate a program to domesticate more indigenous fruits, which can contribute to the reduction of poverty and livelihood enhancement (Poulton and Poole, 2001) and diversify farming systems (Gockowski et al., 2001).

Participatory Domestication

In contrast to the widely cultivated agricultural and horticultural crops of the world that have been domesticated for millennia, the initiatives to domesticate some of the indigenous fruit trees of different ecoregions of the tropics (Leakey and Simons, 1998) are starting now with wild, or virtually wild, gene pools. This imposes responsibilities on the scientists involved to develop an understanding of the potential of the species, to ensure that domestication proceeds wisely, efficiently and within the constraints imposed by the Convention on Biological Diversity, and to maintain and protect the diversity of the genetic resource.

Tree improvement and breeding has usually been the prerogative of national and international research institutes, because of its long-term nature and the emphasis on timber production by government forestry departments. In agroforestry, however, with the much greater emphasis on the social, cultural, and economic needs of resource-poor subsistence farmers, there has been a recent shift toward domesticating trees producing valuable non-timber forest products (NTFPs) with the people and for the people (Sanchez et al., 1997; Tchoundjeu et al., 1998). This requires a very different approach to tree improvement, one based more on horticultural than forestry techniques (Leakey and Jaenicke, 1995); and one situated on the farm rather than in a research station.

The model, which has been developed in Cameroon and Nigeria by ICRAF and partners (Tchoundjeu et al., 1998; Kengue et al., 2002a), is based on involving the farmers in all stages of the process. This starts with asking the farmers about which of the trees from the natural forest they would like to cultivate on their farms (Franzel et al., 1996), and progresses to the development of simple, low-technology plant propagators (Leakey et al., 1990) in the villages. These inexpensive and effective propagators, made from readily available products (wood, sand, and polythene) for the rooting of stem cuttings, do not require running water or electricity. This simple and appropriate technology has many benefits for rural development projects over more complex propagation systems, especially micropropagation, as with the involvement of NGOs, villagers are trained in the basic principles of vegetative propagation so that they can themselves produce and bulk up "cultivars" from the trees that they know and like best in their area. This emphasis on the empowerment of the community and its use of indigenous knowledge about superior phenotypes in the forest allows rapid progress to be made, as it overcomes the need to do expensive and time-consuming mass propagation and selection from populations of seedlings with unknown potential. This is particularly important, when there are a number of different fruit characteristics that together form a "plus-tree" or ideotype (Atangana et al., 2002a; Leakey et al., 2002), as the more traits for which selection is desired, the larger is the number of trees that would need to be screened in a research station approach to tree improvement.

Participatory domestication also allows farmers to be the beneficiaries and guardians of the use of their indigenous knowledge about inter- and intraspecific variation in the population, and germplasm derived from it. This approach conforms to the aims of the Convention on Biological Diversity, which seeks to protect the rights of local people to their indigenous knowledge and germplasm. It is, thus, in stark contrast to the research station model of tree domestication. It does, however, require that the farmers be informed about, and understand, their rights and know how to maintain and protect these rights.

Are Subsistence Farmers Interested in Domestication?

To determine the relevance of agroforestry tree domestication to subsistence farmers in West and Central Africa, a socioeconomic study to examine both the constraints and potential benefits of bringing indigenous trees into cultivation was carried out in Cameroon and Nigeria. The detailed results of this study will be reported elsewhere (Schreckenberg et al., 2002; Degrande et al., 2006). The overall conclusions of this study, obtained through participatory community-level research, household surveys and whole-farm fruit tree inventories, were that farmers in the study area are very interested in the cultivation of indigenous fruits (Mbosso, 1999). Of particular importance in southern Cameroon is *Dacryodes edulis* (safou/African plum), which is widely planted and constitutes 21−57% of all fruit trees in farmers' fields (Schreckenberg et al., 2002).

In all four Cameroonian communities, safou is very important for home consumption. In two communities, Chopfarm and Elig Nkouma, it was ranked higher than all other tree species for its food value, and in the others it was ranked either second or third. Women particularly like the fact that the boiled or roasted fruit can be eaten with cassava, providing a meal that is quick and easy to prepare at a time when most labor has to be devoted to agricultural activities.

In addition to its use for direct consumption, *D. edulis* provides an important income, being ranked among the top three species for commercial value in all four communities. In terms of value, this is more important for women, for whom the marketing of safou fruit is one of the few relatively independent sources of income they have, but the timing of the income (July−September) is also important for men, coming at a time of year when they have few other income sources and school fees are due. Over 90% of *D. edulis* trees occur in the perennial crop farms (mainly cocoa and coffee), which constitute the predominant land use in the area. In addition to provision of shade, *D. edulis* plays an important role as an income buffer when cocoa and coffee prices fall (Schreckenberg et al., 2002).

Tenure is not an insurmountable constraint to planting safou as most households have at least some land with secure tenure. Nor is labor a particular problem as tree-planting and maintenance work is integrated with that required for other tree crops. Bottlenecks may occur at harvest time, but in communities such as Chopfarm, proximity to flourishing *D. edulis* markets (e.g., Gabon and Douala) means that farmers no longer need to invest much labor in harvesting or marketing as outside wholesalers bring in their own labor to harvest whole trees (Schreckenberg et al., 2002).

The most popular indigenous fruit tree in the southern Nigerian study sites was *Irvingia gabonensis* (bush mango/dika nut), which is widely planted in home gardens and, to a lesser extent, in food crop fields (Degrande et al., 2006). The fresh fruits are eaten as a snack while the dried kernel is ground and added to sauces to make them viscous. The sliminess (or "drawability") of the resulting sauces is particularly valued. In southern Cameroon, very few trees of this species are actually planted at present, although naturally regenerating seedlings are protected. Nevertheless, farmers in Nko'ovos II ranked *I. gabonensis* as the most important species for food and commercial value (Mbosso, 1999). Farmers expressed interest in the cultivation of *I. gabonensis* because of greatly increased interest from traders in recent years (Degrande et al., 2006).

Improved market access would enhance communities' opportunities to cultivate and sell indigenous fruits of all species. Similarly, improved market information systems would improve the opportunities to generate income. These systems should be targeted first and foremost at women, for whom the *D. edulis* trade, for example, is particularly important (Awono et al., 2002).

Characterization of Intraspecific Variation in Fruit and Kernel Characteristics

This section of the paper presents the results of a study in West and Central Africa to quantify the tree-to-tree variation in fruit characteristics in *D. edulis* and *I. gabonensis*. The study was carried out in conjunction with the socioeconomic research described previously and within the context of a participatory tree domestication program managed by the International Centre for Research in Agroforestry (ICRAF, now called the World Agroforestry Centre). Its purpose was to identify combinations of fruit traits that could be brought together through "plus-tree" selection and then captured as a "cultivar" by vegetative propagation.

The assessment of tree-to-tree variation in the *I. gabonensis* and *D. edulis* populations in Cameroon and Nigeria was aimed at the determination of:

- the levels of diversity available to farmers within the area of their communal ownership
- the levels of selection intensity being applied by farmers,
- the level of market recognition of variability in fruit or kernel traits.

METHODS

This study is based on data collected from six villages (Table 27.1), four in Cameroon (Atangana et al., 2001, 2002a; Waruhui, 1999; Waruhui et al., 2004), and two in Nigeria (Ukafor, 2002; Anegbeh et al., 2003, 2005). The sites were chosen to represent a range of ethnic, social and environmental factors found in the region. These sites were separated by 100–350 km and hence are clearly different populations. The use of three geographically distinct sites should reduce the chance of finding correlated traits that may be due to random non-general associations that can occur in a single isolated population. Two of the sites (Nko'ovos II in Cameroon and Ugwuaji in Nigeria) were in fact within the genetic diversity hotspots of *I. gabonensis*, identified by Lowe et al. (2000), using DNA markers, each with genetically distinct populations.

Diameter at breast height (dbh) was measured for each tree, while tree height was estimated. Tree-to-tree variation in fruit and kernel characteristics were assessed in all the trees of a randomly selected, discrete population of up to 100 trees per village, depending on availability. Measurements were made of the following fruit traits in 24 randomly collected ripe fruits per tree of *I. gabonensis* and *D. edulis*, using kitchen scales accurate to 2 g and calipers accurate to 0.1 mm (see Leakey et al., 2000):

- Fresh fruit mass (g),
- Nut mass (g)—for *I. gabonensis* only—after drying the residue flesh,
- Fresh kernel mass (g),
- Fruit length (mm),
- Fruit width (mm),
- Flesh depth in the fruit breadth dimension (mm),
- Taste score (1[bitter] to 5 [sweet]),
- Fibrosity score (1 [low fiber] to 5 [high fiber])—for *I. gabonensis* only,
- Oiliness score (1 [low oil] to 5 [high oil])—for *D. edulis* only.

Taste, fibrosity, and oiliness were assessed by the same people at each site. The owners of the trees were asked whether or not the tree had been planted and questioned about the tree's fruiting behavior. The farmers were also asked about the likely market price of each fruit sample.

These data were used to derive: Shell mass (g) (=Nut mass − kernel mass, for *I. gabonensis* only), and fresh Flesh mass (g) (Fruit mass − Nut mass, in *I. gabonensis*, and Fruit mass − Kernel mass, in *D. edulis*). Since the ease with which nuts can be cracked to allow kernel extraction is seen by farmers as an important trait, a shell brittleness score was derived as 50 minus shell mass (so that the desirable trees for selection had a high score).

The kernels of *I. gabonensis* are used as a thickening agent in traditional soups and stews in West and Central Africa. To assess the tree-to-tree variation in these properties, kernels were stored for analysis (Leakey et al., 2005d). This analysis was done using a Rapid Visco-Analyzer to determine changes in the physical properties of each sample of

TABLE 27.1 The location of study villages and the numbers of trees assessed.

		Latitude °N	Longitude °E	Altitude (m)	No. of trees assessed
Irvingia gabonensis					
Cameroon	Elig Nkouma	4°06'	11°24'	460	31
	Nko'ovos II	2°55'	11°21'	610	21
Nigeria	Ugwuaji	6°25	7°32'	175	100
Dacryodes edulis					
Cameroon	Makenene	4°52'	10°48'	580	100
	Elig Nkouma	4°06'	11°24'	460	57
	Nko'ovos II	2°55'	11°21'	610	12
	Chop Farm	3°57'	9°15'	11	31
Nigeria	Ilile	5°19'	6°55'	54	100

defatted dika nut meal in water, so mimicking the changes occurring during the cooking process. Traces, generated during a 15-minute two-phase temperature profile (a "cooking phase" at 95°C and an "eating phase" at 50°C), recorded the electrical energy consumed to maintain constant stirring speed of a paddle in the paste. Two food-thickening parameters were derived from the final 2 minutes of the trace at eating temperature: (1) the average value, taken as the "viscosity" (magnitude of soup thickening) and (2) the presumed "drawability," which was based on the varying spikiness (width) of the trace. The latter is proposed as the ability of the gum to exert periodic viscoelastic restraining forces on the paddle; this is presumed to reflect its ability to be drawn out into tendrils with a spoon.

RESULTS AND DISCUSSION

The relationships between tree height and dbh indicated that the Cameroon and Nigerian populations of *I. gabonensis* differed in demographic structure (Atangana et al., 2001; Anegbeh et al., 2003, 2005), with the Cameroon population reflecting a mature age, while the Nigerian population was much younger. These differences can be explained by the farmers' information, which revealed that the Cameroon population was made up of unplanted natural trees retained when forest was cleared for agriculture, while the Nigerian population was made up of planted trees.

The Levels of Diversity Available to Farmers Within Their Community

Contrary to the suggestion that there are morphologically distinct "varieties" in the on-farm populations of *D. edulis* (Okafor, 1983), this study found continuous variation in all the traits examined. There was, however, highly significant ($p < 0.001$) variation between individual trees for each trait, and as expected, trees with superiority in one trait (e.g., fruit size) are not necessarily superior in other traits (e.g., fruit taste). Consequently, the chance of finding trees with superiority in two or more traits is considerably lower than for a single trait. Nevertheless, it is highly desirable to identify combinations of traits, which should be brought together for cultivar development. To pursue this objective of defining combinations of desirable traits, an ideotype approach has been developed.

In *I. gabonensis*, an examination of all the data (Fig. 27.1) indicates that there are some trees (Ug10, Ug75, Ug12) with high values for fruit traits (fruit length, fruit width, flesh weight, flesh depth, and taste) that are close to the fruit ideotype (solid black line), and thus superior as fruit for eating fresh. In the same way there are other trees (EN26, Nk28, Nk31, Nk6) with kernel traits (kernel weight, shell brittleness) close to the kernel ideotype (solid black line). Interestingly, however, the study of the physical and chemical properties of the kernels (Leakey et al., 2005d) found that none of the trees assessed had high values for both of the food thickening traits (viscosity and drawability). Furthermore, the viscosity and drawability of the polysaccharide extract were poorly related traits (e.g., $r^2 = 0.336$) and thus probably kernels from different trees have different uses in food preparation. Consequently, depending on the use of the kernels, the kernel ideotype should be subdivided into two food-thickening subideotypes, one with good properties for viscosity, and the other for drawability (Leakey et al., 2005d).

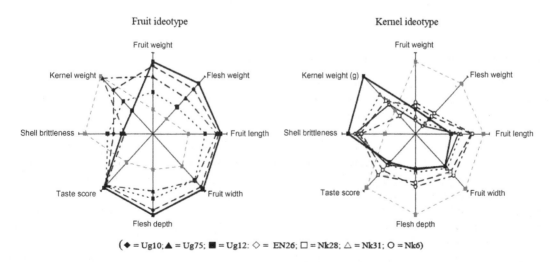

(◆ = Ug10; ▲ = Ug75; ■ = Ug12; ◇ = EN26; □ = Nk28; △ = Nk31; O = Nk6)

FIGURE 27.1 Fruit and kernel ideotypes for *Irvingia gabonensis* compared with data from best trees.

Fat determination and fatty acid profiling of kernels confirmed previous studies (see review by Leakey, 1999a) that fat content ranges from 50% to 70% between samples, although the range in individual tree samples in the present study was greater than this (37.5–75.5%). As also found elsewhere, the study identified myristic acid and lauric acid as the major fatty acid components of the extracted fat. Thus, it appears that kernels for vegetable oil production may have to conform to a third ideotype, depending on the yield and desirable properties of the oils.

To date, the range of nutritional values of *I. gabonensis* kernels has not been reported to vary between samples, although reported protein content of different samples has ranged from 14.3% to 24.1% (Leakey, 1999a). However, it is clear from a protein analysis of defatted kernel samples from six trees, selected for their diverse viscosity properties, that they were similar to the published range. Thus, study of more trees may determine opportunities to further select individual trees for their nutritional value. Electrophoretic analysis of total protein extracts according to molecular weight demonstrated that all six samples had similar protein patterns.

In *D. edulis*, the fruits for eating as a nutritious cooked vegetable would appear to fit a single ideotype characterized by large size, thick flesh and a small kernel. Further refinement of this ideotype may follow once the organoleptic properties of these fruits are better understood. A preliminary study by a trained tasting panel has found that there is variation in acidity, astringency, bitterness, sourness, as well as in fibrosity (Kengni et al., 2001; Leakey et al., 2002). Thus, taking all these traits together, the morphological and organoleptic studies to date suggest that there are opportunities, through ideotype selection, for the development of cultivars that combine large size with good quality attributes for the fresh fruit trade. However, there are also potential industrial uses of these fruits for vegetable oils (Kapseu and Tchiegang, 1996; Silou et al., 2000), which may require further refinement of the fruit ideotype, depending on the oil properties required.

For both species, the preceding definition of ideotypes feeds into the on-farm domestication process, by helping researchers to explain to NGOs and farmers what traits, or combinations of traits (ideotypes), are available for selection and thus their opportunities for cultivar development. The inclusion of this information into the community domestication programs could have very rapid impacts on the level of genetic gains achieved by farmers in the next 10 years. For example, fruit size could probably be increased two- to threefold by creating cultivars that conform to the fruit ideotypes. Since different villages will create different sets of cultivars for each species they wish to cultivate, intra- and interspecific diversity will be maintained at the farm level, at least in the short-to-medium term.

The study has also identified some variation in the fruiting phenology of different trees, illustrating opportunities for selection for the seasonality of production. Seasonality is not a problem in the case of *I. gabonensis* as the storage of dika nuts allows a year-round market, but the fruits of *D. edulis* have a very short shelf life and thus there is a need to extend the productive season, as for example with the recently created "Nöel" cultivar that fruits at Christmas (Tchoundjeu et al., unpublished). Alternatively, research is needed to develop storage and/or processing techniques for the fruits that can be used in the villages or local towns.

An additional advantage of the ideotype approach is that the cultivars may have a broad genetic base in many other characteristics, especially if the cultivars come from unrelated populations. This could make them less susceptible to pest and disease outbreaks (Leakey, 1991). To minimize the risks of narrowing the genetic base and associated disease and pest problems, it would also be wise to ensure that there is a turnover of recommended cultivars arising from an ongoing and continuous program of selection.

The Levels of Selection Intensity Being Applied by Farmers

In an attempt to determine the levels of tree selection by farmers, the frequency distributions of the data for each measured trait were plotted and examined. The results did not provide an answer. However, an assessment of the genetic gain made by farmers through their own selection efforts was achieved.

Typically, trees are outbreeding and genetically very diverse due to the contribution of large numbers of individuals to a shared gene pool and the free segregation of alleles during meiosis (Zobel and Talbert, 1984), typically resulting in normally distributed variation of quantitatively inherited polygenic traits. These patterns of intraspecific variation mean that for any one trait there are relatively rare genotypes that display the desired set of characteristics, so-called plus-trees. In addition, it is well known that tree populations from geographically different locations (provenances) can have different mean values. Leakey et al. (2004) have postulated that when data from different wild populations for a given trait are combined, the overall population will also be normally distributed. In plant breeding, cycles of selecting and crossing between only the best individuals in the population (truncated selection) result in new progenies, which outperform their parents in the selected trait (Futuyma, 1998). The degree of improvement depends on the narrow sense heritability (Stearns and Hockstra, 2000). The domestication of a species must therefore result in changes in the frequency distribution of the values of the selected trait among the members of the population (and typically an increasing reduction in diversity within the selected population, due to an increasing selection intensity). During the course of several generations of truncated selection, the frequency distribution of the trait can thus be expected to change through a progression of stages that ultimately lead to the formation of a variety.

To determine if the stage of farmer-selected domestication reached in different populations of *I. gabonensis* and *D. edulis*, Leakey et al. (2004) have hypothesized that when over a long period of time farmers take and plant seeds from the fruits of their best trees, the frequency distribution of data for the selected trait will change from normal (stage 1), to positively skewed (stage 2), to a flattened normal distribution again (stage 3), to negatively skewed (stage 4) to normal (stage 5)—see Fig. 8.1. With each stage there is also a progressive shift along the x axis, as the progeny is improved for the selected trait. In contrast, the frequency distribution for neutral traits (unselected traits with no correlation to the selected trait) will remain at stage 1.

In *D. edulis* for most traits, the frequency distributions of individual populations were close to normality although there were differences between the traits, with regard to the degree of separation between the populations along the x axis. For example, the peak mean kernel mass of all populations was the same, while for flesh thickness, fruit mass, and fruit width, especially in the Makenene and Elig Nkouma populations from Cameroon, there were considerable differences in the peak mean (i.e., they were distributed along the x axis). Since the kernel of *D. edulis* is usually discarded, as being of little value, while the mass and size of the fruit and the thickness of the flesh are all indicators of a desirable fruit, it seems likely that over time there has been intentional selection by Cameroon farmers for these desirable traits, resulting in domestication progressing to between Stages 2 and 3. Because the mean flesh thickness of the selected population is greater (7.5 mm) than that of the wild population (4.5 mm), these data suggest that farmers have made a 67% genetic gain in flesh depth. Similarly in *I. gabonensis*, evidence for domestication through the selection of large-fruited trees was found in the population from Ugwuaji, in Nigeria. In this species, domestication seems to have advanced to Stage 2, with a genetic gain of 44% in flesh depth. There was no evidence for differences in the stage of domestication between subpopulations in Cameroon. The fact that subsistence farmers have domesticated these fruit to this point emphasizes the importance that they attribute to indigenous fruits for their own consumption and for trade.

The recognition that farmers in Cameroon and Nigeria have initiated the domestication of two of their indigenous fruit trees emphasizes the importance of the current activities to further domesticate the indigenous trees that provide marketable non-timber forest products of importance to local people for food security and income generation. The need now is to go to the next stage of domestication (Fig. 27.2) in which cultivars are developed, using vegetative propagation techniques (Leakey et al., 1990; Shiembo et al., 1996b), from the very best trees available in each village. This is starting in the west African participatory tree domestication program (Tchoundjeu et al., 1998) with the intention of using these cultivars in cocoa and other agroforests to diversify the agroecosystem. In this way, it is envisioned that it may be possible to create land-use systems that enhance the livelihoods of poor subsistence farmers (Leakey, 1999a). In addition, the domestication of these species may lead to the creation of export commodities to diversify both the farmers and national economies. These benefits, together with the international public goods and services (carbon sequestration, biodiversity, etc.) that can be derived from increasing the numbers of trees in agroecosystems, are outcomes that could benefit the global community (Leakey, 2001b).

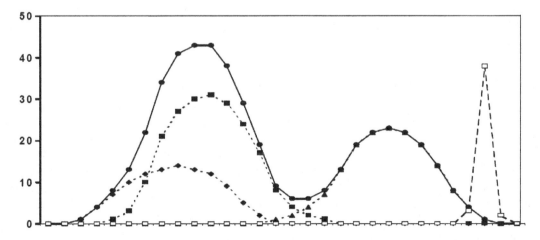

FIGURE 27.2 A fifth stage in the domestication process = Creation of a cultivar by vegetatively propagating a superior individual.

The Level of Market Recognition of Variability in Fruit or Kernel Traits

The importance of indigenous fruits in West Africa is emerging from market studies, which indicate that, for example with *D. edulis*, wholesale traders in Gabon travel to markets like Makenene to buy fruits for importation into Gabon. Similar evidence of regional trade has recently also been documented for *I. gabonensis* and *Ricinodendron heudelotii* kernels, and the nuts of *Cola* spp. (Ndoye et al., 1998; Ruiz Pérez et al., 1999).

To determine if these markets reward farmers for producing fruits with desirable characteristics, fruit samples were purchased at the peak of the season (3–17 August 2000) from urban (Mfoundi, Yaoundé) and rural (Makenene Centre and Makenene East) markets in Cameroon (Atangana et al., 2002b; Leakey et al., 2002). The area around Makenene has a reputation for producing and selling *D. edulis* fruits, leading to the existence of a retail market (Makenene East) and one of the largest wholesale markets in the country (Makenene Centre). The fruits of each sample were characterized as described in the Methods section.

Statistically significant differences in each fruit trait were found between samples for each market, but the relationships between fruit traits and prices were found to be weak in wholesale markets. However, in retail markets, fruit mass, length and width were positively correlated with price per fruit, indicating that small-scale traders can benefit from consumers' preferences for large fruits. Interestingly, the relationship between fruit mass and price was stronger in the urban Mfoundi market than in more rural Makenene East market, suggesting that consumers in urban markets will pay more for large, tasty fruits than they will pay for small or less tasty fruits.

It seems therefore that at present farmers who typically sell their produce in rural wholesale markets are not currently being rewarded for producing superior fruits, although big fruits, which have the most pulp, fetch the highest prices in retail markets. This indicates that retailers, who are in closest contact with consumer demands, take into account phenotypic variation in fruit size when fixing market prices. There is also some evidence that some other qualitative traits (e.g., flavor) are also recognized by urban retailers as Leakey and Ladipo (1996) found that while big fruits tended to have high market prices, some small fruits were also highly priced. Similarly, Waruhiu (1999) found that a relatively uncommon white-skinned fruit type was more expensive than similar sized fruits of the common purple color. Wholesalers, on the other hand, do not appear to take the characteristics of individual fruit types into account when pricing fruits. It is to be hoped that in the future, farmers producing fruits of named cultivars will be rewarded with higher prices.

Impact and Strategic Importance

In these and other indigenous fruits (e.g., *Sclerocarya birrea*), intraspecific variation is typically found to be greatest at the village level, so there seems to be several strategic advantages of tree domestication at this level. Firstly, it gives each community the opportunity for significant genetic gain without empowering one community more than its neighbors. Secondly, this village-level, self-help approach to domestication also helps to maintain a broad genetic base within the species being domesticated, as each village will develop a different collection of cultivars. Furthermore, it allows the farmers to practice their new skills on any other species of interest to them, and so does not restrict the domestication process to the priority species. In the long term, this will promote the species diversity of their farming systems. In terms of ensuring impact from development assistance, this approach of working directly with farmers has the advantage that the project outputs are immediately disseminated into the target population, thus overcoming the delays that often arise from a research station stage in the domestication process.

Potential for Wider Application of Participatory Domestication

The domestication of new species is a major undertaking, one that is a continuous process of improvement and one that has to be justified by the benefits that accrue to the producers and consumers of the products. There is currently some debate among development organizations, focused on poverty alleviation, sustainable livelihoods and food security, about the direction for future research: some favor biotechnology and an expansion of the Green Revolution (McCalla and Brown, 1999; Lipton, 1999), while others see potential for broadening the basket of crops (McNeely and Scherr, 2001)—a Really Green Revolution (Leakey, 2001b). In the case of a number of agroforestry trees, including *I. gabonensis* (Aubry Lecomte ex O'Rorke) Baillon. and *D. edulis* (G.Don) H.J. Lam, it has been argued that their domestication will enhance farmer livelihoods, reduce poverty, and promote economic development (Leakey, 2001b). At the same time, this domestication of agroforestry trees should provide farmers with an incentive to integrate trees into their farming systems, so developing an agroecological succession that can progress to maturity (Leakey, 1996). In this way, it has been suggested (Leakey, 1999b) that the domestication of indigenous fruits could encourage the development of sustainable agroforestry practices that rehabilitate degraded farmland, sequester carbon and other greenhouse gases and enhance both biodiversity and the functioning of agroecosystems. Through the enhanced income generation arising from indigenous fruit trees, farmers have the opportunity to buy fertilizers and other agricultural inputs, and so to raise the productivity of their staple food crops from the normally low yields up toward their biological capacity. In this way, farmers can perhaps reap the benefits from the Green Revolution.

We believe that the experience of domesticating indigenous fruit tree species in West Africa is relevant to many other regions of the tropics, because throughout the tropics:

1. There are many tree species producing edible fruits and other products, which are grown and marketed on a small scale and which are potential candidates for domestication. The range of available species allows for the choice of those that meet labor availability, different markets, systems of tenure, variations in soils and climate, etc.
2. There is a decline in the availability of traditionally important forest products from wild sources, especially near urban markets.
3. There is great interest in agroforestry as a low input, low-risk, sustainable farming system, which supports rural livelihoods.
4. There are many small-scale subsistence farmers who are poor and need opportunities to generate income.
5. There are risks arising from falling cash crop prices, pests and diseases, and increasing environmental pressures, which could be averted by economic and ecological diversification.
6. It could build on and enhances the local social and cultural traditions.
7. It could promote small-scale local processing and entrepreneurial activity at the community level.
8. It could benefit and empower women who are often involved in the labor-intensive harvesting, processing and marketing of forest fruits, and who play a strong role in managing home gardens and village nurseries.
9. It could enhance the health of the rural and urban communities, as indigenous fruits are rich in minerals, vitamins, protein, oils and carbohydrates and so can also meet the needs of poor people for food and nutritional security.

Finally, the participatory approach to domestication is a relatively rapid and low-cost option to development, which given the need to domesticate a wide array of species, cannot be achieved by the Green Revolution approach to agricultural development.

ACKNOWLEDGMENTS

This publication is an output from a research project partly funded by the United Kingdom Department for International Development (DFID) for the benefit of developing countries. The views expressed are not necessarily those of DFID (Project R7190 Forestry Research Programme). We also thank Jean-Marie Fondoun, Joseph Kengue of IRAD, Cameroon; Paul O. Anegbeh, Alain R. Atangana, Ebenezer Asaah, Ann Degrande, Charlie Mbosso of ICRAF, Cameroon and Nigeria; Annabelle N. Waruhiu of Edinburgh University; Cecilia Usoro, Victoria Ukafor of Rivers State University of Science and Technology, Nigeria; Robert C. Munro of CEH Edinburgh and Philip Greenway and Martin Hall of C&CFRA, England, for their contributions to the study.

Chapter 28

Agroforestry—Participatory Domestication of Trees

This chapter was previously published in Leakey, R.R.B., 2014.
In: van Alfen, N. et al., (Eds.), Encyclopedia of Agriculture and Food Systems, vol. 1. Elsevier, San Diego, pp. 253–269, with permission from Elsevier

SUMMARY

Participatory domestication of indigenous tropical trees producing useful and marketable products has been developed to replenish the resource depleted by land clearance for agriculture. The domestication programme which was initiated in the 1990's has now become a global programme aimed at the alleviation of poverty and malnutrition. However when integrated with other aspects of agroforestry in a three-step generic model for food security it becomes an approach to the sustainable intensification of tropical agriculture. This chapter reviews the biological, environmental, social and commercial components of the strategies and techniques involved as well as the outcomes and expected future impacts.

INTRODUCTION

The domestication of agroforestry trees is a technique for the intensification of agroforestry as a low-input farming system delivering multifunctional agriculture for the relief of poverty, malnutrition, hunger, and environmental degradation in tropical and subtropical countries (Leakey, 2010, 2012b).

In the past, tree products were gathered from natural forests and woodlands to meet the everyday needs of people living a subsistence lifestyle. With the advent of the Industrial Revolution and, more recently, the intensive modern farming systems of the Green Revolution, the resource of these trees has declined. This has been due to increased population pressures for agricultural land and the escalation of deforestation. To rebuild and improve this useful resource the concept of tree domestication for agroforestry was proposed in 1992 (Leakey and Newton, 1994a) and subsequently implemented by the World Agroforestry Centre (ICRAF) as a global initiative from 1994 (Simons, 1996). Great progress has been made in the first two decades of this initiative (Leakey et al., 2005a, 2012; Tchoundjeu et al., 2006) which have encouraged local entrepreneurism in the processing and marketing of agroforestry tree products (AFTPs). This has had beneficial impacts on farmers' livelihoods (Tchoundjeu et al., 2010a).

STRATEGY

Domestication through cultivar development relies on three processes: selection, testing, and breeding. Selection identifies certain genotypes for cultivar development, testing exposes the new cultivars to appropriate environments, while breeding creates new genetic variability.

As in animals, plant domestication is a continuous process, which in crops like wheat, rice, maize, oranges, and apples, started thousands of years ago and continues today. In trees, two basic approaches are used to effect genetic improvement: the seed-based breeding approach typical of forestry, and the clonal approach typical of horticulture, but

Multifunctional Agriculture. DOI: http://dx.doi.org/10.1016/B978-0-12-805356-0.00028-3
© 2017 Elsevier Inc. All rights reserved.

the domestication strategies are not dissimilar. In agroforestry, Leakey and Akinnifesi (2008) have suggested a strategy based on the establishment of three interlinked tree populations (Fig. 16.2):

- Gene Resource Population
- Selection Population
- Production Population

The gene resource population is basically the wild population from which new selections can be derived. The selection population is the collection of selected provenances, progenies, or individual trees which are being tested and used to develop clonal cultivars or integrated within breeding programs to create the next generation of breeding stock. A wide range of genotypes may be kept in this population as long as each one has at least one characteristic of possible future interest. The production population consists of the highly selected clones, progenies, or provenances which are currently being planted by farmers.

In their strategy paper, Leakey and Akinnifesi (2008) specifically highlighted the development of a clonal approach to the domestication of high-value trees, and in particular the traditionally important indigenous fruits and nuts. However, the key element of the three population strategy is equally applicable to less valuable tree species grown for their environmental services and propagated by seed. It is important to remember that the practice of domesticating a species is cyclical and thus continuous. For this reason commercial plantings have to be made with whatever material is best at a given time, knowing that they will be superseded later.

The strategy being implemented by ICRAF and others to domesticate high-value indigenous trees for agroforestry, such as those producing marketable fruits and nuts, is based on participatory processes involving local communities. A participatory tree domestication strategy involves the consultation with, and participation of farmers: (1) to determine their priority species for domestication (Franzel et al., 1996, 2008), (2) to make an inventory of the natural resource, (3) to implement a program of genetic selection and mass propagation aimed at the sustainable production of AFTP for food, tree fodder, medicinals and nutriceuticals, timber, wood, fibers, etc., and (4) to sell and trade the products in local traditional and new emerging markets further afield. Such strategies also recognize the importance of the wise use and conservation of genetic resources; the reduction of deforestation and the restoration of degraded land. Participatory approaches have numerous advantages (Leakey et al., 2003), building on tradition and culture and promoting rapid adoption by growers to enhance livelihoods and environmental benefits (Simons and Leakey, 2004). This participatory approach to domestication therefore differs from the more common scientific approach that has typically been implemented to develop most food crops. We can therefore think of these approaches as the "farm" and the "research station" pathways to domestication. In agroforestry, these two pathways have been integrated (Fig. 16.1) in a way that ensures that the participating farming communities can benefit from the close involvement of researchers as mentors both to the communities and the NGOs implementing the domestication program.

Biological Components of the Strategy

Propagation of Superior Trees

The choice of mass propagation by the clonal approach has a number of advantages stemming from the use of vegetative propagation to capture and fix desirable traits, or combinations of traits, found in individual trees (Leakey and Simons, 2000). The main advantage arises from the fact that by taking a cutting, or grafting a scion onto a rootstock, the new plant that is formed has an exact copy of the genetic code of the plant from which the tissue was taken and is therefore genetically identical to the stockplant or "mother" plant (Leakey, 2014c). The clone so formed can be mass produced by further vegetative propagation.

The capture of the first asexual propagule from proven and mature field trees that have already expressed their genetic traits is typically done in the field by air-layering/marcotting shoots on the mother tree, or by collecting scions or buds from it for grafting or budding onto seedlings in the nursery. Alternatively, the tree can be felled and the stump left to coppice to provide juvenile material for propagation by stem cuttings. The latter is preferable for clonal timber production from trees with a well-formed juvenile stem, whereas the former is more suitable for trees producing fruits on mature branches (Leakey and Akinnifesi, 2008). The advantage of propagating fruit trees from mature branches is that they already have the reproductive capacity to form flowers and fruits. This means that they are productive within a few years, so reducing the time taken to produce economic returns. Plants propagated from mature tissues will also have a lower physical stature, making the harvesting of fruits easier (Fig. 28.1).

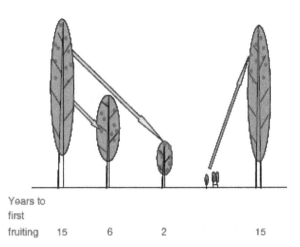

Years to
first
fruiting   15   6   2   15

FIGURE 28.1 Designing trees of different sizes for different uses by selecting the source of propagation material vis-à-vis the state of maturation in the tree. *Reproduced from Leakey R.R.B., 2012b. Living with the Trees of Life — Towards the Transformation of tropical Agriculture, CABI, Wallingford, UK. 200pp.*

Once the clone has been formed by one or other of the described techniques, the best strategy is to mass produce the clone, or cultivar, by rooting cuttings in the nursery. For this process, inexpensive and simple non-mist propagators have been developed (Leakey et al., 1990). This simple, low-technology strategy allows the process to be implemented in remote areas without electricity or running water (Leakey, 2014c). These propagators are also extremely effective and meet the needs of most developing country tree-domestication projects in both the moist and dry tropics.

The preceding strategy of capturing the genetic superiority of mature trees in the field may not always be possible or acceptable, making it necessary to develop clones from seedling plants in the nursery. Basically, this adds an additional genetic selection step to the overall domestication program, which can take many years to complete. Nevertheless, there are three reasons why the use of seedlings may be preferable (Leakey and Simons, 2000):

1. The population of mature trees can be dysgenic (genetically depleted) because the elite specimens have been removed by loggers or farmers. In this case, the seedling progeny derived from such populations will probably have a better array of genetic variation during reproduction due to the remixing of genes during the segregation phase of meiosis.
2. The felling of large numbers of the most desirable mature trees for the purpose of generating cultivars may not be acceptable to the owners. In addition, felling the mature trees may not be environmentally acceptable.
3. The use of seedlings allows the screening of far larger populations, with much more diverse origins, so maintaining genetic diversity among the cultivars.

Through the improvement of tree yield and product quality, tree domestication becomes a strategy for the intensification of agroforestry (Leakey, 2012a), something which is further enhanced when many of the different trees within the farming systems are domesticated in parallel.

Genetic Resource Issues

One of the important requirements of an appropriate strategy is to conserve a substantial proportion of the genetic variability for future use in selection programs and, subsequently, through breeding to broaden the genetic base of the cultivars in the production population. This also serves as a risk-aversion strategy should it be necessary in the future to breed for resistance to pests and diseases. There are three actions, which each contribute to genetic conservation:

- Establishing a Gene Bank (*ex situ* conservation)
- Wise utilization of the genetic resource in cultivation (*circa situ* conservation)
- Protecting some wild populations (*in situ* conservation).

Ex-situ gene banks containing randomly sampled populations from across the natural range of the species should be replicated within a site and across a number of sites. This replication is both for security and to ensure that the sites embrace any climatic or edaphic variation within the region. They should also be located in areas offering long-term protection from external threats—for example deforestation and wild fires.

Regarding *circa situ* conservation and the wise use of genetic resources, it is important to recognize that there is a trade-off between accuracy of genetic value estimation and intensity of selection (i.e., greater accuracy generally comes at the expense of reduced numbers of families, individuals per family, or clones). In long-lived tree crops, there is also the problem that many of the traits on which selection is based are not visible until the trees are nearing their productive age at the end of a long period of growth. Consequently in timber trees, for example, much selection is done on early rates of growth and assumptions that this will relate to yield. In the case of the West African tree *Triplochiton scleroxylon*, a Predictive Test was developed based on apical dominance (the process controlling the formation of branches) in very young plants in the nursery. This was to take advantage of a strong relationship between this physiological process and branching frequency in the nursery and then between branching frequency in the field and stem volume or timber yield (Leakey and Ladipo, 1987). A quick and early test like this allows much larger populations to be screened than would be either practical or economically justifiable through long-term field trials.

The trade-off between the intensity of selection and numbers of plants also affects the diversity of the genetic base—the gene resource population. This diversity should be rigorously enforced as one of the objectives of a domestication program. This therefore involves a good sampling protocol both of the natural population as well as for any existing provenances and progenies from a breeding program. In this respect, it is important to remember that it is not uncommon for there to be a few elite trees in provenances and progenies which may however, on average, be poor. This is especially important if a clonal domestication program is being developed, as these elite individuals from poor populations may be genetically unique.

As the selection process intensifies with time, new traits will be introduced into the domestication program (e.g., seasonality of production, early fruiting, disease and/or pest resistance, drought tolerance, etc.). After the initial phases of clonal selection, the opportunity will arise for controlled pollinations between proven elite clones, as has been done in *Ziziphus mauritiana* (Kalinganire et al., 2012) in the Sahel and *Dacryodes edulis* in Cameroon (Makueti et al., 2012). In this way it should be possible to take advantage of specific-combining ability in unrelated superior clones and to produce progeny with heterosis in desirable traits that exceed what is found in the wild populations. The vegetative propagation of these new genotypes can become a second clonal generation.

Environmental Components of the Strategy

Different agroforestry practices typically involve a range of different species producing either environmental/ecological services such as soil fertility enhancement, erosion control, or different products ranging from low-value fuel wood to high-value marketable products for food, cosmetic, and pharmaceutical industries. Different approaches to domestication will probably be followed for these different species. Likewise these different species will probably fill different spatial and temporal niches in the landscape and be planted in different configurations and densities within farming systems. This is all part of the diversification of the agroecosystem which is an important component of enhancing agroecosystem functions for the rehabilitation of degraded farmland (Leakey, 1999b, Leakey, 2012b).

The domestication of indigenous trees producing high-value products, such as traditional foods and medicines, is one component of a novel strategy for the intensification (Leakey, 2012a) and diversification (Leakey, 2010) of smallholder farming systems in the tropics and sub-tropics through agroforestry. In environmental terms, the diversification with long-lived perennial plants is important because it is the way to rebuild the ecological functions of agroecosystems and landscapes. Soil and land rehabilitation is crucial if agriculture is to use land already cleared of forest rather than abandoning it and cutting down more forest (Leakey, 2012b).

Tree domestication and the commercialization of tree products also create incentives for farmers to plant agroforestry trees as they generate income and promote business, trade and employment opportunities. When added to agroecosystem rehabilitation, tree domestication can be seen as step two in a generic model of how agroforestry can deliver environmentally desirable multifunctional agriculture (Leakey, 2010; Fig. 28.2)—a model with high adaptability to different climatic and edaphic situations (Leakey, 2012b).

Social Components of the Strategy

Many of the biological components of the strategy described previously are not specific to participatory domestication and would apply to more centralized and top-down programs led by research scientists. In contrast, the social components are central to the philosophy of participatory domestication.

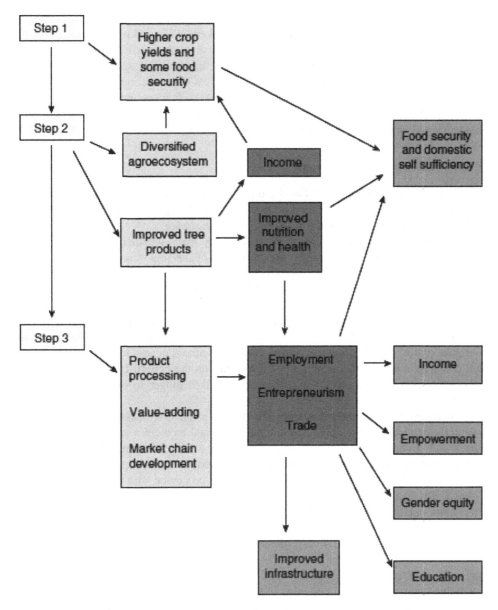

FIGURE 28.2 The delivery of multifunctional agriculture by three agroforestry steps: Step 1. Land rehabilitation, Step 2. Domestication with indigenous trees producing marketable products, and Step 3. Marketing, processing, and value-adding indigenous tree products (see Table 28.1). *Modified from Asaah et al., 2011.*

The prime objective of the participatory approach is to involve the target communities in all aspects of the planning and implementation of the program so that they fully understand it and can buy in and have ownership of the program. In addition, the purpose of this strategy is also to ensure that all engaged members of the community, whether male or female, are empowered by the program and the beneficiaries of the outputs of their own initiatives and labor. This should enhance the livelihoods of the community members in general and promote social equity.

In the longest-running example of participatory domestication in agroforestry trees the researchers have fed their outputs to NGO partners through training-of-trainers courses and by acting as mentors to the NGO-managed farmer training schools (or Rural Resource Centres) established in pilot villages (Tchoundjeu et al., 2002, 2006, 2010a; Asaah et al., 2011). As we will see in the techniques section following, the farmers in this partnership have contributed their knowledge about the use and importance of local species, the range of variation in different traits of relevance to genetic selection and their traditional knowledge (TK) about the role of these species in local culture and tradition. They have also contributed their time and labor. Furthermore and crucially, they have also made available some of their trees for research and for training in domestication techniques.

In implementing this strategy it is of great importance to recognize the legal and socially important communal rights of local people to their TK and local germplasm (Lombard and Leakey, 2010) and to ensure that they benefit from their use and are rewarded for sharing them for the wider good. Because of the sensitivity arising from past commercial exploitation of these rights by individuals, companies, academics, international agencies, and government, it is very clear that the partners in domestication programs have to earn the trust of local communities before TK and germplasm is made freely available. Ideally, to ensure that benefits flow back to the farmers and communities, the recipients of TK and germplasm should enter into formal Access and Benefit Sharing agreements (ICRAF, 2012) in which the rights of the holders of knowledge and genetic resources will be legally recognized.

Commercial Components of the Strategy

With poverty alleviation as one of the objectives of the domestication of agroforestry trees, it is clear that incentives for, and approaches to, income generation are important in the overall strategy. Consequently, improving and expanding the markets for agroforestry trees and their products are central to the strategy.

The first opportunity is the creation of demand for agroforestry trees, whether for environmental and ecological services or for AFTPs. This of course involves the dissemination of the importance of agroforestry for soil fertility replenishment, watershed protection, reduction of erosion, demarcation of boundaries, as well as for the production of wood for fuel and other uses, timber, extractives, medicines, fodder, food and numerous non-food products (Leakey, 2012b). The experience of the last 10–15 years indicates that the first income stream from agroforestry projects is derived from the sales of plants from village nurseries to neighboring communities; and especially the sale of seedlings of nitrogen-fixing or the so-called fertilizer trees (Asaah et al., 2011; Leakey and Asaah, 2013). This is because the loss of soil fertility due to frequent cropping without access to artificial fertilizers is recognized as one of the main constraints to agricultural production. In addition the benefit flows from these trees are obtained relatively quickly (1–3 years). On the other hand, it generally takes longer to obtain returns from the production of AFTPs.

The strategy for increased income generation from AFTPs is, in the first instance, to build on local markets and trade. This is particularly important in the case of traditional foods and medicines as local people are familiar with the use of these products and the demand typically exceeds supply. In the longer term, however, some of these products may have regional and even international markets, firstly with expatriates from tropical countries living in Europe and America, and then as products become more widely known or better processed with global customers.

A secondary reason for focusing on local and regional rather than international markets is that if the demand expands too fast there is a risk that large-scale entrepreneurs may enter the marketplace and out-compete local business people (Leakey and Izac, 1996) so undermining the use of tree domestication as a development strategy. Having said that there are a few interesting and potentially very important initiatives in which multinational companies are becoming engaged with local communities in tropical countries in Public–Private Partnerships (Leakey, 2012b). These are generally operating in a way that is contrary to the trend of globalization in which all the economic benefits flow to industrial countries rather than at least some remaining in the tropical country.

Some tree products, notably fruits, are produced seasonally and have a very short shelf life. To overcome this constraint to year-round marketing it is necessary to investigate opportunities for processing and value-adding. This can take many forms from drying and air-tight packaging; preservation in oil, brine, or sirup; freezing; etc. Generally this involves the need for a level of scale outside the capacity of a smallholder farmer, although "cottage" industries can be a possibility. If this approach to extending the marketing season has to involve industrial companies, then new issues arise. One of these is to involve the companies in the domestication process so that traits which are important in the processed product are included in the genetic selection program (Leakey, 1999a). Later we will consider issues of trading agreements that also result from the involvement of processing companies in marketing.

An alternative approach to processing can be to seek the generally rare plants that flower and fruit outside the normal seasonal pattern and then to develop these as cultivars. While this can be quite simple to achieve, these out-of-season plants may not have quality or yield traits that are as good as those fruiting within the normal season. In this case, developing cultivars which expand the productive season may need to also involve a breeding program.

As the commercialization process involves more players and becomes more complex, so the risks increase that the producers will be exploited and inadequately rewarded for their products and innovations. Entering into any market will expose suppliers to competition. Commentators have suggested that this will undoubtedly jeopardize the rights of farmers or communities in the supply chain. To shut out casual and opportunistic competitors it is obviously important to do as much as possible to ensure that the supply chains leading up to the manufacture of the products are efficient, competitive, and as well protected as possible. Innovative approaches to ensuring that farmers and local communities are rewarded for their innovations have been developed by PhytoTrade Africa (Lombard and Leakey, 2010). It has been engaged in addressing the sustainable commercialization of natural products produced by indigenous plants, especially the trees of the Miombo woodlands. PhytoTrade Africa believes that the risks of exploitation can be minimized if the primary producers are able to secure long-term access to the markets developed for their products. Thus they have worked to ensure that markets can be secured so that supply chains can emerge, so that as wild harvesting leads to domesticated or farmed sources the initial producers are protected as much as possible. The approach involves working with indigenous communities and helping them to secure long-term access to local and even international markets in ways which reward them and protect their intellectual property rights (IPR). Experience to date indicates that these approaches can result in critically important supplementary income of otherwise poor and marginalized farmers and producers. This, in turn, significantly improves their livelihoods.

This partnership approach between producers and the local-to-global cosmetic, food, beverage, herbal medicine, and pharmaceutical industries is based on four areas of intervention aimed at the propoor commercialization of the traditionally important products:

- Product development
- Market development
- Supply chain development
- Institutional development

Such partnerships are developed by carefully constructing commercial agreements with leaders in the relevant sector. Critically this involves the establishment of strong and viable trade associations that are forward thinking and market oriented. Through these partnerships it is possible to ensure long term relationships and supply agreements which ensure that the farmers and local community producers remain in the value chain. The achievement of this can, under certain circumstances, also create a barrier to entry for plantation developers who might wish to out-compete small-scale producers and farmers.

One important consideration in developing these partnerships is the selection of appropriate species. It is critical that the abundance of the resource, and the ownership over the resource by the target producers, is sufficient to ensure sustainable and reliable supply of the products. To date, small-scale producers have depended on wild harvesting to supply the local markets. While this provides a supply-related barrier to large scale trade that favor small-scale producers, it also imposes constraints to the market expansion that is needed to raise these producers out of poverty. For example, wild products are highly variable in quality, often with unreliable levels of production. However, the focus on naturally occurring wild resources is now changing with the recent emergence of highly compatible propoor participatory domestication technologies for indigenous fruit and nut trees. This offers great opportunities to improve product quality through tree selection and cultivar development. Creating these new crops should also greatly increase the supply of very marketable produce. Thus by realizing the importance of "commercialization for domestication" and "domestication for commercialization" there is considerable opportunity for agroforestry to alleviate poverty, malnutrition and hunger in marginalized agricultural communities of developing countries. Thus the symbiotic relationship between domestication and commercialization is a very important aspect of the strategy needed to ensure real impact on the big socioeconomic issues affecting the world. The overall outcome of such developments arising from improved agricultural production and enhanced livelihoods should therefore be improved access to clean water; better diets, health care, and education, etc. Just as we saw above regarding adaptability to different environmental situations, this model case has great adaptability to different socioeconomic situations (Leakey, 2012b).

Thus, in conclusion, when the social and commercial components of the strategy that have been developed for participatory domestication strategy are added to the agroecosystem benefits from the adoption of agroforestry for land rehabilitation we can see the emergence of a three-step generic model for the delivery of a socially and economically desirable multifunctional agriculture by agroforestry (Fig. 28.2, Table 28.1).

Before moving on, however, it is important to recognize that the trade in some new products will require regulatory approval for European Union (EU) and American (USA) markets. It is however possible to tie the approval of these products to the target producers, which has been achieved by PhytoTrade Africa's successful application to have baobab fruit approved as a novel food ingredient in the EU under Regulation (EC) 258/97 and by Unilever Deutschland GmbH for *Allanblackia* seed oil (European Food Security Authority No EFSA-Q-2007-059).

TABLE 28.1 The 3-step model for multifunctional agriculture.

- Step 1: Adopt agroforestry technologies such as two-year "Improved fallows or Relay cropping" with nitrogen-fixing shrubs that improve maize yields from approximately 1 ton ha^{-1} up to approximately 4–5 tons ha^{-1}. This allows the farmers to both improve food security and reduce the area of their holdings planted with maize and thus make space for other crops, perhaps cash crops which would generate income. An additional benefit arising from improved fallows with leguminous shrubs is the reduction of parasitic weeds like *Striga hermontica* and the reduced incidence of insects pests like the stem borers of maize.
- Step 2: Diversify farming system by the inclusion of species producing marketable products or fodder for livestock. The adoption of participatory approaches to the domestication of traditionally important indigenous food plants is a good way to rapidly create new cash crops that generate income, improve nutritional security through diversified diets, enhance gender equity, provide diversified diets rich in micronutrients, empower communities toward self-sufficiency in products of day-to-day domestic importance, and maintain culture and traditions. The sale of these products would allow the purchase of fertilizers and thus, potentially, the increase of maize yields up to 10 tons ha^{-1}. Consequently, the area under maize could be reduced further to allow more cash cropping. The integration of fodder trees and livestock into a farm is one of the elements of diversification that could be part of this step.
- Step 3: Promote entrepreneurship and develop value-adding and processing technologies for the new tree crop products, thus increasing availability of the products throughout the year, expanding trade, and creating employment opportunities. All of these are outputs which should help to reduce the incidence of poverty and enhance gender equity.

Modified from Leakey, R.R.B., 2010. Agroforestry: a delivery mechanism for Multi-functional Agriculture. In: Kellimore, L.R., (Ed.), Handbook on Agroforestry: Management Practices and Environmental Impact. Nova Science Publishers. Environmental Science, Engineering and Technology Series, New York, USA, pp. 461–471, Leakey (2013).

TECHNIQUES

Biological Components of the Techniques

Genetic Selection

The domestication process is basically about the cultivation of plants with superior genetic traits (Gepts, 2014). In trees, two techniques can be used to assist in the identification of superior mature trees (sometimes called elite or plus trees) producing timber and wood, or indigenous fruits and nuts or otherwise useful everyday domestic products. In forestry for timber and wood production, genetic improvement is typically done by selection of a population with inherent genetic quality (a provenance) and by making seed collections from that population. Alternatively, when more is known about the genetic quality of individuals within a population, seeds (a progeny) are typically collected from the best mother-tree(s). The same approach is used in agroforestry when cultivating trees for soil fertility enrichment or for wood products such as poles or wood fuel (i.e., environmental services or low-value products). However, in agroforestry, trees are also cultivated for high-value useful or marketable products like fruits, nuts, medicines or other chemical ingredients. In this case domestication is often done using vegetative propagation to develop cultivars or clones (Leakey, 2014c), and participatory approaches involving local communities are appropriate.

In the participatory domestication of trees producing indigenous foods, two approaches to genetic selection can be adopted. The first is to seek the knowledge of local people about which trees produce the best products (see subsection titled "Social components of the techniques" following as to how to ensure that their traditional rights are recognized). The second, more scientific, technique involves the quantitative characterization of many traits of fruits, kernels, and leaves, which are associated with size, flavor, nutritional value, etc. This characterization also determines the extent of the tree-to-tree variation. This more scientific approach can be used to "back-stop" or enhance the indigenous knowledge approach by adding new knowledge about marketable or commercially important traits to enrich the indigenous knowledge. Over the last two decades much scientific knowledge has been acquired to support the participatory domestication initiatives being implemented around the world (see review by Leakey et al., 2012; Leakey, 2012b).

In the scientific approach to selection, modern laboratory techniques are being increasingly used to examine traits that are not visible to the naked eye: for example, to quantify genetic variation in the chemical and physical composition of marketable products such as polysaccharide food-thickening agents, nutritional content (protein, carbohydrate, oils, fiber, vitamins and minerals, etc.) by proximate analysis, medicinal factors like antiinflammatory properties, the composition of essential oils and fatty acids, the determination of wood density, strength, shrinkage, color, calorific value and other important wood properties correlated with tree growth (Leakey et al., 2012). In addition, molecular characterization of the genetic code is being used to determine the structure of genetic variation in natural, managed and cultivated tree stands, and to devise appropriate management strategies that benefit users (Jamnadass et al., 2009).

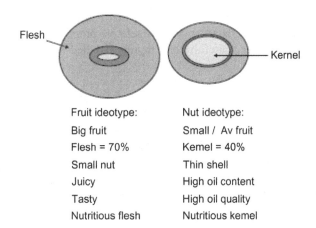

Fruit ideotype: Nut ideotype:

Big fruit   Small / Av fruit

Flesh = 70%   Kernel = 40%

Small nut   Thin shell

Juicy   High oil content

Tasty   High oil quality

Nutritious flesh   Nutritious kernel

FIGURE 28.3   Some traits of different trees which can be combined to form the ideal tree for fresh fruit of kernel production—an ideotype.

This information is used to determine the proportion of a species genetic variation that is available at a local geographic scale, whether or not cultivated stands are of local or introduced origin, and to ensure that domesticated populations have sufficient genetic diversity to avoid future problems from inbreeding (Pauku et al., 2010).

These scientific inputs to the understanding of genetic variation can then inform the process of farmer selection and help to provide guidance as to how best to meet the needs of different market opportunities. To achieve this, the concept of ideotypes developed for the breeding of cereals has been modified to assist tree selection (Leakey and Page, 2006). An ideotype is the ideal combination of traits for a cultivar to produce a product that meets the needs of a particular market (Fig. 28.3). So, for example, an ideotype for a fresh fruit would have a lot of flesh (and small seeds/nuts/kernels), be sweet, juicy, tasty, nutritious, and look attractive. On the other hand, a nut ideotype would have a large kernel(s) (and probably little flesh), have a thin shell so that it is easily cracked, be rich in edible oil with an appropriate fatty acid profile, or have other characteristics meeting the needs of the cosmetic or pharmaceutical industries. In both instances, these quality traits are ideally associated with a high yield of fruits or nuts, so that the cultivar can be said to have a high "harvest index"—a large amount of "ideal" harvestable product.

An obvious trait for inclusion in the selection of all cultivars for food products is their nutritional quality. While we know that the tree-to-tree variation is substantial in these traits, and indeed in antinutritional factors, such as phenolic content, we have a lack of knowledge on the extent of this variation in most species. Instead most researchers have published the average values per species. This allows some comparison between species, but is not useful when seeking to select cultivars from within a species. Likewise, we have very little information about tree-to-tree variation in taste, although again we know it is complex and substantial (Kengni et al., 2001).

Vegetative Propagation and Stockplant Management

The vegetative propagation of elite trees is central to the domestication strategy for indigenous trees producing valuable and marketable products (Leakey and Akinnifesi, 2008). A key element of this strategy is that the domestication is done by local people in their communities so that they are the beneficiaries of their work. This means that the vegetative propagation techniques have to be appropriate for implementation in remote villages by people with little if any formal education. For this reason, the techniques must be simple, robust and must not be dependent on an electricity supply or running water. The achievement of this has been described by Leakey (2004, 2014c). One aspect of this is the way in which stockplants are managed. Thus, the long-term success of a cloning program is heavily dependent on the nursery management and the skills of the nursery staff.

Nursery Management

Good nursery management involves many skills and attention to detail. The process starts with acquisition of high-quality germplasm of known origin (accession records) and recorded genetic quality, supported by documents recognizing the rights of the supplier (Access and Benefit Sharing agreements)—see for example the Agroforestry Tree Genetic Resources Strategy of the World Agroforestry Centre (ICRAF, 2012). Furthermore, this germplasm must have been appropriately stored and handled to avoid loss of viability. Germplasm can be of two types: seeds and vegetative propagules. Seeds also are of two main types: those that can be stored by drying or freezing (orthodox), and those that cannot (recalcitrant). Both the latter and the vegetative propagules are short-lived (from a few days to at most a few weeks). They have to be very carefully and rapidly handled to avoid water or temperature stress and any physical damage.

Accession records must be stored safely and the origin of every plant in the nursery must be labeled or in some other way traceable back to these records. Plants of unknown origin are of zero importance in a domestication program.

Successful plant production is dependent on rapid germination of seeds and propagation of vegetative propagules and their subsequent growth in good, free-draining potting compost well supplied with nutrients and organic matter. During early growth some shade is usually desirable to prevent wilting and water stress. Usually daily watering is required during dry weather. Once growing in the nursery, potted plants must be regularly maintained so that they do not become root-bound and so that the roots do not escape from the pot into the soil below (if this happens it is very difficult to subsequently establish this plant, as its roots will be left in the nursery). Plants should not be kept in the nursery so long that their root systems become coiled. Attention should be paid to weeds and pests.

Plants that are poorly managed in the nursery usually struggle to perform well once planted out in the field. The standards of nursery management are clearly visible to a visitor and are an excellent indicator of the interest and enthusiasm of the nursery staff and manager. Consequently, they are also a good indicator of the likely success of the subsequent planting program.

Predictive Test for Domestication of Timber Trees by Vegetative Propagation

One of the difficulties facing anyone involved in tree domestication is their longevity and long regeneration cycle. Many trees do not reach sexual maturity for 10–20 years and especially in the case of timber they do not display their yield potential and commercially important qualities until they are large trees. This makes tree breeding a slow business, but it also means that clonal forestry is difficult as juvenile trees that have not had time to display their potential are the easiest to propagate by cuttings (Leakey, 2014c). To get around this problem, large trees can be felled and allowed to develop coppice shoots, although this too is less likely from very old trees.

In *T. scleroxylon*, a West African timber tree conforming to Rauh's Model of branching architecture, a technique to predict which seedlings in a given population were likely to grow to form the trees with the best form (branching frequency) and yield has been developed (Leakey and Ladipo, 1987; Ladipo et al., 1991a). This is based on the fact that the trees developing the greatest stem volume had been found to be those producing a few branches in each branch whorl and no branches between whorls—so allocating biomass to stem rather than to branches. The technique is based on an assessment of the "strength" of apical dominance, the process regulating the formation of branches, by decapitating young seedlings and following the pattern of sprouting and the reassertion of dominance by the upper shoot (Ladipo et al., 1992). This nursery-based screening technique in seedlings on 3–6 months old allows the rapid selection of those likely to produce the best stems to enter a program of clonal propagation. Conversely, since the leaves of this species are eaten as a vegetable by local people in Nigeria, the trees producing the most branches can be selected and managed as a hedge for leaf production.

Nursery Management Strategy

Farmers who manage tree nurseries can adapt their management to best meet their needs. For example, if their priority is to meet their own domestic food demands, they can plant the majority of the trees they produce in their own farm rather than selling them or multiplying them up by harvesting successive batches of cuttings or scions. Alternatively, if income generation through the development of a tree nursery business for the sale of plants to other farmers is the priority, they can take three to four crops of cuttings off every plant per year for a few years and so very rapidly build up a large stock of plants for sale. This strategy should result in the greatest income generation in the long term, but has the disadvantage that there are no financial returns in the first few years. An even longer-term commercial strategy would be to do a long period of plant multiplication prior to large-scale planting. In addition to these individual strategies there are, of course, many possibilities of intermediate or mixed options.

Environmental Components of the Techniques

As a low-input farming system based on long-lived perennial trees, agroforestry provides many environmental benefits (Nair, 2014; Garbach et al., 2014; Ong et al., 2014; Lavelle et al., 2014; Sileshi et al., 2014). Tree domestication is normally an approach to make these farming systems more productive and more profitable—in other words to intensify the agroforestry system. Many aspects of this will also have indirect environmental benefits in more rapid cover of the soil, greater root penetration, and lateral spread with consequently better protection and improvement of the soil and agroecosystem function. Rapid tree growth is associated with high water use. While this may have negative impacts on the growth of associated crops, it can also reduce the impacts of salinization by lowering the water table (Ong et al., 2014).

Recent data in *D. edulis* shows that vegetatively propagated cultivars have lower fine root density in the crop rooting zone and hence are likely to be less competitive with crops (Asaah et al., 2012). In addition, the perennial nature of trees also assists in carbon sequestration for the mitigation of climate change, and emerging evidence suggests that vegetatively propagated cultivars store more carbon than seed propagated individuals in their primary roots and shoots—an unexpected benefit of tree domestication (Asaah, 2012).

One important aspect of these environmental impacts is their quantification. This requires baseline surveys based on randomized controlled trials, described by Barrett et al. (2010), as providing solutions to development economics issues with weak causal factors. If applied to agroforestry this approach should help to determine the scale of the impact at the plot, farm, landscape, and regional level.

Social Components of the Techniques

To maximize the social and economic benefits flowing from tree domestication, a participatory approach has been taken by the World Agroforestry Centre and some other organizations and their local partners to identify the farmers' preferences in terms of the tree species they would like to cultivate. Interestingly, in most of the places around the world where this has been done, the farmers have identified indigenous fruits, nuts and leaves for uses as foods or medicines (Franzel et al., 2008). As mentioned previously, under social components of the strategy, participatory domestication has typically followed the identification of priority species, with farmers contributing their knowledge about the traits they would like to see incorporated within a selection program, and the usefulness of the tree products to local households.

To encourage and assist the farmers the agroforestry scientists and NGO partners in Cameroon have provided training and knowledge through specially constituted Rural Resource Centres in pilot villages in different regions of the country (Tchoundjeu et al., 1998, 2006, 2010a; Asaah et al., 2011). These Rural Resource Centres provide opportunities for hands-on training in critical nursery management, vegetative propagation, agroforestry practices, and enterprise development in a community nursery (Fig. 28.4). The focus of this training is that it is appropriate for implementation in often remote village nurseries without piped water and electricity. The farmers are then encouraged to set up similar facilities in their own farm or community, so creating satellite nurseries which help to spread the techniques and concepts to neighboring communities. To encourage this kind of farmer-to-farmer technology transfer, exchange visits, demonstrations, and competitions, covered by radio and TV, are organized. In time these village and community nurseries become self-sufficient and independent.

In recent years the range of topics presented in training programs has been expanded to include the wise use of microfinance and financial management, product marketing, business development, infrastructure development and community organization. In parallel to this, artisans in neighboring towns have been trained in the fabrication of simple equipment and tools for product drying, processing and packaging, while entrepreneurs and women's groups have been helped to develop businesses in food product processing and value-adding aimed at improving the quality of products for the marketplace and increasing market demand. As these processes expand, it is expected that some people will cease to be farmers and producers and instead enter the cash economy as business men and women providing employment to others and developing new enterprises.

While much progress has been made in developing, implementing, and expanding the participatory domestication of a wide range of agroforestry trees and commercializing their products, there is still much to do to quantify the socioeconomic impacts. A critical component of this is the establishment of statistically viable baseline surveys so that accurate assessments can be made in years to come as the concepts are disseminated and scaled up to new communities. This process has begun, but to date the results are preliminary. Likewise, it is still too early to say much about the implementation of work to recognize and protect the rights of farmers and rural entrepreneurs to their innovations.

Commercial Components of the Techniques

Ultimately, as we can see from Fig. 28.1 and Table 28.1, it is the commercialization of the sustainably grown products that potentially delivers the really important impacts from agroforestry and multifunctional agriculture. However, we have

FIGURE 28.4   A village satellite nursery in Bangoua, Cameroon.

also recognized that it is commercialization that can pose the greatest risks affecting the success or failure of the overall initiative. One study (Wynberg et al., 2003) has examined the "Winner or Loser" qualities of different approaches to the commercialization of an indigenous fruit (*Sclerocarya birrea*, Marula) in southern Africa (Table 13.1). It basically found that bottom-up community initiatives had the greatest chance of being Winners, although top-down commercialization involving large multinational companies could also be Winners, if the company recognized the importance of buying raw products from local smallholder producers, rather than from large-scale plantation growers.

Focusing first on how tree domestication can encourage the positive aspects of commercialization from producers practicing agroforestry, the expansion of markets beyond the traditional market or roadside stall (Fig. 28.5) requires increased product quality, uniformity and regular supply. The longer the value chain from local to global, the more important are the attributes of quality, uniformity and regularity of supply (Fig. 15.2). Hence, the development and cultivation of vegetatively propagated cultivars selected for year-round production and other commercially desirable traits makes a quantum leap in the marketability of the products, as it means that traders and wholesalers can purchase a large volume of uniform, high-quality product from a recognized and named cultivar. In return, hopefully the producer will receive a higher price, as it is clear that consumers are willing to pay more for the more desirable varieties (Figs. 28.6 and 28.7).

This demand for uniformity and quality makes the ideotype concept mentioned earlier more important—especially at the upper end of the value chain where processing and value addition increase the value of the marketable products (Fig. 28.8). With the increasing importance of the ideotype the identification of the specific traits that confer market acceptability and market exclusivity and distinctiveness become more and more critical. This, therefore, is the reason for increasingly sophisticated research, mentioned in the earlier subsection titled "Biological components of the technique," to determine the genetic variation found in different tree populations in the chemical, physical, and medicinal properties of the raw products. This of course will lead to the need for stronger linkages between agroforestry researchers and partners in industry (Leakey, 1999a).

Food crop domestication over thousands of years has been credited with the advance of civilization as witnessed in industrialized countries (Diamond, 1997). However, domestication and commercialization are both part of this force for development, as one is considerably weakened without the other. Leakey (2012b, 2012e) has recognized that in the tropics there is a need for a "new wave of domestication" but this too has to be supported by market growth along the value chain. Thus one of the important components in this new wave is the processing and value addition that extends the

FIGURE 28.5 A market stall in Makenene in Cameroon, selling *Dacryodes edulis* fruits.

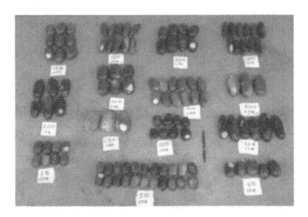

FIGURE 28.6 Fruits of *Dacryodes edulis* from market stalls showing the retail price in Central African francs (CFA) and illustrating consumer preference fruits from certain trees. *Reproduced from Leakey R.R.B., 2012b. Living with the Trees of Life — Towards the Transformation of tropical Agriculture, CABI, Wallingford, UK, 200pp.*

shelf-life of products, expands market demand and makes them more valuable. In Cameroon, the development of cottage industries to dry and package AFTPs has started (Asaah et al., 2011), but this enterprise development and its upscaling needs considerable expansion—recognizing the characteristics of Winners in the development process (Table 28.1). In this regard, one interesting development in recent years has been the involvement of a few multinational companies in public—private partnerships with rural communities engaged in production of agroforestry products in tropical countries (Jamnadass et al., 2010; Leakey, 2012b). Although associated with risks, this also offers great opportunities for the future development of agroforestry tree crops if the strategies and practices can be developed appropriately.

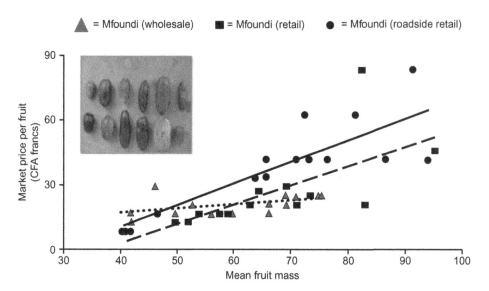

FIGURE 28.7 Relationship between mean *Dacryodes edulis* fruit mass and market price per fruit ($1 = 750CFA francs). *Modified from Leakey, R.R.B., Atangana, A.R., Kengni, E., Waruhiu, A.N., Usuro, C., Anegbeh, P.O., et al., 2002. Domestication of Dacryodes edulis in West and Central Africa: characterisation of genetic variation. Forests Trees Livelihoods, 12, 57–71.*

FIGURE 28.8 Value-added products developed from agroforestry tree species. *Reproduced from Leakey R.R.B., 2012b. Living with the Trees of Life – Towards the Transformation of tropical Agriculture, CABI, Wallingford, UK, 200pp.*

OUTCOMES

To date studies on the environmental, socioeconomic impacts of tree domestication have not been based on randomized controlled trials. Nevertheless, comparisons based on household surveys have indicated that integrated rural development based on agroforestry, tree domestication and local market initiatives with small-scale enterprises having positive impacts on the lives of farming households in participating communities (Table 13.1). Together these impacts illustrate that agroforestry can deliver what the International Assessment of Agricultural Knowledge, Science and Technology for Development (IAASTD) called multifunctional agriculture (Table 28.2). Currently the impacts are only on a local/household/farm/village scale and there is much to be done to upscale and outscale the agroforestry initiatives before the scales are extended (Table 28.3).

Biological Components of the Outcomes

The major outcome of participatory tree domestication, as expected, is new tree crop cultivars with improved product quality and greater market demand. This is happening in many places around the tropics in over 50 species, but is most advanced in Cameroon, where these cultivars have been developed by local farmers for cultivation in their own farms to meet household needs. However, as the supply of product increases it is expected that both these cultivars and their products will be sold—at first locally and then more widely (Tchoundjeu et al., 2010a; Asaah et al., 2011; Leakey, 2012b). The cultivation of these new cash crops is leading to the diversification of farming systems and this is expected to result in healthier agroecosystems. Another biological finding with likely impacts on agroecosystems is that vegetatively propagated trees allocate their dry matter differently, with less in fine root and more in primary roots and shoots. Consequently, clonal cultivars are likely to be less competitive with annual crops (Asaah, 2012; Asaah et al., 2010, 2012).

Environmental Components of the Outcomes

One important outcome expected from more sustainable land-use systems that rehabilitate degraded land is better access to productive land for staple food crop production and hence the opportunity to diversify into other crops with market potential. This diversification either as mixed cropping or as land-use mosaics should improve the agroecosystem functions and so lower the need for inputs such as pesticides. The increased tree cover, especially if arranged along the contours of hillsides, also protects the soil and reduces the risks of serious erosion and can protect watersheds. The hydrological impacts are most fragile in dryland environments (Ong et al., 2014).

The perennial nature of trees also assists in carbon sequestration for the mitigation of climate change, and emerging evidence suggests that vegetatively propagated cultivars store more carbon in their primary roots and shoots than seed-propagated individuals—an unexpected benefit of tree domestication (Asaah, 2012).

Social Components of the Outcomes

One of the original purposes of initiating participatory tree domestication was to improve the livelihoods of poor farmers, especially income generation to reduce poverty; food security to reduce hunger; better nutrition and diet to reduce malnutrition; and better equity to improve the lot of women and children in society. It was envisioned that, indirectly, this would also improve health and education opportunities, and that the overall package would empower individuals and communities allowing them to be more self-sufficient and so to transform their lives, giving them hope for the future. Early indications (Table 13.1) are that these outcomes are starting to emerge in the participating communities. Furthermore, when associated with microfinance, business training and access to simple equipment for the processing and packaging of raw products, there is now evidence of rural people engaging in small businesses either as entrepreneurs or employees.

Significantly, one of the outcomes mentioned by young people in the participating communities is that this now means that they can see a future for themselves if they remain in the village rather than feeling that they have to migrate to towns and cities for a better life (Table 13.1).

Commercial Components of the Outcomes

The important commercial outcomes have been the development of Rural Resource Centres that deliver both education and training in agroforestry and tree domestication, as well as in community development and business management so that the villagers can earn money from the sale of plants and raw products. This money is then being used to make infrastructure developments, such as roads and clean water supplies, as well as to reduce the drudgery of the women (Table 13.1). The consequence of this has been that poor farmers are starting to generate income and enter the cash economy and so begin the climb out of poverty (Fig. 28.9). In the bigger villages and small towns, local people are also

TABLE 28.2 The impacts reported by farmers engaged in a participatory domestication program for agroforestry trees in Cameroon.

Positive impacts

Increased number of farmers adopting agroforestry and the domestication of indigenous trees

Increased production of tree products

Increased income from tree sales by nurseries

Increased income from sale of tree products

Increased income from better farming practices

Increased income from eligibility for microfinance

Increased income used for schooling and school uniforms

Increased income used for medicines and healthcare

Increased income used for home improvements—for example, installation of water and electricity in the home, new buildings.

Increased income used for farm improvement—for example, installation of water and electricity, livestock, wells, agricultural inputs.

New employment opportunities from nurseries

New employment opportunities from processing both agricultural crops (such as cassava) and new markets for processed agroforestry products (fruits, spices, herbs, and medicinal)

New employment opportunities in the emerging workshops producing small tools and appropriate mechanized equipment to service the need for food processing equipment

New employment opportunities from marketing as traders of processed products and the food processing equipment

New employment opportunities in transport from producers to markets and to the processors of agricultural produce

Retention of youths in the villages due to career opportunities by domesticating trees in their village nurseries

Tree domestication has led to better diets and improved nutrition

Luxury food items consumed

Improved health from potable water

Piped water supplied for irrigation and use in nurseries

Increased livestock rearing due to tree fodder

Increased use of traditional medicines and better health

Increased honey production and processing

Reduced drudgery in women's lives from not having to collect water from rivers and farm produce from remote farms, as well as from mechanical processing of food crops

Reduced drudgery in gives more time to look after their families and engage in farming or other income-generating activities

Improved marketing for food and agroforestry products

Improved soil fertility from improved fallows has increased crop yields 2- to 3-fold with better weed control

Improved tree fodder for goats and cattle

Having more time as a result of better farming methods, farmers had more time for marketing and new farming activities

Community feeling empowered, stronger, and optimistic for the future in ways that they could sustain

Knowledge has empowered the Rural Resource Centers as an agent of change

Negative impacts

Increased jealousy and theft

New roads lead to deforestation and land degradation as a result of the expansion of farming activities to more remote areas

Reproduced from Leakey R.R.B., 2012b. Living with the Trees of Life – Towards the Transformation of tropical Agriculture, CABI, Wallingford, UK, 200 pp.

TABLE 28.3 The characteristics of the participatory domestication of agroforestry trees important for the delivery of multifunctional agriculture.

The features that agroforestry brings to multifunctional agriculture to enhance social, economic, and environmental resilience in agriculture and rural development are:

- Based on TK and culture
- Based on participatory techniques to ensure relevance to local people
- Based on integrated natural resources management and sustainable land use
- Based on knowledge of the natural resource

These features mean that multifunctional agriculture can:

- Empower subsistence farmers to control their destiny
- Enhance food security and rural/urban livelihoods, reducing hunger
- Enhance nutrition security and health, reducing malnutrition and diseases
- Enhance opportunity for income generation, reducing poverty
- Diversify farming system at the local and landscape scale, enhancing watershed services and sustainable production
- Create new agricultural commodities
- Diversify market economy and buffer commodity price fluctuations
- Decentralize business opportunities to the villagers
- Create employment in processing and marketing
- Build social responsibility from the "grassroots"
- Enhance international public goods and services, reducing climate change and loss of biodiversity
- Offer opportunities for new policy interventions to combat deforestation, desertification, and land degradation
- Breakdown the disconnects between disciplines and organizations responsible for policy and its implementation in rural development

Reproduced from Leakey R.R.B., 2012b. Living with the Trees of Life – Towards the Transformation of tropical Agriculture, CABI, Wallingford, UK, 200 pp.

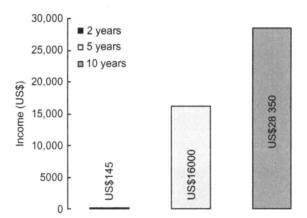

FIGURE 28.9   Income generated from plant sales at Rural Resource Centres in Cameroon after 2, 5, and 10 years. *Reproduced from Leakey, R.R.B., Asaah, E.K., 2013. Underutilised species as the backbone of multifunctional agriculture – the next wave of crop domestication. Acta. Hortic. 979, 293–310.*

developing cottage industries and engaging more in marketing and trade. This relationship between enhanced farm production and urban life is important for the rural economy and the overall alleviation of poverty. It is an example of farm production being the "engine of growth."

FUTURE IMPACT

In the long term the application of these strategies should lead to benefits, which encompass many of the rural development goals of development agencies, as specified in the Millennium Development Goals of the United Nations. At the community level some of these goals have been achieved, but to date we are still only talking about several thousand participating farmers. Achieving these benefits will, however, require the large-scale adoption of the techniques and strategies presented here in ways that will meet the needs of the farmers and those of new and emerging markets (Leakey et al., 2012). This makes it important to get the messages about good domestication strategies to policy makers (e.g., Wynberg et al., 2003).

Clearly, the challenge is to scale up the application of agroforestry and tree domestication to the huge numbers needed to have meaningful impact of national, regional, and global scales.

Biological Components of the Impact

The vision of what could be achieved with the widespread adoption of the domestication of trees and the associated trade of the products in numerous different industries is perhaps best exemplified by the domestication of the wolf to give us large numbers of breeds of dogs of all sorts of shapes, sizes, and appearances, to be both pets and working partners with farmers, policemen, entertainers, the blind, etc. These dog breeds have arisen by breeders teasing out the genetic traits present in the wolf genome. If we did the same for trees, we could have a large number of different cultivars meeting the needs of many food, cosmetic, medicinal, perfume, fiber, and wood markets—not just in one species, but in hundreds of species. This in turn would lead to highly diversified, intensified and more productive multifunctional agriculture (Leakey, 2012b).

Integrated Environmental, Social, and Economic Components of the Impact

The restoration of degraded land by multifunctional agriculture could make productive land available for an expansion of overall farm production to fill the Yield Gap (Beddow et al., 2014; Leakey, 2012b) and meet the needs of a growing human population without the need for further deforestation, and without future negative impacts on erosion, flooding, climate change, etc. A healthier environment could lower the risks of disasters due to flooding, drought, landslides, pest, disease epidemics, etc.

The achievement of even greater success than foreseen by the Millennium Development Goals could lead to greatly improved local and national economies and the gradual transition of developing countries toward developed countries.

Section 6.2

Intellectual Property Rights

Chapter 29

Protecting the Rights of Farmers and Communities While Securing Long Term Market Access for Producers of Non-timber Forest Products: Experience in Southern Africa

This chapter was previously published in Lombard, C., Leakey, R.R.B., 2010. Forests, Trees and Livelihoods, 19, 235–249, with permission of Taylor & Francis

SUMMARY

The participatory domestication of agroforestry trees as an incentive to alleviate poverty, malnutrition, hunger and land degradation has to be linked to the commercialization of the products in ways that ensure that the farmers are the beneficiaries of their germplasm improvement activities, as well as from the marketing of the products. Currently, international law is deficient in providing adequate protection of the rights of poor farmers and their communities, as the legal instruments of Intellectual Property legislation are focused more on the protection of commercial companies and entrepreneurs. PhytoTrade Africa is engaged in addressing the sustainable use and commercialization of natural products produced by indigenous plants, especially trees of the Miombo woodlands in southern Africa. Initially the market focus has been on wild-harvested naturally occurring resources using innovative approaches to protecting the Intellectual Property Rights of poor communities and the businesses they work with, including Patents, Trademarks, and Geographical Indicators, with the intention of securing long-term strategic market access and to be able to influence commercial strategy. The approach which has been developed is to work with indigenous communities and local companies and to help them to secure long-term access to these markets through the protection of their intellectual property rights. Experience to date indicates that, by enabling market opportunities for these local resources, significant livelihood options for otherwise marginalized farmers and producers can be facilitated. Partnerships between producers and the local-to-global cosmetic, food, beverage, herbal medicine and pharmaceutical industries are developed by carefully constructing commercial agreements with leaders in the relevant sector. Critically this involves the establishment of a strong and viable trade association that is forward thinking and market oriented. Through these partnerships it is possible to ensure long-term relationships and supply agreements. Such agreements ensure that the target producers remain in the value chain. This paper also explores opportunities for protecting farmer-improved germplasm through the registration of Plant Breeders Rights in compliance with the International Union for the Protection of New Varieties of Plants (UPOV) and proposals for an affiliated African Intellectual Property Organization (OAPI).

INTRODUCTION

Throughout the world, hunter-gatherers have used tree products such as fruits and nuts for food; wood, timber, and bark to make utensils and homes; leaves for fodder; and all these different components for medicines. Since the advent of agriculture, these natural common-property resources (non-wood forest products—NWFPs) have become progressively

Multifunctional Agriculture. DOI: http://dx.doi.org/10.1016/B978-0-12-805356-0.00029-5
© 2017 Elsevier Inc. All rights reserved.

317

more scarce as forests and woodlands have been cleared for farming. In recent decades agriculture has focused on livestock and a small number of highly domesticated crops, to the detriment of the resource of these useful wild species. Thus poor people have less access to these resources. In addition, increasing pressures on rural livelihoods has led to some local and even regional trading of wood and NWFPs for the generation of small amounts of cash income. Thus marketing and trade are important elements of poor people's livelihood strategies, although they have little access to formal markets. However, the development of commercial opportunities for new marketable products from traditionally and culturally important indigenous species creates tensions between entrepreneurs and local communities that need to be addressed if long-term livelihood benefits are to be achieved.

In recent years two initiatives have tried to help poor people to make more money from these natural resources. Firstly there has been some growth in local markets for both cultivated crops and natural products. However, poor people are disadvantaged in formal markets, especially when NWFPs have potential markets in local, regional, or even international commerce (Welford and Le Breton, 2008). In this situation, PhytoTrade Africa is now playing an important role in helping local communities to formulate and implement partnership agreements with entrepreneurs and industries, so that their capacity to improve their livelihoods is enhanced (Welford and Le Breton, 2008). Secondly, there is a global initiative to involve local communities in the domestication of the species producing NWFPs and thereby to develop them as new crops (Leakey et al., 2005a, 2006). The products of these new cultivated crops have been called AFTPs to distinguish them from the wild resource of NWFPs (Simons and Leakey, 2004). These domestication programs have been innovative and involved the communities directly in the genetic improvement so that they are the beneficiaries of their own activities (Leakey et al., 2003; Tchoundjeu et al., 2006). Exciting and very important benefits, generating income and other livelihood benefits, are now starting to flow from this approach, so reducing poverty and malnutrition (Tchoundjeu et al., 2010a).

The need for simultaneous approaches to the domestication of new crops and the commercialization of their products is well recognized in agroforestry (Leakey et al., 1996; Leakey and Izac, 1996) but all too often domestication is implemented without commercialization of the products, or commercialization is done without domestication. The participatory domestication of agroforestry trees as an incentive to alleviate poverty, malnutrition, hunger, and land degradation has to be linked to the commercialization of the products in ways that ensure that the farmers are the beneficiaries of their germplasm improvement activities, as well as from the marketing of the products (Leakey et al., 2003, 2005a). Currently, international law is deficient in providing adequate protection of the rights of poor farmers and their communities as the legal instruments of Intellectual Property legislation are focused more on the protection of commercial companies and entrepreneurs. Consequently, there is a need for innovative thinking about how to ensure that poor farmers developing their own cultivars of species-producing ATFPs are protected, firstly from unscrupulous entrepreneurs mass propagating and marketing their cultivars, and secondly ensuring that these farmers and those engaged in trading the AFTPs are ensured of their market niche (Leakey and Izac, 1996). Furthermore, there is also a vital need for those industries who are the buyers of AFTPs to collaborate with the agroforesters who are supporting the participatory domestication process, so that the genetic selection is improving the traits and characteristics that are of commercial importance (Leakey, 1999a).

Interestingly, the first such project, which has Unilever as the industrial partner, is in progress and concerns the domestication of *Allanblackia* species for their unique kernel oils (Attipoe et al., 2006; Jamnadass et al., 2010).

PHYTOTRADE AFRICA'S EXPERIENCE

PhytoTrade Africa is the Southern African Natural Products Trade Association. It was formed in 2001 and has some 60 members drawn from Botswana, Malawi, Mozambique, Namibia, South Africa, Swaziland, Zambia, and Zimbabwe. It is a registered Fair Trade Organisation, a Type II partnership under the World Summit on Sustainable Development, a founding member of the UNCTAD supported Union for Ethical Biotrade and has been nominated as a SADC Centre of Excellence in the field of Access and Benefit Sharing. In 2003, PhytoTrade Africa produced Bio-Prospecting Guidelines for its members to advise them and African producers how to respond to the unethical bioprospecting approaches of some entrepreneurs. The activities of PhytoTrade Africa always include the use of Material Transfer Agreements with the companies engaged in testing biological materials. This is done with the Prior Informed Consent of the community providing the material, recognizing that equitable benefit-sharing should result from any commercial developments.

PhytoTrade Africa has experience with wild harvested NWFPs that could greatly benefit the initiatives in participatory domestication, by helping to bring better market access and more secure marketing for the AFTPs. PhytoTrade Africa is engaged in addressing the sustainable use and commercialization of natural products produced by indigenous plants, especially the trees of the Miombo woodlands in southern Africa, where agriculture is constrained by low and unpredictable rain fall, poor soil quality, and poor infrastructure to support agricultural production. PhytoTrade Africa work with indigenous communities and local companies to help the communities to secure long-term market access through the protection of their IPR. Experience to date indicates that by enabling market opportunities for these local resources can facilitate significant livelihood options for otherwise marginalized farmers and producers.

The overall objective of this initiative is to promote both the sustainable use of indigenous resources and to reduce poverty by generating critically important supplementary income and other livelihood benefits for the poorest rural producers. In this endeavor PhytoTrade Africa is engaged in a partnership approach with industry based on four inter-linked areas of intervention aimed at the propoor commercialization of the traditionally important products derived from indigenous trees (Welford and Le Breton, 2008). These areas of intervention, which take considerable financial investment, are:

- Product development to transform a raw product to a marketable commercial product. This can involve the raw product, but it often means developing techniques to improve the shelf life of the product through drying, processing, bottling/canning, or freezing. These steps then involve packaging, labeling, and other forms of presenting the product to markets.
- Market development in areas outside the product's traditional importance to raise market recognition. In addition to locating potential buyers, this involves the need to comply with increasingly stringent production standards, such as food safety systems and Good Agricultural Practices (Woods, 2004), particularly if the products are destined for markets in industrial countries where international compliance codes is audited (Wheatley et al., 2004). Further, global markets increasingly seek to protect consumers and many AFTPs are subject to, for example, the Novel Foods Regulation in the EU. Market development for AFTP requires that these regulatory hurdles be addressed.
- Supply chain development to increase the capacity of primary producers to meet the demand for new products. This is an exercise to facilitate the marketing process from the smallest rural producer to the largest global corporation and at all levels in between, and can involve support to organic and fair trade certification. This also requires effort to ensure that the product meets the standards required for market access. Species domestication can assist this process by increasing the uniformity of the product through the replacement of highly diverse biological material with high-quality products from improved cultivars (Bunt and Leakey, 2008). Domestication can also increase the length of the productive season by the selection of germplasm that flowers and fruits outside the normal season, so allowing the trade and marketing to occur throughout the year.
- Institutional development refers to the development of the trade association as an entity that can act on behalf of the producers and member companies. A viable trade association is a prerequisite to producers being supported in their numerous challenges as they seek to enter global markets. Trade associations also lobby policymakers and regulators to try to ensure that appropriate policies and regulations are in place and that the playing field is not further stacked against small-scale producers. Partnerships between producers and the local-to-global cosmetic, food, beverage, herbal medicine and pharmaceutical industries are developed by carefully constructing commercial agreements with leaders in the relevant sector.

Critically this involves the establishment of a strong and viable trade association that is forward thinking and market oriented. Through these partnerships it is possible to ensure long-term relationships and supply agreements. Such agreements ensure that the target producers remain in the value chain.

PhytoTrade Africa's strategy is to focus initially on wild-harvested, naturally occurring resources. This takes advantage of the easy access that poor producers can have to wild resources. The approach also ensures that supply chains for industry can be more rapidly developed. This provides an advantage to the commercialization process by avoiding long selection and domestication processes. Critically, this strategy uses market forces and tools in support of poor producers. PhytoTrade has been engaged in some innovative approaches to protecting the IPR of poor communities and the business they work with, including Patents, Trade Marks, and Geographical Indicators, with the intention of securing long term strategic market access and to be able to influence commercial strategy. Once markets have been secured the strategy will, where necessary, evolve to establishing supply chains from domesticated or farmed sources in a way that keeps the initial producers in the value chain as far as possible.

Several important factors need to be considered in the selection of the species and the products that can be derived from them. It is critical that the abundance of the resource, and the ownership over the resource by the target producers, is sufficient to ensure sustainable and reliable supply. This approach can, under certain circumstances, also create a barrier to entry for plantation developers who can be seen to be competitors to small-scale producers and farmers. To date, as mentioned, the focus has been on naturally occurring wild resources, but the recent emergence of highly compatible propoor participatory domestication technologies for indigenous fruit and nut trees in developing countries offers considerable opportunity for the better integration of domestication and commercialization for the alleviation of poverty, malnutrition, and hunger in marginalized agricultural communities. Entering into any market will expose suppliers to competition, hence there are concerns that the rights of farmers or communities in the supply chain will be jeopardized in the commercialization process. Thus it is important that the supply chains, and the manufacture of the products, are efficient and competitive to shut out casual and opportunistic competitors. PhytoTrade Africa believes that this risk can be managed and the primary producers are able to secure long-term access to the markets developed for their products.

CASE STUDIES

Marula (*Sclerocarya birrea* subsp. *caffra*)

Marula oil is produced from the seed kernels of the marula tree (*Sclerocarya birrea* (A. Rich.) Hochst. subsp. *caffra* (Sonder)). It has a long history of traditional use encompassing food and cosmetic uses. Now years of innovative research by producers and institutions in Namibia has been combined with the "green chemistry" lipid science of Aldivia S.A. to bring to world markets the first in a new range of African active botanical ingredients Maruline®. This was developed under a groundbreaking partnership between PhytoTrade Africa and Aldivia S.A. of France. Maruline is 100% marula oil with enhanced antioxidant properties obtained through a patented process.

Marula nuts contain an oil with a fatty acid composition comparable to that of olive oil, but with much greater stability ($\times 10$) due to its tocopherol/sterol composition. The oil also has an amino acid profile that is similar to that for human milk and whole hens' eggs, although deficient in lysine (Leakey, 1999a). Consequently, the oil is considered to have great potential in the food industry where it could be used as a coating of dried fruit, as a frying oil or as a substitute for high-oleic safflower oil in baby foods. Marula oil contains nine fatty acids of which palmitic, stearic and arachidonic acids are the most dominant. Oil content (%) and oil yield (g/fruit) of marula kernels has been found to differ significantly between trees (44.7–72.3% and 8.0–53.0 g/fruit) offering great opportunities for cultivar development for oil production through tree selection (Leakey et al., 2005c).

Maruline could be the world's first active botanical ingredient developed through scientific collaboration between traditional resource users and a specialized international R&D company. Since 1996, a Namibian NGO, CRIAA SA-DC, with support from Namibian government institutions and other organizations, have been working with the Eudafano women to develop marula oil for use in the cosmetics sector. The production of Maruline is based exclusively on principles of fair trade and environmental sustainability. The patent is co-owned by The Southern African Natural Products Trade Association, who have in turn secured benefit sharing agreement between Aldivia, the Marula producers in Namibia (Eudafano Women's Cooperative in North Central Namibia) and other relevant institutions in Namibia. This co-ownership with Aldivia S.A. represents a unique partnership that sets new standards for the benefit sharing provisions contained in Article 8(j) of the Convention on Biological Diversity.

When consumers buy products containing Maruline, they can also be sure that they are making a meaningful contribution to the local livelihoods. By creating viable markets for marula in this way, local value is added, traditional culture is preserved, food security is enhanced, and marula trees will be conserved for generations to come.

Eudafano Women's Cooperative and CRIAA SA-DC are both members of PhytoTrade Africa, which has helped them to develop genuine partnerships with Aldivia S.A. This market has created a new income opportunity for local farmers, and especially for rural women. The Maruline story is a great example of how smart partnerships based on indigenous resources, modern science and collaborative commercial strategies can help to alleviate poverty while also increasing public interest and participation in sustainable use of Africa's biodiversity. The relationship between PhytoTrade Africa and Aldivia S.A, linking rural producers, industry and consumers, is a fundamental principle of the founders and shareholders of our company.

Baobab (*Adansonia digitata*)

The baobab tree, *Adansonia digitata* L., is indigenous to much of Africa including Malawi, South Africa, Botswana, Namibia, Mozambique, and Zimbabwe, but can now be found in many African and Asian countries. The fruits, which are typically wild harvested by local people, contain a soft whitish fruit pulp with a distinctive and pleasant flavor due to the presence of organic acids. Nutritional analysis has shown that baobab fruits are an excellent source of pectins, calcium, vitamin C, and iron. Vitamin C content of baobab fruit pulp is three times greater than that of oranges (Manfredini et al., 2002), while the leaves are rich in vitamin A. Sidibé et al. (1996) assessed the tree-to-tree variation in vitamin C content of fruits from three areas of Mali across a rainfall gradient of 450–850 mm and found threefold between individual trees, but this variation was not related to rainfall or tree type. This range of diversity has been confirmed in a genetic study, which found no patterns of genetic relatedness between trees with similar characteristics (Assogbadjo et al., 2009), suggesting that as with other AFTPs, such as *Dacryodes edulis* (G. Don.) H.J. Lam (Waruhiu et al., 2004; Anegbeh et al., 2005), *Irvingia gabonensis* (Aubry-Lecomte ex O'Rorke) Baill. (Anegbeh et al., 2003), *S. birrea* subsp. *caffra* (Leakey et al., 2005b,c), *Barringtonia procera* (Miers) R. Knuth (Pauku, 2005), *Canarium indicum* L. (Leakey et al., 2008), and *Santalum austrocaledonicum* Viell. (Page et al., 2010a), this variability is normal for wild populations.

Traditionally, Africans have eaten the fruits and made refreshing drinks from the juice (Bosch et al., 2004). The pulp in particular is consumed by pregnant women and young children for its calcium content. The pectins are important as dietary fiber (Wilkinson, 2006). The fruit pulp is also cooked and the dried pulp can be used as an alternative to cream of tartar in baking. The unique chemical properties of baobab fruit pulp have led to its sale in US dispensaries and as cosmetics and care products in Canada. In Italy the products have been endorsed by cyclists, a Formula One driver, and AC Milan football players (Wilkinson and Hall, 2007). Medicinally, the tree is important in traditional folklore for the treatment of dysentery and dehydration and is used to reduce fever (Watt and Breyer-Brandwijk, 1962).

In June 2006, PhytoTrade Africa filed for EU regulatory approval so that the use of dried baobab fruit pulp could be legally consumed within the EU. Two years later the EU approved the use of baobab dried fruit pulp as a food ingredient. The approval refers only to those producers who are members of PhytoTrade Africa, and who supply the market through PhytoTrade Africa's partner in South Africa, Afriplex. Other producers or companies who wish to have their baobab fruit legally used as a food ingredient in the EU either have to carry out their own full application, or they can apply on the basis of their product being "substantially equivalent" to PhytoTrade Africa's baobab fruit pulp. The approval in the EU has resulted in significant commercial interest in the product. This has been further increased by PhytoTrade Africa's recent successful notification to the Food and Drugs Administration to have baobab dried fruit pulp accepted as "Generally Recognized as Safe," with significant value adding to the baobab fruit taking place at Afriplex in South Africa. The partnership agreement between PhytoTrade Africa and Afriplex ensures benefit sharing arrangements between the producers and the value-adders in South Africa. Baobab fruit is being targeted at Europe's growing market for health foods, in products such as fruit/cereal bars and smoothie drinks. The fruit's high pectin content is also being used to produce chili sauces and jams. This success is due to finding innovative ways to achieve the EU safety objectives without unnecessary costs. Baobab dried fruit pulp was approved as a novel food ingredient in the EU partly on the basis of its history of safe use in Africa. This is a landmark case, as previously the history of safe use could only be demonstrated by a use in the EU. This saved PhytoTrade Africa significant additional costs in the approval process. Other applications for tree products have, unfortunately, not been successful. For example, to date, approval for Galip nuts from Melanesia has not been granted. But hopefully other indigenous fruits from Africa, South America, the Pacific and Asia will be able to build on the baobab experience and gain access to international markets, opening the way for greater trading opportunities for other AFTPs.

With interest for baobab fruit now on the rise in Europe and the United States, African producers—who are mostly women—are beginning to generate income from baobab fruits. The primary producers (harvesters) are organized into supply associations by Tree Crops Malawi, a member of PhytoTrade Africa. Membership of PhytoTrade Africa provides these supply associations with ongoing technical training to ensure that baobab is harvested, processed and stored sustainably and assists them to attain quality control and traceability standards required by international markets. Support is also provided for members and their producers and suppliers to achieve organic and fair trade status. Phytotrade Africa now plans to introduce a number of other novel foods into the EU.

ISSUES FOR FURTHER ATTENTION

Link the Success of Opening New Markets for NWFPs to the Domestication of AFTPs

As mentioned earlier, developments in agroforestry tree domestication have not been adequately linked to the commercialization processes. This creates issues that need to be addressed. For example, without market expansion domestication could increase the supply so much that it starts to outstrip demand, leading to lower market prices that would reduce the incentive to grow these new crops. On the other hand, when NWFP marketing relies on wild harvesting, there can be issues of the surety of supply as harvesting is time consuming, especially if it entails travel to woodlands and forests remote from present-day villages. In addition, fruiting in wild populations is very susceptible to climatic variability resulting from drought, storms, and unseasonal rainfall, etc. Year-to-year variation in fruiting is also normal in tree species, making it much more difficult in poorer years to meet the needs of the marketplace. Furthermore, as we have already seen, wild populations are characterized by very substantial tree-to-tree variation in fruit size and fruit color as well as in taste and other important attributes likely to affect market prices. For example, studies in *D. edulis* (Safou) in Cameroon have illustrated up to tenfold variation in the size of fruits from different trees (Waruhiu et al., 2004; Anegbeh et al., 2005). Consequently, when traders visit rural areas to buy a truckload of fruits, the purchased fruits have the full range of variation and the wholesale trader pays a low price. When these same fruits are sold in the urban markets by retailers, the fruit are arranged in piles that reflect both their variation in size, but also the quality and desirability of different fruit types (Leakey and Ladipo, 1996) and there is then a wide range of market prices reflecting how much the consumer is willing to pay. In this situation market price is related to the different attributes of the fruit types (Atangana et al., 2002b). For the farmer to benefit from the domestication of his/her best trees, the trader will have to be able to purchase a large quantity of fruits with a uniform quality—in other words, of a single cultivar. The scale of this situation increases as the market opportunities for the product increases, and demand rises above that of local markets to regional and international markets.

Low-quality products from wild populations may be rejected by buyers for these more lucrative markets. Large-scale markets are also less tolerant of a seasonal supply of product, so either the periodicity of production has to be overcome by developing a range of different cultivars (e.g., "Noel") of *D. edulis* which fruits at Christmas (Tchoundjeu et al., 2006), or processing techniques are needed to increase the shelf life of the product.

Participatory domestication of AFTPs is most advanced for *D. edulis* (Safou) in Cameroon. Its highly nutritious, oily fruits have a short shelf life and are difficult to preserve. Fresh fruits are typically roasted and consumed as vegetables, so cultivar selection is based on fruit size, fruit color, oiliness, and taste. Recognition of the need of the marketplace for particular combinations of traits has given rise to the concept of an ideotype that meets a particular market need, in this case a fresh fruit (Leakey and Page, 2006). However, in the Congo there is commercial interest in safou oil (Mbofung et al., 2002), and its extraction for use in the food industry (Kapseu et al., 2002). In this case the ideotype would obviously be based on oil yield and oil quality, leading to a very different set of cultivars and very different market opportunities. Interestingly, the development of safou oil might also benefit the enhancement of the storage life of fresh fruits, as one opportunity might be to preserve either fresh or cooked fruits in sealed containers of their own oil. This would have the benefit of creating two new market niches simultaneously, one for the fruits and the other for the oil.

Intellectual Property Rights for Protection of Farmer's Germplasm

The development of the markets increases the risk that unscrupulous entrepreneurs will seek to undermine the success of poor farmers who are developing their own superior cultivars of agroforestry trees producing novel products for regional and worldwide markets (Leakey and Izac, 1996). Globally there are ongoing international negotiations on how to improve the protection of genetic resources and specifically how to assist poor farmers and indigenous rural communities to protect their germplasm and traditional knowledge from commercial exploitation, as can be seen from the comments of the Secretariat of the Convention on Biological Diversity (Medaglia, 2009) on the "Study on the relationship between the Access and Benefit Sharing International Regime and other international instruments which govern the use of genetic resources: The World Trade Organization (WTO); the World Intellectual Property Rights Organization (WIPO); and the International Union for the Protection of New Varieties of Plants (UPOV)." One of the issues brought to the table by developing countries concerns the disclosure of the origin of genetic resources or associated traditional knowledge, particularly with regard to amendments to the TRIPS Agreement (WTO intellectual property agreement). There are suggestions that this agreement should be amended to ensure that patent applicants disclose the source and origin of any genetic material, indicate any relevant existing traditional knowledge and provide evidence of prior informed consent from a competent authority in the country of origin; together with evidence of fair and equitable benefit sharing agreements.

Plant Breeders Rights (PBR) are the most relevant form of legal protection for intellectual property relating to new plant germplasm developed as varieties through the processes of genetic selection and breeding. PBR aim to encourage the development of new plant varieties by providing plant breeders with a limited monopoly (in terms of duration, research exemption, and farmer's privilege) that allows them to recoup their investment in research and development (ACIPA, 2008). Applications for PBR are made to legally empowered entities, usually in the Ministry of Agriculture of a nation, and require compliance with the criteria of novelty, distinctness, uniformity, and stability. Thus, to be registered under PBR legislation, a variety has to be new (usually commercially available for less than one year), clearly distinguishable from other varieties in field trials and be uniform across individuals within the variety. These requirements must be true-to-type over two generations. Once acquired, PBR give exclusive rights (with some limitations and exceptions) to produce and reproduce the material; condition for the purpose of propagation; and to offer for sale, selling or other marketing, exporting, importing, and stocking (UPOV, 1991). This process, which is oriented at protecting commercial companies in industrial countries, is expensive, time-consuming, and complex. Consequently, it is totally inappropriate for individuals or rural communities in developing countries and some more appropriate form of conferring similar rights to disadvantaged individuals in developing countries has to be found and implemented globally.

In 2000, the WIPO Intergovernmental Committee on Intellectual Property and Genetic Resources, Traditional Knowledge and Folklore (IGC) was established as a forum to address the issues relating to the protection of genetic resources vis à vis concerns about intellectual property, traditional knowledge and culture.

To date much of the debate has concerned the ability of existing and new forms of IPR, particularly *sui generis* protection systems ("of its own kind"—in other words, systems developed on a case-by-case basis), but no globally acceptable outcome has emerged from these various negotiations. Many things are unclear. For example, under the Convention on Biological Diversity, individual nations have rights to their indigenous germplasm, but of course the natural ranges of individual species do not neatly respect the political borders of a single country, so what happens when a species with potential as a new crop plant spans two or more nations? Thus, currently, the rights of farmers and rural communities in developing countries to their germplasm and especially their intellectual property on plant breeding innovations remain inadequately protected. This is an unacceptable situation that needs to be resolved as soon as possible.

What action can be taken to protect the rights of, for example, the smallholder farmers in Cameroon engaged in Participatory Domestication, pending a legally binding, globally acceptable, outcome of international negotiations? It is encouraging that 14 West African nations (Benin, Burkina Faso, Cameroon, Central African Republic, Chad, Congo, Côte d'Ivoire, Equatorial Guinea, Guinea Bissau, Mali, Mauritania, Niger, Senegal, and Togo) have come together to apply to join UPOV, creating the African Intellectual Property Organization (OAPI).

The formations of groupings like OAPI with a common legislative protocol that is compliant with the other nations of UPOV will certainly assist in providing a clearer way forward, and also allow some of the administrative procedures to be shared by the partners. However, until such time as these procedures become legally binding, it is only going to be possible to position Cameroon communities in a way that will allow them to be ready to submit PBR applications once the legislation exists. Thus it is recommended that the following steps should be taken now:

1. establish a register of named varieties developed through participatory domestication in association with the name and address of the farmer/"breeder," the GPS location of the mother tree, and a DNA profile ("genetic fingerprint") of the cloned variety. The entry in this register should also record the origin and history of the selection/breeding of the putative variety to be protected. Maybe this register could be established under the aegis of the World Agroforestry Centre,

2. define the species descriptors, based on published data characterizing the tree-to-tree morphological/seasonal variation of traits likely to be used to precisely define the distinctiveness, uniformity and stability (DUS) of the putative varieties,

3. in parallel with clonal tests of the putative varieties being proposed for PBR protection, establish field trials of a number of different clones representative of the full spectrum of intraspecific variation. Data from these clones will be used as comparators to define the unique distinctiveness of the proposed varieties. At a later date any varieties with PBR must be added to these trials. While a single trial at the most relevant Rural Resource Centre (the nominated Genetic Resource Centre [GRC] which should include a nursery set aside for the purpose of maintaining stockplants) may be sufficient for PBR registration, it would be wise to establish trials at several locations so that source population or "varieties of common knowledge" (VCK) are not lost by incidents such as a bush fire.

Following PBR registration the "breeder" will have exclusive rights to his/her germplasm for the period of PBR registration (usually 25 years for trees), make royalty agreements with commercial partners, and settle disputes with regard to ownership of the variety on the basis of "informed consent."

CONCLUSION

Great progress has been independently made over the last 10 years to help poor and disadvantaged farmers in developing countries to improve their livelihoods by trading (Welford and Le Breton, 2008) and cultivating (Tchoundjeu et al., 2010b) the traditionally important products of indigenous trees. However, it is clear that these processes of domestication and commercialization are mutually beneficial and need to become more closely integrated. In addition, rapid progress is needed to develop acceptable means of protecting the rights of farmers and rural communities to their germplasm innovations if these livelihood-enhancing developments are not to be abused by unscrupulous entrepreneurs.

Commercialization of Agroforestry Tree Products

Chapter 30

The Role of Tree Domestication in Value Chain Development

This chapter was previously published in Leakey, R.R.B., van Damme, P., 2014. Forests, Trees and Livelihoods, 23, 116–126, with permission of Taylor & Francis

SUMMARY

Internationally, there is interest in increasing the trade in "green" market products, such as organic, fair trade, reduction of forest degradation for reduced deforestation and mitigation of climate change, and environmental goods and services. This crucially needs to be extended to the many poor, hungry and marginalized smallholder farmers in developing countries. In this context, agroforestry tree domestication has made great progress over the last 20 years, especially in Africa with the emergence of many new tree crops for food, cosmetic and pharmaceutical industries across many agroecological zones. Tree domestication is important for the enhancement of economic returns as value chains from local to global become more sophisticated and demand higher quality, greater uniformity and a regular and continuous supply. Local farmers are now developing cultivars creating direct benefits from the marketing of food and nonfood products in local and regional markets. This creates business and employment opportunities in local cottage industries. Likewise, through the indirect environmental and ecological services provided by trees, food security can be greatly enhanced by closing the yield gap (the difference between the potential and actual yield) of modern crop varieties. In this way, agroforestry is adding income generation to agroecological approaches which together reverse the cycle of land degradation and social deprivation and transform the lives of poor farmers. However, these benefits do not come without some risks from the loss of genetic diversity, local rights over genetic resources and exploitation by unscrupulous entrepreneurs. Agroforestry developments are therefore focusing on better access to "green" business opportunities for poor smallholder farmers in Africa by maximizing the benefits and minimizing the risks.

INTRODUCTION

Agroforestry tree domestication is entering its third decade. In the previous 20 years, great progress has been made (Leakey et al., 2012). This progress has been on two distinct levels. First, domestication consisted of a phase of field and laboratory research which developed new strategies and techniques and also explored the biology and genetics of trees to provide a better understanding of the potential for genetic selection to meet many different and new domestic and market opportunities (Leakey and Akinnifesi, 2008; Leakey et al., 2012). Second, it was implemented as a participatory process in which local communities are helped by Rural Resource Centres to engage and innovate in ways that ensure that they are the beneficiaries of their own initiatives (Tchoundjeu et al., 2010a; Asaah et al., 2011; Leakey, 2012c). Through the process of domestication and cultivation, the products of little-known wild trees producing common property non-timber forest products are transformed into agricultural crops producing AFTPs (Leakey, 2012d).

The domestication process involves steps to evaluate the ethnobotanical and socioeconomic potential of the wild resource of useful species, to characterize and quantify the extent of genetic variation, to effectively capture and make wise use of genetic variation, and finally to determine how best to integrate the domesticates within farming systems for social, economic, and environmental benefits (Leakey, 2012b). More specifically, the domestication of indigenous

Multifunctional Agriculture. DOI: http://dx.doi.org/10.1016/B978-0-12-805356-0.00030-1
© 2017 Elsevier Inc. All rights reserved.

trees producing traditionally and culturally important products for cultivation in low-input agroforestry systems produce foods, medicines and other goods which are marketable in niche markets that are socially, economically and environmentally sustainable. These sustainably produced AFTPs and their associate environmental/ecological services can therefore be seen to contribute to "green" markets, such as organic and locally grown food, health and natural personal care products; sustainably produced building materials (e.g., timber); complementary, alternative and preventive medicine (naturopathy, Chinese medicine, etc.); while also contributing to "fair trade," ecocertification and sustainable ecotourism.

Local, regional, and international markets for AFTPs are crucial for promoting adoption of agroforestry and agroforestry species on a sufficiently large scale to have meaningful economic, social, and environmental impacts. Importantly, making high-quality germplasm available to farmers opens the way for new niche market developments, creating opportunities for rural communities to enter the cash economy. Agricultural policies have traditionally had their primary focus on promoting the supply side. However, it is important to make sure that supply does not exceed demand. Therefore, the expansion of the market for organic ATFP foods will only be effective if the institutional set-up is adequate, i.e., policy networks including organic farmers' associations, certification agencies and public authorities, and if they include or are based on the provision of targeted R&D as well as educational efforts (know-how).

Ecosystem services markets (source: http://www.green-markets.org/context.htm) include many things, such as watershed protection, the conservation of threatened species, etc., but currently there is great interest in the reduction of measurable and verifiable greenhouse gas emissions and carbon markets, due to the reduction of deforestation and forest degradation (REDD) and the conservation, sustainable management of forests and enhancement of forest carbon stocks (REDD+). REDD consists of a set of steps designed to use market and financial incentives in order to reduce the emissions of greenhouse gases, while REDD+ is about avoiding deforestation. These can all be promoted through adoption of agroforestry. Both REDD and especially REDD+ operate in novel market environments which generate funding and political will. They involve mechanisms that both combat climate change and improve human wellbeing in developing nations. Consequently, they represent a suite of policies, institutional reforms, and programs that provide monetary incentives for developing countries to reduce greenhouse gas emissions and sustain economic growth by halting or preventing the destruction of their forests (source: http://www.conservation.org/learn/climate/solutions/mitigation/pages/climate_redd.aspx).

Agroforestry offers the potential to add tree-based REDD/REDD+ income generation to product marketing. In order to foster the latter, agroforestry embraces a holistic approach to rural development (Leakey and Asaah, 2013). This starts with the prioritization of the most important species through ranking exercises with key stakeholders and using ethnobotany (van Damme and Kindt, 2011). This is followed by farmer-led identification and characterization of superior germplasm, and progresses through participatory tree domestication to tree planting and the sale of AFTPs (Leakey, 2012b).

As it is, many plants and animals have been domesticated by humankind through successive generations of selection and breeding. However, in the case of trees, the conventional crop breeding approach takes a long time and gains are small in each generation. This is because of the long period from seed germination to the attainment of sexual maturity and the outbreeding nature of trees that results in genetic segregation during sexual reproduction. Fortunately, however, plants are amenable to asexual regeneration, which means that it is possible to produce multiple copies of a selected individual plant by rooting stem cuttings, or by grafting, budding, or marcotting (air layering) techniques (Leakey, 2004). Using these relatively simple techniques, it is possible to rapidly capture and multiply selected individuals as clones or cultivars (cultivated varieties). While these techniques have been known and used for thousands of years, they have only recently been applied to tropical and subtropical tree species that have for ages provided local people with traditionally important foods, medicines, and other products for everyday use through gathering from wild stands of trees. The success of these asexual regeneration techniques has highlighted the potential and also been the start of agroforestry tree domestication (Leakey, 2012b).

From the outset of this domestication program, a key element of the strategy has been to ask local, often resource-poor, farmers which species they would have liked to domesticate and what characteristics they would have liked to improve (Tchoundjeu et al., 2010a; Asaah et al., 2011). In response to the first question, the majority of farmers, especially in Africa, say that they would like to domesticate local fruit and nut species which are important nutritionally and which have local (and sometimes regional) markets. Regarding the second question, a common response is that they prefer earlier fruiting (shorter time to sexual maturation) and plants of short stature so that harvesting is made easier. Fortunately, both of these outcomes are easy to achieve when using vegetative propagation techniques by propagating material from the already mature crown of the selected trees.

Likewise, vegetative propagation techniques also make it easy to meet the needs of new and traditional commercial markets for uniformity, quality and regularity of supply—characteristics which enhance the market price charged to the consumer, but also the prices that can be obtained by producers. Uniformity and quality are a direct response to clonal propagation and regularity of supply should be the outcome of large-scale propagation of cultivars. The achievement of these outcomes is of great importance to maximizing the outcomes of the value chain, as these characteristics are of increasing importance with each successive step along the value chain: from local to global (Fig. 15.2), and also for meeting the expectations of green and niche markets.

The specificity of agroforestry systems is that they build on and use symbiotic natural equilibria and mutualistic relationships to counter both abiotic and biotic stresses and to make the most of environment and natural (soil) resources. Agroforests that make no use of external (chemical) inputs are production options appropriate to green marketing but are also highly beneficial as approaches to sustainable intensification of tropical agriculture, as they generate income and boost food and nutritional security; restore and maintain aboveground and belowground biodiversity, as well as corridors between protected forests; serve as CH_4 sinks; maintain watershed hydrology; promote soil conservation (Pandey, 2002); and also have social benefits such as greater security over land tenure, enhanced gender equity, and employment/business opportunities (Leakey, 2013). Agroforests also mitigate demand for wood and reduce pressure on natural forests. The promotion of the woodcarving industry facilitates long-term carbon sequestration in artifacts as well as new sequestration through intensified tree growth. The potential delivery of so many benefits means that there is need to support the implementation of agroforestry approaches to tropical agriculture through the development of suitable policies. This can be assisted by, and based on, robust country- and continent-wide scientific studies aimed at better understanding the potential of agroforestry and ethnoforestry for climate-change mitigation and human well-being.

The Domestication—Commercialization Continuum

The value chain starts with the producer who can ensure that the quality of the raw product sold at the "farm gate" is as high (and uniform) as possible. This runs counter to the natural heterogeneity of products from wild tree populations. Many studies of tree-to-tree variation in a range of marketable AFTPs (fruits, nuts, kernel oils, food-thickening agents, pharmaceuticals, perfumed essential oils, and so on) have demonstrated three- to tenfold variation (Anegbeh et al., 2003; Waruhiu et al., 2004; Leakey et al., 2005b,c,d, 2008; Assogbadjo et al., 2006; Page et al., 2010a; Abasse et al., 2011; Atangana et al., 2011).

Interestingly, and as illustrated by the continuous nature of this intraspecific variation, there has been no indication of specific varieties within these populations (Leakey et al., 2012), and furthermore, it is only recently that work formally describing and characterizing specific ideotypes, or the so-called plus germplasm, has been done (Leakey and Page, 2006). This variability which is seen in any bulk load of fruits is one of the major reasons that wholesale traders pay low prices to farmers, even though retail markets recognize this variability and charge customers more for the better (bigger and tastier) fruits (Leakey et al., 2002; De Caluwé et al., 2010a, 2010b; De Caluwé, 2011). It is not until traders can buy bulk loads of a standard quality (i.e., of a single cultivar) that there is the prospect of farmers receiving a better price for their "farm gate" sales. This can be seen in products such as apples, pears, plums, etc., sold in shops and supermarkets in industrialized countries, but is also increasingly valid for lesser-known, underutilized species such as cherimoya (*Annona cherimola*) as shown by the Cumbe case in Peru (Vanhove and van Damme, 2013). These authors found that value-chain features such as market channels, chain governance, quality performance, and the distribution of added value over the component links in the chain differ significantly between cherimoya fruits that are traditionally produced and marketed versus those that are registered by a collective trademark, such as the Cumbe variety. The latter is exported from its production area (Lima province in Peru) to neighboring Andean countries, as graded and highly selected fruits with greater quality. This creates significant value for both producers and traders, while the former local cherimoyas have lower value due to their uneven and unpredictable quality. Studies on the genetic diversity of cherimoya in the countries of origin have stressed the necessity for conserving highly diverse (southern Ecuador and northern Peru) or rare (Bolivia) cherimoya germplasm.

There are some concerns that this commercial success of superior cultivars may lead to a loss of genetic diversity when farmers who believe that quality is exclusively linked to a certain genotype purchase grafts from each other. Potentially, this is a risk, but currently the evidence suggests that this risk may not be as great as feared, as the genetic erosion has been minimal in species such as apples and grapes, which have been cultivated and traded for hundreds of years (Gross and Miller, 2014). In cutnut (*Barringtonia procera*), indigenous to the Pacific, more than 70% of the tree-to-tree variation in the Solomon Islands was found at the level of a village population (Pauku et al., 2010). This is similar to high person-to-person variation seen in people at the village/town level, although genetic variability also includes elements attributable to specific races, tribes and families: in people we are familiar with, the recognition of elite individuals as sporting heroes, beauty queens, Nobel Prize winners, Oscar winners, etc. In crops, similar recognition can be accorded to individual plants that meet the specific needs of a particular market due to their particular combination of a number of specific genetic traits—an ideotype (Leakey and Page, 2006). To understand all these variations, studies have been done in a number of agroforestry trees to evaluate multitrait variation using characterization web diagrams (Leakey, 2005; Simbo et al., 2013). To appreciate the potential, Leakey (2012c) has likened the opportunities to those captured by dog breeders from the genetic diversity of the wolf (von Holdt et al., 2012).

With regard to the risk of genetic erosion, the current approach in participatory domestication of agroforestry trees results in a number of cultivars of each species in every participating community. This means that the diversity in unselected traits will remain high across the production population of each species. This conclusion is supported by the findings from molecular studies of *B. procera* (Pauku et al., 2010) and *Adansonia digitata* (Assogbadjo et al., 2006) showing that trees with particular morphotypes are not closely related. Thus it seems that the risk of genetic erosion is small, provided a wise domestication strategy is followed (Leakey and Akinnifesi, 2008).

Already there are over 50 agroforestry tree species under some level of domestication (Leakey et al., 2012) and potentially there are hundreds, if not thousands, more that could be domesticated in this way. As many tree species produce more than one useful or marketable product (e.g., *Garcinia kola*; Leakey, 2012b; or *Vitex doniana*; Dadjo et al., 2012), it is also possible to identify a number of different trait combinations (e.g., fruit or nut ideotypes) within a single species, and sometimes within a particular product (e.g., oils for food, cosmetics, and medicinal products from a nut). Thus, kernel oils with different chemical components or physical traits may have potential in pharmaceutical, cosmetic or food industries, leading to a hierarchy of ideotype selections (Leakey, 2012b). Taking all this inter- and intraspecific diversity into account, the potential scale of the value chains emanating from the domestication of agroforestry trees is enormous and offers scope for new industries to dramatically change the economies of tropical societies and nations (Leakey, 1999a). This agroforestry tree domestication program has now become a global program that is being implemented in Africa, Latin America, Asia, and Oceania. It thus represents a new wave of crop domestication, focused on improving the livelihoods of farmers and the national economies of developing countries (Leakey, 2012c; Leakey and Asaah, 2013).

Another potential risk arising from successful domestication is that unscrupulous entrepreneurs seeing the potential of cultivars will exploit the opportunity to undermine the initiatives of poor rural communities engaged in tree domestication (Leakey and Izac, 1996). To try to minimize this risk, Lombard and Leakey (2010) have proposed the development of a register of farmer-derived cultivars, with a GPS location for the mother tree and a DNA "fingerprint" of the clone.

The need to match supply (tree domestication and cultivation) with demand (tree product commercialization) means that domestication initiatives need to be matched with marketing initiatives all along the value chain from the local to the global scales. Currently, the domestication initiatives in agroforestry are mostly at a local scale and consequently much focus is on the development of cottage industries drying and packaging tree products for local and some regional markets (Asaah et al., 2011; Leakey and Asaah, 2013). These initiatives are aimed at moving the place of many tree products from the traditional street markets toward new business opportunities. This process is trying to engage local community members as new entrepreneurs and initiate small- and medium-scale enterprises, thereby creating employment in value addition and associated activities, such as the local fabrication of simple processing equipment (Leakey and Asaah, 2013). However, in addition, there are also some marketing initiatives that are running ahead of domestication, in which the trade in processed tree products to regional and international markets has preceded domestication. These initiatives face the risk that demand may be restricted by low quality, lack of uniformity, and unreliable supply.

At the global level, there are also a few commercial initiatives involving tree products. These are public—private partnerships in which multinational companies are working directly with local communities in developing countries (Leakey, 2012b). One of these involves Unilever plc. and communities in Ghana, Nigeria, and Tanzania who are domesticating Allanblackia spp. as a new oil crop for margarine production (Jamnadass et al., 2010) on account of its unique fatty acid content, which displays considerable tree-to-tree variation in oleic and stearic acid composition (Atangana et al., 2011).

Toward Sustainable and Multifunctional Agriculture

It has recently been suggested that agroforestry can be used to close the Yield Gap (the difference between the potential yield of modern crop varieties and the yield actually achieved by poor smallholder farmers in the tropics and subtropics;

Leakey, 2012b,c). This is achieved by a three-step process involving: (1) the restoration of soil fertility and the reha-bilitation of agroecological functions, (2) tree domestication and (3) the commercialization of AFTPs. Moreover, it is seen that in this way agroforestry is capable of improving food and nutritional security, poverty alleviation, and rehabilitation of degraded land so that new areas of forest do not need to be cleared for the expansion of agriculture to feed the growing human population. In other words, agroforestry can deliver multifunctional agriculture, in which the outputs are the enhanced production of crops and livestock in ways that are environmentally, socially and eco-nomically much more sustainable than conventional farming practices (Leakey, 2012b,e). Seen in this way, the role of tree domestication in value-chain development takes on very high priority in the rural development of tropical and subtropical countries.

Toward a Resilient Green Market for Agroforestry Tree Products and Agroforestry Systems

Developing good-quality germplasm material for a broad range of green markets is one thing. To make those markets function is another. The REDD and REDD+ approaches to marketing for reduced carbon and greenhouse gas emission are becoming more and more global, and are being promoted by national governments and international organizations. At local levels, however, there is some skepticism and dissent based on the tendency for inequitable delivery of the ben-efits to the wrong stakeholders. Nevertheless, the general thrust of current negotiations is toward increasing acceptance by producer groups, and the improvement of current compensation schemes that involve most or all of the stakeholders. To avoid these negative outcomes seen in earlier reforestation schemes, REDD+ must incorporate the following: rights-based spatial planning; equitable and accountable distribution of financial incentives; improved financial gover-nance to prevent corruption and fraud; policy reform to remove perverse incentives for forest conversion; and the strengthening of economic benefits and safeguards for smallholders, e.g., such as the use of multipurpose tree species (Barr and Sayer, 2012).

As aforementioned, AFTPs occurring in resource-poor farming systems are by default—for socioeconomic reasons—generally organic, and thus well-fitted for introduction into green and fair-trade markets. There are, however, a number of elements that may constrain sustainable ATFP market development. These, for example, are the inade-quacy of transportation facilities, communication systems, financial capital or access to credit, market information and linkages, and limited knowledge about ATFP market and market information among households. The latter is limited, deficient and significantly influenced by socioeconomic factors such as household members' education, gender, income level and ethnicity, and the distance to market and road access.

However, local AFTP markets can offer many inspiring and motivating advantages.

Local markets exist and may be relatively large, while export markets often have to be developed. Local markets are relatively stable and guaranteed while export markets are often fickle, uncertain and frequently demonstrate "boom and bust" characteristics. Participants in local markets are often independent, whereas export markets may suffer from dependencies that increase the risk of the loss of benefit by poor producers and the collapse of demand if any of the actors withdraw. Another problem can be the sophisticated requirements of export markets for levels of processing, quality control and grading that are out of the reach of local producers. By comparison, local markets are relatively unregulated and have less bureaucracy. In addition, the lower value of goods sold in local markets poses a lower risk of a takeover by wealthy businessmen or displacement by large-scale, capital-intensive producers. All told, therefore, local markets have lower entry barriers as compared to export markets, as there is a minimal requirement for intervention and capital investment to support local trade and enhance livelihood benefits. This allows poor, unskilled, and marginal-ized community members to engage in the trade.

The cultural value of many local and traditionally traded products also provides market stability and can be used to expand markets among urban communities with strong rural roots. Many of these products have value in local markets which may be unknown in export markets, which tend to be socially and geographically foreign. In addition, the econo-mies of scale of local markets can be appropriate in remote areas where supply and demand are in better balance when products are produced locally. Participators in these local markets may also have greater control, setting their own prices, selling where and to whom they wish, and determining their own work pace to fit in with other household activi-ties. Local producers and traders therefore understand the needs of the market and its quality standards and expecta-tions. Last but not least, local markets are accessible and close to producers/traders, reducing transaction costs relative to export markets (Shackleton et al., 2007).

Local markets also have some disadvantages. For example, they may show limited potential for growth or grow more slowly than export markets and can quickly become saturated. This limits the opportunities for new entrants and so can constrain the expansion of individual businesses and hence limit income generation. Local markets may also have poor external visibility and so are often neglected by policymakers and development planners. Low visibility can also result in inadequate research and development support (e.g., extending shelf life, resource ecology, and management) relative to emerging internationally marketed products. Other constraints can include a lack of the technology, credit, contacts, or skills to develop business opportunities.

Rural areas may have scant access to market intelligence and may be beholden to historical trade patterns with less potential for product diversification to reduce risk of market collapse in the long run. Producers supplying these local markets are often dispersed over large areas, making it difficult to target interventions and build collaboration. In these areas, informal traders may face problems establishing themselves in the marketplace and frequently encounter harassment; furthermore, the conditions under which they operate are often poor. In conclusion, producers supplying local markets may be constrained from performing all or most functions along the trade chain (Shackleton et al., 2007). Nevertheless, the horizontal integration of a value chain offers opportunities for more control, realization of more benefits and lower dependency.

On the other side of the coin, the consumers located near local markets are often poor and have limited buying power, keeping prices low. Products in specialized export markets can often fetch high prices. There may be few buyers in local markets for producers who are creative and produce high-quality, unusual goods. Local markets are often located in marginalized areas characterized by poorly developed transport and communication.

Given all these potential benefits, but also in the face of the constraints faced by the commercialization of AFTPs, there is a need to raise the status of local and national AFTP trade. Shackleton et al. (2007) suggest that this can be done by:

- integrating tree products into national surveys for statistical documentation of volumes and values generated by agricultural and forest goods, and into household income and expenditure surveys;
- communicating trade statistics to increase awareness of the size, value, and significance of the trade among key stakeholders such as traditional authorities, local government structures and municipalities, conservation agencies, forestry officials, retailers, consumers, and the general public;
- seeking political backing for the local and national trade in important indigenous products;
- raising the status of collectors/producers/extractors, and remove associated stigmas;
- recognizing, affirming and facilitating development based on existing/traditional knowledge;
- identifying and supporting cultural links to forest products;
- promoting locally produced products through, e.g., special markets, fairs;
- facilitating multistakeholder fora to support development of AFTP markets;
- seeking to integrate AFTPs with other development sectors to form part of a holistic approach to development and poverty alleviation—AFTPs on their own are often limited in their potential for livelihood support and other forms of income generation are also necessary.

CONCLUSION

In conclusion, important strides are being made in agroforestry to initiate integrated rural development approaches which transform many tree-based food and non-food products, which were formerly harvested from the wild, into new sophisticated market commodities for local, regional, and sometimes even international markets. Furthermore, this is being done in ways which also improve the sustainability of tropical agriculture by reversing the complex set of interacting environmental, social, and economic factors, which cause the downward spiral of land degradation and social deprivation that traps millions of farmers in poverty, malnutrition, and hunger. By focusing on all the links in the value chain, this approach is also creating opportunities for poor rural communities to get onto the bottom rungs of the ladder into the cash economy by creating opportunities for the development of cottage industries and the service industries which support them. This approach also opens up the opportunity to benefit from payments for the environmental services, such as carbon markets and the environmental and social product certification schemes, that flow from more environmentally and socially sustainable agriculture.

Section 6.4

Development and Impact

Chapter 31

Underutilised Species as the Backbone of Multifunctional Agriculture—The Next Wave of Crop Domestication

This chapter was previously published in Leakey, R.R.B. and Asaah, E.K. 2013. Acta Horticulturae, 979, 293–310, with permission of ISHS

SUMMARY

The International Assessment of Agricultural Knowledge, Science and Technology for Development (IAASTD) defined multifunctional agriculture as the inescapable interconnectedness of agriculture's different roles and functions: namely the production of food and nonfood commodities; delivery of environmental services; the improvement of rural livelihoods; and the upholding of traditional crops and local culture. Together these outputs should create greater environmental, social and economic sustainability. These goals mirror those of agroforestry, which has been described as a significant mechanism for the delivery of multifunctional agriculture. Agroforestry outputs are delivered in three steps: i) rehabilitation of degraded land; ii) the domestication of underutilized plant species, and iii) the commercialization of agroforestry tree products (AFTPs). Interestingly, past crop domestication has been credited with being a *"perquisite for the development of settled, politically centralized, socially stratified, economically complex and technologically innovative societies."* While there is good evidence of this, the benefits of modern agriculture based on staple food crops have not been equitably distributed and developing country farmers have been marginalized. In the mid-1990s a new wave of participatory crop domestication was initiated. This second wave of domestication, led by the World Agroforestry Centre, has focused on underutilized tropical trees producing highly nutritious fruits and nuts which provide the everyday needs of smallholder farmers. Recent evidence from Cameroon indicates that the domestication of these new tree crops, within an integrated approach to rural development delivering multifunctional agriculture, can transform the lives of poor farmers. It also has positive impacts on the environment and creates new business and employment opportunities in rural communities. Thus it seems that, if widely implemented, this new propoor wave of domestication could have large impacts on global food production and the alleviation of malnutrition, hunger and poverty in developing countries.

INTRODUCTION

The International Assessment of Agricultural Knowledge, Science and Technology for Development (IAASTD) has recently reviewed the state of global agriculture *vis-à-vis* sustainable rural development (McIntyre et al., 2009). It concluded that to achieve economic, social and environmental sustainability it was necessary to redirect agriculture toward multifunctionality (Kiers et al., 2008) in recognition of the "inescapable interconnectedness of agriculture's different roles and functions." IAASTD recognized the functions of multifunctional agriculture to be: the production of food and non-food commodities; delivery of environmental services; the improvement of rural livelihoods; and the upholding of traditional crops and local culture. Thus, if multifunctional agriculture is adopted, future agriculture will be as much about enhancing the livelihoods, health and nutrition of rural households, and restoring natural capital, as about increasing food security and economic growth.

Why is this redirection of agriculture toward multifunctionality necessary? Land degradation is one of the most serious problems facing agriculture as it affects 2 billion hectares (38% of world's cropland) and consequently many

Multifunctional Agriculture. DOI: http://dx.doi.org/10.1016/B978-0-12-805356-0.00031-3
© 2017 Elsevier Inc. All rights reserved.

smallholder farmers in the tropics are trapped in poverty and hunger, together with malnutrition. With little, if any, land remaining for the expansion of farming, the only option is to use what land we have more efficiently. This therefore means that existing farmland has to be made more productive. The options are either to increase yields of existing crops or to rehabilitate degraded farmland and bring it back into production. In effect this means either further expanding Green Revolution technologies or seeking another solution. In temperate countries the former is appropriate. However, in the tropics there is a problem because although the Green Revolution has hugely improved the yield potential and quality of staple food crops, poor farmers have often been unable to have access to seeds, fertilizers, and pesticides, which are critical components of the overall package of modern intensive agriculture. This inability to implement the whole package is a substantial part of the land degradation problem in the tropics and its associated social deprivation. Several recent reviews of agriculture (IAASTD, 2009; Royal Society, 2009) and the role of agriculture in global environmental issues (e.g., MEA, 2005; GEO, 2007; CAWMA, 2007) have suggested that due to the scale of the problems and the constraints facing poor farmers, the further intensification of Green Revolution technologies—business as usual—is no longer the appropriate option.

So, what are the possible alternatives? Let's start by looking at the problem. It is not a simple problem, with a simple solution. It is important to recognize that the degradation of farmland is intimately interconnected with increasing population densities, declining livelihoods, and the social deprivation that is also associated with malnutrition, hunger, and poverty, creating a cyclical problem (Leakey, 2010). Thus to reverse this situation requires simultaneous interventions at different points in the cycle that can both rehabilitate degraded land and reduce social deprivation. Rehabilitation requires soil fertility replenishment and ecological restoration. This can be achieved by the diversification of agroecosystems at the plot and landscape level with species that can generate income. Many wild and underutilized species producing domestically important and marketable products are suitable for this role and are candidates for a new wave of crop domestication. Some of these appropriate species are large perennials—trees, shrubs, or woody vines—while others are herbs that can productively fill ecological niches under the trees. Establishing productive and useful perennial plants helps to restore agroecosystem function and enhances the provision of environmental services while also generating income (Leakey, 1999b). Meanwhile, filling the niches under the trees with useful and marketable herbs and other species is important for the further improvement of the livelihoods of the rural population, and the overall profitability of the farming system. All this creates a more sustainable and multifunctional approach to land use (Leakey, 1996).

Multifunctional Agriculture and the Role of Agroforestry

There are many examples from around the world of low-input, propoor approaches to rural development that enhance production, livelihoods, and ecosystem service functions. Some of these approaches are based on integrated management systems such as reduced- or no-tillage, conservation agriculture, ecoagriculture, agroforestry, permaculture, and organic agriculture. Of these, agroforestry seems to be particularly relevant to the delivery of multifunctional agriculture. Like the other systems, it addresses the issues of soil fertility management; the rehabilitation of degraded farming systems; loss of biodiversity above- and belowground; carbon sequestration; and soil and watershed protection. However, in addition, agroforestry also provides three crucial outputs that are not provided by the other systems, namely: (1) useful, underutilized, and marketable indigenous tree products for income generation, fuel, food, and nutritional security/health and the enhancement of local livelihoods; (2) complex, mature, and functioning agroecosystems akin to natural woodlands and forests; (3) linkages with culture through the food and other products of traditional importance to local people (Leakey, 2010). Thus, the aims of agroforestry are to simultaneously restore: biological resources and natural capital (soil fertility, water, forests, etc.); livelihoods (nutrition, health, culture, equity, income); and agroecological processes (nutrient and water cycles, pest and disease control, etc.).

In agroforestry, the domestication of underutilized and indigenous trees was initiated in the mid-1990s by the World Agroforestry Centre (ICRAF) and its partners aimed at improving the quality and yield of products from traditionally important species that used to be gathered from forests and woodlands. Since then other groups, such as the Agroforestry and Novel Crops Unit of James Cook University and its partners in Oceania, have joined the initiative. As well as meeting the everyday needs of local people, these products are widely traded in local and regional markets and so have the potential to become new cash crops for income generation and to counter malnutrition and disease by diversifying dietary uptake of micronutrients that boost the immune system. These indigenous tree species also play an important role in enhancing agroecological function and, through carbon sequestration, help to counter climate change.

Agroforestry practices are especially numerous in the tropics and are used by more than 1.2 billion people. They produce the products that are important for the livelihoods of millions of other people in developing countries. The area under agroforestry worldwide has not been determined, but over 1 billion hectares (46%) of farmland have more than 10% tree cover, affecting about 30% of rural people (Zomer et al., 2009). Like organic farming, conservation agriculture and ecoagriculture, agroforestry addresses soil fertility management issues for the rehabilitation of degraded farming systems; loss of biodiversity above- and belowground; carbon sequestration; and soil and watershed protection. On the down side, trees are competitive with crops (Cooper et al., 1996) and the net benefits of agroforestry can be slow to materialize due to the longevity of trees. However, techniques such as the vegetative propagation of ontogenetically mature tissues speed up the benefit flows by creating cultivars from parts of the tree that already have the capacity to flower and fruit without going through a long juvenile phase.

Domestication of Plant Species

Crop domestication is human-induced change in the genetics of a plant to conform to human desires and agroecosystems (Harlan, 1975). Crop domestication has been limited to less than 0.05% of all plant species and about 0.5% of edible species (Leakey and Tomich, 1999) and the process goes back thousands of years; for example, the domestication of oranges and apples goes back about 3000 years in China and central Asia, respectively (Simmonds, 1976). According to Diamond (1997), the domestication of useful species has been "the precursor to the development of settled, politically centralized, socially stratified, economically complex, and technologically innovative societies."

Domestication is also said to be stimulated when demand exceeds supply. The latter would explain this recent interest in domesticating tree crops from wild forest species in the tropics, as deforestation has increased in proportion to population growth. This has made the underutilized wild species a scarce resource, which is much in demand. Interestingly, some poor smallholder farmers have reacted to deforestation by starting to select useful trees for growth within their farms (Leakey et al., 2004; Leakey, 2005). These farmers can probably be said to be practicing "commensal" domestication as they have retained natural seedlings in their fields and home gardens and cut down those that do not have desirable characteristics when they clear land for other crops. Secondly, they also sow and disperse the seeds of the tastier fruits that they eat, close to the homestead. This commensal approach to domestication provides a good foundation for the "direct" pathway to domestication that is now being taken by agroforesters working to empower local communities through participatory approaches.

The definition of domestication used for agroforestry trees (Leakey and Newton, 1994a) encompasses the socioeconomic and biophysical processes involved in the identification and characterization of germplasm resources; the capture, selection, and management of genetic resources; and the regeneration and sustainable cultivation of the species in managed ecosystems. This definition therefore stresses that domesticates will be compatible with sustainable land-use systems and have beneficial socioeconomic and environmental impacts. Consequently, the domestication of agroforestry trees is an incentive to promote sustainable agriculture through diversification with species which generate income, improve diets and health, meet domestic needs, and restore functional agroecosystems, as well as empowering local communities (Leakey, 2012c).

The recent history of agroforestry tree domestication has been reviewed by Leakey et al. (2005a, 2007, 2012), and the products of these cultivated trees have been named AFTPs to distinguish them from the extractive resource of NTFPs (Simons and Leakey, 2004).

Strategies of Domestication for Agroforestry Trees

The tree domestication strategy involves the maintenance and use of three interlinked populations (Leakey and Akinnifesi, 2008):

- Gene resource population, for genetic conservation,
- Selection population, for the development of improved cultivars, and
- Production population, for farmers to plant and grow.

The strategy is equally appropriate for the domestication of species producing fruits and nuts; medicinal products; leafy vegetable and animal fodder; timber and wood; and extractives like essential oils, resins, etc. (Table 31.1). For the purpose of efficiency and speed, the domestication strategy adopted by agroforestry has been a clonal one. This is based on well-known horticultural techniques of vegetative propagation (Leakey, 2004), applied in a simple, robust, and low-tech manner (Leakey et al., 1990) so as to be appropriate for implementation in remote areas of tropical countries which

TABLE 31.1 Tree species being domesticated clonally that have potential as components of agroforestry systems.

Species	Use	Reference
Irvingia gabonensis and *Irvingia wombulu*	Kernels and fruits	Okafor (1980), Shiembo et al. (1996a), Atangana et al. (2001c, 2002a), Anegbeh et al. (2003), Leakey et al. (2005d), Tchoundjeu et al. (2010a).
Dacryodes edulis	Fruits and oils	Okafor (1983), Kengue et al. (2002b), Tchoundjeu et al. (2002a), Waruhiu et al. (2004), Anegbeh et al. (2005), Asaah et al. (2010).
Prunus africana	Bark for medicinal products	Simons et al. (2000), Leakey (1997), Tchoundjeu et al. (2002b), Simons and Leakey (2004).
Pausinystalia johimbe	Bark for medicinal products	Ngo Mpeck et al. (2003a), Tchoundjeu et al. (2004)
Ricinodendron heudelottii	Kernels	Shiembo et al., 1997; Ngo Mpeck et al., 2003b, Tchoundjeu and Atangana, 2006.
Gnetum africanum	Leafy vegetable	Shiembo et al. (1996b), Mialoundama et al. (2002).
Barringtonia procera	Nuts	Pauku et al. (2010)
Inocarpus fagifer	Nuts	Pauku (2005)
Santalum austrocaledonicum and *S. lanceolatum*	Essential oils	Page et al. (2010a, 2010b)
Canarium indicum	Nuts	Nevenimo et al. (2007), Leakey et al. (2008)
Sclerocarya birrea	Fruits and nuts	Leakey et al. (2005b,c), Leakey (2005)
Triplochiton scleroxylon	Timber	Longman and Leakey (1995), Ladipo et al. (1991a, 1991b, 1992)
Chlorophora excelsa	Timber	Ofori et al. (1996a, 1996b, 1997)
Swietenia macrophylla and *S. mahogani*	Timber	Newton et al. (1993)

lack reliable supplies of running water or electricity. Vegetative propagation is a uniquely powerful means of capturing existing genetic traits and fixing them so that they can be used as the basis of a genetic variety or "cultivar." The advantage of using clonal propagules outweighs those of seedlings when the products are valuable, or when the tree has a long generation time and when the seeds are scarce or difficult to keep in storage (Leakey and Akinnifesi, 2008). The consequent uniformity in the crop is advantageous in terms of maximizing quality, meeting market specifications and increasing productivity, but it also increases the risks of pest and disease problems. Consequently, risk aversion through the diversification of the clonal production population is a crucial component of the strategies used. Agroforestry enhances this risk aversion by diversifying agroecosystems in ways that improve agroecosystem function (Leakey, 1999b).

Secondly, to benefit the target population of poor smallholder farmers, the strategy is based on participatory approaches to both decision making and implementation (Tchoundjeu et al., 1998; Leakey et al., 2003). This foundation in participatory processes ensures that domestication is a *farmer-driven* process that also has an eye on the local market to ensure that farmers will be able to sell their products (Simons, 1996; Leakey and Simons, 1998; Simons and Leakey, 2004). The first participatory step involved an exercise in priority setting, in which farmers listed their preferred species for domestication (Franzel et al., 1996, 2008). This was to ensure that the outputs of the program were relevant to farmers' needs and so to encourage their active interest and involvement. Interestingly, almost everywhere in the world where this priority setting has been done, farmers have selected familiar and locally marketed indigenous fruits and nuts as their top priority. This is because these traditionally important products are no longer readily available in the wild and are important domestically to rural people because of their cultural and nutritional value. The second step is a participatory approach to project implementation aimed at empowering local communities, promoting food self-sufficiency, generating income and employment, and enhancing nutritional benefits. By providing knowledge and training, the program assists farmers to develop the skills to set up village nurseries and apply simple and adoptable approaches to nursery management; the horticultural techniques of vegetative propagation and tree selection;

agroforestry and community development. The participatory approach has been adopted in order to provide the incentive for farmers to raise themselves out of poverty, malnutrition, and hunger through enhanced livelihoods, and food and nutritional security. Together these two steps to participatory domestication probably also explain the rapid adoption by rural communities (Tchoundjeu et al., 2006, 2010a, 2010b). After about 12 years the number of engaged communities had grown from two pilot villages in Cameroon to 485 villages centered on five Rural Resource Centres and involving about 7100 farmers (Asaah et al., 2011). The concept has also spread to neighboring countries (Tchoundjeu et al., 2006): 11 villages in Nigeria (about 2000 farmers), 3 villages in Gabon (about 800 farmers), and 2 villages in Equatorial Guinea (about 500 farmers).

This strategy of village-level participatory domestication is also important because it conforms to the Convention on Biological Diversity (Tchoundjeu et al., 1998; Leakey et al., 2003; Simons and Leakey, 2004), by recognizing the rights of local people to their indigenous knowledge and traditional use of native plant species. Protection of the farmers' intellectual property is needed to ensure that participatory domestication by local farmers can be recognized as a good model of biodiscovery; an alternative to biopiracy by expatriate or local entrepreneurs. However, until global negotiations create an effective means of protecting the intellectual property of farmers they remain at risk of being exploited, although some procedures to register farmers' cultivars have been proposed as an interim measure (Lombard and Leakey, 2010).

Constraints to Domestication

One constraint to tree domestication is that agroforestry trees are notoriously difficult to domesticate because they are predominantly outbreeding. This means that gains in selected traits are on average small because of the wide range of intraspecific variation in the progeny arising from controlled pollinations. Additionally the long generation time of many trees (10−20 years) means that an individual geneticist does not produce many generations within his/her career. These problems can be overcome by the horticultural approach to domestication, by using vegetative propagation to mass produce individual trees with superior characteristics. Until recently, however, trees have had the reputation of being very difficult to propagate by stem cuttings. This perception has arisen from poor understanding of the principles determining success, mainly the result of poor experimental reporting leading to a confused research literature (Leakey, 2004). The techniques of grafting and budding and marcotting have provided an alternative means of capturing phenotypic variation, but they too have some technical difficulties (e.g., graft incompatibility, dominance by the rootstock, etc.), as well as requiring some special skills. Currently, the numbers of people in many developing countries with appropriate skills in all approaches to vegetative propagation may be a constraint to the widespread upscaling and adoption of participatory domestication in the future (Simons and Leakey, 2004). Nevertheless, perhaps the overriding factor that has constrained the "direct" approach to tree domestication has been the disinterest of colonists and development agencies in products that did not appeal to "western" tastes. Consequently they were neither promoted by early European settlers, nor were they funded by the Green Revolution.

Putting Tree Domestication Into Practice

A fundamental requirement of the clonal approach to domestication is a good understanding of the intraspecific variation in all traits of importance for selection and improvement. Consequently, quantitative studies have been made of the tree-to-tree variation in a range of fruit and nut traits to determine the potential for highly productive and qualitatively superior cultivars with a high Harvest Index (reviewed by Leakey et al., 2005a). This information is needed in order to identify the elite trees with the desirable combinations of different traits that would be appreciated by different markets (e.g., edible fruits, edible nuts, nuts for food oil, nuts for cosmetic oils, or fruits and nuts for medicinal products). The practical approach is to seek trees, which have particular, market-oriented, trait combinations—such as big, sweet fruits (even seedlessness) for the fresh fruit market (a fruit ideotype); big, easily extracted kernels for the kernel market (kernel ideotype), etc. Ideotypes can then be subdivided into those meeting the demands of different markets (Leakey and Page, 2006), such as food-thickening agents conferring drawability and viscosity (Leakey et al., 2005d), or other products, such as pectins or oils for the food or cosmetic industries (reviewed by Leakey, 1999a). Likewise a preliminary analysis of *Sclerocarya birrea* fruits has identified considerable tree-to-tree variation in protein and vitamin C (Thiong'o et al., 2002), while preliminary studies in *Canarium indicum* from Papua New Guinea have evaluated tree-to-tree variation in fatty acid profiles, protein and vitamin E contents (Leakey et al., 2008). Fatty acid profiles have also been the subject of studies in *Allanblackia* species (Atangana et al., 2011), species which are being developed as a new oil crop for Africa (Jamnadass et al., 2010).

One of the key findings of these characterization studies is that each trait shows very considerable and continuous variation from low to high values. Interestingly, this is greatest at the village level, while the variation between villages is only modest. Importantly, it is also found that high values of one trait are not necessarily associated with high values of another trait: thus large fruits are not necessarily sweet fruits, and do not necessarily contain large nuts or kernels. Consequently the more trees that are examined, the greater are the opportunities for creating exciting new cultivars.

A start has been made to look at the genetic variability in sensory and medicinal traits in a few AFTP producing crops. Kengni et al. (2001) have made a preliminary examination of the variability in flavor, astringency, taste, and aroma in samples of *Dacryodes edulis*, while Leakey et al. (2008) have assessed antioxidant activity and phenolic content in *C. indicum*. Interestingly, the latter study also identified very considerable tree-to-tree variation in the antiinflammatory property of kernel oil between just 10 trees. This finding illustrates the very real opportunity to develop cultivars for medicinal properties. Other evidence of tree-to-tree variation in medicinal value has been recorded in the major sterol component, B-sitosterol, from the bark of *Prunus africana*, which is of importance for the treatment of benign prostatic hyperplasia (Simons and Leakey, 2004). This variability in nutritional quality and medicinal properties is likely to affect the potential for different markets, so there is an urgent need for agroforesters to work closely with the food, nutraceutical, and pharmaceutical industries to optimize the domestication/commercialization partnerships (Leakey, 1999a). One aspect of the potential health benefits of agroforestry is the fortification of the immune systems of HIV/AIDS sufferers through the selection of especially nutritious cultivars of indigenous fruits and nuts (Barany et al., 2001), something that requires further investigation as an output from agroforestry (Villarreal et al., 2006).

Up to this point, the focus of tree domestication has been almost exclusively on food, fodder, and medicinal products, but is the approach relevant to other tree products, such as extractives? To test this, a recent study of four essential oils in two sandalwood species (*Santalum austrocaledonicum* and *Santalum lanceolatum*) has shown that these fragrant oils extracted from heartwood have very marked but similar patterns of continuous tree-to-tree variation (Page et al., 2010a). This result therefore offers opportunities to use these techniques to rebuild the sandalwood industry of the Pacific using participatory domestication and agroforestry (Page et al., 2010b).

In addition to the previous qualitative traits, there is also the opportunity for cultivars to capture variation in quantitative traits and in phenology, such as yield, seasonality, and regularity of production, reproductive biology and reduction of susceptibility to pests and diseases which can reduce productivity or quality (Kengue et al., 2002b). High yield is obviously a desirable trait, but in the early stages of domestication it may be even more important economically, nutritionally, etc., to expand the fruiting season from 2−4 to 6−8, or even 12 months. Emerging evidence suggest that in many species there are rare individuals that flower and fruit outside the main season. This seasonality of production offers important opportunities to increase the duration of the productive season and so reduce the periodicity of income generation for the farmer, as well as to make it easier to provide commercial markets with year-round supply.

Having identified which are the elite trees worthy of becoming cultivars, they are propagated vegetatively to capture the specific combination of genetic traits as a clone (Leakey et al., 1990). To ensure that optimal use of the genetic resource is achieved, the clonal approach is integrated with others to ensure that a wise genetic improvement strategy is adopted (Leakey and Akinnifesi, 2008).

Retention and Protection of Genetic Diversity

Typically only the best plants are brought into domestication programs, so domestication is generally considered to reduce the genetic diversity of the species being domesticated, creating the so-called domestication bottleneck (Cornelius et al., 2006). This is probably true in situations where the domesticated plant replaces or dominates the wild origin, but is probably not the case at the current level of domestication of agroforestry trees. So, for example, in most of the trees currently being domesticated there is still a robust wild population. Evidence from molecular studies of *Barringtonia procera* in the Solomon Islands (Pauku et al., 2010) found that the trees with the largest kernels were found in many different populations and were not closely related. Thus selected cultivars produced by different communities will all have large kernels but they will be genetically diverse in all the unselected traits, such as pest and disease resistance, etc. Similar results have been obtained by Assogbadjo et al. (2009) in baobab (*Adansonia digitata*).

The high frequency of intraspecific variation in village populations (about 80%) indicates that cultivar development at the village level also minimizes loss of genetic diversity, especially when wild populations are also present. This therefore is another advantage of implementing a participatory domestication strategy independently in different villages (Leakey et al., 2003). Modern molecular techniques are useful in the development of a wise strategy for the maintenance of genetic diversity, as within the geographic range of a particular species they can be used to identify the 'hot-spots' of intraspecific diversity (e.g., Lowe et al., 1998, 2000), places which should, whenever possible, be protected for in situ genetic conservation, or be the source of germplasm collections if ex situ conservation is required.

Social, Economic, and Environmental Benefits of Domestication

Crop domestication has been credited with being one of the major stimulants of agricultural development and hence the diversification of civil society and economic development, and even the evolution of civilization (Diamond, 1997). This illustrates the close linkage between domestication and the commercialization of the products. Recognizing this linkage and deliberately promoting the parallel development of domestication and commercialization is a very important part of the domestication strategy for agroforestry trees (Leakey, 1999a; Leakey and Akinnifesi, 2008; Bunt and Leakey, 2008). In West and Central Africa, a number of indigenous fruits and nuts, mostly gathered from farm trees, contribute to regional trade. In Cameroon, the annual trade of the products of five key species has been valued at US$7.5 million, of which exports generate US$2.5 million (Awono et al., 2002). Perhaps because of this trade, evidence is accumulating that AFTPs do contribute significantly to household income and to household welfare (Schreckenberg et al., 2002; Degrande et al., 2006).

In terms of social benefits, women, who are the main retailers of NTFPs (Awono et al., 2002), are often the beneficiaries of this trade and they have especially indicated their interest in marketing *D. edulis* fruits because the fruiting season coincides with the time to pay school fees and to buy school uniforms (Schreckenberg et al., 2002). The role of women in trade and marketing of AFTPs is being enhanced by domestication, and hopefully children will also benefit, not only from improved nutrition, but by greater access to education. Similar trends are emerging in southern Africa, where indigenous fruits have relatively new local and international markets (Shackleton et al., 2002). Because the production and trading of AFTPs are based on traditional lifestyles, it is relatively easy for new producers to enter into production and trade with minimal skills, low capital requirement, and with little need for external inputs. Together these things make this approach to intensifying production and enhancing household livelihoods very easy and adoptable by poor people.

Integrating Domesticates Into the Cropping System

Domesticated trees for the production of AFTPs can be integrated into farming systems in many ways, either in home gardens, or as shade in cash-cropping systems such as cocoa or coffee (Leakey and Tchoundjeu, 2001), as scattered trees in food crop fields, or as boundary trees, to generate income, provide products for domestic use as well as to provide environmental services. Trees also used to maintain tenure of customary land which would otherwise have to be forfeited if not seen to be in use.

Agroforestry often creates opportunities for shade-adapted species to fill shady niches and increase the benefits derived from mixed-cropping systems. In this connection, most existing food crops have been selected and bred for cultivation in full sun, so there are opportunities for plant breeders and domesticators to develop new crops or crop varieties that are better adapted to partial shade—e.g., Eru (*Gnetum africanum*) in Cameroon. When new agroforestry crops are integrated with other agroforestry practices, such as improved fallows for soil fertility management, the combined impacts can reduce the crop Yield Gap (the difference between potential yield of a food crop and the actual yield achieved by farmers) and result in many economic, social and environmental benefits (Asaah et al., 2011) that go a long way toward meeting the goals of multifunctional agriculture.

To date, one tree domestication project—Agricultural and Tree Products Program in west and northwest regions of Cameroon—has been outstanding as an example of how the participatory domestication of underutilized species can be a catalyst for the adoption and delivery of multifunctional agriculture within an integrated rural development program (Asaah et al., 2011). This success has been the outcome of the dissemination of knowledge and skills to neighboring communities, via Rural Resource Centres to break the cycles of land degradation and social deprivation that have kept nearly half the world's population in poverty. These activities are now steering the participating villages down a path toward social, economic, and environmental sustainability. Part of the package on new techniques and knowledge is access to and use of microfinance. This has helped over 1000 farmers to obtain short and small-scale loans (about $200 over a period of a few months) for the purchase of inputs such as seeds, fertilizers and hired labor. This has enhanced crop production and had additional benefits such as releasing children from farm work so that they can attend school.

To rebuild the forest resource of useful indigenous trees the Agricultural and Tree Products Programme has taken an innovative three-step approach (Asaah et al., 2011):

- to mitigate environmental degradation that constrains food production through the use of nitrogen-fixing "fertilizer" trees to restore soil fertility and raise crop yields;
- to create income generation opportunities through the establishment of village tree nurseries and then through the production of indigenous fruits and nuts in agroforestry systems for local and regional trade; and
- to promote local processing and marketing of food crops and tree products in order to create employment and entrepreneurial opportunities for community members.

The village nurseries have produced 1,508,000 fertilizer trees over 3 years for use in improved fallows which have more than doubled crop yields. These leguminous trees and shrubs are also popular with beekeepers, who have significantly increased their honey production. They have also produced over 159,060 trees of indigenous fruit and nut trees for home use, for sale and for integration into the tree improvement program (Table 31.2).

Together these village nursery activities have enriched local farms and generated significant income. Typically, this income stream gathers momentum after about 3 years (Fig. 28.9). For example, after 10 years, plant sales from MIFACIG Rural Resource Centre at Belo, and its 35 satellite nurseries in the Northwest region of Cameroon, were valued at US$28,350. Impressively, income was especially high after only 5 years (US$40,000) at the GIC PROAGRO Rural Resource Centre in Bayangam, which has eight satellite nurseries in the West region of Cameroon. This was due to its strong focus on fruit trees. Soon, these communities will also be able to further increase their income by selling fruits from their named cultivars.

To encourage the processing of both food crops and AFTPs, the project, through the activities of WINROCK International, has encouraged the development and fabrication of simple equipment in local towns. Nine local metal workers have been trained and are making equipment for drying, and grinding products. At least 150 discharge mills and 50 dryers have been sold, generating income in excess of US$120,000. This has created employment opportunities for about 200 machine operators. Profits from this enterprise are about 10–20%. Local entrepreneurs and producers are benefiting from the use of this equipment to improve the quality and shelf life of their produce. For example, one trader in Bamenda market, North West Region, Cameroon is selling sealed packages of Nyangsang (*Ricinodendron heudelottii*), Bitter leaf (*Vernonia* spp.), and Eru (*G. africanum*). His trade increased threefold in four months as he gained a reputation for quality products. These small businesses have also created employment for local people. In addition, many women's groups are setting up businesses for grinding crops like cassava and producing "gari" which is substantially increasing their income.

Putting all this together, the farmers have reported over 30 life-changing positive impacts, which are now being verified and quantified. These impacts range from increased income from the sale of plants and AFTPs, improved diet, improved health, ability to send children to school, improved buildings and other infrastructure, such as wells (Asaah et al., 2011). In addition these farmers are also reporting a feeling of empowerment from increased knowledge and success and they are recognizing that they have a pathway out of poverty. However, perhaps the most important impact has been the decision of young men to stay in the community rather than migrating to town, because they now could see a future in the village. Overall, therefore, this program is delivering a suite of impacts as part of a much bigger package, which is improving the social, economic and environmental sustainability of rural life in Cameroon. It therefore seems that this is the pathway to more widely applied rural development for the alleviation of hunger, malnutrition and poverty. As such it is an example of agroforestry delivering multifunctional agriculture (Leakey, 2010).

TABLE 31.2 Summary of overall production of plants from indigenous fruit and nuts species.

Propagule type	2007	2008	2009	2010	Total
Cutting	650	17,600	28,250	29,500	76,000
Marcots	800	2,600	7,260	7,400	18,060
Grafts	5,250	16,300	21,450	22,000	65,000
Seedlings		350,000	605,000	553,000	1,508,000
Total					1,667,060

Taken together, all these impacts achieved in just 12 years strongly suggest that the domestication of indigenous fruit and nut trees is promoting self-sufficiency through the empowerment of individuals and community groups by disseminating new skills in agroforestry, food production and processing. This project therefore addresses the key socioeconomic and biophysical problems facing smallholder farmers in Cameroon. This success can be attributed to the relevance of the work to the farmers' needs and interests and the fact that the program builds on traditional knowledge, local culture, local species and local markets. This initiative has hit the right set of buttons to appeal to farmers and rural communities. Impressively, this process also snowballs, as each community draws in new neighboring communities in a continuous progression of adoption and knowledge dissemination. As a consequence, these communities are on a path toward a better standard of living that is based on the recognition of traditional knowledge of locally important underutilized species. This approach therefore builds on their traditional "life-support systems" derived from indigenous species formerly ignored by agricultural science (Asaah et al., 2011).

What is needed now is to disseminate these skills and experience to millions of other poor people in Africa and other tropical countries. There are many ways of doing this, but one very interesting and outstanding lesson from this project has been the importance of building rural development from the grassroots, using technologies that are simple, practical and easy to implement without spending large amounts of money. The nurseries are a good example; the facilities needed are well within the reach of most farmers once they have had training in the simple technologies developed by the World Agroforestry Centre for soil fertility management and tree domestication. Once established, these activities are self-supporting. Additionally, the philosophy of self-help integrated rural development promulgated by the Rural Resource Centres has been proven to encourage very strong local participation, and ensured the sustainability of the diverse set of activities.

The challenge now is to scale this project up from 7100 farmers in 485 villages to hundreds of millions of rural people across the tropics, and to help some of them to find employment and business opportunities in the rural economy, but outside farming.

CONCLUSION

Future progress in the domestication and cultivation of underutilized species in integrated rural development will probably come from innovative combinations of domestication with the processing and commercialization of their products, especially within developing countries. As the domestication of underutilized species becomes more sophisticated, commercial companies may become increasingly involved. The ideotype approach to formulating trait combinations will be needed to target products that meet a wider and wider range of commercial markets and so encourage commercial companies (Leakey, 1999a) to enter into public—private partnerships. This increasingly commercial approach will ensure the expansion of agroforestry's role in multifunctional agriculture, but to be successful in achieving agroforestry's mission to improve the livelihoods of poor smallholder farmers the commercial partners must be committed to working with local communities based on a recognition of the farmers' intellectual property and their long-term involvement as producers under formal trade agreements (Lombard and Leakey, 2010). This upscaling of agroforestry tree domestication could be the start of a Second Wave of Domestication aimed at the development of a new generation of "settled, politically centralized, socially stratified, economically complex, and technologically innovative societies in the tropics" (Leakey, 2012c).

ACKNOWLEDGMENTS

I thank the World Agroforestry Centre and University of Nottingham Malaysia Campus for sponsoring my (RRBL) participation in this conference.

Chapter 32

Trees: Ensuring That Farmers Benefit From Domestication: An Update

R.R.B. Leakey

Farmers have been found to have started the process of cultivating culturally-important indigenous fruits and nuts (Chapter 8 [Leakey et al., 2004]), as well as expressing their desire to take this further and domesticate them as new crops (Franzel et al., 1996, 2008; Leakey et al., 2003, Leakey, 2014a). The papers presented in this section now place the outcomes resulting from these new tree domestication activities (Chapter 14 [Leakey et al., 2012]) in their social and economic context, and focus on the benefits accruing to the participating communities (Table 28.2). They also present the substance of the social and economic impacts of the strategy to decentralize the domestication activities (Chapter 16 [Leakey and Akinnifesi, 2008]). These aspects build on the work of other members of the tree domestication team. The process started with participatory priority setting (Franzel et al., 1996), and the importance of indigenous fruits and nuts to rural communities (Schreckenberg et al., 2002, 2006; Degrande et al., 2006). It was clear that these species were socially important, even though the economic returns were limited (Ayuk et al., 1999a, 1999b; Schreckenberg et al., 2002; Cosyns et al., 2011), due to the variability in quality and to constraints in the value chain and from informal marketing (Facheux et al., 2006, 2007). These constraints differ from product to product and country to country, as well as from differences in product collection techniques, postharvest pest infestations and storage techniques (Degrande et al., 2014). Thus, to develop the value chain, it is important to diagnose the problems and gain a better understanding of these issues and hence to stimulate income production (Degrande et al., 2014).

Against the background in which extension services were in decline, the concept of Rural Resource Centres (RRCs) was established, involving a partnership between researchers and local NGOs and CBOs (Relay Organizations), to lead a participatory and bottom-up extension program (Tchoundjeu et al., 2006, 2010a; Degrande et al., 2014; Takoutsing et al., 2014). Initially, these Centres focused on the establishment of community tree nurseries and tree planting, and especially: (1) the use of nitrogen-fixing trees and shrubs for soil fertility enhancement (Degrande et al., 2007; Chapter 4 (Cooper et al., 1996), (2) the use of domestication techniques to improve the quality and uniformity of indigenous fruit and nut trees by vegetative propagation for cultivar development (Chapters 15, 20 and Chapter 26 [Leakey, 2017d, 2014c, 2017e]) and (3) as "diffusion hubs" gathering production and market information and building linkages between farmers and traders to facilitate marketing. In this way farmer-to-farmer dissemination was encouraged, leading to the development of satellite tree nurseries servicing neighboring communities (Tchoundjeu et al., 2006, 2010a). As seen in Chapter 31, (Leakey and Asaah, 2013), the involvement of communities and farmers has grown over 12 years from 10 farmers in two villages to about 10,000 farmers in 500 villages over the North and Northwest Provinces (Tchoundjeu et al., 2002a; Asaah et al., 2011; Degrande et al., 2012).

In addition, specific studies were also initiated as follows:

- Enhanced linkages and partnerships between producers and traders are being developed by promoting group sales, so pooling resources such as credit, information, transportation, and labor (Facheux et al., 2012).
- Promoting group sales and the facilitation of a village-level stabilization fund for better storage methods to promote off-season sales found that the coupling of improved storage and guarantee funds helps enhance farmers' capacity to capture higher prices (Facheux et al., 2012).
- Created opportunities for new income streams: for example, honey production (Degrande et al., 2007).
- Arranged capacity-building sessions to reinforce and strengthen trusting relationships among producers and traders (Cosyns et al., 2013); to promote market information to create awareness of supply and demand, prices and opportunities for transportation; to stimulate negotiation in the value chain (Foundjem-Tita et al., 2011; Degrande

Multifunctional Agriculture. DOI: http://dx.doi.org/10.1016/B978-0-12-805356-0.00032-5
© 2017 Elsevier Inc. All rights reserved.

et al., 2014); to encourage collective action (Gyau et al., 2012); and to reduce transaction costs (Foundjem-Tita et al., 2012a).

- Examined the issues around access and rights to land which are commonly considered to be a severe constraint to the adoption of new farming practices (Schreckenberg et al., 2002; Gyau et al., 2014).
- Examined the linkages between income generation and the evolution of businesses engaged in food storage, processing and value-adding (Asaah et al., 2011; Degrande et al., 2014).
- Developed processing and value-adding, such as the development of a simple nut cracker to facilitate the extraction of marketable kernels traditionally done by hand (Mbosso et al., 2015).
- Discussed national policy and law reform regarding the sale of cultivated non-timber tree products cultivated in agroforestry systems (Foundjem-Tita et al., 2012b).

In addition to setting the domestication activities in their social context, several of these studies promote the commercialization of the tree products. This intervention builds on the finding that local retail traders recognized the commercial value of the tree-to-tree variation in tree food products (Figs. 5.2; 28.6 and 28.7). This led to the recognition (Chapters 5 and 13 [Leakey and Simons, 1998; Leakey et al., 2005a]) that tree domestication should be implemented "hand-in-hand" with initiatives to add value and market AFTPs, in order to promote and expand their commercialization (Leakey and Izac, 1996; Leakey et al., 1996). Then this concept was further expanded in Chapter 30 (Leakey and van Damme, 2014), in recognition of the importance of forming a domestication-commercialization continuum to create a viable value chain to expand and diversify markets locally, and potentially regionally and internationally (Fig. 15.2, see also Fig. 28.8).

In Chapters 34, 35 and 39 (Leakey, 2013; 2017g; Leakey and Prabhu, 2017), we see that ultimately the importance of commercialization is to stimulate meaningful impact on poverty—one of the serious constraints to production—that contributes in the creation of a Yield Gap. Expanded trade, new businesses to add value by local-scale processing and packaging, and other income-generating activities, then come together in Step 3 of a generic model for multifunctional agriculture to close the Yield Gap and reverse the cycle of land degradation and social deprivation (Fig. 28.2). This clearly illustrates the importance of ensuring that the domestication-commercialization continuum is fully functional. This is because improved product quality, uniformity, and reliability of supply all become more important as the scale of trade expands.

From all this, it is clear that community-based Rural Resource Centres are effective as a means to: (1) trigger and enhance a self-help philosophy to agricultural development (Asaah et al., 2011; Degrande et al., 2012); and (2) to ensure that production, harvest, and postharvest interventions are effectively associated with the commercialization process (Degrande et al., 2014). Furthermore, when commercialization and domestication occur in parallel, their integration importantly leads to greatly increased well-being: health, business and employment opportunities, social justice, equity, enhanced self-sufficiency, and empowerment (Tchoundjeu et al., 2006, 2010a; Asaah et al., 2011). Thus we can see the new crops formed by this bottom-up approach to domestication as 'socially modified organisms'—a new tool in agricultural development.

However, one very important issue remains to be resolved. It is how to protect these innovative farmers and their communities from exploitation by unscrupulous entrepreneurs. To try to minimize this undesirable outcome in the short term, some interim actions were suggested in Chapter 29 (Lombard and Leakey, 2010). To help the global IPR community to identify a long-term solution, a recent study has made a detailed examination of the special case of agroforestry and the participatory domestication of indigenous trees by poor smallholder farmers (Santilli, 2015). This report has examined all the relevant international and regional instruments and concluded that none of the existing instruments meets the needs of smallholder farmers practicing participatory tree domestication for the enrichment and intensification of agroforestry systems. Further, it considers that national laws are currently the best mechanism for achieving both defensive and positive protection of native genetic resources and associated traditional knowledge. This is in recognition that, in the short term at least, the threats to farmer innovations come from local entrepreneurs, rather than international companies. These recommendations therefore seem to offer the best opportunity to ensure that farmers and local communities get some protection for their innovations in creating elite clonal cultivars, as well as from the marketing of their products. This is especially pertinent in Cameroon as it is currently reviewing its 1994 Forestry Law.

Other possible approaches to protecting farmers' intellectual property include the use of geographical indications (GI) protection or trademark registry. GIs are protected by the TRIPS Agreement, to which 148 countries are signatories. GIs must first be established at the national level before being extended to the international level. Forty-seven African countries have national legal frameworks for GI protection, but very few of them use this legal instrument. The main challenges are the costs of establishing and administering a GI regime and the lack of availability of technical assistance and capacity building.

In the longer term, the Santilli Report (2015) proposes the development of a new and specialized international Access and Benefit Sharing agreement for agroforestry research and development, justified by its potential to address the global challenges facing agriculture, such as poor nutrition and low health status, climate change, limited agricultural productivity, and forest and biodiversity loss.

To conclude, experience in Cameroon has found that diversified farming systems with 'socially modified organisms' producing traditional tree food products for local markets, together with improvements along the value-chain, have increased household income across many wealth classes, and had numerous positive impacts on the lives and well-being of participating communities (Tchoundjeu et al., 2010a; Asaah et al., 2011; Leakey and Asaah, 2013; Degrande et al., 2014). This approach to tree domestication is also being adopted for *Allanblackia* species being developed as a new oil crop for Africa (Jamnadass et al., 2010), as well as for agroforestry trees in Peru (Weber et al., 2001), southern Africa (Akinnifesi et al., 2006), Solomon Islands (Pauku et al., 2010), and Indonesia (Narendra et al., 2013). It seems that this is a model which could be effective in many other parts of the tropics as appropriate species for domestication are widespread. Gepts (2014) sees this approach being particularly significant as it leads to decentralized domestication activities boosting the genetic improvement of trees, addressing the needs of local farmers as dictated by local adaptation and consumer preferences.

In Chapters 34 and 35 (Leakey, 2013, 2017g) we will see how participatory tree domestication and the marketing of tree products can create a three-step approach to multifunctional agriculture (Figs. 2.3 and 28.2), when combined with the use of trees and shrubs capable of biological nitrogen fixation.

Section 7

Agroforestry: A Delivery Mechanism for Multifunctional Agriculture

In 1997, the first paper presented here recognized that the three main determinants for overcoming rural poverty in Africa were (1) reversing soil fertility depletion, (2) intensifying and diversifying land use with high-value products, and (3) providing an enabling policy environment for the smallholder farming sector. So far in this book we have seen that agroforestry practices can improve food production in a sustainable way through their contribution to soil fertility replenishment, and that the large untapped resource of high-value trees can make farming systems both ecologically more stable and economically more rewarding, while improving food security. Now, in the second paper we have seen steps towards the third activity—policy change. The need for this has been widely recognized by numerous major international reports (MEA in 2005; IAASTD and the Royal Society in 2009; etc.) all calling for a change away from the conventional paradigm for intensive agriculture ("business as usual") which stems from the success of industrial agriculture in temperate climates. This raises the question: Is this temperate model appropriate for the tropics and subtropics where the biophysical, social, and economic environments are very different from those in industrialized economies?

This section is about trying to develop a relatively simple "conceptual framework" for a new paradigm for agricultural policy based on appropriate crop husbandry—a framework for multifunctional agriculture. As we have seen in earlier chapters, we do actually have simple and appropriate technologies in Africa that can increase actual crop yields three- to sixfold and make big advances towards food security for all—with better nutrition and enhanced livelihoods for all. Is this a practical proposition, and could these refinements to the concepts and policies underlying modern conventional agriculture actually improve global agricultural production? If so, would they also increase the economic returns on the substantial global investment in the Green Revolution?

Chapter 33

Trees, Soils and Food Security

This chapter was previously published in Sanchez, P.A., Buresh, R.J., Leakey, R.R.B., 1997. Philosophical Transactions of the Royal Society Series B, 352, 949–961, with permission of The Royal Society

SUMMARY

Trees have a different impact on soil properties than annual crops, because of their longer residence time, larger biomass accumulation, and longer-lasting, more extensive root systems. In natural forests nutrients are efficiently cycled with very small inputs and outputs from the system. In most agricultural systems the opposite happens. Agroforestry encompasses the continuum between these extremes, and emerging hard data is showing that successful agroforestry systems increase nutrient inputs, enhance internal flows, decrease nutrient losses and provide environmental benefits—when the competition for growth resources between the tree and the crop component is well managed. The three main determinants for overcoming rural poverty in Africa are (i) reversing soil fertility depletion, (ii) intensifying and diversifying land use with high-value products, and (iii) providing an enabling policy environment for the smallholder farming sector. Agroforestry practices can improve food production in a sustainable way through their contribution to soil fertility replenishment. The use of organic inputs as a source of biologically fixed nitrogen, together with deep nitrate that is captured by trees, plays a major role in nitrogen replenishment. The combination of commercial phosphorus fertilizers with available organic resources may be the key to increasing and sustaining phosphorus capital. High-value trees—"Cinderella" species"can fit in specific niches on farms, thereby making the system ecologically stable and more rewarding economically, in addition to diversifying and increasing rural incomes and improving food security. In the most heavily populated areas of East Africa, where farm size is extremely small, the number of trees on farms is increasing as farmers seek to reduce labor demands, compatible with the drift of some members of the family into the towns to earn off-farm income. Contrary to the concept that population pressure promotes deforestation, there is evidence that demonstrates that there are conditions under which increasing tree planting is occurring on farms in the tropics through successful agroforestry as human population density increases.

INTRODUCTION

The continuing threat to the world's land resources is exacerbated by protracted rural poverty and food insecurity in the Third World, and wider climatic variations resulting from global warming. During the last decade food security was not a global priority, but studies such as the 2020 Vision (IFPRI, 1996) show that rural poverty in the Third World is one of the main global concerns of our time, and that food insecurity is a major factor in rural poverty. Access for all to sufficient and nutritious food is the key to poverty alleviation—this was one of the main outcomes of the 1996 World Food Summit (FAO, 1996). Food security encompasses both food production and the ability to purchase food. However, calories and protein are not the only factors: nutritional security includes overcoming deficiencies of vitamin A, iron, zinc, iodine, and selenium (IFPRI, 1996). It is also recognized that the attainment of food security is intrinsically linked with safeguarding the natural resource base (IFPRI, 1996). Therefore, the three interlinked factors for reversing rural poverty are (1) income generation, (2) increasing food and nutritional security, and (3) protecting the environment.

Although food insecurity occurs throughout the developing world, it is most acute in sub-Saharan Africa—hereinafter referred to as Africa—where per capita food production continues to decrease, in contrast with increases in other parts of the developing world (FAO, 1996). Africa has the highest rate of population growth of any region in the world

Multifunctional Agriculture. DOI: http://dx.doi.org/10.1016/B978-0-12-805356-0.00033-7
© 2017 Elsevier Inc. All rights reserved.

(2.9% per year) and the highest rate (30%) of degradation of usable land (Cleaver and Schreiber, 1994). Deficiencies in vitamin A and micronutrients are also acute on this continent (IFPRI, 1996). The Malthusian nightmare, although unrealistic at the global scale, could become a reality in Africa.

The bulk of food in Africa is produced on small-scale farms by women. The three main determinants for overcoming rural poverty under these conditions are (1) an enabling policy environment for the smallholder farming sector; (2) reversing soil fertility depletion; and (3) intensifying and diversifying land use with high-value products (Sanchez and Leakey, 1997).

Attaining these three goals can only be achieved in Africa with modern agricultural practices based on traditional rock-phosphate approaches (Borlaug and Dowswell, 1994; Borlaug, 1996) if fertilizers and other farming inputs are available at a price affordable by resource-poor farmers. They can also be achieved with agroforestry—the deliberate use of trees on farms as a low input system—a common feature of small-scale farming throughout the tropics. The purpose of this contribution is to discuss the added value of tree-based agricultural systems and to link them to the three determinants for poverty alleviation.

IMPACT OF TREES ON SOIL FUNCTIONS

Trees have different impacts than annual crops on soil properties, because of their longer residence time, larger biomass accumulation, and continuous and more extensive root systems. In natural forest stands, nutrients are efficiently cycled with very small inputs and outputs from the system, and the soil surface is continuously protected by one or more plant canopies. In most agricultural systems, the opposite happens: nutrient cycling is limited, while inputs and outputs are large, and the soil is not continuously protected by a plant canopy. Agroforestry encompasses the continuum between these two extremes, and emerging hard data show that specific agroforestry systems provide added value to soil processes when the competition for growth resources between the tree and the crop component is adequately managed (Ong and Huxley, 1996). Such added value occurs more commonly in sequential, as opposed to simultaneous, agroforestry systems, because the competition for water, nutrients, and light between the crop and tree component is separated over time (Sanchez, 1995). While the effects of trees on soil functions in agroforestry systems are generally positive, the effects on crop production are often negative. This happens when the competition for light, water or nutrients is intense (Sanchez, 1995). In such cases, trees decrease crop yields (van Noordwijk et al., 1996). Before considering the effects of agroforestry trees on soil properties it is imperative to deal with agronomically successful agroforestry systems.

There are four ways in which trees can have beneficial effects on soil properties, crop production, and environmental protection. Trees in effective agroforestry systems (1) increase nutrient inputs to the soil, (2) enhance internal cycling, (3) decrease nutrient losses from the soil, and (4) provide environmental benefits. These ways are summarized in the following paragraphs, based largely on reviews by the authors (Sanchez et al., 1985, 1997; Leakey and Newton, 1994c; Sanchez, 1995; Leakey et al., 1996; Buresh and Tian, 1997; Sanchez and Leakey, 1997). We focus on nitrogen (N) and phosphorus (P), because these are the main limiting nutrients in smallholder farms in Africa. In contrast to other continents, soil acidity and aluminum toxicity are not widespread constraints in cultivated areas of Africa (Sanchez and Leakey, 1997).

Increased Nutrient Inputs

Trees can provide nutrient inputs to crops in agroforestry systems by capturing nutrients from atmospheric deposition, biological nitrogen fixation (BNF), and from deep in the subsoil, and storing them in their biomass. Biomass transfers from one site to another also provide nutrient inputs. These nutrients become inputs to the soil when the tree biomass is added to and is decomposed in the soil. The main processes are BNF, deep nitrate capture, and biomass transfer.

Biological Nitrogen Fixation

Although the magnitude of BNF is methodologically difficult to quantify, overall annual estimates are in the order of $25-280$ kg N ha^{-1} year^{-1} for leguminous trees (Giller and Wilson, 1991). Woody and herbaceous legumes can provide practical means of capturing nitrogen via BNF when grown as fallows in rotation with annual crops, taking advantage of the dry season in subhumid environments when no crops can be grown. Two years of *Sesbania sesban* fallows in Zambia overcame nitrogen deficiencies for three subsequent maize crops (Kwesiga and Coe, 1994).

There is high genetic variability within tree species in their effectiveness at BNF (Sanginga et al., 1990, 1991, 1994). Phosphorus deficiencies can limit N_2 fixation and growth of N_2-fixing trees. Sanginga et al. (1994, 1995) found large differences in early growth and P-use efficiency among and within N_2-fixing tree species. These results highlight the merit of selecting provenances of N_2-fixing trees that are tolerant to low available P at an early growth stage.

Deep Nitrate Capture

The uptake of nutrients by tree roots at depths where crop roots are not present can be considered an additional nutrient input in agroforestry systems. Such nutrients become an input upon being transferred to the topsoil via tree litter decomposition. Tree roots frequently extend beyond the rooting depth of crops. An exciting dimension has recently been discovered in nitrogen-deficient Nitisols of western Kenya, where mean nitrate levels in six farmers' fields ranged from 70 to 315 kg N ha^{-1} at 0.5–2.0 m depth (Buresh and Tian, 1997). The accumulation of subsoil nitrate is attributed to greater formation of nitrate by soil organic matter (SOM) mineralization in the topsoil than the crop can absorb (Mekonnen et al., 1997). The excess nitrate then leaches to the subsoil where it is sorbed on positively charged clay surfaces, retarding the downward movement and leaching loss of nitrate (Hartemink et al., 1996). Nitrate sorption is well documented in subsoils rich in iron oxides (Kinjo and Pratt, 1971). *S. sesban* fallows deplete this pool, thus capturing a resource that was unavailable to the maize crop (Mekonnen et al., 1997). These relationships are shown in Fig. 33.1.

In soils with high quantities of subsoil nitrate, a N_2-fixing tree should, ideally, be able to rapidly take up the subsoil nitrate before it can be leached. When the tree has depleted subsoil nitrate, it should then ideally meet a substantial proportion of its N requirements through BNF.

Under such conditions, agroforestry trees become a biological safety net. How extensive are these soils? There are 260 million hectares of Nitisols (oxic or rhodic Alfisols and Oxisols) and similar soils in Africa that have anion exchange capacity in the subsoil, where roots of Sesbania and similar agroforestry trees can penetrate (Sanchez et al., 1997). Assuming that one-tenth of them are under cultivation, the magnitude of this resource could be in the order of 3 million tonnes (Mt) of nitrate nitrogen, much more than the annual nitrogen fertilizer consumption rate, 0.8 Mt of nitrogen in sub-Saharan Africa, excluding South Africa (FAO, 1995). We do not yet know the extent to which this resource is renewable. Nevertheless, the utilization of this hitherto unrecognized nitrogen source via its capture by deep-rooted trees is an exciting area of research in Africa, as well as in other regions with similar oxidic subsoils.

Biomass Transfer

The leafy biomass of trees is frequently cut from hedges or uncultivated areas and incorporated into crop fields as a source of nutrients in Africa. While the quantities of biomass farmers are able to apply are often sufficient to supply N to a maize crop with a moderate grain yield of 4 t ha^{-1}, they seldom can supply sufficient P to that crop (Palm, 1995).

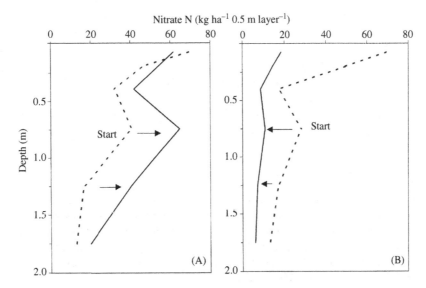

FIGURE 33.1 Nitrate accumulates in the subsoil of this Oxisol, near Maseno, Western Kenya. (A) Maize is unable to access this pool, while *Sesbania sesban* (B) depletes it. *Adapted from Hartemink et al. and Buresh, unpublished data.*

Leguminous trees are most frequently used as biomass transfer systems, but there is increasing evidence that some non-leguminous shrubs may also accumulate high concentrations of nutrients in their biomass. *Tithonia diversifolia*, a common hedge species found at middle elevations throughout East Africa and Southeast Asia has unusually high nutrient concentrations (3.5% N; 0.38% P and 4% K) in its leaf biomass (Gachengo, 1996; Niang et al., 1996). These P and K (potassium) levels are higher than those of commonly used legumes in agroforestry (Palm, 1995). Reasons for such high concentrations remain speculative but members of the Compositae family, to which *Tithonia* belongs, have a reputation for being nutrient scavengers.

The processes involved are not presently identified, but may involve the dissolution of inorganic phosphorus, desorption of fixed soil phosphorus by root exudates, organic acids, and/or extremely effective mycorrhizal associations. Woody species grown in hedges outside the cultivated fields, therefore, may be able to transform less available inorganic forms of phosphorus into more available organic forms, as well as supply significant quantities of N and K when their leaves are incorporated into the soil as biomass transfers.

Enhanced Nutrient Cycling

Trees in agroforestry systems can increase the availability of nutrients in the soil through the conversion of nutrients to more labile forms of SOM. Plants convert inorganic forms of N and P in the soil solution into organic forms in their tissues. The addition of in situ-grown plant material to the soil as litterfall, root decay, green manures, crop residue returns (or animal manures in grazing systems), and its subsequent decomposition results in the formation of organic forms of soil N and P. Mineralization of soil organic N or P converts them once again to nitrates and orthophosphate ions in the soil solution which are readily available to plants. This is the process of cycling.

It is important to distinguish organic cycling from organic inputs. Cycling involves organic materials grown in situ, such as those described in the previous paragraph. They do not add N or P to the soil–plant system, except for additional biological N_2 fixation and capture from below the crop rooting depth, and therefore do not constitute inputs from outside the system. Biomass transfers, composts, and manures produced outside the field are the true organic inputs.

Total SOM generally does not relate to crop yields (Sanchez and Miller, 1986). Nutrient release from SOM is normally more dependent on its biologically active fractions than on total SOM quantity. Microbial biomass P, light fraction organic N and P, and NaOH extractable organic P appear to be relevant fractions in agroforestry systems (Buresh and Tian, 1997).

Soil Organic Nitrogen

Agroforestry tree species vary greatly in their quality, usually measured by the (lignin + phenolics)/N ratio of their leaves (Palm and Sanchez, 1991; Constantinides and Fownes, 1994; Schroth et al., 1995; Tian et al., 1995; Jonsson et al., 1996). High-quality materials are readily mineralized, while low-quality ones decompose slowly and may eventually form part of soil organic pools. For example, Barrios et al. (1997) found that N availability, as determined by inorganic soil N, N in light fraction SOM, and N mineralization in topsoil, was higher in maize plots following improved fallow species with the lowest (lignin + polyphenol/N) ratios in leaf litter in an N-deficient Alfisol in eastern Zambia. *S. sesban* fallows and fertilized maize monocultures resulted in similar inorganic soil N levels, but N mineralization and light fraction N were greater after *S. sesban*. The amount of light fraction N appears to be a sensitive measure of SOM differences among cropping systems and is correlated with N mineralization of the whole soil (Barrios et al., 1996a, b). Light fraction SOM can be increased by addition of tree biomass to maize (Barrios et al., 1996a) and by rotation of maize with planted tree fallows (Barrios et al., 1997). Appropriate agroforestry systems, therefore, seem to enhance internal N flows.

Soil Organic Phosphorus

Most studies have found little or no benefit of trees in agroforestry systems on inorganic soil P tests (Drechsel et al., 1991; Siaw et al., 1991; Kang et al., 1994, 1997). Methods related to labile soil organic P fractions seem more appropriate for agroforestry systems with little or no inorganic P inputs. For example, *S. sesban* fallows, as compared to continuous unfertilized maize, increased soil P availability, measured by chloroform extractable P and P in light fraction SOM, but had no effect on extractable inorganic soil P (Maroko et al., 1999). *S. sesban* fallows, compared with continuous unfertilized maize, increased maize yields when P was the limiting nutrient, but they did not eliminate the need for external P inputs to completely overcome the P deficiency.

Some trees and shrubs, but apparently few crop species, have the ability to exude organic acids from their roots or mycorrhizal associations and dissolve inorganic soil phosphates not otherwise available to roots of crop plants

(Lajtha and Harrison, 1995). Pigeon pea (*Cajanus cajan*) secretes pisidic acid in calcareous soils (Ae et al., 1990; Otani et al., 1996), increasing the plant's phosphorus uptake, while *Inga edulis* is believed to have access to phosphorus not available to maize and beans (Hands et al., 1995). Both these species are legumes, which are known to acidify their rhizosphere in the process of nitrogen fixation. In such cases, organic cycling has the advantage of transforming otherwise unavailable inorganic soil phosphorus into more available organic forms.

Agroforestry will not eliminate the need for P fertilizers on P-deficient soils (Buresh et al., 1997). The integration of organic materials with inorganic P fertilizers is likely to enhance the availability of P added from inorganic fertilizers (Palm et al., 1997). There are, at present, no methods for quantifying nutrient cycling efficiency in agroecosystems and its effects on productivity and sustainability. This is an area that requires further conceptualization, and a start has been made by van Noordwijk (1999) who describes possibilities at various spatial and temporal scales.

Decreased Nutrient Losses From the Soil

Losses caused by runoff, erosion, and leaching account for about half of the N, P, and K depletion in Africa (Smaling, 1993). Agroforestry systems have been found to decrease nutrient losses by runoff and erosion to minimal amounts (Lal, 1989a; Young, 1989).

The evidence for decreased leaching losses is less comprehensive. Horst et al. (1995) reported that *Leucaena leucocephala* hedgerows reduced nitrate leaching as compared with a no-tree control on a sandy Ultisol in the Benin Republic. Lower subsoil water provided indirect evidence of reduced leaching loss of nutrients under trees in agroforestry systems of western Kenya (ICRAF, 1996). Subsoil water in *S. sesban* fallows seldom exceeded field capacity in a clayey Oxisol despite a mean annual rainfall of about 1800 mm. Subsoil water in the natural uncultivated fallow and maize monoculture at the same site occasionally exceeded field capacity, indicating that mobile water was present to transport nitrate downward. Low subsoil water and nitrate content under *S. sesban* were attributed to high water and N demand by the fast-growing tree.

Environmental Benefits

Trees protect the soil surface via two canopies: the litter layer and the leaf canopy, thereby decreasing runoff and erosion losses, dampening temperature and moisture fluctuations and, in most cases, maintaining or improving soil physical properties (Sanchez et al., 1985; Lal, 1989b, c; Hulugalle and Kang, 1990; Hulugalle and Ndi, 1993; Rao et al., 1998). In agroforestry systems, the beneficial effects of protecting the soil surface depend on the spatial and temporal coverage of the tree component. Also, tree roots can loosen the topsoil by radial growth, and improve porosity in the subsoil when roots decompose. The perennial nature of tree root systems provides a dependable source of carbon substrate for microorganisms in the rhizosphere; microbial mucilage binds soil particles into stable aggregates, which results in improved soil structure (Tisdall and Oades, 1982). These two processes, surface soil protection and root penetration, take place continually in agroforestry systems instead of temporarily, as in agricultural systems. Due to these, three major kinds of environmental benefits ensue: soil conservation, biodiversity conservation, and carbon sequestration.

Soil Conservation

Many agroforestry systems help keep the soil in place by biological instead of engineering means (Lal, 1989a; Young, 1989; Kiepe and Rao, 1994; Juo et al., 1995; Rao et al., 1998). While contour hedges do require management, although certainly less than earth terraces, they also become a productive niche on the farm while conserving the soil. Controlling soil erosion biologically has an additional advantage: the slope between the hedges becomes less steep and even flat in some cases (Kiepe and Rao, 1994; Garrity, 1996). These "biological terraces" are produced by taking advantage of the erosion process within the contour hedges, with the vegetative growth keeping up with the higher soil surface at the lower end, something nonbiological terraces cannot do. Trees, however, do not conserve the soil until they are well established and have developed a litter layer (Sanchez et al., 1985). Once established, most trees protect the soil constantly, provided they are healthy and the litter layer is not removed. Biomass transfer of tree leaf litter to cropped fields undermines this process (Nyathi and Campbell, 1993).

Biodiversity Conservation

All agroforestry systems are more diverse than crop or forest plantation monocultures, while some, such as the complex agroforests of Southeast Asia, are nearly as diverse as natural forests (Thiollay, 1995). But, importantly, agroforestry also helps to conserve plant and animal biodiversity by reducing the further clearance of tropical forests through viable alternatives to slash-and-burn agriculture (Sanchez, 1994; Schroeder, 1994). Precise estimates of these substitution values do not exist for agroforestry systems, although figures of 7.1 and 11.5 ha saved for each hectare put into successful agroforestry have been reported (Schroeder, 1993).

Multistrata or complex agroforests are one such alternative to slash-and-burn. In these systems, annual food crops are planted along with trees, and cover the ground quickly until they are shaded out by these trees, which in turn eventually occupy different strata and produce high-value products such as fruits, resins, medicinals, and high-grade timber (de Foresta and Michon, 1994; Michon and de Foresta, 1996). Plant diversity is in the order of 300 species ha^{-1} in the mature, complex rubber agroforests of Sumatra, Indonesia. This level of plant biodiversity by far exceeds that of rubber plantations (5 species ha^{-1}) and approximates to that of adjacent undisturbed forests, with 420 plant species ha^{-1}. The richness of bird species in mature damar (*Shorea javanica*)-based agroforests is approximately 50% that of the original rainforest (Thiollay 1995), and almost all mammal species are present in the agroforest (Sibuea and Herdimansyah 1993). This is possible because such agroforests, composed of hundreds of small plots managed by individual families, occupy contiguous areas of several thousand hectares in Sumatra. Tracks of the rare Sumatran rhino (*Dicerorhinus sumatrensis*) were recently discovered in one of these rubber agroforests, implying that they may provide a habitat similar to the natural rainforest (Sibuea, 1995). Such high biodiversity levels, however, cannot be expected of shorter duration agroforestry systems, such as improved fallows, or in systems that are less geographically extensive.

Agroforestry plays a major role in the reclamation of degraded and abandoned lands, and is generally considered the most workable approach to mimic natural forest succession and increase biodiversity (Anderson, 1990). Hard data on increasing biodiversity in degraded lands through agroforestry, however, are practically nonexistent (Sanchez et al., 1994).

Belowground biodiversity is also higher in agroforestry systems than in crop monocultures, approximating the levels of the natural forest in the Amazon (Lavelle and Pashanasi, 1989). Soil macrofauna and microflora are key regulators of the basic decomposition processes that provide nutrients to higher plants and animals. While they are not as attractive as "furry and feathered creatures," soil communities are a major component of biodiversity conservation and ecosystem functioning.

Carbon Sequestration

Agroforestry systems help keep carbon in the terrestrial ecosystem and out of the atmosphere by preventing further deforestation and by accumulating biomass and soil carbon (Schroeder, 1994). As with biodiversity conservation, the main contribution of improved agroforestry systems to terrestrial carbon conservation comes from its preventive effect, i.e., the area of natural forests that will not be cleared because farmers can make continuous use of already cleared land through improved agroforestry systems (Schroeder, 1993; Unruh et al., 1993; Sanchez, 1994). One hectare of humid tropical forests contains on average 160t C (carbon) ha^{-1} in the aboveground biomass (Houghton et al., 1987). When it is slashed and burned, most of it is emitted to the atmosphere, either immediately during the burn, or gradually through the decomposition of unburned logs and branches. Keeping this carbon resource (some 96 billion tonnes of C in the remaining humid tropical forest biomass) in situ is of critical importance.

Complex agroforestry systems of long duration, such as the jungle rubber and damar agroforests of Sumatra and multistrata systems throughout the humid tropics, can sequester carbon in their tree biomass, where it remains for decades. In addition, complex agroforests act as sinks for methane emitted by adjacent paddy fields, thereby neutralizing these greenhouse gas emissions at the landscape scale (Murdiyarso et al., 1996).

The greatest potential for carbon sequestration is probably in soils that have been depleted of carbon and nutrients and have the potential to regain their original SOM levels. Woomer et al. (1997) estimate that 66 tonnes ha^{-1} of carbon can be sequestered in woody biomass and nutrient-depleted soils in Africa over a 20-year period by a combination of nutrient recapitalization, erosion control, boundary tree plantings, and woodlot or orchard establishment.

The overall magnitude of carbon sequestration by agroforestry is considered among the highest compared with other land use systems by climate change researchers. Unruh et al. (1993) performed complex calculations of agroforestry systems in Africa, their biomass accumulation, and their potential distribution using GIS techniques. Their results suggest that a huge amount of carbon can be sequestered, ranging from 8 to 54 Gt (billion tonnes) of C in a total of 1.55 billion hectares where agroforestry could potentially be practiced. This represents the theoretical upper limit. Above- and belowground carbon sequestration values, however, need to be generated locally, taking into account the duration of each agroforestry system, and extrapolated geographically in a realistic fashion, based on actual rates of agroforestry adoption.

TREES AND OVERCOMING RURAL POVERTY IN AFRICA

While agroforestry trees may improve soil fertility, nutrient-use efficiency, and provide major environmental benefits, they are not likely to have a significant impact on food security or alleviate poverty by themselves. Successful agroforestry can contribute to (1) food security from the production point of view through soil fertility replenishment, along with fertilizers, (2) poverty alleviation and access to enough and nutritious food through the domestication of indigenous trees, and (3) enabling policies. This section examines these possibilities.

Soil Fertility Replenishment

Soil fertility depletion in smallholder farms in Africa is beginning to be recognized as the fundamental biophysical limiting factor responsible for the declining per capita food production of the continent (IFPRI, 1996; Sanchez et al., 1996, 1997). The magnitude of nutrient mining is huge, as evidenced by nutrient balance studies. An average of 660 kg of N, 75 kg of P and 450 kg of K ha^{-1} has been lost during the last 30 years from about 200 million ha of cultivated land in 37 African countries. The total annual nutrient depletion in sub-Saharan Africa is equivalent to 7.9 Mt $year^{-1}$ of N, P and K, six times the amount of annual fertilizer consumption to the region, excluding South Africa (Sanchez et al., 1997). Nutrient capital has gradually been depleted by crop harvest removals, leaching, and soil erosion. This is because farmers did not sufficiently compensate for these losses by returning nutrients to the soil via crop residues, manures, and inorganic fertilizers. The consequences of nutrient depletion are felt at the farm, watershed, and on national and global scales, and include major economic, social, and environmental externalities. Sanchez et al. (1996, 1997) suggested that soil fertility replenishment should be considered as an investment in natural resource capital.

Phosphorus replenishment strategies are mainly fertilizer-based, with biological supplementation, while N-replenishment strategies are mainly biological, with chemical supplementation. Replenishing phosphorus capital can be accomplished by large applications of P fertilizers in high P-fixing soils. Africa has ample rock-phosphate deposits that could be used directly or as superphosphates to reverse phosphorus depletion.

One of the problems is the need to add acidifying agents to rock-phosphates, in order to facilitate their dissolution in many P-depleted African soils that have pH values above 6.0, which are too high for acidification to occur at a rapid rate. Decomposing organic materials produce organic acids that may help acidify rock-phosphate. Mixing rock-phosphates with compost has shown promise in increasing the availability of rock-phosphate at sites in Burkina Faso (Lompo, 1993) and Tanzania (Ikerra et al., 1994). Organic acids produced during the decomposition of plant materials may temporarily reduce the P-fixation capacity of the soils by binding to the oxides and hydroxide surfaces of clay particles (Iyamuremye and Dick, 1996). Through this process P availability and nutrient-use efficiency are temporarily increased. Research in western Kenya with Minjingu rock-phosphate and triple superphosphate indicates higher maize yields following incorporation of P with *T. diversifolia*, rather than urea, at an equivalent N rate (Fig. 33.2). The benefit

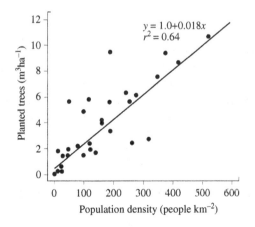

$$y = 1.0 + 0.018x$$
$$r^2 = 0.64$$

Planted trees ($m^3 ha^{-1}$) vs Population density (people km^{-2})

FIGURE 33.2 The effect of nitrogen source, as either urea or *Tithonia diversifolia* biomass transfer (1.8 t ha^{-1} of dry mass), with Minjingu rock-phosphate (RP) and triple superphosphate (TSP). Both applied at a recapitalization rate of 250 kg ha^{-1} of P, on maize grain yield on an acid soil near Maseno, Kenya. The amounts of N supplied by urea and *T. diversifolia* were the same, 60 kg ha^{-1} of N. *Adapted from Buresh, R.J., Smithson, P.C., Hellums, D., 1997. Building up soil P capital in sub-Saharan Africa. In: Replenishing Soil Fertility in Africa. ASA-SSSA Special Publication (In the press).*

from *T. diversifolia* was partially attributed to the addition of K and about 5 kg of P ha^{-1} (Buresh et al., 1997). Subsequent research confirmed higher maize production with sole application of *T. diversifolia* biomass than with an equivalent rate of NPK mineral fertilizer on a P- and K-deficient soil (Bashir Jama et al., unpublished data). The integration of available organic resources, such as *T. diversifolia*, with commercial P fertilizers may be important in increasing and sustaining soil phosphorus capital (Palm et al., 1997).

Given the largely biological nature of the nitrogen cycle, the use of organic inputs, as a source of biologically fixed nitrogen and deep nitrate capture, plays a crucial role in N-replenishment. Agroforestry trees and herbaceous leguminous green manures play a major role in internal cycling. Organic inputs have an important advantage over inorganic fertilizers with regard to fertility replenishment; they provide a carbon source for microbial utilization, resulting in the formation of soil organic N. Inorganic fertilizers do not contain such carbon sources; therefore, most of the fertilizer N not used by crops is subject to leaching and denitrification losses, while much of the N released from organic inputs and not utilized by crops could build soil organic N capital (Sanchez and Palm, 1996). Nitrogen fertilizers are likely to be needed to achieve high crop yields on top of the nutrient contributions of agroforestry (Sanchez et al., 1996).

Accompanying technologies and enabling policies are needed to make recapitalization operational. Soil conservation technologies must be present in order to keep the nutrient capital investment in place, and to avoid polluting rivers and ground waters. Policy improvements are needed to provide the timely availability of the right types of fertilizers at reasonable cost, better infrastructure, credit, timely access to markets, adaptive research and extension education—particularly in the combined use of organic and inorganic sources of nutrients. The issue of who should pay for this recapitalization is based on the principle that those who benefit from a course of action should incur the costs of its implementation. On-farm maintenance costs should be borne by farmers, whereas national and global societies should share the more substantial costs of actual phosphorus applications. This sharing should reflect the ratio of national to global benefits (Sanchez et al., 1997).

Intensifying and Diversifying Land Use Through Tree Domestication

Soil fertility replenishment can go a long way in boosting agricultural production in Africa. However, although it is necessary, it is not sufficient for attaining food security and eliminating rural poverty—particularly considering the economic constraint on farmers' affording fertilizers. Numerous other factors have to come together as well, such as postharvest losses, pests, and disease attacks, the declining size of land holdings and declining human health. The last two factors have an impact on the availability of field labor, which is also a consequence of family members moving to the town to secure off-farm income. What is needed is a paradigm shift from policies directed only at increasing yields of the few staple food crops to one geared at "putting money in farmers' pockets." This rock-phosphate approach has played, and will continue to play, an important part in meeting the needs of the rural poor, but additional steps must also be taken. It is in this vein that Sanchez and Leakey (1997) suggest that a further transformation is needed in the long run: intensifying and diversifying land of smallholder farms in Africa in ways that generate income for farmers so that they have the option to invest in farm inputs.

President Yoweri Museveni of Uganda, in his opening address to a Special Program for African Agricultural Research (SPAAR) meeting in Kampala, 6 February 1996, articulated this idea very clearly. He stated: "It does not make sense to grow low-value products (maize and beans) at a small-scale; instead, high-value products should be grown at a small-scale, while low-value products should be grown on a large-scale."

The obvious implication is that small-scale farming in Africa must diversify by producing a combination of high-value, profitable crops along with the basic food crops. Examples of this strategy occur in western Kenya, where small patches—in the order of 100 m^2—of French beans are grown by smallholders contracted by an exporting company for fresh consumption in Europe. The market is assured, and farmers intensively water, fertilize, and weed these islands of wealth among their lower-value crops. But the largest opportunities for farm diversification come from trees producing an array of marketable products.

Traditionally, people throughout the tropics have depended on indigenous plants for fruits and everyday household products, from medicines to fibers. These products have also provided the essential vitamins and minerals for family health, and through local and regional trading have generated cash to meet household needs for purchased products and services. May be it is here, in peoples' own backyards, that the solution lies. But sadly, through deforestation, the forest or woodland that used to be in the farmers' backyard has now all but disappeared for the vast majority of people in Africa. This is where tree domestication as part of agroforestry becomes so important. Already there is a body of biophysical information on the techniques available to domesticate a wide range of wild tree species (Leakey and Newton,

TABLE 33.1 Examples of "Cinderella" tree species with high potential for domestication (Leakey et al., 1996).

Species	Common names	Ecoregion	Products
Irvingia gabonensis	Bush mango, Mango sauvage	Humid West Africa	Fruit, kernels
Uapaca kirkiana	Mahobohobo	Miombo of Southern Africa	Fruit
Sclerocarya birrea	Marula	Miombo of Southern Africa	Fruit, beverage
Bactris gasipaes	Peach palm, pejibaye, pupunha pijuayo chontaduro	Western Amazonia	Fruit, heart of palm, fibers, parquet floors,
Vitellaria paradoxa	Karite, shea nut	Sahel	Edible oils Cosmetics
Prunus africana	Pygeum	Montane tropical Africa	Medicinal
Pausinystalia johimbe	Johimbe	Humid West Africa	Medicinal

1994b,c; Newton et al., 1994; Franzel et al., 1996; Leakey et al., 1996). Furthermore, guidelines have been developed for determining the species priorities of farmers (Franzel et al., 1996; Jaenicke et al., 1996).

These "Cinderella" species—so called because their value has been largely overlooked by science although appreciated by local people—include indigenous fruit trees and other plants that provide medicinal products, ornamentals, or high-grade timber. Some examples are shown in Table 33.1.

Techniques being developed to convert some of these wild species into domesticated crops in agroforestry systems include vegetative propagation and clonal selection designed to capture genetic diversity (Leakey and Jaenicke, 1995). Domestication involves the formulation of a genetic improvement strategy for agroforestry trees and a strategy for the use of vegetative propagation to capture the additive and nonadditive variation of individual trees in tree populations (Simons, 1996). The domestication strategy for these indigenous fruit tree species, as well as for *Prunus africana* and *Pausinstalia johimbe*, two priority trees for medicinal products, is to conserve their genetic resource in living-germplasm banks and subsequently to develop cultivars for incorporation into multistrata agroforests (Leakey and Simons, 1997).

High-value trees can fit in specific niches on farms, making the system ecologically stable and more rewarding economically, thus diversifying and increasing rural incomes and improving food security. Timber trees can also be grown on-farm boundaries with leguminous fodder trees under them. Similarly, fuelwood trees can be grown on field boundaries or as contour hedges on sloping lands. In such a scheme, improved fallows become a crucial part of the crop rotation. The result is that farm income is increased and diversified, providing resilience against weather or price disruptions, soil erosion is minimized, nutrient cycling is maximized and above- and belowground biodiversity is enhanced. The farm truly approximates a functioning ecosystem. The latest definition of agroforestry summarizes this approach: a dynamic, ecologically based, natural resource management system that, through the integration of trees in farms and in the landscape, diversifies and sustains smallholder production for increased social, economic, and environmental benefits (Leakey, 1996).

Through domestication these tree crops could become higher yielding, produce higher-quality products, be more attractive commercially, and diversify diets (Leakey et al., 1996). Such progress could improve household welfare by providing traditional food and health products, boosting trade, generating income and diversifying farming systems, both biologically and economically, beyond the production of basic food crops. Generally, tree crops have lower labor requirements than basic food crops, and could thus allow farmers time for off-farm income generation. A new paradigm for smallholder farming in Africa emerges: one that instead of being based on a limited number of highly domesticated crops, often grown in monoculture, is based on a much greater diversity of commercially important plants that together produce food and high-value products (Leakey and Izac, 1996).

Enabling Policies

Current policy recommendations place a high priority on the revitalization of the agricultural sector in Africa (FAO, 1996; IFPRI, 1996), and some success stories are beginning to emerge (Cleaver and Schreiber, 1994). The fact that most food in Africa is produced by smallholders, often female farmers, is frequently considered a major constraint to agricultural development. In contrast, we believe that small-scale farms can be an asset rather than a liability when supported by appropriate policies. The agricultural production boom in Asia is a product of smallholder farms and not of a shift from small- to large-scale farming. The policies include improvements in land tenure, infrastructure, marketing information, credit, research, extension, and access to inputs and markets at reasonable prices (Place, 1996). Public investment to increase access to education of girls and improve public health services in rural areas also plays an important role in this transformation process. Policy reform to seize opportunities for smallholder development and to eliminate policies that discriminate against the smallholder agricultural sector therefore remains a top priority. Indeed, policy reform is a necessary, but not a sufficient, condition for food security and environmental conservation. In order for enabling policies to work in most of Africa, the twin issues of soil fertility depletion and land use intensification and diversification have to be tackled.

Therefore, the vision now is of agroforestry as an integrated land use policy that combines increases in productivity and income generation with environmental rehabilitation and the diversification of agroecosystems. Such a vision can be fitted to the range of situations found in the major ecoregions of Africa. According to Cooper et al. (1996) and Sanchez et al. (1997), the realization of this vision, however, is going to be dependent on (1) the appreciation by the international community of the importance of soil fertility replenishment and high-value indigenous species in the lives and welfare of local people, as well as incentives (or the removal of disincentives) for local people to plant trees on their farms; (2) replenishment of plant nutrients, that can also be viewed as an investment in natural resource capital, similar to investments in dams and irrigation; (3) the domestication of commercially important indigenous tree species producing high-value products; and (4) the development of processing infrastructure at the rural scale and a dynamic market perspective at the national and global scales.

Commercialization is both necessary and potentially harmful. It is necessary because without it the market for products is small, and the opportunity for rural people to make money would not exist. A degree of product domestication is therefore essential. On the other hand, commercialization is potentially harmful to rural people if it expands to the point where outsiders with capital to invest come in and develop large-scale monocultural plantations. However, from the experience of the complex agroforests in Southeast Asia (de Foresta and Michon, 1994; Michon and de Foresta, 1996), smallholder units producing nontimber forest products that are also biologically diverse and economically viable, indicate that the intensification and diversification of land use is not a pipe-dream.

THE WAY FORWARD

While land use intensification caused by demographic pressure is generally associated with environmental degradation, the long-term relationship between land resource degradation and demographic pressure is not necessarily negative and linear (Harwood, 1994; Scherr and Hazell, 1994). With further increases in population pressure, however, a point is

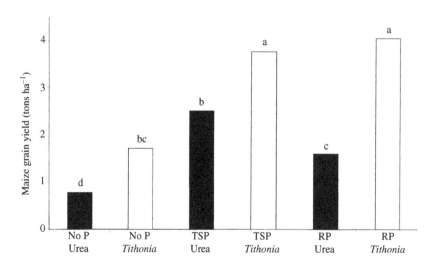

FIGURE 33.3 Correlation between high population density and planted woody biomass for districts in high potential areas of Kenya (Holmgren et al., 1994).

reached where degradation is reversed, with further land intensification and incorporation of trees within the farm. This has happened in the semiarid Machakos District of Kenya, where, despite increasing population pressure since the 1930s, farmers were able to reverse land degradation through an indigenous soil conservation technology that improved both crop and livestock productivity (Pagiola, 1994; Tiffen et al., 1994). This technology did not have a major agroforestry component, but recent evidence in eastern Africa indicates that the same is true with agroforestry. In the more heavily populated areas of Burundi (Place, 1995), Kenya (Holmgren et al., 1994; Bradley et al., 1995; Patel et al., 1995) and Uganda (Place and Otsuka, 1997) where farm size is extremely small, the number of trees on farms is also expanding as farmers increasingly recognize their value (Fig. 33.3). In fact, much of the reforestation in the tropics is taking place on farms, though agroforestry, and not as plantations (J. Spears, personal communication). Most of the planted trees are generally of low-value and used for fodder, fuelwood, boundary delineation and exotic fruits like avocado and mango. The next step is to incorporate high-value domesticated trees into these farms. If the three determinants are realized—replenished soils, high-value trees and enabling policies—Africa will be facing a win-win-win situation (socially, economically, and ecologically) where poverty alleviation, food security, and environmental protection go hand in hand.

Chapter 34

Addressing the Causes of Land Degradation, Food/Nutritional Insecurity and Poverty: A New Approach to Agricultural Intensification in the Tropics and Sub-Tropics

This chapter was previously published in Leakey, R.R.B., 2013. In: Hoffman, U. (Ed.), Wake Up Before It Is Too Late: Make Agriculture Truly Sustainable Now for Food Security in a Changing Climate, UNCTAD Trade and Environment Review 2013. UN Publications, Geneva, Switzerland, pp. 192–198 (Chapter 3), with permission from United Nations

The productivity of conventional high-input agriculture has been greatly increased by the achievements of the Green Revolution, saving millions of people from starvation. However, this achievement came at a high environmental cost in terms of land conversion from forest (deforestation), land degradation, and the overexploitation of natural resources—especially soil and water. It is now also recognized as being a major contributor to climate change. Furthermore, despite the success of improved productivity of major food staples, there are still billions of people suffering from poverty, malnutrition and hunger. Consequently, there have been many calls for a new approach to food production, especially in the tropics and subtropics where the problems and issues are most urgent and prevalent. The key issues to be addressed are land rehabilitation, food and nutritional security and income generation—all within sustainable land-use practices. The overriding questions are: How can the land be used to feed a growing population without further damage to the local and global environment? How can food and nutritional security be achieved on a declining area of available land? How can the land be used to enhance the livelihoods and income of those in poverty?

Answers to these questions fall into two main camps: those that insist that the only way forward is an intensification of the high-energy input Green Revolution model involving further productivity improvements through new and exciting approaches to crop and livestock genetics, against those that think more ecologically based approaches to low-input agriculture are the way forward. To consider the merits of these two contrasting and highly polarized views, let's look at the environmental and socioeconomic problems arising from land conversion to agriculture and then seek some solutions.

Current land-use practices in the tropics have led to deforestation, overgrazing and overexploitation of soils and water resources (Fig. 34.1), causing a cascade of negative impacts: land degradation, loss of soil fertility, loss of biodiversity, the breakdown of agro-ecosystem function, declining yields, hunger and malnutrition, and declining livelihoods. Associated with this is reduced access to traditional wild foods, loss of income, and the increased need for costly (often unaffordable) agricultural inputs. The response of proponents of intensive, high-input industrial farming is to redouble efforts to increase the yield of staple food crops by enhancing their capacity to withstand biotic and abiotic stress.

Multifunctional Agriculture. DOI: http://dx.doi.org/10.1016/B978-0-12-805356-0.00034-9
© 2017 Elsevier Inc. All rights reserved.

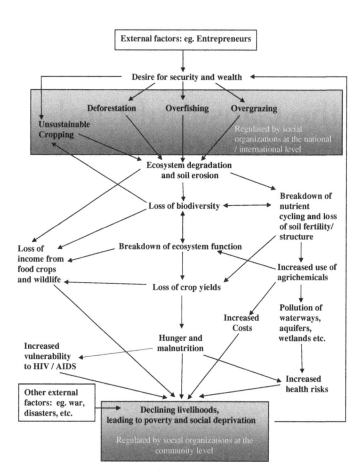

FIGURE 34.1 The cycle of biophysical and socio-economic processes causing ecosystem degradation, and increased social and economic deprivation. *Modified from Leakey, R.R.B., Tchoundjeu, Z., Schreckenberg, K., Shackleton, S., Shackleton, C., 2005a. Agroforestry tree products (AFTPs): targeting poverty reduction and enhanced livelihoods. Int. J. Agric. Sustain. 3, 1–23.*

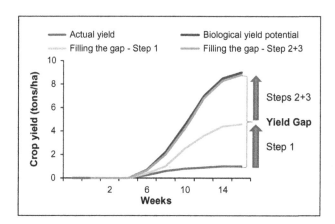

FIGURE 34.2 A diagrammatic representation of the Yield Gap (the difference between potential yield per hectare and actual yield achieved by farmers) in agriculture and the steps needed to close the Gap.

This approach fails to recognize three important points: (1) farmers are failing to grow staple foods anywhere near their existing biological potential, creating what is called the Yield Gap (Fig. 34.2), so increasing the biological potential will not help; (2) poor smallholder farmers locked in the Poverty Trap cannot afford to buy the fertilizers and pesticides (even if they had adequate access to them) that would allow them to practice monoculture agriculture; and (3) the over-riding dominance of starchy food staples in modern agriculture may provide adequate calories for survival, but they lack the protein and micronutrients for healthy living, not to mention the sensory pleasures of traditional and highly nutritious foods which used to be gathered from the forest. In addition, the widespread clearance of forest from the landscape, especially from hillsides, exposes soils to erosion and increases run-off, resulting in landslides and flooding that destroy property and cause the death of large numbers of people. Loss of perennial vegetation also contributes to climate change. In conclusion, therefore, an alternative approach to agricultural intensification is required, as several recent reviews of agriculture (IAASTD, 2009; Royal Society, 2009) and the role of agriculture in global environmental issues (e.g., MEA, 2005; GEO, 2007; CAWMA, 2007) have suggested that "business as usual" is no longer the appropriate option due to the scale of the problems and the constraints facing poor farmers.

So, is there a better alternative? The answer is "yes." Let's go back to basics and look at the cycle of land degradation and social deprivation (Fig. 34.1). Clearly a focus on crop yield is important, but rather than trying to increase yield potential, let's focus on closing the Yield Gap. In the worst cases farmers growing maize are achieving only $0.5-1.0 \, t \, ha^{-1}$ when the potential is around $10 \, t \, ha^{-1}$. In this situation, closing the Gap could increase food production by 15- to 20-fold; but even if it was only 2- to 3-fold on average, this is well over the 70% increase that might be required to feed the 9 billion people predicted to populate the world by 2050, according to IFPRI (2011).

The primary cause of the Yield Gap is poor crop husbandry, which leads to the loss of soil fertility and agroecosystem functions, such as the cycling of nutrient, carbon and water; the progress and operation of life cycles and food webs that maintain the natural balance between organisms; pollination and seed dispersal, etc. Typically, soil nitrogen is the prime constraint to crop growth in degraded soils. This can be restored by harnessing the capacity of certain legumes to fix atmospheric nitrogen in root nodules colonized by symbiotic bacteria (*Rhizobium* spp.). Numerous techniques have been developed to integrate appropriate legume species within farming systems. Probably the most effective and adoptable are high-density improved fallows with species like *Sesbania sesban* and *Tephrosia vogelii* or relay cropping with *Gliricidia sepium* (Cooper et al., 1996; Buresh and Cooper, 1999). Leguminous crops like beans and peanuts can also contribute to this process. Together the legumes will increase soil nitrogen to a level that will give maize yields of $4-5 \, t \, ha^{-1}$ within 2–3 years. In other words the Yield Gap is partially filled and food security is greatly increased. At this point, however, other soil nutrients are generally limiting and so the complete closure of the Yield Gap would require another approach involving the provision of inorganic nutrients, such as rock phosphate, or chemical fertilizers, which have to be purchased. So, the need now is to generate income.

However, before addressing the need for income, agroecosystem function has to be restored. The legumes will start this process. For example, one of the serious weeds of cereal crops like maize, millet and sorghum is *Striga hermonthica*. It is a root parasite on these cereals and its seeds germinate in response to root exudates from the young cereal plants. Interestingly, however, *S. sesban* and the fodder legumes *Desmodium intortum* and *Desmodium uncinatum* also trigger *Striga* germination, so they can be used to promote suicide germination in the absence of the cereal hosts (Khan et al., 2002). *Desmodium* spp. also acts as a repellent to insect pests of cereals, for example the stem borers *Busseola fusca* and *Chilo partellus*. Likewise, simple agroecological benefits can also be attained by planting Napier grass (*Pennisetum purpureum*) as an intercrop, or around small fields, as it attracts the pests away from the crops (Khan et al., 2006).

The two preceding interventions can therefore be used to restore soil fertility and initiate an agroecological succession, so rehabilitating farm land and reversing some of the land degradation processes. We can think of this as the first step towards closing the Yield Gap (Fig. 34.3).

Going the next step to a fully functional and more productive agroecosystem involves the integration of trees within the farming systems. Some trees are of course cash crops like coffee, cocoa and rubber, which in the past were either grown as large-scale monocultural plantations or as a two-species mixture, such as cocoa under the shade of coconuts or *G. sepium*. Increasingly, however, these are becoming smallholder crops grown in much more diverse species mixtures, such as bananas with fruits trees like mango, avocado and local indigenous trees producing marketable products (Leakey and Tchoundjeu, 2001). This is well developed in Latin America and Asia and is becoming widely recognized as a way to restore the biodiversity normally found in natural forests (Schroth et al., 2004; Clough et al., 2011). Certainly the replacement of shade trees with trees that also produce useful and marketable products is a good strategy for farmers wanting to maximize output from the land and to minimize the risks associated with reliance on a single crop species.

There has also been another silent farmer-led revolution in the tropics, especially in Southeast Asia. In Indonesia in particular, many farmers who used to practice shifting agriculture have replaced the natural fallow with a commercial fallow (agroforest) based on tree crops (Plate 34.1). They grow rice in the valley bottoms and plant a wide range of useful and commercially important tree species among the other food crops which they have planted on the valley slopes (Michon and de Foresta, 1999). These trees became productive in a succession in later years, creating a continuous supply of marketable produce (cinnamon, tung nut, damar, duku, rubber, etc.) for several decades—often ending in a

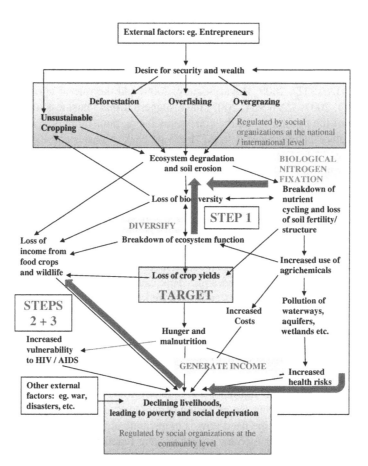

FIGURE 34.3 Procedures for closing the Yield Gap.

PLATE 34.1 Intensive rice cultivation in the valley bottom, with hillsides planted with a diverse array of commercially important trees for income generation and environmental benefits on the sloping farm land. There are some 3 million hectares of these "agroforests" in Indonesia alone.

timber crop. This diversification of the farming system with perennial crops therefore achieves several important outcomes. It protects sloping land from erosion, improving water infiltration into the soil; it sequesters carbon and so mitigates climate change; it generates income; it enhances biodiversity and promotes agroecosystem function—in other words it does all the things that large-scale monocultures fail to do and the livelihoods of the farmers are orders of magnitude better than those locked in poverty growing a failed maize crop in Africa. This approach to agriculture achieves high crop yields close to the biological potential on the best and most fertile land, and income generation from tree crops on the more marginal land, creating a land-use mosaic with many environmentally beneficial impacts (Plate 34.2). Importantly, there is also some evidence that complex perennial vegetation, such as natural forest or an agroforest, is better than an herbaceous crop at recycling moisture to the atmosphere to be advected downwind to fall as rain. Thus agroforests are likely to be beneficial to rainfed agriculture in dry and drought-prone areas of the world.

In a further initiative, over the last 20 years, agroforesters have been working to take this strategy to a higher level by starting to domesticate some of the very wide range of tree species that have been the source of locally important food and nonfood products traditionally gathered from the forest (Leakey, 2012b; Leakey et al., 2005a, 2012). The approach here has been to apply well-known horticultural techniques of vegetative propagation for cultivar development (Leakey, 2004; Leakey and Akinnifesi, 2008). Unconventionally, this has been implemented at the village level as a participatory process with local communities, rather than on a research station (Leakey et al., 2003; Tchoundjeu et al., 2006; Asaah et al., 2011). This participatory approach was implemented to ensure that the farmers are the instant beneficiaries of the domestication and that they are empowered by the development of their indigenous and local knowledge. Because wild populations of tree species contain 3- to 10-fold variation in almost any trait of commercial interest (Leakey et al., 2005a), the potential for substantial improvements in fruit/nut size, quality, chemical content, etc. is very large. This means that new, highly productive cultivars yielding good quality produce with the product uniformity required by markets is easily obtainable. Furthermore, because the multiplication process is implemented by vegetative propagation based on mature tissues with the capacity to flower and fruit, the long unproductive period usually associated with tree crops is circumvented, and trees are productive in 2−3 years.

Proof of concept has recently been demonstrated by the implementation of a participatory tree domestication project in Cameroon (Tchoundjeu et al., 2006, 2010; Asaah et al., 2011). In 12 years it grew from 4 villages and a small number of farmers to over 450 villages with 7500 farmers. The flow of benefits, such as income generation, started within less than 5 years (Fig. 28.9), while the farmers are reporting many other ways in which the project has also improved their lives (Asaah et al., 2011). Perhaps the most signifcant outcome has been the fact that young men and women in these communities now see a future for themselves within the community rather than from migration to local towns. In addition, the processing and value-addition of produce from domesticated trees and other crops have been found to provide off-farm employment and to stimulate local enterprise and trade.

PLATE 34.2 A multifunctional agriculture landscape mosaic in Vietnam with many income-generating tree-based production systems on hillsides surrounding an area of intensive food production on the most fertile soils.

Historically crop domestication has been implicated in the rise of civilizations that are settled, politically centralized, socially stratified, economically complex and technologically innovative societies (Diamond, 1997). As the first wave of crop domestication primarily benefited the industrial countries of northern latitudes, it seems that the time is now ripe for a second wave of domestication to favor tropical and subtropical countries and so enhance social equity and environmental rehabilitation worldwide (Leakey, 2012c; Leakey and Asaah, 2013).

The creation of new cash crops from the domestication of traditionally important, highly nutritious and useful species can be seen as the second step towards closing the Yield Gap, because they can generate the income needed to purchase fertilizers and other agricultural inputs (Fig. 34.2). These trees can be used to enrich and improve the farming systems, whether as shade for commodity crops, agroforests on hillsides, orchards, field and farm boundaries, fodder banks, or woodlots. However, farmers have many other competing demands for their money, whether for local ceremonies, health care, children's education, farm infrastructure, market transport, etc. Consequently, the third step to closing the Yield Gap is to further expand the commercialization of these new tree crops, so creating business opportunities and employment.

Most of the traditionally important products from tropical forests have been marketed locally for centuries. Over the last decade an increasing number of these have been processed as new food, medicinal, nutraceutical, or cosmetic products, based on the fruits, nuts, gums, resins and fibers. Some of these have entered regional and international markets. The marketing and trade of commodities from tropical producers has often been exploitative. So with the emergence of this new trade there has been a parallel initiative to ensure that the producers receive a fair price—see the Fair Trade Foundation (www.fairtrade.org.uk). In addition, ways have been sought to develop marketing partnerships aimed at the propoor commercialization of the traditionally important products derived from indigenous trees (Lombard and Leakey, 2010). These partnerships work to develop the products to a marketable standard and establish strong and viable trade associations that are forward thinking and market oriented. Through these partnerships it is possible to ensure long-term relationships and supply agreements that ensure that the producers remain in the value chain.

We should not forget the importance of livestock in agriculture. The 2020 projections of the International Food Policy Research Institute suggest that we will need 40% more grain and will eat a lot more meat. As we have just seen, we can greatly increase grain production by closing the Yield Gap. Recent developments have also demonstrated that fodder trees can be used to increase the productivity of cattle and goats. The integration of fodder trees and livestock into a farm is one of the elements of diversification that could be part of Step 2.

Another recent development has been the establishment of public-private partnerships between multinational companies, national and international research teams and local producer communities, to promote and produce new products for international trade. Examples include Daimler AG in Brazil who are manufacturing components for the motor industry based on products produced in agroforestry systems by local communities (Panik, 1998), as well as Unilever plc. which is developing a new oil crop for margarine production with communities in Ghana and Tanzania using kernel oil from *Allanblackia* spp. (Jamnadass et al., 2010).

All of these developments offer a new approach to agriculture, delivered by agroforestry practices (Leakey, 2010), which is more sustainable—environmentally, socially and economically—than current conventional approaches. This model conforms to the concepts of Multifunctional Agriculture promoted by the International Assessment of Agricultural Knowledge, Science and Technology for Development (IAASTD, 2009) which was ratified by over 60 countries in 2009.

Chapter 35

Trees: A Call to Policy Makers to Meet Farmers' Needs by Combining Environmental Services With Marketable Products: An Update

R.R.B. Leakey

Chapter 34 (Sanchez et al., 1997) takes us back to some of the first ideas about how to develop agroforestry as the integration of applied agroecology (Chapters 1 and 2 [Leakey, 1996; 2017a]) with income generation from indigenous food trees (Chapter 7 [Leakey, 1999]). Chapter 33, (Sanchez et al., 1997) identified the three main determinants for overcoming rural poverty in Africa as: "(1) reversing soil fertility depletion, (2) intensifying and diversifying land use with high-value products, and (3) providing an enabling policy environment for the smallholder farming sector." As we have seen in Chapters 10, 15, 26 and 32 (Leakey, 2017c,d,e,f), great progress has been made with regard to the development of strategies, techniques and community-based interventions to address the first two of these important problems. Sadly, however, much less progress has been made on the third activity—policy change—despite calls for a new approach (IAASTD, 2009; Kiers et al., 2008).

Why has policy lagged behind? The answer, I think, is that policymakers are hooked on the conventional paradigm of intensive modern agriculture, in which trees are seen as a hindrance to mechanical operations, rather than as a key component of healthy agroecosystem function. This concept of what is "right" stems from the success of industrial agriculture in temperate climates where the biophysical, social and economic environments are all totally different from those in the tropics or subtropics, especially in Africa. So agricultural interventions involving the planting of trees with crops are counterintuitive. Recognizing this dilemma, Leakey and Tomich (1999) posed some questions, answers to which might improve understanding for the need to have a different mindset. Since then, many of the questions have been addressed and, as we saw earlier (Chapters 28 and 31 [Leakey 2014a; Leakey and Asaah, 2013]), the Cameroon case study illustrates the large number of factors that need to be considered in implementing a more appropriate agricultural paradigm, especially when current activities have caused land degradation and trapped farmers in poverty.

The need for policy change has been widely recognized by numerous major international reports, such as the International Assessment of Agriculture, Science, Knowledge and Technology for Development (IAASTD, 2009; Leakey et al., 2009); Millennium Ecological Assessment (Hassan et al., 2005), The Royal Society's *Reaping the Benefits: Science and the Sustainable Intensification of Global Agriculture* (Royal Society, 2009) and other international reports, which all called for a change away from "business as usual." This lack of progress is despite remarkable consensus among the hundreds of IAASTD authors, selected for their skills and knowledge and who recommended a relatively simple package of interventions that would enhance agricultural production while reducing damage to the environment (Kiers et al., 2008).

The main criticisms of the IAASTD Report *Agriculture at a Crossroads* (IAASTD, 2009) were from agribusiness and academia, with the former saying that it was "antibusiness" and the latter that it was "antiscience." As we have just seen, closing the Yield Gap involves overcoming the environmental, social, and economic constraints by a combination of better crop husbandry and income creation for the purchase of agricultural inputs. This needs to be done on a massive scale and is actually a process that would create millions of new customers for the products of agribusiness. Furthermore, as we have seen, the new tree products arising from the domestication of indigenous food and nonfood

Multifunctional Agriculture. DOI: http://dx.doi.org/10.1016/B978-0-12-805356-0.00035-0
© 2017 Elsevier Inc. All rights reserved.

tree crops (see Chapters 14 and 15 [Leakey et al., 2012; Leakey, 2017d]) create opportunities for new rural industries (Chapter 25 [Leakey and Page, 2006]). With a change of mindset away from processing in industrial countries, toward local processing supported by multinational companies, an economic revolution could occur in the tropics and subtropics that would go a long way toward the alleviation of global poverty: a "pro-business" scenario, based on a new way of doing business.

Likewise, the charge of being "antiscience" is perhaps more a matter of perception. What is good science? Well, in my view it should address serious problems, and few could be so great as those we have been discussing. Second, it should be relevant and appropriate to the needs of the clients: in this case millions of poor people throughout the tropics and subtropics. As we have seen, the failure of agriculture in Africa to resolve food and nutritional insecurity lies in environmental degradation and lack of access to agricultural inputs due to poverty. Yet, the major focus of agricultural research is on upstream biotechnology rather than on agroecology. As we saw in Chapter 2 (Leakey, 2014e), modern science is only just starting to understand the highly complex interactions that drive the proper functioning of agroecosystems. Agroecology is indeed one of the most exciting frontiers for future science. Furthermore, as we saw in Chapter 14 (Leakey et al., 2012) and Chapter 15 (Leakey, 2017d), some areas of biotechnology are playing an important role in the domestication of new crops from wild fruit and nut trees. However, the benefits of much of the current focus on the genetic modification of conventional food crops will not resolve the land degradation and social deprivation constraints to agricultural production in the tropics, the main causes of the Yield Gap. This is because they do not offer solutions to the broader socioeconomic problems faced by developing countries (Kiers et al., 2008). Indeed, it seems there is little point in producing new conventional crop varieties for Africa with greater potential yield, when it is the "actual yield" achieved, rather than "potential yield," which is the problem. The focus now needs to be on improving actual yield by better crop husbandry. In the longer term, once the Yield Gap in Africa is closed, new crop varieties with higher potential yield will indeed be pertinent.

To address the Yield Gap, Chapter 34 (Leakey, 2013) is about trying to develop a relatively simple "conceptual framework" for a new paradigm for agricultural policy based on appropriate crop husbandry. To understand the problem, this initiative is first based on a further analysis of the "cycle of land degradation and social deprivation" (Figure 34.1) that arises from the loss of soil fertility and the lack of income to purchase conventional farm inputs like fertilizers and pesticides. Then from an understanding of the multiple interacting biophysical, social and economic factors that lie behind this cycle, to formulate integrated interventions to underpin future policy options. To address the Yield Gap, Chapter 34 (Leakey, 2013) that food insecurity arising from poor yields from cereal and other crops is the consequence of the Yield Gap (Figure 34.2)—the difference between actual and potential yield. The challenge for tropical agriculturalists is therefore to identify a package of simple and low-cost crop husbandry interventions forming a delivery mechanism for the upscaling multifunctional agriculture that is appropriate to the economic, social, and market constraints faced by subsistence households. In particular these need to address the key problems identified by farmers during "Diagnosis & Design" surveys (Raintree, 1987)—soil infertility, degradation and erosion; a lack of tree products, hunger, and a lack of income (see Fig. 4.4)

To improve crop husbandry and to generate new income streams, a generic model for multifunctional agriculture, delivered by agroforestry, has been presented in Chapter 34 (Leakey, 2013). It is based on three steps (see Fig. 28.2): Step 1 uses biological nitrogen fixation to alleviate soil nitrogen deficiencies and recreate agroecological function; Step 2 then diversifies and intensifies the farming system with new tree crops developed from local flora to generate income and enhance agroecological function; then in Step 3, income is further increased by adding value to the tree products and using this income to purchase farm inputs (fertilizers, pesticides, casual labor, etc.). Step 1 partially closes the Yield Gap and so enhances food security; it is the combination of Steps 2 and 3—the added value accumulated down the domestication-commercialization continuum—which can allow farmers to reverse the cycle of land degradation and social deprivation (Figure 34.3) and so provide the means to close the Yield Gap. However, potentially, the positive impacts on poverty can be even greater, leading to sufficient income generation to enhance opportunities for education, health treatments, social equity, business and employment; and develop local infrastructure (Leakey, 2012b).

From the pilot project in Cameroon (Chapters 28 and 31 [Leakey, 2014a; Leakey and Asaah, 2013]), evidence is emerging that at the level of several hundred villages, the cycle of land degradation and social deprivation can be reversed. The remaining challenge is, therefore, to resolve the upscaling issues, which in effect brings us back to the slow progress being achieved in bringing about a new paradigm for tropical agricultural policy discussed earlier. It has to be hoped that the private sector will play a role in this (Kiers et al., 2008). Sadly, it seems that big business has been blind to what could be very lucrative global economic growth if they were to think outside the box and work with local entrepreneurs, to grow truly tropical enterprises in-country.

A new approach to global economic growth would help to address the issues of our divided world—half rich and half poor. We need to recognize that the cultures, the environment, soils and ecology of temperate and tropical regions affecting agriculture are totally different (Table 35.1), and thus that farming systems do not need to be identical around the world. Thus, I believe that by using socially-modified tree crops it is perfectly possible to intensify agriculture that address the constraints to productive smallholder agriculture in ways that are appropriate to the environment and the economic growth of tropical and subtropical countries. Chapters 36–40 (Leakey, 2012a; 2014f; 2017h; 2017i; Leakey and Prabhu, 2017) try to provide pointers in this direction, while also offering unconventional ideas about how to progress toward the achievement of the new Sustainable Development Goals.

To conclude, the problems of the Yield Gap mean that the great potential advances of the Green Revolution are not delivering the outcomes that are needed to feed the world and to lift almost half the world population out of low incomes and poverty. Basically, this is because we are not growing crops sustainably. If, on the other hand, we were to scale-up the approaches to multifunctional agriculture described here, we could potentially increase actual yields three- to sixfold and make big advances toward food security for all—with better nutrition and livelihoods for all. This would also greatly increase the economic returns on the substantial global investment in the Green Revolution. Surely, it's time to take action!

TABLE 35.1 Comparison of shifting agriculture, modern intensive agriculture and the application of agroforestry toward the transition to multifunctional agriculture.

	Traditional shifting agriculture	Modern intensive agriculture	Agroforestry – step 1	Agroforestry – steps 1 + 2	Agroforestry – steps 1 + 2 + 3
Land resource	Access to 15 + ha per household	Access to economies of scale – up to industrial scale	Access restricted to 2–5 ha per household	Access restricted to 2–5 ha per household	Access to 2–5 ha with possible expansion as some households transfer to other sectors
Farming activity	Nonsedentary – rotation around ten plots of land each cultivated for 2–3 years	Sedentary farming with continuous cropping. 1–5% of population in agriculture	Sedentary farming with wild trees providing environmental services and some products. 80% + of population in agriculture	Diversified sedentary farming with improved trees yielding more and quality marketable products	Enriched and diversified farming with improved tree products processing opportunities outside agriculture
Agroecological functions	Natural processes of soil fertility restoration and ecosystem function in natural fallows	Artificial fertilizers, herbicides, and pesticides to replace natural agroecosystem function	Biological nitrogen fixation (BNF) hamessed for soil fertility enrichment by improved fallows	BNF hamessed for soil fertility enrichment by improved fallows and crop diversification for better agroecosystem function	BNF hamessed for soil fertility enrichment by improved fallows and crop diversification for better agroecosystem function
Livelihoods	Subsistence lifestyle	High-input farming with commercial lifestyle	Subsistence lifestyle with improved food security	Rising from subsistence with greater access to income	Further improved lifestyle with greater access to income, employment, and business
Impacts	Unsustainable with declining livelihoods when population pressures rise	Unsustainable without access to capital, income, and natural resources	Barely sustainable with low standard of living	Slowly increasing living standards	Growing opportunities for income generation and rising living standards—in and outside agriculture

Section 8

Sustainable Intensification of Tropical Agriculture

Despite all the good news presented so far, the reality is that the progress to date barely scratches the surface of the global problems. Over 35% of farm land worldwide (around 2 billion hectares) can be categorized as degraded and nearly 50% of the population (about 3 billion people) live on less than US$2 per day. These problems are essentially ones found in the tropics and subtropics, especially in Africa. Analysis of the factors leading to these and their associated problems of food insecurity and malnutrition, low crop yields, loss of biodiversity, and agroecological dysfunction identifies a downward spiral of land degradation and social deprivation which create a large gap between actual and potential yield that is attributable to conventional approaches to agriculture. Thus in the tropics the conventional approach to agricultural intensification seems to be the perpetrator of a food crisis, rather than the provider of food for all. Furthermore, the failure to address the issues of African agriculture may also be putting a brake on the global economy, as these issues in effect leave about half the world's population outside the cash economy.

The previous chapter presented three simple steps to a locally appropriate and multifunctional form of intensification that represents a generic and adaptable model for more sustainable agriculture in the tropics. This refinement builds on the success of, and enhances the outputs of, the Green Revolution and can reverse the complex downward cycle of land degradation and social deprivation through ecological and socioeconomic interventions at several different "pressure-points" within this spiral. This chapter advocates 12 important principles that rehabilitate degraded land and improve the livelihoods of poor smallholder farmers. This multifunctional approach, centered on "agroecology + income," involves the combination of many annual and perennial crop species to diversify and enrich agricultural systems as a way to make them more productive and beneficial to rural communities, while also being climate smart and wildlife-friendly. This is not an alternative to current agricultural systems, but is a way to break the deeply engrained mold of current policy for tropical agriculture that stems from the relative success of temperate agriculture developed under a totally different social and environmental context. The previous suggestions resonate well with recent discussion about ecological intensification and sustainable intensification.

Chapter 36

The Intensification of Agroforestry by Tree Domestication for Enhanced Social and Economic Impact

This chapter was previously published in Leakey, R.R.B., 2012. CAB Reviews: Perspectives in Agriculture, Veterinary Science, Nutrition and Natural Resources, 7 (035), 1–3

SUMMARY

Agroforestry can be used to improve the productivity of staple food crops by improving soil fertility and promoting agroecosystem function. The domestication of agroforestry trees producing marketable products then intensifies the agroforestry system and leads to marketing, trade, and business opportunities that improve farmers' livelihood. Together, these steps provide a generic and adaptable model for more sustainable agriculture in the tropics, which builds on the success of, and enhances the outputs of, the Green Revolution.

INTRODUCTION

There are some common misconceptions about agroforestry that need to be addressed. Agroforestry is a low-input approach to agriculture, but it does not run counter to the Green Revolution. Instead it is an approach to correct some of the mistakes of the Green Revolution and to increase the productivity of modern crop varieties. It thus aims to improve the returns on the investment in the Green Revolution. If widely adopted, it should then open new windows of opportunity for agribusiness and help to achieve the original objectives of the Green Revolution to overcome hunger, malnutrition and rural poverty.

The need for more sustainable, but productive, agriculture has been widely recognized (World Summit on Sustainable Development in Rio de Janeiro in 1992; Millennium Ecosystem Assessment in 2005; Global Environmental Outlook 4 and the Comprehensive Assessment of Water Management in Agriculture in 2007; International Assessment of Agricultural Knowledge, Science and Technology for Development and The Royal Society in 2009 [Royal Society, 2009]). Several of these international reports have stressed that, because of the scale of global environmental and social problems, "business as usual" in agriculture is no longer an acceptable option. The difficulty, however, has been to see how to intensify farming systems while also alleviating poverty, malnutrition, hunger, and reducing environmental degradation—including climate change. In other words, how can we rehabilitate degraded farm land in order to feed the growing world population and so prevent further deforestation and land conversion?

Fundamental to this multifunctional objective is the recognition that the constraints to implementing improved food production are not simply overcome by increasing the potential yields of existing food crops. Instead the need is to understand the causes of poor yields and to address them. The prime constraints to improve crop yield are in fact social and environmental and not associated with the biological potential of modern varieties of the key staple food crops (Leakey, 2010).

Multifunctional Agriculture. DOI: http://dx.doi.org/10.1016/B978-0-12-805356-0.00036-2
© 2017 Elsevier Inc. All rights reserved.

THREE STEPS TO INTENSIFICATION

To address these constraints, it is necessary to find ways of improving soil fertility with the minimum use of artificial fertilizers as, in addition to being inaccessible, these are unaffordable by poor farmers living with an income of only US$1−2/day. For more than 25 years, agroforestry research has been developing systems based on "fertilizer trees"— leguminous trees and shrubs, which fix atmospheric nitrogen and release it to the soil. This low level of diversification in the fields can also initiate improved agroecological functions, especially those controlling some of the aggressive weeds such as *Striga hermonthica* and key pests of cereals such as the stem borers *Busseola fusca* and *Chilo partellus* (Khan et al., 2006). In effect this is the first of three steps to filling the yield gap (Leakey, 2010), the difference between the potential crop yield and that actually achieved by farmers. In much of Africa, this gap can be very substantial, owing to land degradation and the poverty of the farmers. The returns on the investment in the Green Revolution will be greatly increased by filling the yield gap.

The second step is to further diversify the farming system with trees and other perennial plants producing traditionally important and highly nutritious food, as well as nonfood, products. This diversification will further enhance the restoration of agroecological functions by creating more ecological niches in the farming system.

It will also create the opportunity to attract payments for environmental services—things such as watershed protection and the reduction of landslides, erosion, and flooding; carbon sequestration and biodiversity conservation (Swallow et al., 2009). The important social and economic benefits from this diversification, which improve the livelihood of poor farmers, can then be increased by domesticating them as new crop plants in partnership with local communities.

This participatory domestication intensifies agroforestry systems by improving the yield, uniformity, and quality of the products. It is thus a start to addressing the poverty constraint that is part of the cause of the yield gap by generating income from the sale of plants and products in local markets. This international program of agroforestry tree domestication began in the early 1990s and has been widely reported (see reviews, Leakey et al., 2005; Leakey, 2012b,c,e).

The third step is the commercialization of agroforestry tree products, so that the market is expanded and there are greater employment opportunities within trade and new business opportunities in developing countries (Leakey and Asaah, 2013). This step is where the rural population can start to enter the cash economy and continue the climb out of poverty.

Together, these three steps create a highly adaptable generic model—a new approach to agricultural intensification—which fits a very wide range of climates, agroecosystems and socioeconomic settings (see Table 35.1). Further development of this model also opens new frontiers of science in agroecology, postharvest technology and food science.

The agroforestry tree domestication initiative is seen to be the start of a second wave of crop domestication (Leakey, 2012c; Leakey and Asaah, 2013). The first wave started 1000s (thousands) of years ago and is currently manifest in the very productive set of staple food crops from the Green Revolution. These staple food crops have been credited with being fundamental to the creation of "civilizations that are settled, politically centralized, socially stratified, economically complex and technologically innovative societies" (Cribb, 2010)—such as those found primarily in temperate and Mediterranean latitudes. The second wave, now at the end of its second decade, is targeted at creating a similar impact in the tropics and subtropics.

In this respect, it seeks to create new crops to meet the needs of smallholder farmers currently suffering from poverty, malnutrition, and hunger (Leakey and Asaah, 2013; Asaah et al., 2011). This was the vision foreseen at the start of the initiative to promote the domestication of tropical trees (Leakey and Newton, 1994a,b), subsequently described as the "Really Green Revolution" (Leakey and Tomich, 1999; Leakey, 2001b).

CONCLUSION

The participatory domestication of new tree crops based on traditionally important trees from natural forests and woodlands enhances the multifunctionality of agroforestry systems that rehabilitate degraded land, restore agroecosystem function, and provide a sustainable pathway to lift poor smallholder farmers out of poverty, malnutrition, and hunger.

ACKNOWLEDGMENTS

I thank all my former colleagues at the World Agroforestry Center (ICRAF) and James Cook University in Australia and Center for Ecology and Hydrology in UK; our collaborators in many countries; and the farmers who allowed us to work on their farms.

Chapter 37

Twelve Principles for Better Food and More Food From Mature Perennial Agroecosystems

This chapter was previously published in Leakey, R.R.B., 2014. In: Proceedings of Perennial Crops for Food Security FAO Workshop, (Chapter 22), 28–30 August 2013, Rome, Italy

SUMMARY

An analysis of the factors leading to unsustainable agriculture and its associated problems of food insecurity, malnutrition and poverty, identifies a downward spiral of land degradation and social deprivation which is associated with lower crop yields, loss of biodiversity and agro-ecological function, and declining farmer livelihoods. This spiral is responsible for the Yield Gaps (the difference between the potential yield of a modern crop varieties and the yield actually achieved by farmers) found in many modern farming systems. To reverse this complex downward cycle and close the Yield Gap requires simultaneous crop and soil husbandry, ecological and socio-economic interventions at several different 'pressure-points' within this spiral. This paper advocates 12 important principles for the achievement of food security, which including the adoption of a simple, yet highly adaptable, three-step generic model involving perennial crops to kick-start the reversal of the spiral and so the closure of the Yield Gap. This agroforestry approach involves both the use of biological nitrogen fixation from trees and shrubs, as well as the participatory domestication and marketing of new highly nutritious cash crops derived from the indigenous tree species that provide poor people with the traditionally and culturally important foods, medicines and other products of day-to-day importance. Closing the Yield Gap improves food security by improving the yields of staple crops, but also has beneficial social, economic and environmental impacts. Agroforestry involving the combination of many annual and perennial crop species is, therefore, not an alternative to current agricultural systems, but is a way to diversify and enrich them, making them more sustainable. It does this by increasing food and nutrition security, increasing social and environmental sustainability, generating income, creating business and employment opportunities in rural communities and mitigating climate change. Agricultural policy currently tends not to appreciate these outcomes delivered by tropical and sub-tropical production systems which are based on perennial species and meet the requirements of 'sustainable intensification'.

INTRODUCTION

Agriculture faces a very complex set of social and biophysical issues associated with economic, social, and environmental sustainability. This paper examines the role of perennial species, especially trees, in the attainment of improved staple crop yields; provision of nutritious traditional food; the reduction of poverty, hunger, malnutrition, and environmental degradation; the improvement of rural livelihoods; as well as the mitigation of climate change—all with increased economic growth with a program of Integrated Rural Development (Leakey, 2010, 2012b, 2013). It therefore provides a model, or policy roadmap, for the delivery of the sustainable intensification of productive tropical and sub-tropical agriculture which is propoor and multifunctional—i.e., enhancing agriculture economically, socially, and environmentally (Leakey, 2012b). This paper is based on 12 interconnected Principles (see Box 37.1).

Multifunctional Agriculture. DOI: http://dx.doi.org/10.1016/B978-0-12-805356-0.00037-4
© 2017 Elsevier Inc. All rights reserved.

Box 37.1 Twelve principles for improved food security within multifunctional agriculture and enhanced rural development

1. Ask, do not tell
2. Do not throw money at farmers, but provide skills and understanding
3. Build on local culture, tradition, and markets
4. Use appropriate technology, encourage diversity and indigenous perennial species
5. Encourage species and genetic diversity
6. Encourage gender/age equity
7. Encourage farmer-to-farmer dissemination
8. Promote new business and employment opportunities
9. Understand and solve underlying problems: The Big Picture
10. Rehabilitate degraded land and reverse social deprivation: Close the Yield Gap
11. Promote Multifunctional Agriculture for environmental/social/economic sustainability and relief of hunger, malnutrition, poverty, and climate change
12. Encourage integrated rural development

PRINCIPLES

Principle 1: Ask Farmers What They Want, Do Not Tell Them What They Should Do

As the human population has grown, shifting cultivation has become less and less sustainable as deforestation has made new productive land scarcer. One consequence of this has been that farmers have been forced to become more sedentary. With this their crop yields have declined and farmers have struggled to feed their families, let alone generate income from surplus production. These families have therefore become increasingly trapped in hunger, malnutrition and poverty and are in need of help and substantial policy reform to free them from the circumstances that they are in. The problem originates with the advent of colonialism and the Industrial Revolution, because there has been a tendency for leaders in developed countries to think that agricultural developments that have worked in the temperate zone must be applicable in the tropics, despite big differences in the climate, soils, ecology, and socioeconomic conditions. As a result, agricultural policy in developing countries has often been based on a model that is not well adapted to local conditions.

Recognizing this issue, the work reported here began with a participatory approach to priority setting (Franzel et al., 1996, 2008) that sought the ideas of farmers on what they needed. These farmers identified their desire to grow the forest species from which, as hunter-gatherers and subsistence farmers, they had formerly gathered wild fruits, nuts, and other products of everyday value (Leakey, 2012b). This has led to an unconventional approach to agricultural development that focuses on the domestication of indigenous fruit and nut trees using a participatory approach.

From this initiative the following principles have emerged (Tchoundjeu et al., 2002, 2006, 2010; Leakey et al., 2003; Asaah et al., 2011; Degrande et al., 2006; Leakey and Asaah, 2013).

Principle 2: Provide Appropriate Skills and Understanding, Not Unsustainable Infrastructure

Many agricultural and other rural development projects provide funding for communities to implement new and "improved" technologies—often the ones based on concepts which are foreign to the farmers. While the funds are flowing, these projects can be successful, but very often when the project comes to an end the new approaches are not sustained. Typically this is because the stakeholders are still dependent on a continuing stream of finance, but this is often exacerbated by a lack of "buy-in" to the new approach. To try to overcome these problems, the work reported here first asked farmers what they wanted and then, once that was agreed, went on to assist by providing skills and understanding through training, but without direct financial assistance. Thus project funds were spent on training and mentoring the participating communities, with only the provision of minimal facilities. Then, as the concepts were adopted and the program grew, these facilities were improved by both donor funds and by community contributions. In this way, pilot village nurseries grew into Rural Resource Centers (RRCs) staffed by village members with support from local NGOs and community-based organizations (CBOs) (Tchoundjeu et al., 2006, 2010; Asaah et al., 2011). This has been found to be an effective strategy for the dissemination of agroforestry innovations (Degrande et al., 2012).

Principle 3: Build on Local Culture, Tradition, and Markets

In the past, tree products were gathered from natural forests and woodlands to meet the everyday needs of people living a subsistence lifestyle. Nontimber forest products gathered from the wild in this way have played an important role in the lives and culture of local people, as is recognized by the study of local flora (e.g., Abbiw, 1990) and ethnobotany (Cunningham, 2001). With the application of intensive modern farming systems, this resource has declined. To rebuild and improve this useful resource, the concept of tree domestication for agroforestry was proposed in 1992 (Leakey and Newton, 1994a,b) and subsequently implemented by the World Agroforestry Center (ICRAF) as a global initiative from 1994 (Simons, 1996). Great progress was made in the first two decades of this initiative (Leakey et al., 2005a, 2012b, 2013) which has encouraged local entrepreneurism in the processing and marketing of agroforestry tree products. This has had beneficial impacts on farmers' livelihoods (Tchoundjeu et al., 2010; Leakey, 2014a).

To capitalize on this tradition and culture, the domestication of indigenous fruit and nut trees for integration into farming systems through agroforestry is based on participatory processes involving local communities. The prime objective of the participatory approach is to involve the target communities in all aspects of the planning and implementation of the program so that they have ownership of the program, while also benefiting from the close involvement of researchers and NGOs as mentors in the domestication program. By building on tradition and culture in this way, participatory tree domestication has stimulated rapid adoption by growers and has enhanced the livelihoods of the households and communities involved (Leakey et al., 2003; Simons and Leakey, 2004; Asaah et al., 2011).

In implementing this strategy it is of great importance to recognize the legal and socially important communal rights of local people to their traditional knowledge and local germplasm (Lombard and Leakey, 2010) and to ensure that they benefit from their use and are rewarded for sharing them for the wider good. Because of the sensitivity arising from past commercial exploitation of these rights by individuals, companies, academics, international agencies and government, it is very clear that the partners in domestication programs have to earn the trust of local communities. That is, to ensure that benefits flow back to the farmers and communities, the recipients of traditional knowledge and germplasm should enter into formal Access and Benefit Sharing agreements (ICRAF, 2012) in which the rights of the holders of knowledge and genetic resources will be legally recognized.

With poverty alleviation as one of the objectives of the domestication of indigenous trees, it is clear that incentives for, and approaches to, income generation are important in the overall strategy. Consequently, improving and expanding the markets for agroforestry trees and their products are central to the strategy. The experience of the last 10−15 years indicates that this is transforming the lives of the participating farmers and helping them to break into new business and employment opportunities (Leakey and Asaah, 2013).

In many countries land tenure systems are complex with a combination of community customary rights and individual legal rights based on land purchase. In addition, government attempts to regulate logging and deforestation make the sale of tree products illegal. These issues can affect farmers' decisions about the growth of tree crops. In Cameroon, a study of formal policies found that regulations do not clearly distinguish between products from trees found in the wild and those gathered from farmers' fields (Foundjem-Tita et al., 2012). This finding supports the need to distinguish between common-property wild forest resources (e.g., nontimber/wood forest products) and private domesticated tree resources (agroforestry tree products) growing in farm land (Simons and Leakey, 2004) and to recognize that the exploitation, transport, import, and export of indigenous fruit crops from farmers' fields do not pose any threat to conservation (Schreckenberg et al., 2006b). Defining agroforestry tree products (timber and nontimber) as conventional farm products in this way should increase farmers' incentives to formally cultivate trees and harvest their products, with beneficial impacts on farmers' income, national revenues, rehabilitation of degraded land and the environment (Schreckenberg et al., 2006a).

A strategy to increase income generation from the sale of tree products in local markets is particularly important as local people are familiar with the use of these food and medicinal products and the demand typically exceeds supply. In the longer term, this trade often has potential to expand regionally and even internationally as the products become more widely known or better processed for global customers. However, as the commercialization process involves more players and becomes more complex, so the risks that producers will be exploited increases. To counter this risk, innovative approaches to ensure that farmers and local communities are rewarded for their marketing innovations have been developed by PhytoTrade Africa and are being extended to tree domestication (Lombard and Leakey, 2010; Leakey, 2014a). Again, the approach involves working with indigenous communities and helping them to secure long-term access to markets in ways that reward them and protect their intellectual property rights.

Principle 4: Use Appropriate Technology and Indigenous Perennial Species

Principles 1 and 3 mentioned the relevance of indigenous trees and their products to tropical and subtropical farmers. To capture, harness and improve the flow of benefits from these trees, recent approaches to their domestication have focused on the large opportunity for genetic selection and clonal propagation as horticultural cultivars. This is based on the capacity of vegetative propagation to capture and fix desirable traits, or combinations of traits, found in individual trees (Leakey and Simons, 2000). This approach to clonal propagation also has the benefit that selected trees can be propagated from mature tissues so that the cultivar has a lower physical stature and early fruiting—making early returns on effort and the harvesting of fruits easier.

The simplest technique for mass clonal propagation is the rooting of leafy stem cuttings.

Studies over the last 50 years have greatly enhanced the understanding of basic principles for robust and efficient techniques (Leakey, 2004, 2014c), as well as the development of simple, low-cost propagation systems for implementation in remote village nurseries without access to running water and electricity (Leakey et al., 1990). With only a little training, these propagators made from locally available materials have been widely and successfully adopted around the tropics by unskilled and illiterate farmers and have opened up the opportunity to develop improved clones/cultivars of over 50 tree species for local planting, as well as for sale to others. Without this appropriate technology, participatory tree domestication would probably not have been possible.

To decide which trees have potential for cultivar development it is necessary to have an understanding of the tree-to-tree variation within wild populations. Fortunately, farmers who have gathered products from the wild trees in their area are generally well aware which trees have particular traits, such as large fruit or nut size, good taste, or particular elements of seasonality—all desirable traits that attract a good market price (Fig. 28.6). To assist this process of farmer selection, appropriate quantitative techniques have also been developed for the selection of superior trees that meet the needs of local markets and industries. The tree-to-tree variation in hundreds of morphological traits of importance to the development of food, cosmetic, pharmaceutical, and other products have been assessed in the field and used to identify appropriate multitrait combinations that can be easily understood by local farmers. Scientific studies of chemical and physical traits have been done in parallel and the results of these are used to assist farmers to understand the potential for the development of new commercial products. The previous scientific inputs to the understanding of genetic variation can then inform the process of farmer selection and help to provide guidance as to how best to meet the needs of different market opportunities. Based on the concept of ideotypes for tree selection (Leakey and Page, 2006) cultivars can be developed that have the ideal combination of traits for a product to meet the needs of a particular market. So, for example an ideotype for a fresh fruit would have a lot of flesh (and small seeds/nuts/kernels), be sweet, juicy, tasty, nutritious, and look attractive. On the other hand, a nut ideotype would have a large kernel(s) (and probably little flesh), have a thin shell so that it is easily cracked, be rich in edible oil with an appropriate fatty acid profile or have other characteristics meeting the needs of the cosmetic or pharmaceutical industries. In both instances, these quality traits are ideally associated with a high yield of fruits or nuts, so that the cultivar can be said to have a high harvest index—a large amount of "ideal" harvestable product.

To assist the marketing of tree products (especially nuts), simple, low-technology tools are being developed for nut cracking and the pressing of oil from nut kernels (e.g., Mbosso et al., 2015). These are labor saving, better for large-scale processing and safer than many traditional methods, such as the use of a machete to extract kernels.

Principle 5: Encourage Species and Genetic Diversity

Of the 20,000 plant species producing edible products, only about 0.5% have been domesticated as food crops, yet many have the potential to become new crops through the implementation of participatory domestication; indeed research is already in progress in over 50 tree species (Leakey et al., 2012). Adding new crops to small farms reduces risks from crop and market failures, as well as playing an important role in the rebuilding of agroecological functions on degraded farm land (Leakey, 1999b, 2012b). In environmental terms, the diversification with long-lived perennial plants is important because it is the way to rebuild the ecological functions of agroecosystems and landscapes.

Some people are rightly concerned that the domestication of new food crops will result in the loss of their genetic diversity by narrowing the genetic base. This can certainly happen if the domestication process is not based on a wise strategy that is correctly implemented. In the case of agroforestry trees being domesticated by participatory processes implemented at the village level, there is good evidence that both the strategy (Leakey and Akinnifesi, 2008) and the implementation (Pauku et al., 2010) are not creating any serious concerns. About 70–80% of the tree-to-tree variation is found at the village level and selected trees with morphologically desirable traits have been found by DNA analysis to be unrelated. Consequently, development of different sets of unrelated cultivars in different villages ensures that the narrowing of the genetic base is minimal. In other words "decentralized domestication" seems to be a means of ensuring genetic diversity is retained. Furthermore, by gaining an understanding of the tree-to-tree variation and developing different sets of cultivars based on ideotypes formulated to meet the needs of different markets, it should be possible to repackage genetic diversity and develop cultivars which are as different from each other as breeds of dogs are different from each other (Leakey, 2012a), without destroying the wild species.

In the scientific approach to selection, modern laboratory techniques are being increasingly used to examine traits that are not visible to the naked eye: for example, to quantify genetic variation in the chemical and physical composition of marketable products such as polysaccharide food thickening agents, nutritional content (protein, carbohydrate, oils, fiber, vitamins, minerals, etc.) by proximate analysis, medicinal factors like antiinflammatory properties, the composition of essential oils and fatty acids, the determination of wood density, strength, shrinkage, color, calorific value, and other important wood properties correlated with tree growth (Leakey et al., 2012). Molecular DNA analysis is increasingly being used to gain understanding of genetic variation and relatedness (Jamnadass et al., 2009).

Principle 6: Encourage Gender and Age Equity

In many rural communities around the world, women in particular have been engaged in gathering, using and marketing tree products. One of the purposes of a participatory tree domestication strategy is to ensure that all members of the community, whether male or female, are empowered by the program and are the beneficiaries of the outputs of their own initiatives and labor. This has been found to enhance the livelihoods of the community members in general and promote social and gender equity (Kiptot and Franzel, 2012), with exciting long-term benefits for youths (Leakey and Asaah, 2013; Degrande et al., 2012).

Principle 7: Encourage Farmer-to-Farmer Dissemination

Through the development of RRCs as the hubs of participatory tree domestication, there has been a steady growth in the number of communities (from 2 to over 450) and number of people (from 20 to over 10,000) becoming engaged in participatory tree domestication as satellite nurseries have been developed in the areas around the RRCs (Tchoundjeu et al., 2006)—a process which is continually expanding (Asaah et al., 2011). Much of this has been word-of-mouth, neighbor-to-neighbor dissemination, but in addition efforts have been made for longer distance dissemination by community-to-community visits, fairs and competitions, as well as stories in the national media.

Evidence from Cameroon (Degrande et al., 2012) suggests that the involvement of grassroots organizations in the extension of agroforestry through the RRCs has led to a relatively high level of satisfied farmers and been successful in reaching the women and youths often excluded by other extension systems.

Principle 8: Promote New Business and Employment Opportunities

As mentioned earlier, local markets often exist for traditionally important food and nonfood products from trees. Thus local knowledge and acceptance of the products is good. Again as mentioned, through the application of the ideotype concept (Leakey and Page, 2006), tree domestication enhances the quality, uniformity, and marketability of these products as clonal cultivars, selected for commercially desirable traits, and stimulates a quantum leap in the marketability of the products. This means that traders and wholesalers can purchase a large volume of uniform, high-quality product from a recognized and named cultivar. In return, hopefully the producer will receive a higher price, as it is clear that consumers are willing to pay more for the more desirable varieties. To ensure that these price benefits are passed back to the small-scale community producers, the development of trade associations, business partnerships, and agreements arc cssential (Lombard and Leakey, 2010). Interestingly, the benefits from tree domestication become increasingly important as the value chain progresses from local to global (Leakey and van Damme, 2014). In the case of marketing Njangsang (*Ricinodendron heudelottii*) kernels in Cameroon, more kernels were traded, with faster integration and greater financial benefits, when interventions to enhance commercialization were implemented (Cosyns et al., 2011). Other relevant evidence from Cameroon suggests that the adoption of collective action in kola nut production is influenced by its ease of use, absence of entry barriers and emphasis on social activities that serve as an intrinsic motivator for farmers (Gyau et al., 2012).

Much work remains to be done to select cultivars for year-round production and to develop postharvest technologies for the extension of the shelf life of agroforestry tree products and processing for added value. Interestingly, there are a growing number of processed tree products on regional and international markets—for example, there are over 410 baobab products (PhytoTrade Africa, www.phytotradeafrica.org). Many of these products rely on wild harvesting for their supply; this supply can be of very variable (nonuniform) and mixed quality, as well as irregular across seasons and producers.

With the increasing importance of market acceptability, exclusivity, and distinctiveness, the use of ideotypes for the identification of the specific trait combinations becomes more and more critical. To meet this demand increasingly sophisticated research to determine the genetic variation in the chemical, physical, and medicinal properties of the raw products is underway (Leakey et al., 2012). This also leads to the need for stronger linkages between agroforestry researchers and partners in industry (Leakey, 1999a), as can be seen in the case of Allanblackia oil (Jamnadass et al., 2010).

Principle 9: Understand and Solve Underlying Problems—The Big Picture

Over the last 60 years, agricultural intensification has resulted in substantial gains in crop and livestock production. These are due to advances in breeding (e.g., genetic gain, stress resistance), husbandry (e.g., fertilizer, irrigation, mechanization), policy (e.g., intellectual property rights, variety release processes), microfinance (e.g., credit, provision of inputs), education and communication (e.g., farmer-field schools), and market and trade (e.g., demand, incentives).

World cereal production, for example, has more than doubled since 1961, with average yields per hectare also increasing around 150% (with the notable exception of sub-Saharan Africa). Likewise, modern agriculture has led to great improvements in the economic growth of many developed countries, with concomitant improvement in the livelihoods of many farmers. In real terms, food has become cheaper (although currently prices are increasing) and calorie and protein consumption have increased. Thus, on a global scale, the proportion of people living in countries with an average per capita intake of less than 2200 kcal per day has dropped from 57% in the mid-1960s to 10% by the late 1990s. However, these benefits have come with a high environmental cost and only marginal improvements in reduced poverty, malnutrition and hunger in developing countries. Some of the major issues affecting global agriculture are:

- The scale of natural resource degradation (affecting 2.6 billion people and 2 billion ha of farm land), the depletion of soil fertility (nitrogen, phosphorus, and potassium deficiencies affecting 59%, 85%, and 90% of crop land, respectively), loss of biodiversity (valued at US\$1542 billion/year), depletion of water resources (2664 km^3/year) and agroecosystem function, against a background in which new land for agriculture is increasingly scarce. This situation, which has arisen from the overexploitation of natural capital, makes the rehabilitation of farm land, and its associated natural assets, an imperative.
- The incidence of poverty (3.2 billion people with an income of less than US\$2/day), malnutrition, and nutrient deficiency (2 billion people) and hunger (0.9 billion people) remain at unacceptable levels, despite the very significant improvements in agricultural production. In addition, 1 billion people are affected by obesity due to poor diet.
- There are numerous organizational and conceptual "disconnects" between agricultural disciplines and organizations, especially those responsible for environmental services and sustainable development. Agricultural production and governance have focused on producing individual agricultural commodities rather than seeking synergies and the optimum use of limited resources through technologies promoting integrated natural resources management and multifunctional agriculture.
- Modern public-funded agricultural knowledge, science, and technology research and development has largely ignored the improvement of traditional production systems based on "wild" resources which, traditionally, have played an important role in peoples' livelihoods.
- Agriculture is responsible for 15% of greenhouse gas emissions.
- Since the mid-20th century, the globalization pathway has dominated agricultural research and development as well as international trade, at the expense of the "localization" benefits of many existing small-scale activities of farmers and traders that are aimed at meeting the needs of poor people at the community level.

Together, these issues contribute to the formation of a downward cycle of land degradation and associated social deprivation (Fig. 34.1) that drive down crop yields and suppress farmers' livelihoods, which together are responsible for a yield gap (Fig. 34.2) between the biological potential of modern crop varieties and the yield that poor farmers typically manage to produce in the field (Leakey, 2010, 2012b).

An analysis of the cycle of land degradation and associated social deprivation recognizes that the cycle is driven by a desire for security and wealth, which in turn drives deforestation, overgrazing, and unsustainable use of soils and water, all of which cause agroecosystem degradation (Leakey, 2010, 2012b). In farmers' fields this is seen as soil erosion, breakdown of nutrient cycling and the loss of soil fertility and structure. The consequence of this degradation is the loss of biodiversity, the breakdown of ecosystem functions and the loss of crop yield. Low crop yields result in hunger, malnutrition, increased health risks and a loss of income, all of which are manifest as declining livelihoods and so return the cycle to a desire for security and wealth. It is recognized that at all of the steps within this conceptual diagram, there are a range of socioeconomic and biophysical influences that will determine the speed of the downward progress at any particular site. Such factors include: access to markets, land tenure, and local governance—not to mention external factors such as natural disasters, conflict and war, and economic drivers such as international policy and trade agreements.

Principle 10: Rehabilitate Degraded Land and Reverse Social Deprivation: Close the Yield Gap

To be productive, conventional approaches to modern agriculture typically require large inputs of fertilizers, pesticides, mechanization and, in dry areas, irrigation. However, the dependence of this type of agriculture on income and financial capital makes it inaccessible to hundreds of millions of poor farmers due to their high cost and local availability. As it is clear that cutting more forest down for agriculture is not an acceptable option, it is crucial to find ways of making degraded land productive again. Unfortunately, agricultural research and development has focused more on increasing potential yield than on addressing the cycle of land degradation and social deprivation that creates the yield gap.

To close the yield gap, Leakey (2010, 2012b) has suggested the following three-step approach as a way forward, using the example of maize (*Zea mays* L.) production in eastern and southern Africa. The approach is based on the use of agroforestry fallows, perennial crops, tree domestication, and the marketing of agroforestry tree products as a way to deliver multifunctional agriculture:

- *Step 1*. Adopt agroforestry technologies such as two-year improved fallows or relay cropping with nitrogen-fixing shrubs that improve food security by raising maize yields fourfold from around 1 Mg ha^{-1} (Buresh and Cooper, 1999; Sileshi et al., 2008). Likewise, stands of *Faidherbia albida* (Del.) A. Chev. trees play a similar role in the so-called Evergreen Agriculture (Garrity, 2012; Swaminathan, 2012). This allows the farmers to reduce the area of their holdings planted with maize and so make space for other crops, perhaps cash crops which would generate income. This diversification could also include the establishment of perennial grains. An additional benefit arising from improved fallows with leguminous shrubs like *Sesbania sesban* (L.) Merr. and *Desmodium* spp. is the reduction of parasitic weeds like *Striga hermonteca* Benth., and the reduced incidence of insect pests like the stem borers of maize (Cook et al., 2007).
- *Step 2*. Adopt the participatory domestication of indigenous trees producing marketable products, so that new, locally important and nutrient-rich cash crops are rapidly developed as a source of income and products of day-to-day domestic importance, and help empower women and maintain culture and traditions (Cooper et al., 1996; Sanchez and Leakey, 1997). Sale of these products would allow the purchase of fertilizers and so, potentially, the increase of maize yields up to 10 Mg ha^{-1}. Consequently, the area under maize could be reduced further to allow more cash cropping. Filling the Yield Gap will also maximize returns on past investments in food crop breeding.
- *Step 3*. Promote entrepreneurism and develop value-adding and processing technologies for the new tree crop products, so increasing availability of the products throughout the year, expanding trade and creating employment opportunities—outputs which should help to reduce the incidence of poverty.

This approach, which is based on good land husbandry to rebuild natural soil fertility and health, therefore increases food security by improving crop yields. However, it does more than that. The inclusion of trees and other perennial crops within farming systems increases the number of niches in the agroecosystem. These are filled by a wide range of organisms (the unplanned biodiversity) in ways that improve nutrient, carbon, and hydrological cycles; enrich food chains and meet the needs of more complex food cycles; and reduce the risks of pest and disease outbreaks. As the trees increase in size and the ecosystem progresses toward maturity, the numbers of niches for further ecosystem diversity continues to increase, further enhancing agroecosystem function and services. This diversification makes these farming systems less damaging and more sustainable. The high species diversity of moist and dry tropical forests and woodlands means that there are many species available to play these important ecological roles in a developing agroecological succession (Leakey, 1996). The domestication of indigenous trees as new crop plants offers opportunities to increase the numbers of cultivated plants (the "planned biodiversity") in these systems in ways that increase the wild organisms (the "unplanned biodiversity") that fills the niches in the diversified farming system. The new crops of course also provide products to meet the social and economic needs of poor farmers (70% of the 3.2 billion people living on less than US$2 per day) for food self-sufficiency, micronutrients, medicines, and all their other day-to-day needs not provided by modern monocultures. An important part of this approach is therefore to "hedge" against environmental and ecological risk and provide the livelihood needs of the local communities.

By including the domestication of traditional food species and the marketing of their products, this approach also meets the needs of the community for micronutrients that mitigate malnutrition and boost immunity to diseases (Leakey et al., 2012; Leakey, 2012b, 2013). Concomitantly, the commercialization of the tree products matches the product value chain to the needs of traders for more uniform and higher-quality products with improved shelf life. This emphasis on enhanced trade is then being found to open up a pathway out of poverty based on new sources of employment and new local business opportunities (Leakey, 2012b). So, as a package, this combination of social and economic advancement with the environmental restoration creates a generic model for closing the Yield Gap—a model that is highly adaptable to a very wide range of climatic and edaphic environments and to numerous socioeconomic situations, on account of the very large numbers of candidate tree species appropriate to all environments (Leakey, 2010; Leakey, 2012b, 2013).

Principle 11: Promote "Multifunctional Agriculture" for Environmental/Social/Economic Sustainability and Relief of Hunger, Malnutrition, Poverty, and Climate Change

Multifunctional agriculture, as described by International Assessment of Agricultural Science and Technology for Development (IAASTD) (McIntyre et al., 2009), has the objective of simultaneously promoting the social, economic, and environmental benefits of farming systems. In other words, agriculture is very much more than just the production of food (Fig. 37.1).

Agroforestry is particularly relevant to the delivery of multifunctional agriculture as it addresses: (1) environmental issues: (a) soil fertility management, (b) the rehabilitation of degraded farming systems, (c) loss of biodiversity above and below ground, (d) soil and watershed protection, (e) carbon sequestration, and (f) energy needs through the provision of wood fuel; (2) Economic issues: (a) income generation through trade in useful and marketable tree products, (b) the creation of business and employment opportunities in trade and value-adding through the processing of tree and nontree products, and (c) the creation of new cottage industries for diversification and enrichment of the rural economy; (3) Social issues: (a) lack of gender equity and the need for community empowerment, (b) urban migration, (c) poverty and health related problems, (d) loss of cultural identity and of traditional knowledge, (e) loss of food sovereignty, (f) the lack of income for better education and training, provision of essential skills, and (g) the lack of income for community projects such as the supply of potable water, community infrastructure developments, and transport.

Together, these benefits help to resolve the higher level livelihood issues of: (1) a lack of food and nutritional security—and associated poor health, (2) extreme and widespread poverty, (3) the loss of self-esteem arising from the marginalization of poor communities by the social elite and the consequent vulnerability to exploitation arising from a lack of self-sufficiency, (4) deforestation and overexploitation of natural resources, (5) the lack of available productive land due to the degradation of complex mature and functioning agroecosystems and the fragmentation of agricultural landscapes (Perfecto and Vandermeer, 2010; Leakey, 2010; van Noordwijk et al., 2012).

With the increasing recognition of the need to address climate change, the integration of trees in farming systems is being recognized as crucial for the reduction of greenhouse gas emissions and climate smart agriculture (Nair, 2012; van Noordwijk et al., 2011). Large perennial trees have a high volume of standing biomass and through litter fall and root turnover they also enrich the soil with carbon (Minang et al., 2012). Studies suggest that the conversion of degraded farm land to mature agroforest could increase carbon per hectare from 2.2 to 150 mg over a potential area of 900 million ha worldwide (World Agroforestry Center, 2007).

So, we see that by using agroforestry to resolve the production, food and nutritional security and poverty issues causing the yield gap, we simultaneously move farming systems toward the objectives of multifunctional agriculture and create an approach to tropical agriculture which both builds on the positive outcomes of the last 60 years of the Green Revolution and addresses some of its negative outcomes. As a consequence, tropical agriculture becomes more productive—a process of intensification—yet environmentally, socially, and economically more sustainable than the current conventional approach to modern agriculture (Leakey, 2012a).

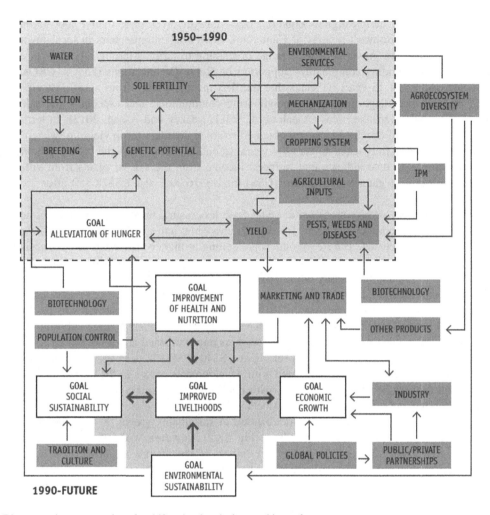

FIGURE 37.1 Diagrammatic representation of multifunctional agriculture and its goals.

Principle 12: Encourage Integrated Rural Development

So far, we have seen that agroforestry has two important roles in the development process relating to agriculture and the rural economy: (1) it provides techniques for the implementation of a highly adaptable set of three steps for the closure of the yield gap that includes value-adding within the marketing of a wide range of indigenous tree products from mixed farming systems, and (2) it is a delivery mechanism for intensified multifunctional agriculture. While these are big steps toward more sustainable rural development, they need to be set within an even wider context in which agroforestry and multifunctional agriculture are part of a regional program of integrated rural development.

To pull the previous 11 principles together into a single project, the ICRAF in Cameroon initiated a development program in 1998 centered around the provision of training in agroforestry for the rehabilitation of degraded land and the domestication/commercialization of fruits and nuts from indigenous trees. This was implemented in a participatory manner through RRCs, which in addition provided training in nursery management, entrepreneurism and the use of microfinance, community organization and infrastructure development, fabrication of simple tools and equipment for value-adding tree and nontree food products and the expansion of the value chain for traditional food products. In this longest running example of participatory domestication in agroforestry trees, the researchers fed their outputs to NGO partners through training-of-trainers courses and by acting as mentors to the NGO-managed RRCs established in pilot villages (Tchoundjeu et al., 2002, 2006, 2010; Asaah et al., 2011). The farmers in this partnership contributed their knowledge about the use and importance of local species, the range of variation in different traits of relevance to genetic selection and their traditional knowledge about the role of these species in local culture and tradition. They have also contributed their time and labor. Furthermore and crucially, they also made available some of their trees for research and for training in domestication techniques.

This case study—a winner of the prestigious Equator Prize—now involves more than 10,000 farmers and over 200 communities in the West and Northwest regions of Cameroon, as well as entrepreneurs in local towns. The project is centered on five RRCs, which are providing a wide range of training to farmers through the growth of more than 120 satellite tree nurseries in surrounding communities supported by Relay Organizations (NGOs, CBOs, etc.) in the villages. The experience of the last 15 years indicates that the first income stream from agroforestry projects is derived from the sales of plants from village nurseries to neighboring communities; and especially the sale of seedlings of nitrogen-fixing or the so-called fertilizer trees (Asaah et al., 2011; Leakey and Asaah, 2013). In terms of soil fertility replenishment, the benefit flows from these trees are obtained relatively quickly (crop yield up two- to threefold in 2−3 years). On the other hand, it generally takes longer (>4 years) to obtain returns from the production and sale of the tree products. On average, results to date indicate that farmers' income from the sale of plants from village nurseries has risen dramatically as the project gathers momentum (US$145, US$16,000, and US$28,350 after 2, 5, and 10 years, respectively).

In addition, to overcome one of the constraints to better food processing, local metal workers in nearby towns have been supported to develop appropriate equipment for drying, chopping, and grinding a range of foodstuffs, including tree products not previously processed. The tree products are selling at higher than usual prices and in a few cases are being sent abroad. This component of the program has created employment for metal workers and allowed local entrepreneurs to extend the shelf life and the quality of the produce they sell in local markets.

For example, the fabrication of about 150 discharge mills and 50 dryers has generated income in excess of US$120,000 (Asaah et al., 2011; Leakey and Asaah, 2013). In parallel, women in nearby towns have set up businesses for grinding crops like cassava (*Manihot esculenta*) and have also increased their income substantially. The largest of these groups was run by 10 women who employed eight workers and processed about 66 180 kg bags of dried cassava flour per day throughout the year. Profits from bags selling at US$40 to US$54 per bag, depending on the season, were said to be more than US$2.5 per bag. When integrated with developments across in the agricultural sector, small business developments such as these benefit from linkages with microfinance, business training and better access to simple equipment for the processing and packaging of raw products.

From this discussion, it is clear that the commercialization of sustainably grown products delivers really important impacts from agroforestry and multifunctional agriculture (Fig. 28.2). However, we have to recognize that commercialization can also pose great risks affecting the success or failure of the overall initiative. One study has found that bottom-up community initiatives like those described here have the greatest chance of being "winners," although if the companies involved recognize the importance of buying raw products from local smallholder producers, top-down commercialization can also be effective (Wynberg et al., 2003).

One important and exciting thing about the Cameroon project has been the wide range of positive livelihood impacts that the farmers are saying have truly transformed their lives (Leakey and Asaah, 2013). These require further quantification and verification, but include substantially increased income, new employment opportunities, improved nutrition, improved health from potable water and better diets, and the ability to spend money on children's schooling, home improvements, wells, etc. Significantly, one of the outcomes mentioned by young people in the participating communities is that this now means that they can see a future for themselves if they remain in the village rather than feeling that they have to migrate to towns and cities for a better life. In addition, women have indicated that improved infrastructure (wells, roads, etc.) has reduced the drudgery in their lives as a result of not having to collect water from rivers and carry farm produce from remote farms. These benefits, like the mechanical processing of food crops, have meant that they had more time to look after their families and engage in farming or other income-generating activities.

It is encouraging that the levels of income generation achieved in Cameroon, albeit on a very small-scale, exceed those proposed in the Millennium Development Goals. This and the other impacts presented here strongly suggest that by promoting self-sufficiency through the empowerment of individuals and community groups through the provision of new skills in agroforestry, tree domestication, food production and processing, community development, and microfinance, it is possible for communities to climb the entrepreneurial ladder out of poverty, malnutrition, and hunger. What is needed now is to disseminate this approach to millions of other poor people in Africa and other tropical countries.

To conclude, through the integration of rural development activities, farmers in Cameroon are intensifying their farming systems in ways that are environmentally, socially and economically more sustainable, while people in local villages and small towns are developing cottage industries and engaging more in marketing and trade. The consequence of this has been the start of the climb out of poverty and entry into the cash economy. This relationship between enhanced farm production and urban life is important for the rural economy, as it is an example of farm production being the "engine of growth." This is perhaps the start of a new approach to rural development in the tropics—one that perhaps replicates what happened thousands of years ago in the Near East and Europe as cereals and other staple food crops were domesticated and brought into cultivation. Interestingly, Diamond (1997) has credited the domestication of food crops with the advance of western civilization. Recognizing this power of crop domestication, Leakey (2012b,c) has called for a "new wave of domestication" to benefit people in developing countries who did not greatly benefit from the first wave. In this regard, one interesting development in recent years has been the involvement of a few multi-national companies in public−private partnerships with rural communities engaged in production of agroforestry products in tropical countries (Jamnadass et al., 2010; Leakey, 2012b). Although associated with risks, this also offers great opportunities for the future development of agroforestry tree crops if the strategies and practices can be developed appropriately.

SUSTAINABLE INTENSIFICATION

Currently, there is great interest internationally in seeking "sustainable intensification" (Garnett and Godfray, 2012; Garnett et al., 2013). This paper, presenting 12 principles for achieving both better and more food from mature perennial agroecosystems, seeks to contribute to this debate and illustrate how the domestication of indigenous trees producing high-value products, such as traditional foods and medicines, can be a catalyst for sustainable and integrated rural development. This paper also emphasizes that an important strategy within this approach to sustainable intensification is the implementation of steps to restore productivity to degraded land and close the yield gap and meet the needs of a growing human population without the need for further deforestation (Fig. 28.2; Leakey, 2012b).

Clearly, the challenge for the future is to scale up the application of the principles outlined here to have meaningful impact on national, regional and global scales. A key to achieving this will be the attainment of political will. Toward this end, the IAASTD (McIntyre et al., 2009) placed a need for greater emphasis on:

- Integrated approaches to land-use management involving participatory approaches to planning and implementation.
- Less exploitative approach to natural resources, especially soils and water, and a lower dependence on inorganic inputs and fossil energy.
- Good husbandry to support agroecosystem health, restoration of degraded land and the reduction of the yield gap.
- Increased involvement of local user groups in actions to improve natural resources management.
- Diversification of agriculture for improved soil amelioration, pest and disease control, and new marketable products.
- The domestication of new nutritious and marketable crops from local species, especially trees, to diversify diets and the local economy.
- Enhancement of rural livelihoods by meeting the needs of local people and supporting culture and tradition.
- Better integration of agricultural sectors, government departments and institutions, communities, and stakeholders to overcome "disconnects" in policy and practice.
- Public−private partnerships involving diverse stakeholder groups at the local level to support sustainable production, and in-country processing and value-adding.
- There is strong accord between these pointers to a better future for agriculture from IAASTD and the principles outlined in this paper.

Chapter 38

Trees: Delivering Productive and Sustainable Farming Systems: An Update

R.R.B. Leakey

Despite all the good news about agroforestry, the reality is that it barely scratches the surface of the problems. Over 35% of farm land worldwide (around 2 billion hectares) can be categorized as degraded and nearly 50% of the population (about 3 billion people) live on less than US$2 per day. Set against this, more than 43% of agricultural land (about 1 billion hectares) has 10% or more tree cover (Zomer et al., 2014), but it is unknown what proportion of these trees have been planted, or deliberately regenerated, rather than retained from forest or woodland during land clearance. It is very likely that the majority are essentially retained natural trees, and furthermore, that only a very small proportion of any planted trees are an output of a genetic selection program. Thus the extent that this area that can be described as "intensified agroforestry" is probably minute. This emphasizes the potential for dramatic improvement. However, this does not belittle the fact that 0.9 billion (Zomer et al., 2014) to 1.2 billion people (World Bank, 2004) are engaged in agroforestry, while even more, an estimated 1.5 billion, are thought to use agroforestry tree products (Leakey and Sanchez, 1997). These statistics illustrate the importance of trees outside forests, while another 0.5 billion people are thought to use tree products from forests. Many of the people using these products are part of the urban population.

As mentioned in Chapter 36 (Leakey, 2012a), agroforestry is an approach to increase the productivity of modern crop varieties produced by the Green Revolution and thus aims to improve the returns on the investment in the Green Revolution and so help to achieve its original objectives to intensify production, and to overcome hunger, malnutrition, and rural poverty. Today, the model for conventional intensive agriculture is still to use "high inputs for high outputs" based on improved crop varieties and livestock breeds, mechanization, artificial fertilizers, irrigation, etc. and the results from industrialized countries in temperate latitudes have been very impressive. As we have seen in Chapter 34 (Leakey, 2013), the social, economic and environmental situations in Africa, and in some countries elsewhere in the tropics and subtropics, are very different. The consequence of trying to transfer this form of intensive agriculture to Africa has been the emergence of the yield gap and average actual yields that are only about 10−15% of potential yield (Fig. 34.2).

In Chapter 4 (Cooper et al., 1996), and Chapter 34 (Leakey, 2013), we saw that agroforestry interventions, such as the use of nitrogen-fixing "fertilizer" trees, can partially close the yield gap and partially alleviate food insecurity. In this scenario, "low biological input" importantly raises yield from "extremely low" (e.g., about 1.5 tonnes of maize per hectare) to "low" (e.g., about 3−5 tonnes of maize per hectare). Unfortunately, achieving this hugely important improved level of food security does little to raise farmers out of a subsistence lifestyle. Again, we have already seen income generation (from the sale of high-quality tree products) can improve farmers' lives and allow them to purchase inorganic fertilizers, pesticides, etc. and so complete the closure of the yield gap. Thus "agroecology + income" seems to be an alternative and more appropriate approach to the intensification of tropical/subtropical agriculture than the industrialized agriculture model (Chapters 34 and 36 [Leakey, 2013; 2012a]).

The previous suggestions resonate well with recent discussion about ecological intensification (Tittonell and Giller, 2012; Tittonell, 2014) and sustainable intensification (Pretty et al., 2011; Garnett et al., 2013). These are all concepts that recognize that conventional intensified agriculture is not truly sustainable even under the near ideal conditions found in temperate latitudes, and that it falls far short of being sustainable under more severe environments. Specifically in the tropics and subtropics, as we will see in Chapter 39 (Leakey and Prabhu, 2017), real progress at the farm level depends on achieving the combination of environmental sustainability linked with social and economic

Multifunctional Agriculture. DOI: http://dx.doi.org/10.1016/B978-0-12-805356-0.00038-6
© 2017 Elsevier Inc. All rights reserved.

livelihood factors. The concept of Total Factor Productivity (the efficiency with which all factors of production are utilized for overall output), is normally applied at the national level (Coelli and Prasada Rao, 2005), but could perhaps be usefully applied at the farm level.

The question now is how to scale up this form of intensification in ways that are appropriate to the needs and aspirations of millions of people, while maintaining the environment? To assist this process, Chapter 37 (Leakey, 2014f) draws together some lessons learned in Cameroon into a set of 12 principles (Box 37.1). Interestingly, there are many overlapping thoughts between this set of 12 principles and the key policy requirements for sustainable intensification (Pretty, 2008: Pretty et al., 2011; Pretty and Bharucha, 2014), as shown in Table 39.2. Likewise, there is strong congruence between the outputs of agroforestry projects described in this book and the challenges for sustainable agricultural development identified in Chapter 3 of the Global Report *Agriculture at a Crossroads* (IAASTD, 2009). The latter examined 297 impacts of agricultural knowledge, science and technology on development. It found that, although positive impacts exceeded negative impacts by about 4:1, there were 10 major challenges for sustainable development (Table 38.1). At the local level, these challenges arise because "the great improvements in productivity had been achieved at the expense of environmental and social sustainability." In particular this report recognized the overexploitation of natural resources and the loss by some societies of their traditions and individuality, as well as a failure in the agricultural sector to adequately appreciate socially relevant, propoor approaches to agriculture for the improvement of productivity in degraded farming systems.

TABLE 38.1 The key challenges (abridged) to the implementation of sustainable agriculture (IAASTD, 2009) and the capacity of the approaches presented in this book to address them.

Challenges identified by IAASTD	Addressed
1. To develop and use agricultural knowledge, science and technology to reverse the misuse and ensure the judicious use and renewal of water bodies, soils, biodiversity, ecosystem services, and atmospheric quality.	✓
2. To find ways to close the yield gap by overcoming the constraints to innovation and improving farming systems in ways that are appropriate to the environmental, economic, social, and cultural situations of resource-poor small-scale farmers.	✓
3. To acknowledge and promote the diversification of production systems through the domestication, cultivation, or integrated management of a much wider set of locally important species for the development of a wide range of marketable natural products—as well as provide ecosystem services.	✓
4. To meet the needs of poor and disadvantaged people—both as producers and consumers, and to reenergize some of the traditional institutions, norms and values of local society that can help to achieve this.	✓
5. To enhance the nutritional quality of both raw foods produced by poor small-scale farmers, and the processed foods bought by urban rich from supermarkets; and overcome health and food safety problems arising from land clearance, food processing and storage, urbanization, use of pesticides, etc.	✓
6. To reverse the degradation of environmental assets due to practices involving land clearance, soil erosion, pollution of waterways, inefficient use of water, and dependency on fossil fuels by the promotion and application of more sustainable land-use management. Given the impacts of climate change, farming systems need to become closer to carbon-neutral.	✓
7. To mainstream technologies promoting Integrated Natural Resources Management using a range of biological, ecological and sustainable development frameworks and tools. New approaches to the upstream and downstream transfer of this knowledge will be needed to address the decline of formal agricultural extension—including engagement in participatory activities with a wider range of farmers.	✓
8. To address the numerous organizational and conceptual "disconnects" between agriculture and other sectors of the rural economy. Future agriculturalists will need to be better trained in "systems thinking" across many disciplines.	✓
9. To recognize all the household/community livelihood assets (human, financial, social, cultural, physical, natural, informational) that are crucial to the multifunctionality of agriculture, and to build systems and capabilities to adopt an appropriately integrated approach to rural development.	✓
10. To redress the balance between Globalization and Localization by combining the technological efficiency of modern globalized agriculture with more environmentally and socially friendly local systems.	✓

Currently, however, there is a problem with putting the upscaling of agroforestry research outputs into practice in development projects. This is because agroforestry, unlike the Green Revolution research outputs of the commodity centers of the CGIAR, does not have a delivery agency. The research outputs of commodity crop and livestock breeding were picked up by agribusinesses. They developed fertilizers, pesticides, irrigation, mechanization, etc. and the dissemination of crop varieties and livestock breeds were supported by national extension services. These typically do not exist now in many developing countries. Thus, the decentralized, low-cost, low-technology RRCs, described in Chapter 28 (Leakey, 2014a) and Chapter 30 (Leakey and van Damme, 2014), are proposed as a replacement (Franzel et al., 2015). Nevertheless this bottom-up, community-based approach needs investment if it is to expand and provide the necessary capacity-building and technology-dissemination services on a scale that has meaningful impact.

Finally, the successful achievement of upscaling this agroforestry approach to multifunctional agriculture will probably also depend on resolving the important issue of how to protect the IPR of the farmers who have invested time and effort in the development of innovative, clonally propagated, elite cultivars of the new tree crops (Chapters 25 and 26 [Leakey and Page, 2006; Leakey, 2017e]). Without this, there may not be sufficient incentive for widespread adoption of participatory domestication, one of the keystones to this approach.

Finally, Chapters 39 and 40 (Leakey and Prabhu, 2017; Leakey, 2017i) will try to see how agroforestry can be implemented as an Integrated Rural Development package to address most of the new 2030 Sustainable Development Goals.

Section 9

Integrating Rural Development to Deliver Multifunctional Agriculture

In the context of sustainable intensification, this final chapter attempts to pull together all the data, information, and concepts presented in the previous chapters, combining elements of agroecology, organic farming, permaculture, and modern industrial inputs within agroforestry, as the delivery mechanism for multifunctional agriculture. This approach recognizes the fact that many African smallholders operate on very small farms as self-sufficient units virtually outside the cash economy—and unlike their industrialized country counterparts, they do not function as a business.

This chapter draws heavily on a 12-year case study in Cameroon where many of the strategies, techniques and concepts presented in previous chapters were developed and tested. It is truly a multi- and transdisciplinary study focused on the concepts of multifunctional agriculture, and has delivered on many of these. It is, however, just the start, a beginning toward a different way of thinking about the needs of Africa—needs that span from putting food onto people's plates and money into their pockets and go all the way up the value chain to the development of new industries. Interestingly, many of these new industries could be founded on a wide array of new crop products developed from species that most people in industrialized countries have never even heard about. This approach was advocated by farmers in Cameroon when they were asked what they would like from agriculture. Subsequently, the program that emerged was developed, led, and implemented by African researchers and their partners in NGOs and local communities and farming households.

The untapped potential of Africa is enormous; I just hope and pray that future upscaling of what is presented here will be done in a way that benefits Africans. We must learn from past mistakes and engage a very different mindset: one that levels up the inequalities which currently divide our world into rich and poor. This new mindset should also eliminate the "silo" mentality that tries to squeeze tropical smallholders into economic theories designed for market economies in which farms are large and have access to income, capital and/or credit, land ownership, and price information. This, it could be said, would integrate *Homo sapiens* within the management and functioning of intensified agroecosystems around the world.

The final commentary to this book concludes that developing a sustainable and multifunctional approach to the intensification of agriculture is much less expensive, but more multidimensional, than that of annual monocultures. However, for policymakers and development agencies it is knowledge intensive, and requires an understanding of techniques and strategies that cut across a number of academic disciplines. This calls for agricultural science education to broaden the range of disciplines taught to students.

Chapter 39

Toward Multifunctional Agriculture — An African Initiative

R.R.B. Leakey and R. Prabhu*

SUMMARY

Sustainable Intensification is especially important in Africa where the need is greatest. We present eleven targets for action, paying specific attention to the needs of poor smallholder farmers in Africa. We describe multi-cropping systems integrating new crops developed from culturally-important traditional food species that intensify and enhance the productivity of smallholder farms by reducing the yield gap and providing multiple environmental, social and economic benefits. These include energy security and the creation of new local business and employment opportunities off-farm. We describe the integration of relatively simple activities to promote better soil fertility and income generation at the community level through the diversification of farming systems with trees, other food crops and livestock that increase total production. This approach, which is best developed in Cameroon, has been shown to meet the needs of food and nutritionally insecure smallholder farmers who also suffer poverty and social injustice at multiple, nested scales from farm through to landscapes and regions. This form of intensification integrates numerous different concepts of enhanced agricultural sustainability with conventional modern agricultural technologies and reverses the Cycle of Land Degradation and Social Deprivation. This leads to integrated rural development and visions of a better world. Consequently, we recommend that those wishing to address the complex and interacting issues which are the targets of the 2030 Sustainable Development Agenda should rethink the interventions needed by African farmers.

INTRODUCTION

Agronomy in the developing world has become a polemic issue in which agreement focuses around wide recognition that "business as usual" is not an option for the future (e.g., MEA, 2005; IAASTD, 2009). After many years of discussion about the sustainability of agriculture, especially in the tropics and subtropics, the concept of sustainable intensification is gaining traction as a result of being increasingly well elucidated and justified (Royal Society, 2009; Garnett et al., 2013; Pretty and Bharucha, 2014). Additionally, awareness is increasing that progress in this direction will depend on the integration of biological, environmental, and social sciences to deliver benefits of practical value for food security, poverty alleviation, and mitigation of climate change, etc. Toward this objective Garnett et al. (2013) have provided underpinning premises and areas of interfacing policy (Table 39.1), while Pretty and Bharucha (2014) have identified requirements for policy change. Within this wider debate there are also calls for strategies aimed at resilience (Bennett et al., 2014) and other holistic concepts to ensure that food security is achieved with environmental and social benefits conferring land-use efficiency (Loos et al., 2014), and the creation of "green economies" respecting imperatives such as Rockström's Planetary Boundaries (Rockström, 2009). While we agree with much of the general thrust of these analyses, we believe the debate so far is not putting sufficient emphasis on the risk and path dependency of the options seminal to sustainable intensification of smallholder agriculture in Africa. Pretty et al. (2011) have, however, surveyed 40 existing projects in 20 African countries and found examples that demonstrate that food outputs can be improved without harm to the environment. Typically these also fostered social, economic, and cultural relations, and enhanced rural and urban livelihoods. While the study identified some common constraints affecting African agriculture, and listed seven key requirements for

*World Agroforestry Centre, Nairobi, Kenya.

Multifunctional Agriculture. DOI: http://dx.doi.org/10.1016/B978-0-12-805356-0.00039-8
© 2017 Elsevier Inc. All rights reserved.

TABLE 39.1 Premises and interfacing policies for multifunctional landscapes (Garnett et al., 2013).

Premises for sustainable intensification

1. Conventional modern agriculture has a supply side problem based on underproduction to meet growing demand for food.
2. Increased production should be achieved without the use of an increased area of land, because of the environmental cost of clearing more land for agriculture.
3. Food security requires attention to environmental sustainability.
4. SI merits diverse approaches (conventional, high-tech, agroecological, and organic) in different biological and social contexts.

Policies interfacing with sustainable intensification.

1. Biodiversity and land use
2. Animal welfare
3. Human nutrition
4. Rural economics
5. Sustainable development

TABLE 39.2 Key requirements for policies to support the scaling up of sustainable intensification to larger numbers of farms (Pretty et al. 2011; Pretty and Bharucha, 2014).

1. Scientific and farmer input into technologies and practices that combine crops and animals with appropriate agroecological and agronomic management.
2. Creation of novel social infrastructure that both results in flows of information and builds trust among individuals and agencies.
3. Improvement of farmer knowledge and capacity through the use of farmer field schools, videos and modern information communication technologies.
4. Engagement with the private sector to supply goods and services (e.g., veterinary services, manufacturers of implements, seed multipliers, milk, and tea collectors) and development of farmers' capacity to add value through their own business development.
5. A focus particularly on women's educational, microfinance and agricultural technology needs, and building of their unique forms of social capital.
6. Ensuring that microfinance and rural banking is available to farmers' groups.
7. Ensuring public sector support to lever up the necessary public goods in the form of innovative and capable research systems, dense social infrastructure, appropriate economic incentives (subsidies, price signals), legal status for land ownership, and improved access to markets through transport infrastructure.

policy interventions (Table 39.2), it did not seek innovative practical interventions aimed at new approaches to maximizing productivity, farmer well-being, and environmental rehabilitation.

The problems of African agriculture are a special case. There are intricately enmeshed issues of land degradation and poverty (Vosti and Reardon, 1997; Scherr, 2000); thus, interventions to achieve sustainable intensification must combine good land husbandry and functioning agroecosystems, with steps to improve the income, livelihoods and well-being of extremely poor farmers (Leakey, 2012b). Critically, it is important to help these farmers to gain entry into the cash economy and be able to adopt more productive land-use practices as well as acquire the ability to purchase food and other day-to-day goods and services (Leakey, 2010, 2012a). In other words, sustainable intensification is also about improving the lives of people so that more intensive and productive farming is associated with both better land-use management, empowered farmers with better livelihoods, and enhanced economic development. In this way, some members of the rural population will be able to get out of the subsistence agriculture that is driving land degradation, and into either new local business development or local employment. To broaden the debate we therefore present some action-oriented targets for sustainable intensification offering a more developing country/low-input, smallholder agriculture view, involving agroforestry. This is based on experience in the tropics, and particularly Cameroon, and has an income generation and agroecological/landscape perspective in which the importance of diversity (social, economic, ecological and environmental) is stronger (Leakey, 2012b,e). We believe that this wider perspective is needed to illustrate interventions that can "tick the boxes" of some expected outputs and then formulate policies more relevant to smallholder agriculture in the tropics and to enhance investment in support of sustainable intensification.

To start by stating the obvious, the purpose of agriculture is to meet realistic food and nutrition needs of all people cost-effectively, equitably and with resource efficiency. To this goal, the International Assessment of Agricultural Science and Technology for Development (IAASTD, 2009) added multifunctionality embracing livelihood and sustainability goals. Despite the obvious successes over the last 60−70 years in crop and livestock breeding, there is clear evidence from an analysis of 297 impacts of Agricultural Knowledge, Science and Technology that we are not currently living up to these wider expectations (Leakey et al., 2009). This is further emphasized by the enunciation of the new

2030 Sustainable Development Goals which embrace the need for agricultural impacts across a wide range of interrelated social, economic and environmental outcomes, and by calls for better communication to promote advances in policy that link biodiversity and ecosystem services with human well-being (Bennett et al., 2015).

WHAT ARE THE ISSUES?

In the early days of agriculture, yields were sustained by implementing periods of fallow that harness natural processes of soil fertility replenishment and the maintenance of agroecosystem function. Then, with increasing intensification, monocultures were developed involving improved crop varieties and artificial inputs such as chemical fertilizers and pesticides, irrigation, and mechanization (Pingali, 2012); but its implementation has come with an environmental cost. This industrial approach to agriculture is most effective in industrial countries where only a small proportion of the population is engaged in farming and the farmers can use capital-intensive technologies on large farms with fertile soils. In this situation, the high financial costs of industrial agriculture are absorbed by economies of scale. While maintaining high crop yields artificially is biologically possible in the tropics and subtropics—as is seen on research stations (Sileshi et al. 2008a, 2014)—economies of scale are not available to poor smallholder farmers in the tropics and subtropics. These farmers typically have about 2 hectares (1–5 ha) of land (sometimes in small, isolated, and scattered parcels), and do not have access to cash income to purchase modern technologies (African Development Bank, 2015). In addition, in Africa, modern conventional agriculture commonly comes with high resource use, inefficiency and often with considerable social and environmental "externalities" (Pretty, 2008). Life in rural areas is also hampered by poor roads and local transport, lack of local infrastructure, issues of land tenure and ownership, poor market information, and poor extension services for technical support. For these farmers, desperately trying to support their dependent households by self-sufficiency, without the safety net of insurance and social services, mixed-cropping is a strategy for risk aversion. However, with declining staple crop yields, the area committed to them has to increase. Together, these problems combine and seem to make Africa the least responsive to the Green Revolution (Evenson and Gollin, 2003) and in need of special consideration (Nin-Pratt and McBride, 2014).

The reality is that in nearly half the world, food production by extremely poor farmers is severely constrained by poverty and the impacts of deforestation, overgrazing, and nutrient mining, which lead to land degradation and soil infertility (Leakey, 2013), outcomes which can be exacerbated by tillage, especially mechanical tillage. Together these impacts create a complex downward spiral in crop productivity that involves the loss of soil fertility and above- and belowground biodiversity. Together these cause a breakdown of the agroecological functions that are an important part of ecological sustainability. Together, the consequence of these negative impacts is lower crop yields and subsequently a decline in livelihoods, which then traps farmers in poverty, hunger, and malnutrition—the cycle of land degradation and social deprivation (Fig. 39.1) (Leakey, 2010, 2012b,e). A critical consequence of this is that farmers in Africa are not able to implement the Green Revolution technology package, and hence farm yields are well below the biological potential (Leakey, 2012a). For example, average maize yields in Africa are about 1.5 tonnes per hectare, while the biological potential of the varieties being used is around 7 tonnes per hectare (Sebastian, 2014). This very substantial yield gap is the result of the inaccessibility of technology due to low income, and hence is an efficiency gap (van Noordwijk and Brussaard, 2014) that could be filled by appropriate technology that addresses the supply side issues (Garnett et al., 2013). Closing this yield gap across Africa would substantially enhance food security in the continent as well as having substantial environmental, economic, and social benefits (Leakey, 2012b,e). Furthermore, in addition to addressing food insecurity in a world with a rapidly expanding population, sustainable intensification has to address the other Sustainable Development Goals. This will require complementary multidisciplinary perspectives and solutions to resolve issues of food availability, the stability of supply, income generation, social injustice, access to markets, and the nutritional utility and safety of food (Poppy et al., 2014). These are the result of complex and challenging issues, now being exacerbated by climate change, which together exemplify the need to examine the inextricable links between ecosystem services and human well-being in novel and inter- and transdisciplinary ways (Leakey, 2013, 2014f; Poppy et al., 2014). To guide this process we identify 11 critical targets for transformative action (Table 39.3). Some of these targets interact, and so have overlapping impacts, but are presented separately here for emphasis of their importance, and because of the failure of many agencies to engage in integrated approaches to development.

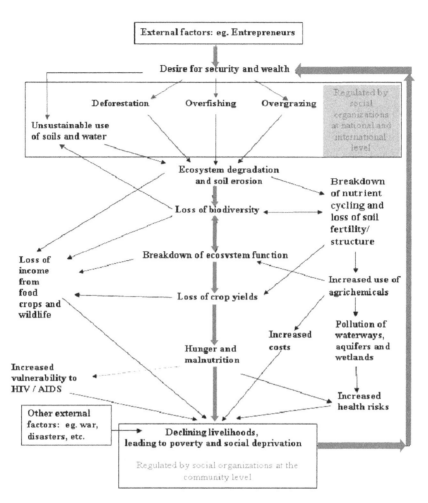

FIGURE 39.1 The cycle of land degradation and social deprivation. *Modified from Leakey et al., 2005a.*

TABLE 39.3 Eleven targets for action to transform the productivity and sustainability of tropical agriculture and associated rural development.

No.	Target
1.	Recognize need for different approaches, both agronomic and economic.
2.	Restore and maintain soil fertility for sustained high-level production.
3.	Restore and maintain agroecological processes for sustained and resilient production.
4.	Domesticate and improve indigenous species as new crops for: (1) better nutrition and (2) income generation.
5.	Close the yield gap by addressing local supply-side and livelihood problems.
6.	Provide training in rural communities to enhance their capacity to implement technologies relevant to sustainable intensification.
7.	Achieve energy security without environmental damage.
8.	Reduce and eliminate waste.
9.	Promote integrated livestock management in drylands, and animal welfare.
10.	Maintain landscape functions.
11.	Maintain global functions.

Target 1: Recognize Need for Different Approaches—Both Agronomic and Economic

Many advocates of change in agriculture have called for more sustainable systems (e.g., MEA, 2005; IAASTD, 2009). The suggestions often show remarkable polarity, at one extreme by greater use of molecular science to create genetically modified crops with resistance to herbicides and biotic/physical stresses, and at the other end certified organic agriculture free of pesticide use and genetically modified organisms. We consider that a more appropriate target embraces elements of both approaches.

Julian Cribb (2011) in his book, *The Coming Famine*, also recognizes the need to embrace the philosophical divide as these different approaches are appropriate in different places and circumstances. The requirement for diverse approaches in different biological and social contexts for the achievement of sustainable intensification is also recognized by Garnett et al. (2013). We concur with these assessments and particularly want to emphasize that what is appropriate in industrial nations of temperate latitudes almost inevitably is not relevant to developing nations in the tropics and subtropics. For example, we have to recognize that many tropical farmers moving from shifting cultivation to sedentary smallholder systems have been unable to sustainably implement the latest wave of agricultural intensification—the Green Revolution—due to their lack of income for the purchase of the inputs. It is therefore premature to try to implement the biotechnology revolution while they face severe land degradation and socioeconomic constraints that prevent the achievement of the already improved biological potential of modern crop varieties.

Typically, under conventional approaches to agriculture in the tropics, farmers can generate income by growing commodity products such as tea, coffee, cocoa, cotton, tobacco, and rubber. However, there are two problems here. Firstly, the prices paid to tropical farmers for these commodity crops are regulated in industrial countries, raising issues of "fair trade"—for example, smallholder tea, coffee, and cocoa farmers receive only 3–10% of the retail price (Curtis et al., 2013). Secondly, the average household of 6–8 people only has about 2 ha of land to provide all their needs. When the food crops have very low yields, there is little possibility of available land to grow cash crops for income generation. Another factor in the poverty debate arises from the fact that many youths migrate to urban areas to seek employment. This leaves the old on the farms and a labor shortage. Interestingly, in a number of countries, this is leading to a switch to tree crops that have a lower labor demand (Place, 1995; Place and Otsuka, 2000).

Recognizing all these problems, we believe a stepped approach is needed to resolve the issues constraining the productivity of tropical agriculture. Thus, before further crop breeding to enhance yield potential, there is a need to promote good crop husbandry and land rehabilitation. This can then be followed by a focus on income generation so that hundreds of millions of poor tropical farmers can enter the cash economy and purchase the agricultural inputs needed to intensify their production and completely close the yield gap that is common in the cultivation of staple food crops (Leakey, 2012b,e, 2013). As increasing crop yields will only address one aspect of what makes individuals, households, communities and nations food secure or insecure, we present in what follows an integrated approach to additionally resolving the current issues of land degradation, nutritional insecurity, poverty, social reform, and economic growth— steps toward the sustainable intensification of tropical agriculture, especially in Africa. This involves experience from work not generally recognized as mainstream agriculture, e.g., in agroecology (Altieri, 2002; Gliessman, 1998; Leakey, 2014e) and agroforestry (Leakey, 2012b; Leakey and Asaah, 2013).

Target 2: Restore and Maintain Soil Fertility for Sustained High-level Production

The fertility of the agricultural soils of sub-Saharan Africa have been depleted by years of weathering and nutrient leaching, resulting in highly acidic soils (<5.5 pH) which are vulnerable to aluminum toxicity (Sebastian, 2014) and reduced phosphorus availability and uptake (Rowell, 2014). Soil acidity affects about 32% of cropland. In total about 80% sub-Saharan Africa's cropland is considered highly unsuitable for agriculture, because farmers' yields are limited (Sebastian, 2014). A recent report estimates that nearly 180 million people in sub-Saharan Africa are negatively impacted by this land degradation (Naylor, 2014) and suggests that the economic cost in terms of lost yield is about $68 billion (Naylor, 2014). This depletion of nutrient resources is exacerbated by the frequent or continuous cropping of extremely poor farmers who are desperate to feed their families, but lack the income to purchase fertilizers. This soil degradation also results in agroecological problems due to loss of soil biodiversity, reduced nutrient and carbon recycling, loss of soil structure, poor water infiltration, soil erosion, and consequently problems of water uptake (Sileshi et al., 2014). Early approaches to agroforestry's "Diagnosis and Design" (Raintree, 1987) recognized that a different and more appropriate approach was needed, as financially intensive options of maintaining soil fertility are not available to poor smallholder farmers in the tropics and subtropics.

Unfortunately, inorganic fertilizers are less efficient at ameliorating the physical and biological degradation of soils; thus, organic methods of restoring fertility are especially beneficial on degraded land. However, the use of manure and organic mulching is constrained in large-scale farming systems by the difficulty of producing sufficient organic matter ($10-40$ Mg ha^{-1} year^{-1}) to balance the quantity of nutrients lost (Mafongoya et al., 2006). An alternative option is to use leguminous plants that fix atmospheric nitrogen through a symbiotic association with soil bacteria on their roots (Sprent, 2009). Nitrogen is the soil nutrient most commonly limiting plant growth and can be very effectively replenished by this inexpensive and relatively simple biological process. Unfortunately, the biological replenishment of other major soil nutrients like phosphorus and potassium is only possible by recycling biomass.

Some nitrogen-fixing leguminous plants are food crops like beans; some are fodder species which produce $23-176$ kg N ha^{-1} year^{-1} (Herridge et al., 2008), while others are shrubs and small trees which when grown at high density can produce $300-650$ kg N ha^{-1} year^{-1} (Nygren et al., 2012). Many years of research in Africa (Cooper et al., 1996; Buresh and Cooper 1999; Sanchez, 2002) have demonstrated that $2-3$ year improved fallows with leguminous shrubs and trees will enhance soil nitrogen such that typical yields from cereal crops grown on formerly degraded soils are increased two- to threefold (Sileshi et al., 2008a) in ways that are easily attainable by poor smallholder farmers (Kwesiga et al., 1999). Likewise, in the drylands, stands of *Faidherbia albida* trees play a similar role in the so-called evergreen agriculture (Garrity et al., 2010). These yield responses, which are also reported for vegetables (cabbage, onion, rape, paprika) (Sileshi et al., 2014) and groundnuts (Degrande et al., 2007), in response to biological nitrogen fixation have led to these trees and shrubs being called "fertilizer trees." However, as with the use of inorganic fertilizers, N_2O emissions are associated with increased levels of biologically fixed soil nitrogen (Dick et al., 2006; Hall et al., 2006).

Sileshi et al. (2014) emphasize that both the organic/biological benefits of fertilizer trees and the inorganic replenishment of soil nutrients by artificial fertilizers are important for food production and that decision makers in development agencies should take advantage of the synergy between biological and mineral fertilizers rather than focusing on the "organic vs. inorganic" debate. We support this conclusion in Target 5. However, as will be seen in the following, there are other benefits of fertilizer trees, especially for the rehabilitation of degraded land. Nevertheless, one advantage of organic inputs is that, in contrast to inorganic nitrogen, much of the unused nitrogen from organic inputs is typically incorporated into soil organic matter or taken up by the trees for future recycling.

Soil health is also enhanced by the biological nitrogen fixation of leguminous trees and shrubs as it increases aggregate stability, porosity, and hydraulic conductivity by increasing the soil organic matter and improving both soil structure and water infiltration (Table 39.4). These changes have also been shown to improve water use efficiency and rain-use efficiency (Sileshi et al., 2011). The synergy between soil health and soil fertility is important for the attainment of good crop yields, and for resilience and sustainability.

TABLE 39.4 Some examples from Msekera, Zambia of changes in soil physical properties (0–20 cm) due to fertilizer trees/shrubs (Fert Tree) in improved fallow and the control (sole maize) and the % change.

Variable	Tree species	+ Fert Tree	Control	% Change
Bulk density (Mg m^{-3})	Gliricidia	1.39	1.53	-9.2
	Gliricidia	1.40	1.42	-1.4
Aggregate stability (mm)	Sesbania	8.3	61.2	36.1
Infiltration rate (mm h^{-1})	Gliricidia	16.0	4.0	300.0
Time to run-off (min)	Sesbania	7.0	3.0	133.3
Drainage (mm)	Sesbania	56.4	15.8	257.0
	Sesbania	10.9	1.0	990.0
	Sesbania	61.1	7.6	703.9
	Sesbania	10.7	5.7	87.7
Penetrometer resistance (Mpa)	Sesbania	2.2	3.2	-31.3
	Pigeon pea	2.9	3.2	-9.4

After Sileshi, G.W., Mafongoya, P. Akinnifesi, F.K., Phiri, E., Chirwa, P., Beedy, T., et al., 2014. Agroforestry: Fertilizer trees. In: Encyclopedia of Agriculture and Food Systems, Vol. 1, Neal Van Alfen, editor-in-chief. Elsevier, San Diego, pp. 222−234.

An additional benefit of enhanced yield in cereal crops is increased stover production (up to 2 tonnes ha^{-1}), which can be used as livestock fodder.

These leguminous trees and shrubs can simultaneously be grown for their environmental services, e.g., along contours to reduce soil erosion (Angima et al., 2001); to control serious weeds (Sileshi et al., 2008c, 2014); or they can also double-up as productive crops producing building poles, fuel wood, or livestock fodder, and in some cases edible fruits and seeds. Furthermore, they can treble-up as bee fodder for better crop pollination and honey production (Degrande et al., 2007). On the negative side, fast-growing trees grown in association with food crops can compete in complex ways for light, water, and nutrients, especially early in an agroecological succession (Ong and Leakey, 1999), while in the longer term having numerous agroecological benefits (see Target 3). Nevertheless, depending on complex spatial and temporal interactions between the biological, physical, hydrological, and climatic components of the system, there is potential to derive considerable benefits by optimizing the use of different tree and crop species, appropriate to the type, depth and fertility of the soil, the quantity and distribution of rainfall, and the capture of solar radiation (Ong et al., 2014).

It is also relevant to recall here that overdependency on inorganic artificial fertilizers is one of the environmental "costs" of high-input agriculture which impinges on the sustainability debate due to the high fossil-fuel usage during their manufacture, and the contamination of aquifers and pollution of rivers and the ocean, which results when they enter the groundwater (Molden, 2007).

Target 3: Restore and Maintain Agroecological Processes for Sustained and Resilient Production

In the past 10–15 years, increasing efforts have been made to maintain or increase flows of ecosystem services, while also increasing both yields and farm profitability. These have been promoted under a variety of names and approaches. A review of over 280 such interventions over 37 million hectares found evidence of improved crop yields together with enhanced water use efficiency and other ecosystem services, including carbon sequestration (Pretty et al., 2006).

While fertilizers and pesticides provide a technological "fix" substituting for the natural processes in high-input farming systems, they typically have negative impacts on agroecological processes and so contribute to the loss of soil health. In this situation, harnessing natural processes of fully functional agroecosystems is a more appropriate and effective alternative (Altieri, 2002; Gliessman, 1998), recognized in African traditional knowledge (Boafo et al., 2015). Like natural vegetation, agroecosystems progress through a succession of stages from pioneer to more complex and mature assemblages (Leakey, 1999b). This is the natural process of recovery from severe ecosystem damage, such as land clearance. In farming systems, a crop monoculture is an extreme pioneer stage which can be maintained ad infinitum under perfect situations, or encouraged to mature by diversification of the farming system either at the field level in a landscape mosaic, with diversification with other crops (especially perennials) creating niches above- and belowground for wild organisms to occupy as it matures (Leakey, 2014e). On and below the soil surface, these organisms decompose the biomass, and return the nutrients back to the plants for their continued growth (Barrios et al., 2012). These processes are the driving forces of the nutrient and carbon cycles—the foundations of soil fertility and the reduction of CO_2 emissions to the atmosphere (Leakey 1999b). The greatest opportunity for practices in which trees and crops are grown simultaneously is therefore to fill niches within the landscape where soil and water resources are currently underutilized by crops. In this way, agroforestry can lead to successional development akin to natural ecological succession and mimic the large-scale patch dynamics and successional progression of a natural ecosystem (Ong and Leakey, 1999). Long-lived trees grown for their marketable products maximize these agroecological benefits, while also generating income and other benefits. So far, unrealized opportunities exist to understand how to create multistrata canopies composed of species producing useful and marketable products, which differ in height, form, growth habit, etc., and which could be grown in various different densities and configurations (Leakey, 1998b, 2014e). Such opportunities would also offer niches for conventional staple food crops bred for shady agroecosystems.

One component of soil health is attributed to its physical properties, as expressed by variations in soil depth, bulk density, aggregate stability, infiltration rates, hydraulic conductivity, water-holding capacity, and penetration resistance, time run-off, drainage, and penetrometer resistance, many of which are affected by the presence of organic matter and the turnover of fine root populations of plants, especially long-lived perennials like trees and shrubs, including fertilizer trees. For example, in Zimbabwe and Zambia leguminous fallows lowered soil bulk density by 12% and raised aggregate stability by 18–36% (Table 39.4). Likewise, pore density was raised from about 255 m^{-2} to 285–443 m^{-2}, and the pore density was greater, up from 2689–3938 m^{-2} to 4521–8911 m^{-2} (see review by Sileshi et al., 2014), attributable to larger pore sizes, up from 0.03 mm to 0.07—0.12 mm at 5 cm tension. Improvements in soil structure like these improve soil drainage by 42—600%, and reduce water run-off by 40—133%, which is especially important during wet periods (Table 39.4). Such changes in soil physical properties resulting from fertilizer trees improve water recharge, retention, storage, and availability to associated crops, while the tree canopies intercept rain and release it slowly to the soil, where it is available for crop growth with demonstrated improvements in water-use efficiency/rain-use efficiency in Africa (Sileshi et al., 2011).

Together with enhanced soil fertility, structure and function, all these benefits are important for the restoration of the biodiversity of soils, especially soil fauna—the decomposers, belowground micropredators and soil engineers (Sileshi et al., 2008b,c; Lavelle et al., 2014; Garbach et al., 2014) and the beneficial mycorrhizal fungi that have symbiotic associations with plants (Alexander and Lee, 2005). Gaining a thorough understanding of the complexity of agroecosystems function and their optimization in production systems that also improve livelihoods through approaches to agricultural intensification is probably the greatest challenge for scientific endeavor today (Leakey, 2014e). Progress is especially needed to gain a mechanistic understanding of the organisms, guilds, and ecological communities that provide ecosystem services together with quantitative measures of yield and ecosystem services in the same farming systems, especially at multiple scales (see review by Garbach et al., 2014). In this regard, considerable progress has been made in recent years to understand the categories and roles of the many different types of organisms that live in soils (see review by Lavelle et al., 2014). They classify them in five scales:

1. Microbial biofilms and colonies—occupying the smallest soil habitats found in assemblages of mineral and organic particles of approximately 20 mm in size
2. Micropredators such as nematodes and protoctists that feed on microbial biomass in meso-aggregates, at a scale of approximately 100—500 mm
3. Ecosystem engineers, such as earthworms, living at the scale of decimeters to decameters, which affect the architecture of soils through the accumulation of soil particles into aggregates separated by pores of different sizes. They can have important effects over scales ranging horizontally from decimeters to 20—30 m and vertically from a few centimeters up to a few meters in depth
4. Organisms occupying complex spatial domains such as community structures representing colonies of organisms, like termites and ants
5. Organisms occupying ecosystem and landscape mosaics.

The best practical examples of the agroecological impacts of crop diversification in Africa show that nitrogen-fixing legumes can lead to crops with lower susceptibility to weeds, pests, and diseases (Sileshi et al., 2014). For example, weed control results from shading and smothering aboveground and greater completion belowground, as in the case of the serious weed, Spear grass (*Imperata cylindrica*). Other important effects result from the release of allelochemical compounds that inhibit seed germination (Sileshi et al., 2014); or as in the special case of *Sesbania sesban* and the fodder legumes *Desmonium intortum* and *Desmonium uncinatum* which reduce populations of the parasitic weed of cereals, *Striga* spp., by stimulating "suicide germination" in the absence of the host (Khan et al., 2002). This is affected by the rate of decomposition and nitrogen mineralization of organic residues (Gacheru and Rao, 2001). With regard to pest control, in simple cereal/legume mixtures *Desmodium* spp. has been shown to act as repellents to the cereal stem borers *Busseola fusca* and *Chilo partellus*. On the other hand, Napier grass (*Pennisetum purpureum*), planted as an intercrop or around small fields, attracts the pests away from the crops (Khan et al., 2006; Cook et al., 2007). Thus, a secondary benefit of diversifying farming systems is the provision of habitat for ecologically important components of wildlife (Leakey, 2014e) that regulate pests and diseases. Consequently, by maintaining the "balance of nature," which is lost in monocultures, there is a reduced need for expensive pesticides to protect the crop artificially.

An agroecological approach to maintaining ecological health is therefore a good option for poor smallholder farmers as, without the need for financial expenditure, it sustains resilience to severe weather and other environmental hazards by enhancing the natural processes that maintain soil health. In addition, the noncrop components of the system can provide other useful, edible and marketable products (Leakey, 1999a; Jamnadass et al., 2011). However, on its own, attending to agroecological processes does not solve all the problems, because the lack of income for the purchase of inorganic fertilizers providing phosphorus and potassium salts and trace elements, as well as pesticides, remains a constraint to achieving the full biological potential of modern crop varieties (Leakey, 2010, 2013).

We argue that the multiple biological, ecological, social, economic, and environmental benefits arising from complex agroecosystems are the most important outcome of farm diversification through multicropping. It results in higher outputs from fewer inputs, with the added advantages of ecological resilience and risk aversion, i.e., greater efficiency (synonymous with total factor productivity). There are excellent examples of these diverse and productive mature agroecosystems in Southeast Asia (Leakey, 2001a) and Latin America (Schroth and do Socorro Souza da Mota, 2014), providing a model for Africa (Leakey, 2001b). In Africa, there is a wealth of traditionally important indigenous species providing food, medicinal and other useful products to create both the upper middle and lower strata of such complex systems (Abbiw, 1990; Vivien and Faure, 1996; Leakey, 1999a,b; Nono-Wombin et al., 2012; Awodoyin et al., 2015) that could be domesticated to fill different niches and make these systems highly productive. Multicropping is especially advantageous to smallholder farmers who have to produce all their household food and nonfood products without agrichemicals and with minimal risk of an ecological crash (Leakey, 1999b). Nevertheless, by providing habitat for all forms of local plants and animals, mature ecosystems are also a means of conserving wildlife and genetic resources (Atta-Krah et al., 2004) threatened by habitat loss and at the same time reduce the incidence of pest attacks (see reviews by Leakey, 2014e; Schroth and do Socorro Souza da Mota, 2014). These multiple biological, ecological, social, economic, and environmental benefits contrast with the common view, however, that there are "trade-offs" between production and wildlife conservation and ecosystem services (Godfray and Garnett, 2014). We concur with others who recognize that trade-offs are not inevitable (Maes et al., 2012) and that synergies can be achieved between ecosystem services and production (Nelson et al., 2009). Thus the compensation of farmers for adopting low-input farming systems by greater intensification elsewhere, or by the reallocation of land for biodiversity conservation and other environmental services (such as "set-aside" systems in Europe) is not needed in the tropics where smallholders have to provide all the food and nonfood needs of their families on 1–5 ha of degraded land. In some circumstances, however, payments for environmental services (PES) such as carbon sequestration by trees also meeting household needs for a range of tree products, may be an option (van Noordwijk et al., 2011).

Target 4: Domesticate and Improve Indigenous Species as New Crops for: (1) Better Nutrition and (2) Income Generation

Since the mid-1990s there has been a major research initiative to domesticate traditionally important, indigenous food and nonfood tree species for their nutritional, cultural and income generation benefits (Leakey et al., 2005a; Jamnadass et al., 2011); this initiative is now global (Leakey et al., 2012).

1. Nutrition:

   Modern approaches to agriculture have dramatically enhanced access to a small number of starch-based staple foods and this has allowed the growth of the human population to the current 7 billion. However, despite this success nearly half the world population still suffers from inadequate access to food (food insecurity) and from dietary deficiencies (nutritional insecurity). In contrast, the use and availability of traditional foods which were gathered from forests and woodlands or grown in home gardens have declined (Assogbadjo et al., 2012). Modern agriculture has focused on calorie production at the expense of micronutrient intake and dietary diversity—both elements of healthy living. Of special concern is the fact that about 80% of African farmers suffer hunger and malnutrition at some point in the year—the so-called hungry farmer paradox (FAO, WFP and IFAD, 2012; Bacon et al., 2015). These deficiencies in agriculture raise human rights issues regarding food sovereignty and rights of poor and marginalized people to food; to traditional knowledge and to the germplasm of traditional food species (de Schutter, 2011; Claeys, 2015).

   Malnutrition affects about 3 billion people due to dietary imbalance arising either from undernutrition in poor areas (2 billion), or overeating processed "fast" foods in rich areas (1 billion). To address this there are initiatives to use modern genetic techniques to fortify conventional staple foods with added vitamins (Nestel et al., 2006), but this overlooks the opportunity to also diversify and enrich farming systems with many traditional food species that are rich in vitamins, minerals and other micronutrients (Leakey, 1999a; van Damme and Termote, 2008; Nono-Wombin et al., 2012; Agoyi et al., 2014; Boedecker et al., 2014). This cultural approach has multiple benefits—sociocultural, agroecological, dietary, as well as for the enhanced health and well-being of rural households (Assogbadjo et al., 2012)—and is becoming recognized as an appropriate strategy for national and regional nutrition/health programs (Fungo, 2011). Consequently this approach to addressing the nutritional needs of malnourished people has many advantages over the genetic fortification of cereal crops.

2. Income

   Farm income in the tropics is typically acquired from the sale of cereals and other staple foods surplus to domestic household requirements, or by producing cash crops, generally for export. The domestication of indigenous trees offers several ways to enhance farm income, and so alleviating poverty as a constraint to purchasing agricultural inputs (fertilizers, pesticides, irrigation, better farm infrastructure, etc.). In addition to timber and wood for constructing numerous things, from housing to agricultural implements, trees produce foods (fruits, nuts, edible leaves and edible oils, etc.), medicines, extractives (resins, gums, latex, perfumes, dyes, tannins, etc.) and fibers for paper, crafts, baskets, etc., which are widely recognized for their usefulness as well as being sold in local markets (Abbiw, 1990; Vivien and Faure, 1996). For example, a cocoa farmer with 1.4 ha of land might have about 17 trees providing shade to the cocoa. If these trees are an indigenous fruit tree like *Dacryodes edulis* producing fruits worth $20−$150 per tree, depending on the characteristics of the individual trees, the farmer can make an additional income of about $700 from these fruits in the local market. Evidence from Nigeria, however, has shown that the average annual income from the sale of products from 100 trees of each of three species (*Chrysophyllum albidum*, *Irvingia gabonensis* and *Garcinia kola*) across 10 villages ranged from $US300 to $US1300 and that this contributed 20−60% of annual family income (Onyekwelu et al., 2014). In Cameroon, the kernels of another fruit (*Ricinodendron heudelotii*) contributed 12−15% of household cash generation (Cosyns et al., 2011).

Remembering that over 70% of the population of very large parts of Africa live on less than $1.25 per day (Sebastian, 2014), it is important to appreciate that even small increases in income from tree products can be very important to the livelihoods of rural households. Thus, for example, sums of only $50 to $100 from tree products (e.g., Ayuk et al., 1999a,b; Schreckenberg et al., 2006; Shackleton and Shackleton, 2005; Shackleton et al., 2011) have real significance for expenditures on food, medicines, clothing, schooling, farm inputs, etc. Typically, women are the traders in these tree products and the income derived from them is spent on the needs of the household and the children.

In addition to the sale of raw products on local markets, communities engaged in tree domestication are also generating income from the sale of plants to neighbors and other communities. For example, in this way, communities in Cameroon have built up to an annual income of over $28,000 after 10 years (Asaah et al., 2011; Leakey and Asaah, 2013).

Despite the existence of local markets for tree products, farmers do not receive good prices because of dysfunctional value chains (Ingram et al., 2015), poor infrastructure, limited market information, inadequate processing and storage methods, poor product quality, and lack of uniformity, etc. (Facheux et al., 2006, 2007). To address these issues and maximize and ensure the sustainability of income benefits, domestication has to go hand in hand with commercialization (Leakey and Izac, 1996) to take advantage of any complementarities (Degrande et al., 2014). The role of domestication becomes increasingly important as the products progress up the value chain because the more formal regional and especially international markets insist on the quality and uniformity of products, as well as the regularity of supply (Leakey and van Damme, 2014). Evidence suggests that the greatest benefits are derived when production, harvest and postharvest interventions are linked with activities that promote the effectiveness of local organizations and policies to support the commercialization process (Degrande et al., 2014). Product uniformity has also been found to be important

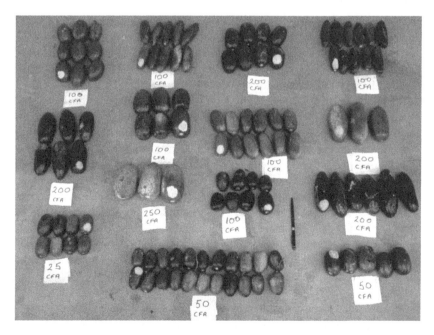

FIGURE 39.2 Safou (*Dacryodes edulis*) fruits in Cameroon. Consumers recognize and are willing to pay for desirable traits—more than just size and color.

in local wholesale trade of *D. edulis* fruits in Cameroon if farmers are to obtain a price greater than that paid for the typical mixture of wild fruits (Leakey et al., 2002). In the retail trade, however, stall holders sell small collections of uniform fruits at different prices, depending on consumer preferences (Fig. 39.2).

Tree nuts often have another inherent constraint to marketing: they are difficult and/or laborious to crack without damaging the kernels—e.g., *I. gabonensis, R. heudelotii, Sclerocarya birrea*. This poses a disincentive to engage in trade, which is now being addressed by the development of simple nut crackers, which are showing promise especially to enhance economic advantages for women (Mbosso et al., 2015). They increase returns to labor and are increasing the number of trees that a household can grow and harvest.

To help farmers generate more income from the sales of AFTPs, linkages and partnerships between producers and traders are being developed by promoting group sales, thus pooling resources such as credit, information, transportation, and labor (Facheux et al., 2012). Despite initial low levels of trust between the AFTP producers and traders, both actors were committed to continue the alliance for strategic reasons such as increased negotiation power, share production and market information, and gain from capacity-building programs (Foundjem-Tita et al., 2011) and reduced transaction costs (Foundjem-Tita et al., 2012a). In another initiative the facilitation of a village-level stabilization fund for better storage methods to promote off-season sales found that the coupling of improved storage and guarantee funds helps enhance farmers' capacity to capture higher prices (Facheux et al., 2012). However, the lower cost of group sales was found to be a better starting point for group interventions.

In general there is not much processing or value adding yet done for these traditional food products, but it has been demonstrated that there is great opportunity to increase the income-generation opportunities by simple drying, packaging, etc. in "cottage" industries (Asaah et al., 2011). This will extend the shelf life, allowing the expansion of markets both geographically and into the off-season period (Leakey, 2012b; Leakey and van Damme, 2014). Indeed, the long-term potential is even greater as many tree products are used in international pharmaceutical, cosmetic, nutraceutical, and in food and beverage industries, and even in car manufacture (Leakey, 2012b). This is an area for future expansion as the domestication process progresses (Leakey et al., 2012). This process will be enhanced if the relevant industries interact with those engaged in the domestication, so that the selection process is driven by the needs of the industry (Leakey, 1999a), based on identification of relevant ideotypes (Leakey and Page, 2006) which, through genetic selection processes, can make use of the three- to tenfold tree-to-tree variation to tease out different strands from the wild resource to create cultivars as different from each other as are the breeds of dogs that have been derived from the wolf (Leakey, 2012b).

Throughout the tropics and subtropics, traditional markets selling local products are essential to the everyday life of people, both vendors and consumers. Nevertheless, these informal markets are seldom recognized in national statistics or policies. However, in recent years, wider recognition of the potential and importance is growing (Beattie et al., 2005; Shackleton et al., 2000; Jamnadass et al., 2010), local processing and value addition are being initiated, (Bille et al., 2013) and new tree products are entering commerce: e.g., baobab fruit and leaf (www.phytotrade.com/products/baobab; www.aduna.com; www.theafricanchef.com); marula fruits and kernel oils (www.phytotrade.com/products/marula; www.wildfruitsofafrica.com); and alcoholic beverages (www.amarula.com) and "Becel" margarine from Allanblackia kernel oil (www.allanblackiapartners.org).

While there is evidence that farmers in Africa (Leakey et al., 2004, 2005b,c) and other continents (Parker et al., 2010) have initiated their own domestication processes for indigenous fruit trees, these have not had the benefits of a scientific approach. To progress this enterprise, there is now a global research program to assist poor farmers to domesticate traditional food and nonfood species and address the needs for better nutrition and greater income (Simons, 1996; Kengue et al., 2002; Akinnifesi et al., 2008; Ræbild et al., 2011; Leakey et al., 2012). The idea behind this initiative arose from discussions with farmers in the 1990s about what they would like to see from agriculture (Franzel et al., 1996, 2008). They indicated that they would like to cultivate the indigenous food tree species that produce products that rural people used to gather from forests and woodlands. A participatory approach is being implemented (Tchoundjeu et al. 2002, 2006, 2010; Asaah et al., 2011) to ensure that farmers are the beneficiaries of this work. The program is based around rural resource centers (RRCs) which assist smallholder farmers with access to resources of indigenous trees to identify elite individuals meeting the domestic food or other day-to-day needs of the household (Tchoundjeu et al., 2006; Takoutsing et al., 2014) and to develop these as cultivars. Many of the products of these trees also have potential to be marketed locally or regionally. Using simple, low-cost horticultural techniques these elite trees are then propagated vegetatively (Leakey et al., 1990) from juvenile or mature tissues, as appropriate, either on-farm or within the village (Leakey and Akinnifesi, 2008). The resulting plants are then out-planted in farming systems so that the benefits accrue directly to the farmer. Contrary to the slow process of tree breeding, this horticultural strategy results in superior cultivars that start to produce fruits or other products within 2–3 years (Leakey and Simons, 1998; Leakey and Akinnifesi, 2008).

The domestication process requires multidisciplinary research inputs (strategies and techniques) to capture and make the best use of the genetic variation present in wild populations; to make certain that the process is sustainable and adoptable;

TABLE 39.5 Multidisciplinary topics studied in Cameroon to develop effective participatory tree domestication within an agroforestry approach to deliver multifunctional agriculture.

Topics	References
Strategy for participatory involvement of communities	Leakey et al. (2003), Tchoundjeu et al. (2010)
Techniques and domestication strategy	Leakey and Akinnifesi (2008)
Farmer livelihood strategies	Degrande et al. (2006)
Characterization of intraspecific variation	Atangana et al. (2001, 2002), Waruhiu et al. (2004), Anegbeh et al. (2003, 2005)
Genetic molecular characterization	Lowe et al. (2000), Muchugi et al. (2008), Assogbadjo et al. (2009), Jamnadass et al. (2009), Mwase et al. (2010)
Sensory evaluation	Kengni et al. (2001)
Genetic resource management	Dawson et al. (2014)
Selection of trees well adapted to climate change	Weber et al. (2008, 2015), Weber and Sotelo Montes (2010), Sotelo Montes and Weber (2009), Sotelo Montes et al. (2011)
Horticultural protocols	Leakey et al. (1990), Leakey (2014c)
Root systems	Asaah et al. (2010, 2012)
Community constraints and benefits	Schreckenberg et al. (2002), Degrande et al. (2012)
Policy	Leakey and Tomich (1999), Simons and Leakey (2004)
National forest laws	Foundjem-Tita et al. (2012b)
Impact	Degrande et al. (2012), Asaah et al. (2011), Leakey and Asaah (2013)
Trade, marketing, and industry development	Jamnadass et al. (2010, 2014), Leakey et al. (2012), Leakey and van Damme (2014)

and to ensure that it has beneficial social and economic outputs and impacts for the growers and local traders (see review by Leakey, 2014a). Research at the World Agroforestry Center and its partners over the last 22 years to develop this participatory approach to tree domestication implemented in the villages by the farmers (Tchoundjeu et al., 2006, 2010; Asaah et al., 2011, Degrande et al., 2006, 2012) has been multidisciplinary and addressed many transdisciplinary issues (Table 39.5). It has achieved many positive impacts, with farmers saying that their lives have been transformed (Asaah et al., 2011, Leakey and Asaah, 2013) as they now have more food, better and more diverse diets, and are healthier.

The participatory approach employed by this multifaceted program is delivering a wide array of social, economic and livelihood benefits (Leakey, 2014a) and avoids concerns about the loss of genetic diversity, as about 70−80% of the intraspecific variation found at the village-level (de Smedt et al., 2011; Pauku et al., 2010) decentralized domestication in scattered villages has the benefit that it allows the capture of elite individuals at each location without seriously narrowing the genetic diversity of the species (Leakey, 2012b). Nevertheless, there are important issues still to be addressed, such as how to create equitable partnerships between producers and entrepreneurs to ensure that the farmers' innovations and cultivars are protected from unscrupulous entrepreneurs (Laird, 2002). Toward this end, interim interventions have been initiated (Lombard and Leakey, 2010) and work is in progress to promote new approaches to protecting the intellectual property of African farmers engaged in participatory tree domestication (Santilli, 2015).

Earlier we recognized that, unlike the situation in industrialized countries where typically less than 1% of the population is engaged in farming, the proportion in developing countries can be more than 75%, each with only a small area of land to provide all the food and other needs of the family. Breaking this demographic scenario is crucial to rural development, especially in Africa. Thus the creation of a pathway out of poverty through the development of new, off-farm, income-generation opportunities for farming households has to be part of the concept of sustainable intensification (Leakey, 2012a)—an outcome that also offers the opportunity to engage in microfinance schemes (Asaah et al., 2011). Experience in Cameroon has shown that developing diversified farming systems, which produce traditional tree food products for local markets, is having numerous positive outcomes and impacts on the livelihoods of participating communities (Tchoundjeu et al., 2010; Asaah et al., 2011; Leakey and Asaah, 2013; Degrande et al., 2014), and

TABLE 39.6 The local level cascade of expected outputs and outcomes from implementing agroforestry to deliver multifunctional agriculture. Thirty-two of these outcomes have already been reported (Tchoundjeu et al., 2010; Asaah et al., 2011; Degrande et al., 2014; Leakey, 2014a).

Intervention ➡️ Impacts

	Results	Outputs	Outcomes	Impact
Step 1 — Harness biological nitrogen fixation by planting leguminous trees	Replenishment of soil nitrogen	Crop yield increased	Partial closure of yield gap	Enhanced food security
	• Enhanced agrobiodiversity in soils	Improved agroecological function below ground	Improved soil health — reduced risk of crop failure	Enhanced food security
		• Increased organic matter	Carbon sequestration	Some mitigation of climate change
		• Reduced erosion	Reduced soil run-off	Enhanced soil protection
		• Enhanced water infiltration	Groundwater recharge	Water table replenished
	• Production of tree fodder	Increased livestock production	Increased consumption of meat, dairy products, etc.	Better dietary health and income generation
	• Production of bee fodder	Enhanced pollination	Beekeeping for honey production	Income generation and improved dietary health
	• Production of fuel wood	Reduced labor on fuel collection	Improved energy self-sufficiency and income	Enhanced well-being
	• Business opportunity	Establish tree nurseries	Sale of tree seedlings	Income generation
Step 2 — Domestication of indigenous food / medicinal trees	Tree planting and replenishment of depleted and threatened resource	Enhanced agrobiodiversity in soils	Improved agroecological function below ground and greater soil health	Reduced risk of crop failure and enhanced food security
	• Production of useful tree products	Domestic consumption	Improved diet and nutrition	Better household health
		• Marketing opportunity	Local trade	Income generation
			• Postharvest processing for wider trade year-round	Income generation
	• Establish participatory domestication process	Community engagement in rural resource centers	Acquire skills and understanding	Community empowerment and self-sufficiency
		• Self-help process	Better self-image	Improved self-esteem
			• Satisfaction	Enhanced well-being
		• Involvement of women and youth	Gender and youth equity	Healthy rural communities

(Continued)

TABLE 39.6 (Continued)

Intervention	Results	Outputs	Outcomes	Impact
	• Selection of elite trees	Production of superior planting stock	Farm diversification	Improved agroecological function below ground
		• Multiplication of superior varieties	Greater uniformity of product quality	Reach more regulated markets
			• Opportunity to match products to industrial market needs using ideotypes	Reach more specialist or niche markets (even export markets)
			• Opportunity to market further up the value chain	Regional trade and income generation
		• Farm intensification	Greater total productivity	Enhanced social and economic lifestyle
				• Opportunities to purchase farm inputs and develop farm infrastructure
Step 3 — Commercialization of tree products	Postharvest processing and packaging	Longer shelf life	Opportunity to market outside production "season"	Increased income generation
			• Opportunity to expand trade geographically	Increased income generation
		• Creation of local business enterprises	New entrepreneurism and job opportunities	Increased income generation
			• Create opportunity for local equipment fabricators	Local employment and income generation
			• Opportunity for microfinance	Greater income generation
			• Opportunity for women and youth	Greater social equity
			• Enterprise diversification	Diversified and healthy rural economy
			• Enhanced wealth	Opportunities to purchase education and health care
				Opportunities to develop local infrastructure

creating hope and optimism (Table 39.6). Indeed, in parallel with these developments, small cottage industries are emerging in which tree and other farm products are being processed, packaged (Asaah et al., 2011, Mbosso et al., 2015) and traded. Value adding in this way is extending the shelf life of tree products, such as the leaves of *Gnetum africanum* and kernels of *R. heudelotii*, expanding the season, and opening up new and more distant markets.

Interestingly, with the decline of former plantation systems for many cash crops, such as those for cocoa, tea, coffee, and rubber, smallholder farmers are developing innovative smallholder agroforestry systems based on complex species mixtures that include cash crops, as illustrated by cocoa grown under indigenous fruit trees instead of unproductive shade trees (Leakey and Tchoundjeu, 2001). Extending these concepts further, opportunities are now arising for local communities engaged in these approaches to produce a range of different tree products in multistrata farming systems—a Win:Win situation (Leakey, 2001a,b). Furthermore, this can be extended with the foundation of larger scale public–private partnerships with innovative and progressive multinational companies (Jamnadass et al., 2010, 2014; Leakey, 2012b). As mentioned earlier (Chapter 32 [Leakey, 2017f]) this initiative can be considered as the 'social modification of useful organisms' existing wild genetic resources by community-based organizations.

In recent years there has been growing interest in traditional food crops and neglected and underutilized species, as is evident from international conferences in Tanzania (2008), Malaysia (2011), Ghana (2013), and Senegal (2014). Together these developments indicate that there is now a need for policies that give greater recognition to the value of underutilized species as producers of nutritious traditional foods when cultivated as new crops for their health benefits, as well as being important components of the ecology of farming systems, which additionally sequester carbon for the mitigation of climate change. Unfortunately, in many tropical countries policies to reduce deforestation by reduced exploitation of forest trees have not differentiated between timber and nontimber forest products (fruits, nuts, leaf, fiber, bark, extractives, etc.) many of which are now gathered from farmland rather than from natural forests and woodlands. Making the sale of nontimber products illegal regardless of their origin provides a strong disincentive for farmers to diversify their cropping systems with local species (Foundjem-Tita et al., 2012b). As a consequence, it has been suggested that terms like "non-timber forest products" should only be applied to common-property forest resources and that "agroforestry tree products" be used to refer to private-property products from farmland (Simons and Leakey, 2004).

Target 5: Close the Yield Gap by Addressing Local Supply Side and Livelihood Problems

As mentioned earlier, the difference between potential yield and actual yield in staple food crops—the yield gap—is the result of the cycle of land degradation and social deprivation (Fig. 39.1), a linked set of complex environmental, social, and economic factors associated with: (1) environmental degradation, (2) a lack of access to natural resources (Ellis and Allison, 2004), and (3) poverty (Tittonell and Giller, 2012). Consequently, the solution goes beyond agronomic interventions and we see a need to adopt an integrated approach to addressing Targets 1, 2, 3, and 4. This is because poor soil fertility, low crop productivity and agroecosystem health, and a lack of income all interact to cause a dysfunctional relationship which, in turn, results from a failure to recognize that there are social and environmental constraints that hinder the adoption of the Green Revolution package of technologies in Africa. As a consequence, unlike temperate agriculture, modern technologies are failing to deliver food security, public health, and wealth creation.

Practical steps to close this yield gap have been proposed by Leakey (2010, 2012b):

Step 1. Adopt 2–3 year improved fallows or relay cropping with leguminous trees and shrubs to restore soil nitrogen fertility—see Target 2—and raise crop yields two- to threefold and so lead to about 50% closure of a yield gap. This also initiates the improvement of soil health (see Target 3). The development of village tree nurseries additionally creates income-generation opportunities. Likewise, income from stakes, fuelwood, livestock, and bee fodder importantly offset any crop yield losses due to tree/crop competition, as well as compensating for any extra labor demand for planting high density tree fallows. These leguminous trees and shrubs also have beneficial agroecological impacts (Target 3).

Step 2. Initiate and adopt the domestication of trees producing nutritious traditional foods (as well as other marketable products such as medicines, essential oils and other extractives, and fuelwoods with high calorific value) for domestic consumption as well as to generate income for the purchase of agricultural inputs and for infrastructure improvements. This step also creates new business and employment opportunities in new rural industries (see Target 4). This also further improves agroecosystem functions (see Target 3). Such diversified and intensified cash cropping systems combine production with ecosystem services and biodiversity conservation (Leakey, 2014e; Cerda et al., 2014), while diversifying income and enhancing food security (Fouladbash and Currie, 2015). Within this step, the domestication of nutritious fodder trees can enhance livestock components of mixed farming systems. The cultivation of new socially-modified tree crops also has the advantage of increasing sinks for greenhouse gases and the consequent mitigation of climate change.

Step 3. Promote entrepreneurism for the expansion of marketing, trade, value adding and processing of traditional and other food and nonfood products to further stimulate income generation and value chain development (see Target 4) to allow rural communities to expand cottage industries.

Smallholder access to markets for higher-value or differentiated agricultural and food products is a vital opportunity for lower-income farm households to improve their cash incomes and food security by shifting their livelihoods from low-value commodities to fruits, vegetables, milk, and other high-value products (World Bank, 2007). Jaffee et al. (2011) have pointed out that constraints related to basic infrastructure, farmer organizations, access to finance, etc. remain as barriers to smallholders' participation in markets linked to value chains. Interestingly, an evaluation of the impact of value chain interventions to lessen these constraints on the commercialization of kola nuts (*G. kola*) in Cameroon found benefits to farmers' livelihoods arising from substantial cost reduction, improved market information, strengthened bargaining position, and improved product conservation (Ferket, 2012; Gyau et al., 2012). Together these led to higher selling prices and increased the average household income from kola as well as its contribution to total income. These benefits spilled over to others in the villages across all wealth classes. However, it is evident that value-chain constraints differ from product to product and country to country, and arise from differences in product collection techniques, postharvest pest infestations, storage techniques, etc. Thus the first step in value-chain development should be a diagnosis of the specific issues (Degrande et al., 2014). Nevertheless, improvements introduced across the value chain have substantial benefits to the livelihoods of rural households.

One logical, but perhaps unanticipated benefit of using the income from the sale of products from trees integrated into farming systems to purchase inputs like fertilizers to intensify food crop production and close the cereal yield gap is that the cultivation of the land and the application of inputs such as fertilizers have additional production impacts through the increased yield of tree products (Khasanah et al., 2015). In South Africa, for example, wild marula (*S. birrea*) trees in natural vegetation only produced an average of 3200−6500 fruits, while those growing in farmers' cultivated fields produced 17,000−115,000 fruits (Shackleton et al., 2003). This indicates that total production per hectare can even exceed that achieved by closing the cereal yield gap.

Overgrazing by livestock has been one of the contributing factors to land degradation, especially around water holes in dry areas. We see opportunities to integrate fodder trees for small-scale livestock and fish more effectively into the farming system as a valuable source of protein and income, in combination with their already mentioned ecosystem and nitrogen-fixing benefits. Trees producing fodder for livestock have many livelihood benefits, including the provision of meat, milk and other useful products for better human nutrition, income generation and general utility (see review by Franzel et al., 2014). In Africa, farmers typically plant locally adapted fodder trees in a wide range of on-farm niches that minimize the loss of space for staple food crops. Fodder is often harvested and fed to tethered or stall-fed livestock, especially during the dry season when herbaceous fodder resources are limited. Yields vary depending on the tree species and the site (see review by Franzel et al., 2014). For reasons of digestibility, it is generally recommended that tree leaf fodder should be 15−30% of the diet. Thus in East Africa, a farmer needs about 500 regularly cut-back trees to feed a dairy cow 2 kg of dry matter per day throughout the year. This equates to approximately 1 kg of dairy meal, and produces about 0.6−0.75 kg milk per kilogram of fodder. Mean net returns from four sites over 2 years in Kenya and Uganda from 500 trees ranged from US$30 to US$114 per year (Wambugu et al., 2006, 2011). In the future, if the yield gap is closed on a large-scale and maize yield increased from 1−2 to 7−8 tonnes per hectare, surplus grains could be used to further enhance the opportunities for livestock and fish production. While we recognize that the suggestion to moderate demand for resource-intensive foods like meat and dairy as a component of sustainable intensification (Garnett et al., 2013) may be appropriate in industrialized countries, we believe that the opportunity for small-scale and more sustainable livestock production in Africa is important for people's livelihoods. However, there is an issue regarding the need for support by extension services. Although fodder trees require relatively little land, labor, or capital, farmers need specialist skills and knowledge for wider adoption (Franzel et al., 2014).

Target 6: Provide Training in Rural Communities to Enhance Their Capacity to Implement Technologies Relevant to Sustainable Intensification

The circumstances of poor smallholder farmers in Africa have often prevented them from having the opportunity to gain more than a basic education. This situation has also been exacerbated by the reduction of funds to support agricultural extension services which used to provide practical training in farming techniques. This has serious implications for agricultural development and especially the introduction of new skills, knowledge and understanding to enhance the productivity and sustainability of farms. Women in Africa are in special need of interventions to ease their burden in farming systems (Kwamina et al., 2015). Several approaches to agroforestry ease the burden of women and are ready for upscaling, having demonstrated their potential for wider adoption and indicated some important characteristics that encourage adoption: a farmer-centered approach, provision of a range of options for farmers, building local capacity, sharing knowledge and information, learning from experience and the development of strategic partnerships through facilitation activities

(Franzel et al., 2004; Kiptot and Franzel, 2015). Issues around access and rights to land can, however, be a problem under some circumstances (Schreckenberg et al., 2002, Gyau et al., 2014).

In association with the implementation of interventions to address Targets 2, 3, 4, and 5, in Cameroon, the concept of RRCs evolved as a partnership between researchers and relay organizations (local NGOs and CBOs that lead the extension program) (Tchoundjeu et al., 2006, 2010; Degrande et al., 2014), to reconcile the needs of farmers with research (Takoutsing et al., 2014). Initially, the RRCs focused on community tree nursery establishment and management for the production of nitrogen-fixing trees and shrubs and their planting in farming systems. These nurseries then expanded into simple tree domestication techniques (elite tree selection, genotype capture, cultivar multiplication by vegetative propagation, and stockplant management). Over the early years this program has involved bottom-up, farmer-to-farmer dissemination so that the number of participating farmers has "snowballed" as RRCs developing satellite tree nurseries servicing neighboring communities (Tchoundjeu et al., 2006, 2010). The media, intervillage competitions and local fairs have been used to publicize the program. In this way, over 12 years, the program has grown from 10 farmers in two villages to about 10,000 farmers in 500 villages over the North and Northwest provinces (Asaah et al., 2011; Degrande et al., 2012).

As this program has grown, so the activities of the RRCs and relay organizations have evolved and expanded to provide new skills and knowledge through the further evolution of the training programs. This has led to the availability of microfinance, marketing, product processing and value adding, community management committees for infrastructure developments, and enhanced trade (Asaah et al., 2011), spawning new opportunities for income generation and employment in the wider community. One role of the relay organizations as "diffusion hubs" is to gather production and market information and to build linkages between farmers and traders so as to facilitate price negotiations. By 2012 a market information system had been developed in Cameroon, which extended in DR Congo, for five tree products spanning 99 producer groups and 71 traders. Evidence showed that, to the satisfaction of both producers and traders, this system was creating awareness of the supply of products and their quality, the current market prices, and the opportunities for transportation (Degrande et al., 2014). In addition, capacity-building sessions have reinforced and strengthened social assets and coherence among villagers by forging new relationships among producers as part of the value chains (Cosyns et al., 2013).

In conclusion, community-based RRCs have been found to be effective as a means to enhance a self-help philosophy for agricultural development with socially-modified organizations (Asaah et al., 2011; Degrande et al., 2012) and are being found to greatly increase well-being: health, opportunity, social justice, equity, enhanced self-sufficiency, and empowerment (Tchoundjeu et al., 2010; Asaah et al., 2011). Thus, for the future they provide a replacement for former national extension services (Franzel et al., 2014), which have been in decline for several decades.

Target 7: Achieve Energy Security Without Environmental Damage

While there are many new and emerging energy technologies from solar, wind and water power, currently about 2.7 billion people depend on woody biomass as fuel for cooking and other uses (WHO, 2014), mostly gathered, unsustainably, from natural vegetation. This fuel wood gathering is one of the drivers of forest and woodland destruction and land degradation, and is closely associated with poverty, hunger, and malnutrition, as well as health issues from inhaling wood smoke. As wood becomes more scarce, women and children have to spend more time and effort gathering it, or make or purchase charcoal. The use of charcoal, often transported long distances, is also on the increase, especially in urban areas. In parallel, simple but more efficient solid-fuel cookers are being increasingly used.

While the cultivation of trees in woodlots as a source of fuel is possible, many families must spend their time and effort to grow food. Nevertheless, trees of many species can be used for fuel wood and thus an agricultural system diversified with trees for soil fertility enrichment, animal and bee fodder, fruits, nuts, and medicines affords the opportunity to use dead wood as a byproduct of farming. In addition, approaches to bioenergy production are becoming an important opportunity for farmers as part of a strategy for clean and sustainable energy source (ICRAF, 2015). For example, wood from agroforestry systems can be used for charcoal production, the development of liquid biofuels and for electricity generation, so creating more renewable sources of bioenergy. In addition, there are opportunities to enhance the fuel quality of wood in terms of its density and calorific value (Sotelo Montes and Weber, 2009; Sotelo Montes et al., 2011). Provision of fuel wood in these ways is much more sustainable than gathering it from natural forests and woodlands.

Target 8: Reduce and Eliminate Waste

In industrial countries food wastage from supermarkets, shops, restaurants and homes is serious (Garnett et al. 2013) and needs to be reduced by changes in patterns of consumption and better market planning. However, in contrast, in developing countries, where food is scarce, the main issues in food wastage are: postharvest losses of raw products due to contamination, rotting and decomposition, and pest attacks. These arise from poor storage and drying facilities for food products and techniques, and inadequate transport infrastructure. The resolution of these issues is constrained by lack of income for village storage infrastructures and facilities. Consequently, there are strong linkages here with Targets 4 and 5 aimed at income generation and the evolution of businesses engaged in food processing and value adding (Asaah et al., 2011; Degrande et al., 2014).

Target 9: Promote Integrated Livestock Management in Drylands and Animal Welfare

Approximately 1 billion head of livestock are held by more than 600 million poor smallholders, comprising approximately 70% of the world's rural poor (IFAD, 2004). Many of these animals are found in the African drylands where nomadic herdsmen use their local knowledge to move their flocks of goats and herds of cattle around areas with natural pasture and waterholes, as has been happening for generations. Probably the biggest issue here is the increasing size of these herds and the degradation of traditional grazing lands, especially close to watering holes. Consequently, sustainable intensification has to be about increasing the availability of feed and fodder, and in worst-case scenarios, the maintenance of wells and water resources to avoid starvation and dehydration, especially in the dry season. As in parts of the Sahel, this problem can raise quite complex social issues in order to avoid confrontation between the resident sedentary farmers growing cereals like sorghum and millet, and the nomads. Customary laws often give the farmers exclusive land rights in the cropping season, but in the dry season nomadic herdsmen often have the right to let their animals graze anywhere (Barrow, 1996). This results in a disincentive for farmers to grow perennial crops or to have irrigated vegetable gardens near wells, as these are likely to be browsed by livestock. Furthermore, as the stock of natural trees is depleted by the need for fuel wood, the tree resource of a wide range of both tree fodder and nonfodder products is threatened by the problems of their regeneration. The serious knock-on environmental effects of overgrazing can increase wind erosion, and they also pose a threat to the availability of species like baobab, which produces nutritious fruits and leaves for human consumption (Leakey, 1999a), as well as bark fibers. Worse still, there is some evidence that the absence of trees to draw nutrients from the lower strata of the soil and to deposit and recycle them as leaf litter on the soil surface is resulting in a downward nutrient pulse that threatens the potability of groundwater through nitrate toxicity (Edmunds and Gaye, 1997). This could have very serious future implications for both humans and livestock dependent on aquifers for their water supply. All of these issues make the deliberate cultivation of tree fodder an important component of tropical agriculture (Franzel et al., 2014). These trees, whether for fodder or other products, would also draw deep nitrogen back to the soil surface.

As seen earlier (Target 5), there is great potential to domesticate trees producing leaves important for livestock fodder (Franzel et al., 2014), as well as human food (Leakey, 2012b). Critically, however, the survival of these young trees depends on their protection from roaming livestock by the establishment of thorny hedges or overenclosures. Such actions to exclude livestock have the potential to bring the sedentary farmers into conflict with the herdsmen (Leakey, 2012b). On the other hand, if sedentary farmers were to produce fodder banks of indigenous trees with edible leaves for herdsmen and their livestock in exchange for money or animal products, then mutual benefits may allow the intensification of these drylands, and also reduce the environmental risks of overgrazing. Tree planting is not the only way to improve the resource of local trees in dry areas, as farmer-managed natural regeneration, through the exclusion of animals while young trees grow to a size that saves them from browsing animals, is also very successful and is now happening on a significant scale (Haglund et al., 2011) with numerous socioeconomic and environmental benefits (Weston et al., 2015). This too, however, involves the evolution of new relationships between farmers and nomadic herdsmen.

The elimination of animal cruelty has been identified as one component of sustainable intensification (Garnett et al., 2013). Although not absent from developing countries, intensive factory farming is more an issue in industrialized countries. In the rural tropics, livestock holdings per farm are typically just a few animals close to the household. In some cases, for example in the African Highlands, these animals are penned and fed by "cut and carry" systems where cruelty may happen, but in general this is not common, as the well-being of these animals is typically very important to the household's own livelihoods.

Target 10: Maintain Landscape Functions

The scale and spatial diversity of landscapes create diverse microenvironments that provide niches for many organisms, as well as physical formations appropriate for agricultural production and the protection and management of natural resources and public goods, such as watershed management or the conservation of biodiversity and genetic resources (Leakey, 1999b; Dawson et al., 2014), aided by remote sensing and geographic information systems.

Currently, there is growing recognition that sustainability, and hence sustainable intensification, is not just a phenomenon of importance at the plot or field scale, and that landscapes aggregate and integrate the range of outcomes at smaller scales in ways that actually provide a more meaningful expression of the state of natural capital and the ecosystem services that regulate them. Furthermore, the benefits of healthy landscapes can be greater than the sum of the smaller scales. As a consequence, landscape management has promising potential to reshape the processes and governance systems affecting land-use by individuals and by society at local, national, and global scales (Scherr et al., 2014; Minang et al., 2015). However, there is a need for much more evidence of the impacts of landscape-scale interventions with regard to farming practices and the level of intensification over time. In addition, there is a need to better understand how local agricultural decision making takes place in reaction to macrolevel economic and price dynamics, especially vis à vis the role of incentives and investments to enhance synergistic interventions across a multifunctional landscape (Scherr et al., 2014; Minang et al., 2015).

Multifunctionality is reflected in the ability of landscapes to provide services that are not normally traded, such as genetic conservation, carbon capture, pollination, and other ecosystem or environmental services (Torquebiau et al., 2013). In development, the integration of these different land-uses is often easier at larger scales. However, it is important to recognize that landscapes are dynamic and constantly changing and indeed can be either degrading or rejuvenating. Thus, at least five classes of landscapes can be appreciated:

- Pristine landscapes (very rare)
- Stable landscapes being effectively managed for production and environmental services
- Senile landscapes (progressing toward degradation: scale 10 − 1)
- Degraded landscapes that are neither productive nor ecologically/environmentally viable
- Rejuvenated landscapes that are being/have been rehabilitated (scale 1−10).

One of the complicating factors is that some senile landscapes can look much better than rejuvenated landscapes. However, the important point is that they are traveling in different directions.

It is at the scale of landscapes and the mosaics within them that we can begin to understand the aggregated effects of the environmental services on the sustainability of smallholder agriculture, which typically exceed the sum of their parts, providing positive multifunctional outcomes spanning agricultural production (food, wood, medicines, etc.), ecosystem and biodiversity conservation, human livelihoods and institutional planning and coordination (Estrada-Carmona et al., 2014), such as public policy and collective action initiatives like infrastructure developments shaping the relationship between society and the environment (Wu, 2013). Landscape approaches are also recommended to identify synergies between competing land uses or to link local initiatives with national and regional policies (Yaap and Campbell, 2012), as well as to reconcile conservation and developmental trade-offs (Peng et al., 2011; Sayer et al., 2013). Work in South Africa has found that, by characterizing landscape performance, it is possible to develop integrated solutions that effectively promote the multiple use of land (Torquebiau et al., 2013). It is important to recognize that the failure to maintain many of these landscape functions can have impacts at both ends of a transect. As an illustrative example regarding groundwater resources, the failure to match the yield potential of the relevant source landscapes to their rural and urban usage has negative impacts for rural and urban communities (MacDonald et al., 2012).

The delivery of the ecological, social, and economic landscape functions is subject to scale effects, history dependence, multiple interactions, nonlinear effects and uncertainty, all of which call for special management attention (Scott, 1998; Walker et al., 2004), such as the use of finance and stakeholder reward schemes for the ecosystem services (PES/RES) that simplify the management of landscape functions (van Noordwijk, et al., 2012). The term *socioecological production landscapes* refers to fostering human well-being, biodiversity, and ecosystem services (Gu and Subramanian, 2012) simultaneously managed at multiple nested scales and levels (Cash et al., 2006). To ignore the challenge of biologically rich farming systems that are both resilient and supporting farmers' livelihoods is to fail to address the real-world complexity of landscape functions and their externalities caused by high-input intensification (Jackson et al., 2012; Tscharntke et al., 2012). This is likely true in many landscapes, but Phalan et al. (2011) have shown that it is unlikely that landscapes can fulfill all use functions and deliver the desired ecosystem services; it will

TABLE 39.7 The landscape and global level cascade of expected outputs and outcomes from implementing agroforestry to deliver Multifunctional Agriculture.

Intervention				Impacts
	Results	Outputs	Outcomes	Impact
Up-scaling of Steps 1–3 for landscape and global benefits	Develop land-use mosaics and biodiversity corridors	Conserve biodiversity by expanding food webs and life cycle functions	Greater resilience to ecological and environmental shocks	Sustainable land-use
			• Integrated pest management to reduce crop and livestock failure	Enhanced food security
		• Protect watershed functions	Maintain groundwater resources	Support rural, urban and industrial water use
		• Increased carbon sequestration in perennial plants and in soils	Mitigation of and adaptation to greenhouse gas emissions	Reduced global climate change
	• Enhance access to finance	Expand supply of products	Create business opportunities and meet customer needs	Income generation, well-being, food security and economic sustainability
	• Enhance access to markets	Expand demand for products	Create business opportunities and employment	Income generation and enhanced food security
	• Enhance access to farmer cooperatives, training schools, credit associations, stakeholder reward schemes, etc.	Expand capacity to innovate, manage and implement better land-use systems	Greater productivity, social collaboration and cohesion; community coordination and well-being	Improved food security, poverty alleviation, social equity and empowerment; and environmental resilience and sustainability
	• Expand involvement of certification schemes	Stakeholder incentives to practice more sustainable practices	Enhance rewards for rural populations	Improved environmental, social, and economic sustainability
	• Enhance access to remote sensing, information systems and communication systems	Better management of resources	Greater productivity, social collaboration and cohesion; community coordination and well-being	Improved environmental, social, and economic sustainability
	• Expand global awareness initiatives for greater environmental, social, and economic sustainability	Greater understanding of global issues and ways of addressing them	Acceptance of need for new initiatives to address sustainability of production systems	Public awareness and agreement with sustainable approaches to the intensification of agriculture
	• Improve global policies for greater environmental, social, and economic sustainability in agriculture	Agree on new approaches to agricultural intensification that enhance production without depleting natural resources	Implementation of new approaches to intensive agriculture that ensure food security and land rehabilitation, as well as personal, national and global economic development	A world more in tune with the needs to meet the sustainable use of natural resources while supporting a growing population

be necessary in some cases to prioritize some uses over others, and especially to consider conservation uses, in order to deliver the required ecosystem services. Nevertheless, experience suggests that many desirable outputs, outcomes and impacts are possible (Table 39.7). Indeed, by restoring the productivity of degraded land it should be possible to reduce the pressures to clear forests and woodlands for agriculture.

Recently debate has focused on "climate-smart agriculture" as an approach to concurrently produce food, conserve ecosystem services, enhance agroecosystem resilience and mitigate climate change by the reduction of greenhouse gas emissions from livestock, rice paddies, and soil denitrification (van Noordwijk, 2014; Rosenstock et al., 2015; Minang et al., 2015). This multifunctional concept in which trees in agricultural landscapes play an important role (van Noordwijk et al., 2011) has some similarities to that for sustainable intensification being described here.

Many opportunities to address climate change issues occur at the landscape level. Landscape approaches present opportunities for sustainable development by enhancing opportunities for synergy between many social, economic, and environmental objectives in landscapes (Minang et al., 2012, 2015). Together these seek to recognize the importance of landscape functions in the mitigation of, and adaptation to, climate change and range from policy interventions such as REDD + (Reducing Emissions from Deforestation and Forest Degradation), and land-use certification schemes (such as for coffee, tea, and cacao), through cooperatives and credit associations to the construction of biodiversity corridors and integrated pest management. In East Africa, a project aimed at identifying, verifying and scaling up climate-smart farm management practices that both increase productivity and emit fewer greenhouse gases (GHGs) per unit of produce has so far found two candidates (Rosenstock et al., 2015). One is the use of leguminous trees to intensify cereal production in Tanzania, and the second is an integrated crop-livestock system combining agroforestry and pasture for smallholder dairy production in Kenya.

Regarding the identification of socially just strategies for enhanced adaptation to climate change, it is necessary to understand the numerous and diverse barriers to adaptation which relate to biophysical, knowledge, and financial constraints on agricultural production and rural development (Shackleton et al., 2015), as well as those political, social, and psychological barriers which are often hidden. The financing of integrated landscape management poses a challenge as national and international planning processes tend to operate sectorially with crops, livestock, fish, forests, wildlife, the environment and rural development individually managed. Consequently, financial strategies for integrated landscape approaches are under discussion (Shames et al., 2014).

To conclude this section, an integrated landscape approach recognizes the importance of the continua of multifunctionality, transdisciplinarity, participation, complexity, and sustainability that exist across a landscape (Freeman et al., 2015). The application of this approach to enhancing sustainable agriculture is still in its infancy, but the combination of the ability to recognize and interpret landscape functions appears to hold out great promise for the effective delivery of integrated production, conservation and livelihood functions, as well as tackling the complex issues affecting climate change, about which there is some controversy (van Noordwijk, 2014).

Target 11: Maintain Global Functions

Concern about the unsustainable trajectory of agriculture globally has been evident since the Club of Rome report (Meadows et al., 1972) and yet subsequent legally binding instruments such as the Rio Conventions have barely affected the trajectory of development. Consequently, the Intergovernmental Panel on Climate Change still recognizes agriculture as a major challenge to the tackling of climate change (IPCC, 2007). Thus, any discussion of sustainable intensification of agriculture that omits the need to address mitigation and adaptation targets is clearly out of touch with reality. There are two challenges in this regard: the first is to adapt, promote and adopt agricultural practices that have been shown to mitigate climate change by improving carbon sequestration in standing biomass and to enhance adaptation through increased soil organic matter, so conferring climate-smart resilient agroecosystems (Harvey et al., 2014), such as diverse mature agroecosystems.

The second challenge is to provide metrics that allow an assessment of progress toward sustainability goals and promote adaptive learning and improvement of governance and management systems. This is even more difficult. Gross domestic product (GDP) is the most widely used metric of development since Kuznets first proposed its precursor gross national product, but it is not highly appropriate in the sustainability context as acknowledged by its inventor who said "the welfare of a nation can scarcely be inferred from a measure of national income" (Kuznets, 1934). A number of alternative metrics have subsequently been proposed (Pretty, 2013; Kubiszewski et al., 2013), but so far none has received the global acceptance required to allow it to be used to assess the sustainability of ecological, economic, and social functions necessary to assure the well-being of the growing number of people on the planet (Bernard et al., 2014). This may change shortly with the adoption of new accounting approaches by the international community that aim to account for ecosystem services as well as their economic performance (European Commission, OECD, UN, World Bank, 2013).

HAVING IDENTIFIED ACTION-ORIENTED TARGETS FOR SUSTAINABLE INTENSIFICATION, WHERE DO WE GO FROM HERE?

Based on the 11 targets for transformative action presented here, along with their associated information, we suggest that sustainable intensification involves the processes that reverse the downward spiral of the cycle of land degradation and social deprivation (Fig. 39.1). Through a combination of applied agroecology and income generation, there are realistic opportunities to rehabilitate hundreds of millions of hectares of degraded/abandoned farmland, while also raising productivity and alleviating the associated complex of social and economic constraints that lead to continuing hunger, malnutrition, and poverty. Building on these outcomes, there are also opportunities for further economic growth. As local income-generating activities grow, the capacity of these rural households to purchase foods and goods produced elsewhere should also expand, so stimulating the local economy. This expansion of the local economy and the ability to purchase imported food does not seem to have been taken into account when predicting the required scale of future agricultural production (Foley et al., 2011). Thus, by delivering opportunities for self-determination, more judicious and equitable use of natural capital, and a wider range of livelihood options, the achievement of sustainable intensification could have very important impacts on global agricultural production and create new and more sustainable horizons for local economic and social growth, as well as for global agriculture.

The approach presented integrates numerous different concepts of more sustainable agriculture (agroforestry, climate-smart agriculture, conservation agriculture, ecoagriculture, integrated landscape management, integrated rural development, and organic agriculture) with some conventional Green Revolution technologies (improved varieties, chemical fertilizers, pesticides, etc.). In particular it uses the domestication of hugely underutilized resources of traditionally important food and nonfood indigenous tree species as the means to intensify these concepts of sustainable agriculture, by creating a new suite of locally important socially-modified cash crops, with potential for creating new businesses and industries (Leakey, 2012b). Furthermore, trees deliver unique environmental benefits in terms of the protection of soils and watersheds, carbon sequestration, and development of fully functional agroecosystems (Leakey, 2014e; Atangana et al., 2015). Thus, in contrast to the concept of trade-offs between production and good environmental management, there is a synergism that goes a long way toward a fuller and more inclusive concept of sustainable intensification, which better addresses the needs of smallholder agriculture in the tropics and subtropics, especially in Africa.

In conclusion, the combination of restored agroecological function and income generation provides a way out of hunger, poverty and environmental degradation, by closing the yield gap and meeting the needs of poor, smallholder farmers. In these ways it meets the key requirements for sustainable intensification identified by Pretty et al. (2011) and Pretty and Bharucha, 2014), by simultaneously addressing its three components (ecological intensification, genetic intensification, and market intensification) (Conway, 2012). The issue, of course, is whether or not there is the political will to develop and implement appropriate policies for its application.

Chapter 40

Trees: Meeting the Social, Economic and Environmental Needs of Poor Farmers—Scoring Sustainable Development Goals: An Update

R.R.B. Leakey

Many questions were raised by Leakey and Tomich (1999) with regard to the possible social and economic outcomes arising from tree domestication within agroforestry systems and its potential to contribute to the achievement of sustainable development goals (see also Chapter 13 [Leakey et al., 2005a]). However, it's very clear from this chapter, and indeed from this book, that developing a sustainable and multifunctional approach to multistrata, mixed-cropping, tropical agriculture has great potential to address the big issues of hunger, malnutrition, poverty, and land degradation. It is also clear that this approach to the intensification of agriculture is much less expensive, but more multidimensional, than that of an annual monoculture. Indeed for policymakers and development agencies it is knowledge intensive, and requires an understanding of techniques and strategies that cut across a number of academic disciplines. This calls for a multidisciplinary/transdisciplinary approach to agricultural science education (Leakey, 2012b) that is not currently part of university curricula. In this chapter special attention is drawn to how this multidisciplinary approach has been developed and put into practice in Cameroon (Table 39.5; Tchoundjeu et al., 2002, 2006, 2010; Asaah et al., 2011; Degrande et al., 2012, 2014), based on the 11 action-oriented targets presented in Chapter 39 (Leakey and Prabhu, 2017).

The Cameroon case study has harnessed the benefits of biological nitrogen fixation to restore soil fertility (Degrande et al., 2007), and then followed it up with the cultivation and genetic improvement of indigenous trees that produce traditionally and culturally important fruits and nuts (Degrande et al., 2006), as well as a wide range of nonfood products from trees deemed as priority species by local farmers (Franzel et al., 1996, 2008). Importantly, this study was done in response to the declared desires of local farmers in Cameroon, a feature probably explaining the good adoption and dissemination of the approach (Leakey, 2012b). It achieves these important outcomes by implementing "appropriate technology" to improve the yields of staple food crops and supplementing them with local fruits and nuts—so together addressing hunger and malnutrition. In addition, farmers' lives have been improved by a steady source of income from their tree nurseries and the sale of a wide range of farm products, also offering the opportunity for microfinance. This income is then being used to further improve their livelihoods by purchasing things like livestock and fertilizers, digging wells and installing piped water, as well as building new infrastructure such as storerooms, bridges and roads, and providing access to child education and health services. The knub of this approach is that through: (1) the diversification of farming systems, local markets and household diets, and (2) the use of simple and appropriate technologies, it is possible to increase total productivity with few external inputs, so maximizing the efficiency of low-input agriculture. Furthermore, this approach builds environmental, social, and economic resilience. Together these outcomes improve livelihoods and minimize risk. Additionally, by processing these products local people, especially women and youths, are creating employment and business opportunities in value adding and the fabrication of simple processing equipment such as driers, grinding machines, and oil presses (Asaah et al., 2011; Leakey and Asaah, 2013). In this way, new industries (Figs. 7.15; 7.20; 28.8) are created using natural resources previously only marketed in informal traditional markets (Figs. 7.1; 28.5; 28.6). Income generated by these business initiatives (US $3000–$4000) is substantial by local standards. Interestingly, some youths in these communities now say that they can

Multifunctional Agriculture. DOI: http://dx.doi.org/10.1016/B978-0-12-805356-0.00040-4
© 2017 Elsevier Inc. All rights reserved.

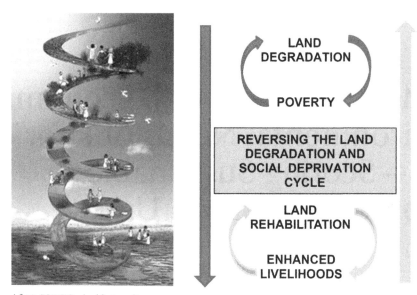

After: CGIAR Dryland Systems Programme

FIGURE 40.1 Sustainable intensification involves reversing the cycle of land degradation and its associated social deprivation issues.

see a future in their villages and so are deciding not to migrate to towns and cities looking for employment, where a life of crime and drugs beckons. Clearly, there is some risk that these domestication and commercial initiatives may be open to misuse (Fig. 13.3), resulting in both "winners and losers"(Table 13.1), but evidence from southern Africa suggests that, when done sensitively, most rural people can be winners (Wynberg et al., 2002; Shackleton, et al., 2003b).

The model to reverse the cycle of land degradation and social deprivation (Figs. 28.2 and 40.1) has the advantage of being highly adaptable, as there are appropriate indigenous tree species adapted to a very wide range of climates and soils that are well known to local people for food, medicinal and other day-to-day uses, many of which are already traded and marketed locally and regionally. Some of these socially-modified organisms, in addition, have the potential for new industries with international markets (Leakey, 2012b; Jamnadass et al., 2014; Leakey and van Damme, 2014). It has been found, however, that even at the local level, smallholder livelihoods are greatly enhanced by initiatives to promote and expand the marketing of traditionally and culturally important agroforestry food products within a fully functional and integrated value chain that extends from tree planting through to product processing and business development (Degrande et al., 2014). The wider vision of multifunctional agriculture within integrated rural development encompasses issues relating to the Right to Food (de Schutter, 2011) and food sovereignty (Claeys, 2015).

All of these benefits from the cultivation of trees within farming systems are captured under the generic term "agroforestry" (ICRAF, 1997; Atangana et al., 2015). Agroforestry combines elements of agroecology, organic farming and modern intensive inputs. Its unique and defining feature is that trees form part of a multicropping enterprise. One of the great benefits of this multiproduction strategy based on the integration of trees in farming systems is that, in addition to producing a wide range of useful and marketable products, they also produce a profusion of new ecological niches above- and belowground for a biodiverse and functioning agroecosystem that nurtures natural life cycles, food chains and the recycling of natural resources (Sileshi et al., 2008). An additional feature of this system of "carbon farming" with perennial trees is that it sequesters carbon in both standing and soil biomass (Toensmeier, 2016). Consequently, in addition to improving food security and rural livelihoods (employment, education, health, empowerment and equity), agroforestry restores the productivity of abandoned farm land, so removing the need to clear forest land for agricultural expansion. Creating an alternative resource of forest resources (including carbon) in the hands of small-scale farmers with livelihood incentives to manage them wisely can be seen to be good for both forest and wildlife conservation (Leakey, 2014a).

Very recent evidence gathered from an agricultural intensification transect in southern Cameroon (Tata epse Ngome, 2015) indicates that land clearance for agricultural intensification has had negative impacts on nutritional security and health in the rural population due to the reduction of indigenous fruits in the diet. Another consequence of this land clearance is that more people at the intensified end of the transect go to bed hungry, have been unable to acquire their preferred foods, and so eat a dull, repetitive and undiversified diet just to fill their stomachs. Interestingly, it also records that local people in Cameroon, even in the periurban population, have a strong affinity and desire for indigenous fruits, recognizing their nutritional value. Furthermore, there was no social stigma or shame attached to consuming wild fruits. This finding supports the conclusion of the Royal Society (Royal Society, 2009), the Millennium Ecosystem Assessment (MEA, 2005) and IAASTD (IAASTD, 2009), which all recognized the importance of ecologically sound intensification as the preferred long-term solution to global environmental issues arising from agriculture (Cassman et al., 2005), rather than "business as usual."

Thus, by scaling up the agricultural and rural development experience of the Cameroon project, African nations could be offered the opportunity to become vibrant economies based on very different, productive, sustainable and environmentally appropriate approaches to agriculture, supporting new local industries[1]. This in turn would expand the global economy, creating a "greener," carbon-neutral world capable of feeding and sustainably supporting a population well in excess of limits currently foreseen. In an ideal world, this perhaps can point us toward an even bigger target. With this in mind, it is pertinent to recognize that a study of urban:rural interactions across four measures of productivity in Zambia found a symbiotic relationship between agricultural growth, nonfarm growth and poverty reduction, as well as confirming that farms with less than 5ha were superior in terms of land:labor and labor:capital efficiencies (Nkonde et al., 2015).

Looking again at the Cameroon case study, where is the point at which we can say that sustainable intensification has been achieved and sustainable rural development begins? In fact, it seems that these two approaches to sustainable living overlap, forming a continuum which should not be separated by an arbitrary line. Maybe we can see this if we follow the trajectory of the projects in Cameroon in which interventions for greater productivity merge with young people saying that they can now see a future for themselves in their home villages and towns, rather than having to migrate to big cities for employment. Interestingly, a recent review of African agricultural development suggests that the current debate about how to remedy the failings of markets that deny smallholder farmers access to inputs and other innovations sees hope for their provision by private (NGO and collective) institutions, rather than public provision (Wiggins, 2014). It seems from evidence like that reported previously that smallholder farmers in Africa will invest and innovate when given the chance.

The failure to address the issues of African agriculture has put a brake on the global economy and is also having serious impacts on the global environment, especially manifested as climate change. The intensity with which the monocropping philosophy has gripped modern agriculture is blinding policy and decision makers, as well as agribusiness and academia, to the opportunities for sustainable intensification based on the net benefits from multiple-output, mixed-cropping systems. This is especially important in Africa, where agriculture is constrained by a complex of interacting social, environmental and economic factors. Addressing these then has spin-off benefits off-farm that enhance developments in other sectors of the national culture and economy (food sovereignty and the protection of traditional knowledge).

Nevertheless, progress toward long-term, holistic solutions to the global challenges seems to be thwarted by a failure to recognize the need to harness and combine multidimensional environment and production benefits within an intensification strategy (Leakey, 2012a,b). Instead, there has been a tendency toward a "silo" mentality and trying to squeeze tropical smallholders into economic theories designed for market economies in which farms are large and have access to income, capital and/or credit, land ownership, and price information—in other words, trying to make subsistence smallholders operate as a "business" and enterprise which can be more efficient by increasing the economies of scale. This overlooks the fact that many Africa smallholders operate on very small farms as self-sufficient units virtually outside the cash economy and so do not function as a business. Consequently, alternative and more appropriate "home production" economic models should be considered which are more risk averse, based on complementarity (such as between leguminous tree fallows and cereal crops) and supplementarity (such as between fruit trees and cereal crops), and concerned with the linkages between time allocation and utility maximization in the home (Ellis, 1993; Ellis and Allison, 2004).

Finally, I reemphasize that, just as food security is about the many factors affecting access to food, and not just about increasing yields (Poppy et al., 2014), so sustainable intensification has to be much more than just increased productivity, especially in Africa. It needs to address the unique social and environmental constraints which are reasons for the failure to intensify African agriculture. Thus, contrary to much conventional wisdom (e.g., Long et al., 2015), those wishing to address the complex and interacting issues which are the objectives of the 2030 Sustainable Development Goals (Table 40.1) should take much more notice of the ideas of African farmers and scientists about how to address the multifaceted needs of African agriculture—an African solution to an African problem (Leakey, 2014b). Interestingly, the action-oriented targets presented in Chapter 39 (Leakey and Prabhu, 2017) are not specific to the

---

1. *"Food crop domestication has been the precursor of settled, politically centralized, socially stratified, economically complex and technologically innovative societies"* according to Jared Diamond (1999) in *Guns, Germs, and Steel: The Fates of Human Societies* (WW Norton & Co., New York).

TABLE 40.1 UN Sustainable Development Goals: Transforming Our World: the 2030 Agenda for Sustainable Development (United Nations, 2015).

Goal	Purpose	Contribution from presented activities (Scored 1–5)
1	End poverty in all its forms everywhere	4
2	End hunger, achieve food security and improved nutrition, and promote sustainable agriculture	5
3	Ensure healthy lives and promote well-being for all at all ages	4
4	Ensure inclusive and equitable quality education and promote lifelong learning opportunities for all	1
5	Achieve gender equality and empower all women and girls	4
6	Ensure availability and sustainable management of water and sanitation for all	2
7	Ensure access to affordable, reliable, sustainable, and modern energy for all	0 – but improved access to fuel wood
8	Promote sustained, inclusive and sustainable economic growth, full and productive employment and decent work for all	4
9	Build resilient infrastructure, promote inclusive, and sustainable industrialization and foster innovation	2
10	Reduce inequality within and among countries	3
11	Make cities and human settlements inclusive, safe, resilient, and sustainable	0 – but produce better food for urban people
12	Ensure sustainable consumption and production patterns	2 – especially in rural population
13	Take urgent action to combat climate change and its impacts (under the auspices of UN Framework Convention on Climate Change)	4
14	Conserve and sustainably use the oceans, seas and marine resources for sustainable development	0
15	Protect, restore and promote sustainable use of terrestrial ecosystems, sustainably manage forests, combat desertification, and halt and reverse land degradation and halt biodiversity loss	5
16	Promote peaceful and inclusive societies for sustainable development, provide access to justice for all and build effective, accountable and inclusive institutions at all levels	2
17	Strengthen the means of implementation and revitalize the Global Partnership for Sustainable Development	1

achievement of different sustainable development goals. This further illustrates the need for integrated solutions to address the interconnected goals.

Ultimately, as we saw in Chapters 2 and 3 (Leakey, 1996; 2017a), agroecosystem functions have to include organisms at all trophic levels and scales; thus we need to recognize that global sustainability has to embrace *Homo sapiens* within the functioning of intensified agroecosystems. To conclude, therefore, we should perhaps add a twelfth and ultimate target of *Security, peace, and social justice* to our list. To create a productive and sustainable world we need to address the issues dividing our world into rich and poor. Poverty, hunger, and disillusionment lead to jealousy and resentment, which probably lie behind the global social and security issues of illegal migration, human trafficking, international drug smuggling, and terrorism. So maybe by resolving this economic and social divide, we can create a more peaceful world. This dream will certainly be impossible if we continue to plunder natural capital, damage the local and global environment, and ignore the needs of disadvantaged people in areas where agriculture is not meeting the social and economic needs of their populations.

References

Abasse, T., Weber, J.C., Katkore, B., Boureima, M., Larwanou, M., Kalinganire, A., 2011. Morphological variation in *Balanites aegyptiaca* fruits and seeds within and among parkland agroforests in eastern Niger. Agrofor. Syst. 81, 57–66.

Abbiw, D., 1990. Useful Plants of Ghana: West African Uses of Wild and Cultivated Plants. Intermediate Technology Publications, Royal Botanic Gardens, Kew, 337 pp.

Abrahamczyk, S., Kessler, M., Dadang, D.P., Waltert, M., Tscharntke, T., 2008. The value of differently managed cacao plantations for forest bird conservation in Sulawesi, Indonesia. Bird Conserv. Int. 18, 349–362.

Achinewhu, S.C., 1983. Ascorbic acid content of some Nigerian local fruits and vegetables. Qual. Plant. Plant Foods Hum. Nutr. 33, 261–266.

ACIPA, 2008. Plant Breeders Rights: A Guide for Horticultural Industries. Australian Centre for Intellectual Property in Australia, University of Queensland, Brisbane, Queensland, Australia, 49 p.

Adejuwon, J.O., Adesina, F.A., 1990. Organic matter and nutrient status of soils under cultivated fallows; an example of *Gliricidia sepium* fallows from south western Nigeria. Agrofor. Syst. 10, 23–32.

Adeola, A.O, Aiyelaagbe, I.O.O., Appiagyei-Nkyi, K., Bennuah, S.Y., Franzel, S., Jampoh, E.L., et al., unpublished. Farmers' preferences among tree species in the humid lowlands of West Africa. In: Boland, D., Ladipo, D.O., (Eds.), Irvingia: Uses, Potential and Domestication, 10–11 May 1994, ICRAF, Nairobi, Kenya.

Adin, A., Weber, J.C., Sotelo Montes, C., Vidaurre, H., Vosman, B., Smulders, M.J.M., 2004. Genetic differentiation and trade among populations of peach palm (*Bactris gasipaes* Kunth) in the Peruvian Amazon – implications for genetic resource management. Theor. Appl. Genet. 108, 1564–1573.

Adu-Tutu, M., Afful, Y., Asante-Appiah, K., Lieberman, D., Hall, J.B., Elvin-Lewis, M., 1979. Chewing stick usage in southern Ghana. Econ. Bot. 33, 320–328.

Ae, N., Akihara, J., Okada, K., Yoshinara, T., Johansen, C., 1990. Phosphorus uptake by pigeon pea and its role in cropping systems in the Indian subcontinent. Science. 248, 477–480.

African Development Bank, 2015. Abuja Declaration on fertilizer for African Green Revolution. www.afdb.org/en/topics-and-sectors/initiatives-partnerships/africanfertilizerfinancingmechanism/abuja-declaration/

Agbessi Dos-Santos, D., 1987. Manuel de Nutrition Africaine: Elements de Base Appliquée, Tome 1, ACCT, IPD et Editions Karthala, Dakar, Senegal.

Agoyi, E.E., Assogbadjo, A.E., Gouwakinnou, G., Okou, F.A.Y., Sinsin, B., 2014. Ethnobotanical assessment of *Moringa oleifera* Lam. In Southern Benin (West Africa). Ethnobot. Res. Appl. 12, 551–560.

Ahn, J.H., Robertson, B.M., Elliot, R., Gutteridge, R.C., Ford, C.W., 1989. Quality assessment of tropical forage browse legumes: tannin content and protein degradation. Anim. Feed Sci. Technol. 27, 147–156.

Aina, J.O., 1990. Physico-chemical changes in African mango (*Irvingia gabonensis*) during normal storage ripening. Food Chem. 36, 205–212.

Aiyelaagbe, I.O.O., Popoola, L., Adeola, A.O., Obisesan, K., Ladipo, D.O., 1996. *Garcinia kola*: its prevalence, farmer valuation and strategies for its conservation in the rainforest of southeastern Nigeria, Workshop on the Rainforest of Southeastern Nigeria and Southwestern Cameroon, 21–23 October 1996, Cross River National Park, Obudu Ranch, Nigeria.

Ajewole, K., Adeyeye, A., 1991. Seed oil of white star apple (*Crysophyllum albidum*) – Physicochemical characteristics and fatty acid composition. J. Sci. Food Agric. 54, 313–315.

Akinnifesi, F.K., Chilanga, T.G., Mkonda, A., Kwesiga, F.K., Maghembe, J.A., 2004. Domestication of *Uapaca kirkiana* in southern Africa: preliminary results of screening provenances in Malawi and Zambia. In: Rao, M.R., Kwesiga, F.R. (Eds.), Agroforestry Impacts on Livelihoods in Southern Africa: Putting Research into Practice. World Agroforestry Centre (ICRAF), Nairobi, pp. 85–92.

Akinnifesi, F.K., Kwesiga, F., Mhango, J., Chilanga, T., Mkonda, A., Kadu, C.A.C., et al., 2006. Towards the development of miombo fruit trees as commercial tree crops in southern Africa. For. Trees Livelihoods. 16, 103–121.

Akinnifesi, F.K., Leakey, R.R.B., Ajayi, O.C., Sileshi, G., Tchoundjeu, Z., Matakala, P. (Eds.), 2008. Indigenous Fruit Trees in the Tropics: Domestication, Utilization and Commercialization. CAB International, Wallingford, UK, 438p.

Akinnifesi, F.K., Mng'omba, S.A., Sileshi, G., Chilanga, T.G., Mhango, J., Ajayi, O.C., et al., 2009. Propagule type affects growth and fruiting of *Uapaca kirkiana*, a priority indigenous fruit tree of Southern Africa. Hortic. Sci. 44, 1662–1667.

Akubor, P.I., 1996. The suitability of African bush mango juice for wine production. Plant Foods Hum. Nutr. 49, 213–219.

Akyeampong, E., 1996. The influence of time of planting and spacing on production of fodder and fuelwood in associations of *Calliandra calothyrsus* and *Pennisetum purpureum* grown on contour bunds in the highlands of Burundi. Exp. Agric. 32, 79–85.

Akyeampong, E., Muzinga, K., 1994. Cutting management *of Calliandra calothyrsus* in the wet season to maximize dry season fodder production in the central highlands of Burundi. Agrofor. Syst. 27, 101–105.

Akyeampong, E.B., Duguma, B., Hieneman, A.M., Kamara, C.S., Kiepe, P., Kwesiga, F.K., et al., 1995. A synthesis of ICRAF's research alley cropping. Paper presented at the International Alley Farming Conference 14–18 September 1992. IITA, Ibadan, Nigeria.

Alexander, I.J., Lee, S.S., 2005. Mycorrhizas and ecosystem processes in tropical rain forest: Implications for diversity. In: Burslem, D., Pinard, M., Hartley, S. (Eds.), Biotic Interactions in the Tropics. Cambridge University Press, Cambridge, UK, pp. 165–203.

Altieri, M.A., 2002. Agro-ecology: the science of natural resource management for poor farmers in marginal environments. Agric. Ecosyst. Environ. 93, 1–24.

Altieri, M.A., Nicholls, C.I., 1999. Biodiversity, ecosystem function, and insect pest management in agricultural systems. In: Collins, W.W., Qualset, C.O. (Eds.), Biodiversity in Agroecosystems. CRC Press, New York, pp. 69–84.

Amin, M.N., Jaiswal, V.S., 1993. In vitro response of apical bud explants from mature trees of jackfruit (*Artocarpus heterophyllus*). Plant Cell Tissue Organ Cult. 33, 59–65.

Aminah, H., Dick, J.Mc.P., Leakey, R.R.B., Grace, J., Smith, R.I., 1995. Effect of indole butyric acid (IBA) on stem cuttings of *Shorea leprosula*. For. Ecol. Manage. 72, 199–206.

Amubode, F.O., Fetuga, B.L., 1984. Amino acid composition of some lesser known tree crops. Food Chem. 13, 299–307.

Anderson, A.B. (Ed.), 1990. Alternatives to Deforestation: Steps Toward Sustainable Use of the Amazon Rainforest. Columbia University Press, New York.

Anderson, L.S., Sinclair, F.L., 1993. Ecological interactions in agroforestry systems. For. Abstr. 54, 489–523.

Andersson, M., Gradstein, S.R., 2005. Impact of management intensity on non-vascular epiphyte diversity in cacao plantations in western Ecuador. Biodivers. Conserv. 14, 1101–1120.

Anecksamphant, C., Bonchee, S., Sajjappongse, R., 1990. Management of sloping land for sustainable agriculture in northern Thailand. Trans. 14th Int. Congr. Soil Sci. 6, 198–203.

Anegbeh, P.O., Usoro, C., Ukafor, V., Tchoundjeu, Z., Leakey, R.R.B., Schreckenberg, K., 2003. Domestication of *Irvingia gabonensis*: 3. Phenotypic variation of fruits and kernels in a Nigerian village. Agrofor. Syst. 58, 213–218.

Anegbeh, P.O., Ukafor, V., Usoro, C., Tchoundjeu, Z., Leakey, R.R.B., Schreckenberg, K., 2005. Domestication of *Dacryodes edulis*: 1. Phenotypic variation of fruit traits from 100 trees in southeast Nigeria. New Forests. 29, 149–160.

Angima, S., Stott, D.E., O'Neill, M.K., Ong, C.K., Weesies, G.A., 2001. Use of calliandra-napier contour hedges to control soil erosion in central Kenya. Agric. Ecosyst. Environ. 97, 295–308.

Aniche, G.N., Uwakwe, G.U., 1990. Potential use of *Garcinia kola* as substitute in lager beer brewing. World J. Microbiol. Biotechnol. 6, 323–327.

Arkcoll, D.B., Aguiar, J.P.L., 1984. Peach palm (*Bactris gasipaes* H.B.K.), a new source of vegetable oil from the wet tropics. J. Sci. Food Agric. 35, 520–526.

Arlet, M.E., Molleman, F., 2010. Farmers' perceptions of the impact of wildlife on small-scale cacao cultivation at the northern periphery of Dja faunal reserve, Cameroon. Afr. Primates. 7, 27–34.

Arnold J.E.M., 1995. Socio-economic benefits and issues in non-wood forest product use. Report of the International Expert Consultation on Non-wood Forest Products. Non-wood Forest Products No. 3, pp 89–123

Arnold J.E.M., 1996. Economic factors in farmer adoption of forest product activities. In: Leakey, R.R.B., Temu, A.B., Melnyk, M., Vantomme, P. (Eds.), Domestication and Commercialization of Non-timber Forest Products in Agroforestry Systems, pp 131–146. Non-wood Forest Products No. 9. FAO, Rome, Italy.

Asaah, EK, 2012. Beyond vegetative propagation of indigenous fruit trees: case of *Dacryodes edulis* (G.Don) H. J. Lam and *Allanblackia floribunda* Oliv. PhD. Thesis. Faculty of Bioscience Engineering, Ghent University, Belgium, 231 pp.

Asaah, E.K., Tchoundjeu, Z., Wanduku, T.N., van Damme, P., 2010. Understanding structural roots system of 5-year-old African plum tree (*D. edulis*) of seed and vegetative origins (G. Don) H. J. Lam). Trees. 24, 789–796.

Asaah, E.K., Tchoundjeu, Z., Leakey, R.R.B., Takousting, B., Njong, J., Edang, I., 2011a. Trees, agroforestry and multifunctional agriculture in Cameroon. Int. J. Agric. Sustain. 9, 110–119.

Asaah, E.K., Tchoundjeu, Z., Ngahane, W., Tsobeng, A., Konodiekong, L., Jamnadass, R., et al., 2011b. *Allanblackia floribunda* a new oil tree crop for Africa: amenability to grafting. New For. 41, 389–398.

Asaah, E.K., Tchoundjeu, Z., van Damme, P., 2012a. Beyond vegetative propagation of indigenous fruit trees: case of *Dacryodes edulis* (G. Don) H. J. Lam and *Allanblackia floribunda* Oliv. Afrika Focus. 25, 61–72.

Asaah, E.K., Wanduku, T.N., Tchoundjeu, Z., Kouodiekong, L., van Damme, P., 2012b. Do propagation methods affect the fine root architecture of African plum (*Dacryodes edulis*)? Trees. 26, 1461–1469.

Ash, J.A., 1989. The effect of supplementation with leaves from leguminous trees *Sesbania grandiflora*, *AlbUia chinensis* and *Gliricidia sepium* on the intake and digestibility of guinea grass hay by goats. Anim. Feed Sci. Technol. 28, 225–232.

Assogbadjo, A.E., Sinsin, B., Codjia, J.T.C., van Damme, P., 2005. Ecological diversity and pulp, seed and kernel production of the baobab (*Adansonia digitata*) in Benin. Belg. J. Bot. 138, 47–56.

Assogbadjo, A.E., Kyndt, T., Sinsin, B., Gheysen, G., van Damme, P., 2006. Patterns of genetic and morphometric diversity in baobab (*Adansonia digitata* L.) populations across different climatic zones in Benin (West Africa). Ann. Bot. (Lond). 97, 819–830.

Assogbadjo, A.E., Kyndt, T., Chadare, F.J., Sinsin, B., Gheysen, G., Eyog-Matig, O., et al., 2009. Genetic fingerprinting using AFLP cannot distinguish traditionally classified baobab morphotypes. Agrofor. Syst. 75, 157–165.

Assogbadjo, A.E., Glèlè Kakaï, R., Vodouhê, F.G., Djagoun, C.A.M.S., Codjia, J.T.C., Sinsin, B., 2012. Biodiversity and socioeconomic factors supporting farmers' choice of wild edible trees in the agroforestry systems of Benin (West Africa). Forest Policy Econ. 14, 41–49.

Atangana, A.R., Tchoundjeu, Z., Fondoun, J.-M., Asaah, E., Ndoumbe, M., Leakey, R.R.B., 2001. Domestication of *Irvingia gabonensis*: 1. Phenotypic variation in fruit and kernels in two populations from Cameroon. Agrofor. Syst. 53, 55–64.

Atangana, A.R., Ukafor, V., Anegbeh, P.O., Asaah, E., Tchoundjeu, Z., Usoro, C., et al., 2002a. Domestication of *Irvingia gabonensis*: 2. The selection of multiple traits for potential cultivars from Cameroon and Nigeria. Agrofor. Syst. 55, 221–229.

Atangana, A.R., Asaah, E., Tchoundjeu, Z., Schreckenberg, K., Leakey, R.R.B., 2002b. Biophysical characterisation of *Dacryodes edulis* fruits in three markets in Cameroon. In: Kengue, J., Kapseu, C., Kayem, G.J. (Eds.), 3ème Séminaire International sur la Valorisation du Safoutier et autres Oléagineux Non-conventionnels Yaounde, Cameroun, 3–5 Octobre 2000. Presses Universitaires d'Afrique, Yaoundé, pp. 106–118.

Atangana, A.R., van der Vlis, E., Khasa, D.P., van Houten, D., Beaulieu, J., Hendrickx, H., 2011a. Tree-to-tree variation in stearic and oleic acid content in seed fat from *Allanblackia floribunda* from wild stands: potential for tree breeding. Food Chem. 126, 1579–1585.

Atangana, A.R., Ngo Mpeck-Nyemeck, M.-L., Chang, S.X., Khasa, D.P., 2011b. Auxin regulation and function: Insights from studies on rooted leafy stem cuttings of tropical tree species. In: Keller, A.H., Fallon, M.D. (Eds.), Auxins: Structure, Biosynthesis and Functions. Nova Science Publishers, Inc., New York, USA, pp. 53–66.

Atangana, A.R., Khasa, D., Chang, S., Degrande, A., 2014. Tropical Agroforestry. Springer Science, Dordrecht, The Netherlands, 380 pp.

Atta-Krah, K., Kindt, R., Skilton, J.N., Amaral, W., 2004. Managing biological and genetic diversity in tropical agroforestry. Agrofor. Syst. 61, 183–194.

Attipoe, L, van Andel, A, Nyame, SK, 2006. The Novella Project: developing a sustainable supply chain for Allanblackia oil. In: Agro-food Chains and Networks for Development, pp. 179–189.

Aumeeruddy, Y., 1994. Local representations and management of agroforests on the periphery of Kerinci Seblat National Park, Sumatra, Indonesia, People and Plants Working Paper No 3. UNESCO, Paris.

Aweto, A.O., Obe, O., Ayanniyi, O.O., 1992. Effects of shifting and continuous cultivation of cassava (*Manihot esculenta*) intercropped with maize (*Zea mays*) on a forest alfisol in southwestern Nigeria. J. Agric. Sci. Camb. 118, 195–198.

Awodoyin, R.O., Olubode, O.S., Ogbu, J.U., Balogun, R.B., Nwawuisi, J.U., Orji, K.O., 2015. Indigenous fruit trees of tropical africa: status, opportunity for development and biodiversity management. Agric. Sci. 6, 31–41.

Awono, A., Ndoye, O., Schreckenberg, K., Tabuna, H., Isseri, F., Temple, L., 2002. Production and marketing of Safou (*Dacryodes edulis*) in Cameroon and internationally: market development issues. Forest Trees Livelihoods. 12, 125–147.

Ayuk, E.T., Duguma, B., Franzel, S., Kengue, J., Mollet, M., Tiki-Manga, T., et al., 1999a. Uses, management and economic potential of *Dacryodes edulis* (Burseraceae) in the humid lowlands of Cameroon. Econ. Bot. 53, 292–301.

Ayuk, E.T., Duguma, B., Franzel, S., Kengue, J., Mollet, M., Tiki-Manga, T., et al., 1999b. Uses, management and economic potential of *Garcinia kola* and *Ricinodendron heudelotii* in the humid lowlands of Cameroon. J. Trop. For. Sci. 11, 746–761.

Ayuk, E.T., Duguma, B., Franzel, S., Kengue, J., Mollet, M., Tiki-Manga, T., et al., 1999c. Uses, management and economic potential of *Irvingia gabonensis* in the humid lowlands of Cameroon. For. Ecol. Manage. 113, 1–9.

Baadsmand, S., Andersen, A.S., 1984. Transport and accumulation of indole-3-acetic acid in pea cuttings under two levels of irradiance. Physiol. Plant. 61, 107–113.

Bacon, C.M., Sundstrom, W.A., Flores Gómez, M.E., Méndez, V.E., Santos, R., Goldoftas, B., et al., 2015. Explaining the 'hungry farmer paradox': smallholders and fair trade cooperatives navigate seasonality and change in Nicaragua's corn and coffee markets. Global Environ. Change. 25, 133–149.

Badifu, G.I.O., 1989. Lipid composition of Nigerian *Butyrospermum paradoxum* kernel. J. Food Comp. Anal. 2, 238–244.

Bai, Z.G., Dent, D.L., Olsson, L., Schaepman, M.E., 2008. Proxy global assessment of land degradation. Soil Use Manage. 24, 223–234.

Bailey, N.J.L., 1959. Statistical Methods in Biology. English Universities, London, p. 200.

Bali, A., Kumar, A., Krishnaswamy, J., 2007. The mammalian communities in coffee plantations around a protected area in the Western Ghats, India. Biol. Conserv. 39, 93–102.

Banda, A., Maghembe, J.A., Ngugi, D.N., Chome, V.A., 1994. Effect of intercropping maize and leucaena hedgerow on soil conservation and maize yield on a steep slope near Ntcheu, Malawi. Agrofor. Syst. 25, 1–6.

Bani-Aameur, Ferradous, 2001. Fruit and stone variability in three argan (*Argania spinosa* (L.) Skeels) populations. For. Genet. 8, 39–45.

Bani-Aameur, F., Ferradous, A., Dupuis, P., 1999. Typology of fruits and stones of *Argan spinosa* (Sapotaceae). For. Genet. 6, 213–219.

Bannister, M.E., Nair, P.K.R., 1990. Alley cropping as a sustainable agricultural technology for the hillsides of Haiti: experience of an agroforestry outreach project. Am. J. Altern. Agric. 5 (2), 51–59.

Barany, M., Hammett, A.L., Sene, A., Amichev, B., 2001. Non-timber forest benefits and HIV/AIDS in sub-Saharan Africa. J. For. 99, 36–42.

Barany, M., Hammett, A.L., Leakey, R.R.B., Moore, K.M., 2003. Income generating opportunities for smallholders affected by HIV/AIDS: linking agro-ecological change and non-timber forest product markets. J. Manage. Stud. 39, 26–39.

Barnes, R.D., Simons, A.J., 1994. Selection and breeding to conserve and utilize tropical tree germplasm. In: Leakey, R.R.B., Newton, A.C. (Eds.), Tropical Trees: The Potential for Domestication and the Rebuilding of Forest Resources. HMSO, London, pp. 84–90.

Barr, C.M., Sayer, J.A., 2012. The political economy of reforestation and forest restoration in Asia-Pacific: critical issues for REDD + . Biol. Conserv. 154, 9–19.

Barrett, C.B., Ashley, J.G., Carter, M.R., 2010. The power and pitfalls of experiments in development economics: some non-random reflections. Appl. Econ. Perspect. Policy. 32, 515–548.

Barrios, E., Buresh, R.J., Sprent, J.I., 1996a. Organic matter in soil particle size and density fractions from maize and legume cropping systems. Soil. Biol. Biochem. 28, 185−193.

Barrios, E., Buresh, R.J., Sprent, J.I., 1996b. Nitrogen mineralization in density fractions of soil organic matter from maize and legume cropping systems. Soil. Biol. Biochem. 28, 1459−1465.

Barrios, E., Kwesiga, F., Buresh, R.J., Sprent, J.I., 1997. Light fraction soil organic matter and available nitrogen following trees and maize. Soil. Sci. Soc. Am. J. 61, 826−831.

Barrios, E., Sileshi, G.W., Shepherd, K., Sinclair, F., 2012. Agroforestry and soil health: linking trees, soil biota and ecosystem services. In: Wall, D. H. (Ed.), The Oxford Handbook of Soil Ecology and Ecosystem Services. Oxford University Press, Oxford, UK, pp. 315−330.

Barrow, E.G.C., 1996. The Drylands of Africa: Local Participation in Tree Management. Initiatives Publishers, Nairobi, Kenya, 268pp.

Beare, M.H., Reddy, M.V., Tian, G., Srivastava, S.C., 1997. Agricultural intensification, soil biodiversity and agroecosystem function in the tropics: the role of decomposer biota. Appl. Soil Ecol. 6, 87−108.

Beattie, A.J., Barthlott, W., Elisabetsky, E., Farrel, R., Kheng, C.T., Prance, I., et al., 2005. New products and industries from biodiversity (Chapter 10). In: Hassan, R., Scholes, R., Ash, N. (Eds.), Ecosystems and Human Well-Being: Vol 1. Current State and Trends. Findings of the Condition and Trends Working Group of the Millennium Ecosystem Assessment, Island Press, Washington DC, USA, pp. 271−296.

Becker, B., 1983. The contribution of wild plants to human nutrition in the Ferlo (Northern Senegal). Agrofor. Syst. 1, 257−267.

Beddow, J.M., Hurley, T.M., Pardey, P.G., Alston, J.M., 2014. Food security: yield gap. In: van Alfen, N., et al., (Eds.), Encyclopedia of Agriculture and Food Systems, Vol 3. Elsevier Publishers, San Diego, pp. 352−365.

Bekele-Tesemma, A., Birnie, A., Tengnäs, B., 1993. Useful trees and shrubs for ethiopia: identification, propagation and mmanagement for agricultural and pastoral communities. SIDA Soil Conservation Unit, Technical Handbook No 5, Swedish International Development Authority, Nairobi, Kenya, 474pp.

Belcher, B.M., 2003. What isn't an NTFP? Int. For. Rev. 5, 161−168.

Belmain, S.R., Amoah, B.A., Nyirenda, S.P., Kamanula, J.F., Stevenson, P.C., 2012. Highly variable insect control efficacy of *Tephrosia vogelii* chemotypes. J. Agric. Food Chem. 60, 10055−10063.

Bennett, E., Carpenter, S., Gordon, L., Ramankutty, N., Balvanera, P., Campbell, B., et al., 2014. Toward a more resilient agriculture. Solutions. 5, 65−75.

Bennett, E.M., Cramer, W., Begossi, A., Cundill, G., Diaz, S., Egoh, B.N., et al., 2015. Linking biodiversity, ecosystem services and human well-being: three challenges for designing research for sustainability. Curr. Opin. Environ. Sustain. 14, 76−85.

Bernard, F., van Noordwijk, M., Luedeling, E., Villamor, G.B., Gudeta, S., Namirembe, S., 2014. Social actors and unsustainability of agriculture. Curr. Opin. Environ. Sustain. 6, 155−161.

Bertomeu, M., 2004. Smallholder timber production on sloping lands in the Philippines: a systems approach. ICRAF World Agroforestry Center, Southeast Regional Office, Bogor, Indonesia, 257 pp.

Bertomeu, M., Roshetko, J.M., Rahayu, S., 2011. Optimum pruning intensity for reducing crop suppression in a Gmelina-maize smallholder agroforestry system in Claveria, Philippines. Agrofor. Syst. 83, 167−180.

Bertomeu, M.G., Sungkit, R.L., 1999. Propagating *Eucalyptus* species − recommendations for smallholders in the Philippines. For. Farm. Comm. Tree Res. Rep. 4, 68−72.

Bignell, D.E., Tondoh, J., Dibog, L., Huang, S.P., Moreira, F., Nwaga, D., et al., 2005. Belowground biodiversity assessment: developing a key functional group approach in best-bet alternatives to slash and burn. In: Palm, C., Vosti, S., Sanchez, P., Ericksen, P. (Eds.), Slash and Burn: The Search for Alternatives. Columbia University Press, New York, pp. 119−142.

Bille, P.G., Shikongo-Nambabi, Cheikhyoussef, A., 2013. Value-addition and processed products of three indigenous fruits in Namibia. Afr. J. Food Agric. Nutr. Dev. 13, 7192−7212.

Boafo, Y.A., Saito, O., Kato, S., Kamiyama, C., Takeuchi, K., Nakahara, M., 2015. The role of traditional ecological knowledge in ecosystem services management: the case of four rural communities in Northern Ghana. Int. J. Biodivers. Sci. Ecosyst. Serv. Manage. Available from: http://dx.doi.org/10.1080/21513732.2015.1124454.

Boedecker, J., Termote, C., Assogbadjo, A.E., van Damme, P., Lachat, C., 2014. Dietary contribution of wild edible plants to women's diets in the buffer zone around Lama Forest, Benin —an underutilized potential. Food Security. 0, 000-000, http://dx.doi.org/10.1007/s12571-014-0396-7.

Boehringer, A., Ayuk, E.T., Katanga, R., Ruvuga, S., 2003. Farmer nurseries as a catalyst for developing sustainable land use systems in southern Africa. Agric. Syst. 77, 187−201.

Boffa, J.-M., Yaméogo, G., Nikiéma, P., Knudson, D.M., 1996. Shea nut (*Vitellaria paradoxa*) production and collection in agroforestry parklands of Burkina Faso. In: Leakey, R.R.B., Temu, A.B., Melnyk, M., Vantomme, P. (Eds.), *Domestication and Commercialization of Non-timber Forest Products in Agroforestry Systems, Non-Wood Forest Products No. 9.* FAO, Rome, Italy, pp. 110−122.

Boland, D., Ladipo, D.O., unpublished. *Irvingia gabonensis*: state of knowledge, strategy for germplasm collection and potential for genetic improvement for agroforestry development in West Africa. In: Boland, D., Ladipo D.O., (Eds.), Proceedings of ICRAF Workshop, May 1994, Ibadan, Nigeria.

Bonkoungou, E.G., 1995. Practiques agroforestieres traditionnelles et gestion des ressources naturelles dans les zones semi-arides de l'Afrique de l'Ouest. In: Proceedings of the International Workshop for a Desert Margins Initiative, ICRISAT, Hyderabad, India.

Booth, F.E.M., Wickens, G.E., 1988.). *Non-timber uses of selected arid zone trees and shrubs in Africa*, FAO Conservation Guide 19. FAO, Rome, Italy, p. 176.

Borchert, R., 1976. The concept of juvenility in woody plants. Symp. Juv. Woody Perenn. Acta Hortic. 56, 21−36.

Borlaug, N., 1996. Mobilizing science and technology for a rock-phosphate in African agriculture. In: Breth, S.A. (Ed.), Achieving Greater Impact from Research Investments in Africa. Sasakawa Africa Association, Mexico City, pp. 209–217.

Borlaug, N., Dowswell, C.R., 1994. Feeding a human population that increasingly crowds a fragile planet. Supplement to Transactions of the 15th World Congress of Soil Science. International Society of Soil Science, Acapulco, Mexico. Chapingo, Mexico.

Bos, M.M., Höhn, P., Saleh, S., Büche, B., Buchori, D., et al., 2007a. Insect diversity responses to forest conversion and agroforestry management. In: Tscharntke, T., Leuschner, C., Guhardja, E., Zeller, M. (Eds.), *The Stability of Tropical Rainforest Margins: Linking Ecological, Economic and Social Constraints of* Land-Use *and Conservation.* Springer Verlag, Berlin, pp. 279–296.

Bos, M.M., Steffen-Dewenter, I., Tscharntke, T., 2007b. The contribution of cacao agroforests to the conservation of lower canopy ant and beetle diversity in Indonesia. Biodivers. Conserv. 16, 2429–2444.

Bos, M.M., Steffen-Dewenter, I., Tscharntke, T., 2007c. Shade tree management affects fruit abortion, insect pests and pathogens of cacao. Agric. Ecosyst. Environ. 120, 201–205.

Bos, M.M., Tylianakis, J.M., Steffan-Dewenter, I., Tscharntke, T., 2008. The invasive yellow crazy ant in Indonesian cacao agroforests and the decline of forest ant diversity. Biol. Invasions. 10, 1399–1409.

Bosch, C, Sie, K, Asafa, B 2004. *Adansonia digitata* L. Fiche de Protabase. In: Grubben, G.J.H., Denton, O.A., (Eds.), Plant Resources of Tropical Africa. Wageningen, The Netherlands.

Botelle, A., du Plessis, P., Pate, K., Laamanen, R., 2002. A survey of Marula fruit yields in north-central Namibia, Report to UK DFID Forestry Research Programme (Project No R7795), CRIAA-SA-DC, PO Box 23778, Windhoek, Namibia, 26 pp.

Bowen, M.R., Howland, P., Last, F.T., Leakey, R.R.B., Longman, K.A., 1977. *Triplochiton scleroxylon:* its conservation and future improvement. FAO For. Genet. Resour. Inf. 6, 38–47.

Bradley, P.N., Chavangi, N., van Gelder, A., 1995. Development research and energy planning in Kenya. Ambio. 14, 228–236.

Brewbaker, J.L., Sorensson, C.T., 1994. Domestication of lesser known species of the genus *Leucaena*. In: Leakey, R.R.B., Newton, A.C. (Eds.), Tropical Trees: the Potential for Domestication. Rebuilding Forest Resources. HMSO, London, pp. 195–204.

Brigham, T., Chihongo, A., Chidumayo, E., 1996. Trade in woodland products from the miombo region. In: Campbell, B.M. (Ed.), The Miombo in Transition: Woodlands and Welfare in Africa. CIFOR, Bogor, pp. 137–174.

Buchanan, M., King, L.D., 1992. Seasonal fluctuations in soil microbial biomass carbon, phosphorus, and activity in no-till and reduced-chemical-input maize agroecosystems. Biol. Fertil. Soils. 13, 211–217.

Bulatao, R.A., Bos, E., Stephens, P.W., Vu, M.T., 1990. World population projections 1989–90 edition. The Johns Hopkins University Press for the World Bank, Baltimore.

Bundersen, W.T., 1992. Final report of the agroforestry technical assistant to the national agroforestry team, MARE project, Chitedze Agricultural Research Station, Malawi.

Bunt, C., Leakey, R.R.B., 2008. Domestication potential and marketing of *Canarium indicum* nuts in the Pacific: commercialization and market development. Forests Trees Livelihoods. 18, 271–289.

Burchi, F., Fanzo, J., Frison, E., 2011. The role of food and nutrition system approaches in tackling hidden hunger. Int. J. Environ. Res. Public Health. 8 (2), 358–373.

Buresh, R.J., Cooper, P.J.M. (Eds.), 1999. The science and practice of short-term fallows. *Agrofor. Syst.* 47, 1–358.

Buresh, R.J., Tian, G., 1997. Soil improvement by trees in sub-Saharan Africa. Agrofor. Syst. 38, 51–76.

Buresh, R.J., Smithson, P.C., Hellums, D., 1997. Building up soil P capital in sub-Saharan Africa. Replenishing Soil Fertility in Africa. ASA-SSSA Special Publication (In the press.).

Burger, A.E.C., de Villers, J.B.M., du Plessis, L.M., 1987. Composition of the kernel oil and protein of the marula seed. S. Afr. J. Sci. 83, 733–735.

Burley J., von Carlowitz P. 1984a. Multi-purpose tree germplasm. Proceedings of a planning workshop at National Academy of Sciences, Washington, DC, June 1983. ICRAF, Nairobi, Kenya.

Burley J., von Carlowitz P., 1984b. Multipurpose Tree Germplasm. International Center for Research in Agroforestry, Nairobi, Kenya, 298 pp.

Campbell-Platt, G., 1980. African locust bean (*Parkia* species) and its West African fermented food product, dawadawa. Ecol. Food Nutr. 9, 123–132.

Cannell, M.G.R., 1989. Food crop potential of tropical trees. Exp. Agric. 25, 313–326.

Carandang, W.M., Tolentino, E.L., Roshetko, J.M., 2006. Smallholder tree nursery operations in southern Philippines – supporting mechanisms for timber tree domestication. Forest Tree Livelihoods. 17, 71–83.

Carsan, S., Stroebel, A., Dawson, I., Kindt, R., Swanepoel, F., Jamnadass, R., 2013. Implications of shifts in coffee production on tree species richness, composition and structure on small farms around Mount Kenya. Biodivers. Conserv. 22, 2919–2936.

Carter, J., 1995. Alley farming: have resource poor farmers benefited? ODI Natural Resources Perspectives No. June 1995.

Cash, D.W., Adger, W., Berkes, F., Garden, P., Lebel, L., Olsson, P., et al., 2006. Scale and cross-scale dynamics: governance and information in a multilevel world. Ecol. Soc. 11, 8.

Cassano, C.R., Kierulff, M.C.M., Chiarello, A.G., 2011. The cacao agrotorests of the Brazilian Atlantic forest as habitat for the endangered maned sloth *Bradypus torquatus*. Mamm. Biol. 76, 243–250.

Cassman, K.G., Wood, S., Choo, P.S., Cooper, D., Devendra, C., Dixon, J., et al., 2005. Cultivated Systems (Chapter 26). In: Hassan, R., Scholes, R., Ash, N. (Eds.), Ecosystems and Human Well-Being: Vol 1. Current State and Trends. Findings of the Condition and Trends Working Group of the Millennium Ecosystem Assessment, Island Press, Washington DC, USA, pp. 745–794.

Catacutan, D., Bertomeu, M., Arbes, L., Duque, C., Butra, N., 2008. Fluctuating fortunes of a collective enterprise: the case of the Agroforestry Tree Seeds Association of Lantapan (ATSAL) in the Philippines. Small Scale For. 7, 353–368.

Cavigelli, M.A., Maul, J.E., Szlavecz, K., 2012. Managing soil biodiversity and ecosystem services. In: Wall, D. (Ed.), The Oxford Handbook of Soil Ecology and Ecosystem Services. Oxford University Press, Oxford, pp. 337−356.

CAWMA, 2007. In: Molden, D. (Ed.), *Water for Food: Water for Life. A Comprehensive Assessment of Water Management in Agriculture*. Earthscan, London, 645 pp.

Cerda, R., Deheuvels, O., Calvache, D., Niehaus, L., Saenz, Y., Kent, J., et al., 2014. Contribution of cocoa agroforestry systems to family income and domestic consumption: looking toward intensification. Agrofor. Syst. 88, 957−981.

Chapell, M.J., LaValle, L.A., 2011. Food security and biodiversity: Can we have both? An agroecological analysis. Agric. Hum. Values. 28, 3−26.

Chavelier, A., 1943. Nouveau procédé de traitment des noix de Karité. Rev. Bot. Appl. 5, 536−537.

Cheikhyoussef, Embashu, 2013. Ethnobotanical knowledge on indigenous fruits in Ohangwena and Oshikoto regions in Northern Namibia. J. Ethnobiol. Ethnomed. 9 (34), 1−12.

Chianu, J.N., Chianu, J.N., Mairura, F., 2012. Mineral fertilizer in the farming systems of Sub-Saharan Africa. A review. Agron. Sustain. Dev. 32, 545−566.

Chirwa, P.W., Ong, C.K., Maghembe, J.A., Black, C.R., 2007. Soil water dynamics in cropping systems containing *Gliricidia sepium*, pigeon pea and maize in southern Malawi. Agrofor. Syst. 69, 29−43.

Chivandi, E., Mukonowenzou, N., Nyakudya, T., Erlwanger, K.E., 2015. Potential of indigenous fruit-bearing trees to curb malnutrition, improve household food security, income and community health in Sub-Saharan Africa: a review. Food Res. Int. 76, 980−985.

Chivenge, P., Mabhaudhi, T., Modi, A.T., Mafongoya, P., 2015. The potential role of neglected and underutilised crop species as future crops under water scarce conditions in sub-Saharan Africa. Int. J. Environ. Res. Public Health. 12, 5685−5711.

Christensen, M.V., Eriksen, E.N., Andersen, A.S., 1980. Interaction of stockplant irradiance and auxin in the propagation of apple rootstocks by cuttings. Sci. Hortic. 12, 11−17.

Claeys, P., 2015. Human Rights and the Food Sovereignty Movement: Reclaiming control. Routledge, London and New York, 197pp.

Clark, D.P., Pazdernik, N.J., 2016. Biotechnology: Applying the Genetic Revolution. Academic Press, New York.

Clark, L.E., Sunderland, T.C.H., 2004. *The Key Non-Timber Forest Products of Central Africa: State of Knowledge*, Technical Paper 122, SD Publication Series. USAID, Washington, p. 186.

Clarke, P.A., 1998. Early Aboriginal plant use in southern South Australia. Proc. Nutr. Soc. S. Aust. 22, 17−20.

Cleaver, K.M., Schreiber, G.A., 1994. Reversing the spiral; the population, agriculture and environment nexus in sub-Saharan Africa. World Bank, Washington, DC.

Clement, C.R., 1988. Domestication of the pejibaye palm (*Bactris gasipaes*): past and present. Adv. Econ. Bot. 6, 155−174.

Clement, C.R., 1989. The potential use of the pejibaye palm in agroforestry systems. Agrofor. Syst. 7, 201−212.

Clement, C.R., 1990. Pejibaye. In: Nagy, S., Shaw, P.E., Wardowski, W.F. (Eds.), Fruits of Tropical and Subtropical origin: Composition, Properties and Uses. Florida Science Source Inc., Lake Alfred, FL, pp. 302−321.

Clement, C.R., Arkcoll, D.B., 1985. The *Bactris gasipaes* H.B.K. (Palmae) as an oil producing crop: potential and priority of investigation. In: Forero P, L.E. (Ed.), Informe del Seminario-Taller Sobre Oleaginosas Promisorias. Programma Interciencias de Recursos Biologicos, Bogota, Colombia.

Clement, C.R., Villachica, H., 1994. Amazonian fruits and nuts: potential for domestication in various agroecosystems. In: Leakey, R.R.B., Newton, A.C. (Eds.), Tropical Trees: The Potential for Domestication and the Rebuilding of Forest Resources. HMSO, London, UK, pp. 230−238.

Clement, C.R., Weber, J.C., van Leeuwen, J., Domian, C.A., Cole, D.M., Arévalo Lopez, L.A.A., et al., 2004. Why extensive research and development did not promote use of peach palm fruit in Latin America. Agrofor. Syst. 61, 195−206.

Clement, C.R., Denevan, W.M., Heckenberger, M.J., Junqueira, A.B., Neves, E.G., Teixeira, W.G., et al., 2015. The domestication of Amazonia before European conquest. Proc. R. Soc. B. 282, 20150813.

Clough, Y., Faust, H., Tscharntke, T., 2009a. Cacao boom and bust: sustainability of agroforests and opportunities for biodiversity conservation. Conserv. Lett. 2, 197−205.

Clough, Y., Putra, D.D., Pitopang, R., Tscharntke, T., 2009b. Local and landscape factors determine functional bird diversity in Indonesian cacao agroforestry. Biol. Conserv. 142, 1032−1041.

Clough, Y., Barkmann, J., Juhrbandt, J., Kessler, M., Wanger, T.C., et al., 2011. Combining high biodiversity with high yields in tropical agroforests. Proc. Natl. Acad. Sci. USA. 108, 8311−8316.

Coe, R., 1994. Through the looking glass: 10 common problems in alley-cropping research. Agrofor. Today. 6, 9−11.

Coelli, T.J., Prasada Rao, D.S., 2005. Total factor productivity growth in agriculture: A Malmquist index analysis of 93 countries, 1980-2000. Agric. Econ. 32, 115−134.

Collins, W.W., Qualset, C.O. (Eds.), 1999. Biodiversity in Agroecosystems. CRC Press, New York.

Constantinides, M., Fownes, J.H., 1994. Nitrogen mineralization from leaves and litter of tropical plants: relationship to nitrogen, lignin and soluble polyphenol concentrations. Soil Biol. Biochem. 26, 49−55.

Conway, G., 2012. One Billion Hungry, Can We Feed the World. Cornell University Press, Ithaca.

Cook, S.M., Khan, Z.R., Pickett, J.A., 2007. The use of "push-pull" strategies in integrated pest management. Annu. Rev. Entomol. 52, 375−400.

Cooper, P.J.M., Leakey, R.R.B., Rao, M.R., Reynolds, L., 1996. Agroforestry and the mitigation of land degradation in the humid and sub-humid tropics of Africa. Exp. Agric. 32, 235−290.

Cornelius, J., Clement, C.R., Weber, J.C., Sotelo Montes, C., van Leeuwen, J., Ugarte Guerra, L.J., et al., 2006. The trade-off between genetic gain and conservation in a participatory improvement program: the case of peach palm (*Bactris gasipaes* Kunth). Forest Tree Livelihoods. 16, 17−34.

Cornelius, J.P., Mesén, F., Ohashi, S.T., Leão, N., Silva, C.E., Ugarte-Guerra, J., et al., 2010. Smallholder production of agroforestry germplasm: experiences and lessons from Brazil, Costa Rica, Mexico and Peru. For. Trees Livelihoods. 19, 201−216.

Cosyns, H., Degrande, A., de Wulf, R., van Damme, P., Tchoundjeu, Z., 2011. Can commercialization of NTFPs alleviate poverty? A case study of *Ricinodendron heudelottii* (Baill.) Pierre ex Pax. Kernel marketing in Cameroon. J. Agric. Rural Dev. Trop. Subtrop. 113, 45−56.

Cosyns, H., van Damme, P., de Wulf, R., Degrande, A., 2013. Can rural development projects generate social capital? A case study of *Ricinodendron heudelotii* kernel marketing in Cameroon. Small-Scale For. 13, 163−182.

Cribb, J., 2010. The Coming Famine: The Global Food Crisis and What We Can Do to Avoid It. University of California Press, Los Angeles, USA, 248 pp.

Cuni Sanchez, A., de Smedt, S., Haq, N., Samson, R., 2011. Comparative study on baobab fruit morphological variation between Western and Southeastern Africa: opportunities for domestication. Genet. Resour. Crop Evol. 58, 1143−1156.

Cunningham, A.B., 2001. Applied Ethnobotany: People, Wild Plant Use and Conservation. People and Plants Conservation Manual. Earthscan, London (300 pp).

Cunningham, A.B., Mbenkum, F.T., 1993a. Sustainability of harvesting *Prunus africana* bark in Cameroon: a medicinal plant in international trade. People and Plants Working Paper 2. UNESCO, Paris, 28 pp.

Cunningham, A.B., Mbenkum, F.T., 1993b. Medicinal bark and international trade: a case study of the Afromontane tree *Prunus africana*. Report of the WWF/UNESCO/Kew 'People and Plants Programme'. WWF, Godalming, England.

Cunningham, A.B., Ayuk, E., Franzel, S., Duguma, B., Asanga, C., 2002. An economic evaluation of medicinal tree cultivation: *Prunus africana* in Cameroon. People and Plants Working Paper 10. UNESCO, Paris, 35 pp.

Cunningham M., Cunningham A.B. and Schippmann U. 1997. Trade in *Prunus africana* and the implementation of CITES. Bundesamt für Naturschutz, Bonn, Germany.

Curtis, M., Crowther, B., Aldred, T., Rangan, V., Gidney, M., Dodgeon, S., et al., 2013. Powering up Smallholder Farmers to Make Food Fair: A Five Point Agenda. Fairtrade Foundation, London, 47pp.

da Silva Moço, M.K., da Gama-Rodrigues, E.F., da Gama-Rodrigues, A.C., Machado, R.C.R., Baligar, V.C., 2009. Soil and litter fauna of cacao agroforestry systems in Bahia, Brazil. Agrofor. Syst. 76, 127−138.

Dadjo, C., Assogbadjo, A.E., Fandohan, B., Glélé Kakaï, R., Chakeredza, S., Houehanou, T.D., et al., 2012. Uses and management of black plum (*Vitex doniana* Sweet) in Southern Benin. Fruits. 67, 239−248.

Dafni, A., 1992. Pollination Ecology: A Practical Approach. Oxford University Press, Oxford, p. 250.

Dawson, I.K., Hollingsworth, P., Doyle, J.J., Kresovich, S., Weber, J.C., Sotelo Montes, S., et al., 2008. Tree origins and conservation on farm: a case study from the Peruvian Amazon. Conserv. Genet. 9, 361−372.

Dawson, I.K., Lengkeek, A., Weber, J.C., Jamnadass, R., 2009. Managing genetic variation in tropical trees: linking knowledge with action in agroforestry ecosystems for improved conservation and enhanced livelihoods. Biodivers. Conserv. 18, 969−986.

Dawson, I.K., Guariguata, M.R., Loo, J., Weber, J.C., Lengkeek, A., Bush, D., et al., 2013. What is the relevance of smallholders' agroforestry systems for conserving tropical tree species and genetic diversity in circa situ, in situ and ex situ settings? A review. Biodivers. Conserv. 22, 301−324.

Dawson, I.K., Leakey, R.R.B., Clement, C., Weber, J.C., Cornelius, J.P., Roshetko, J.M., et al., 2014. The management of tree genetic resources and the livelihoods of rural communities in the tropics: non-timber forest products, smallholder agroforestry practices and tree commodity crops. For. Ecol. Manage. 333, 9−21.

de Beenhouwer, M., Aerts, R., Honnay, O., 2013. A global meta-analysis of the biodiversity and ecosystem service benefits of coffee and cacao agroforestry. Agric. Ecosyst. Environ. 175, 1−7.

de Caluwe, E., 2011. Market potential of underutilized plant species: the case of baobab (*Adansonia digitata* L.) and tamarind (*Tamarindus indica* L.) in Mali and Benin. PhD thesis. Ghent University, Ghent, 188 pp.

de Caluwe, E., Halamová, K., van Damme, P., 2010a. *Adansonia digitata* L. A review of traditional uses, phytochemistry and pharmacology. Afrika Focus. 23, 11−51.

de Caluwe, E., Halamová, K., van Damme, P., 2010b. *Tamarindus indica* L. A review of traditional uses, phytochemistry and pharmacology. Afrika Focus. 23, 53−83.

de Foresta, H., Michon, G., 1994a. Agroforests in Sumatra: where ecology meets economy. Agrofor. Today. 6, 12−13.

de Foresta, H., Michon, G., 1994b. Agroforest: an original agro-forestry model from smallholder farmers for environmental conservation and sustainable development. In: Traditional Technology for Environmental Conservation and Sustainable Development in the Asian-Pacific Region, Japan, pp. 52−58.

de la Mora, A., Livingston, G., Philpott, S.M., 2008. Arboreal ant abundance and leaf miner damage in coffee agroecosystems in mexico. Biotropica. 40, 742−746.

de Schutter, O., 2011. The right of everyone to enjoy the benefits of scientific progress and the right to food: from conflict to complementarity. Human Rights Q. 33, 304−350.

de Smedt, S., Alaerts, K., Kouyaté, A.M., van Damme, P., Potters, G., Samson, R., 2011. Phenotypic variation of baobab (*Adansonia digitata* L.) fruit traits in Mali. Agrofor. Syst. 83, 87−97.

de Wit, C.T., 1960. On competition. Verslagen van landbouwkundige onderzoekingen. 660, 1−82.

Degrande, A., Schreckenberg, K., Mbosso, C., Anegbeh, P.O., Okafor, J., Kanmegne, J., 2006. Farmers' fruit tree growing strategies in the humid forest zone of Cameroon and Nigeria. Agrofor. Syst. 67, 159−175.

Degrande, A., Asaah, E., Tchoundjeu, Z., Kanmegne, J., Duguma, B., Franzel, S., 2007. Opportunities for and constraints to adoption of improved fallows: ICRAF's experience in the humid tropics of Cameroon. In: Bationo, A., Waswa, B., Kihara, J., Kimetu, J. (Eds.), Advances in Integrated Soil Fertility Management in Sub-Saharan Africa: Challenges and Opportunities. Springer, Dordrecht, The Netherlands, pp. 901–910.

Degrande, A., Franzel, S., Yeptiep, Y.S., Asaah, E., Tsobeng, A., Tchoundjeu, Z., 2012. Effectiveness of grassroots organisations in the dissemination of agroforestry innovations. In: Kaonga,, M.L. (Ed.), Agroforestry for Biodiversity and Ecosystem Services – Science and Practice. Elsevier, London, pp. 141–164.

Degrande, A., Gyau, A., Foundjem-Tita, D., Tollens, E., 2014. Improving smallholders' participation in tree product value chains: experiences from the Congo Basin. Forests Trees Livelihoods. 23, 102–115.

Deharveng, L., 1992. Field Report for the Soil Mesofauna Studies. Bogor ICRAF, Indonesia.

Delabie, J.H.C., Jahyny, B., Cardoso Do Nascimento, I., Mariano, C.S.F., et al., 2007. Contribution of cocoa plantations to the conservation of native ants (Insecta: Hymenoptera: Formicidae) with a special emphasis on the Atlantic Forest fauna of southern Bahia, Brazil. Biodivers. Conserv. 16, 2359–2384.

Delwaulle, J.C., 1983. Creation et multiplication vegetative par bouturage d'*Eucalyptus* hybrids en Republique Populaire du Congo. Silvicultura. 8, 775–778.

Delwaulle, J.C., Laplace, Y., Quillet, G., 1983. Production massive de boutures d'*Eucalyptus* en Republique Populaire du Congo. Silvicultura. 8, 779–781.

Dembner, S., 1991. Provisional data from the Forest Resources Assessment 1990. Unasylva. 164 (42), 40–44.

Deriaz, R.E., 1961. Routine analysis of carbohydrates and lignin in herbage. J. Sci. Food Agric. 12, 152–160.

Dewees, P.A., Scherr, S.J., 1996. Policies and Markets for Non-timber Tree Products. EPTD Discussion Paper No. 16. IFPRI, Washington, DC, 78 pp.

Diamond, J., 1997. Guns, Germs and Steel: The Fates of Human Societies. W.W. Norton & Co, London, UK.

Diamond, J., 2002. Evolution, consequences and future of plant and animal domestication. Nature. 418, 700–707.

Diarrasouba, N., Divine, B.N., César, K., Christophe, K., Abdourahamane, S., 2007. Phenotypic diversity of shea (*Vitellaria paradoxa* C. F. Gaertn.) populations across four agroecological zones of Cameroon. J. Crop Sci. Biotechnol. 10 (4), 223–230.

Dick, J., Skiba, U., Munro, R., Deans, D., 2006. Effect of N-fixing and non N-fixing trees and crops on NO and N_2O emissions from Senegalese soils. J. Biogeogr. 33, 416–423.

Dick, J.Mc.P., Dewar, R.C., 1992. A mechanistic model of carbohydrate dynamics during the adventitious root development in leafy cuttings. Ann. Bot. (Lond.). 70, 371–377.

Dick, J.Mc.P., Leakey, R.R.B., 2006. Differentiation of the dynamic variables affecting rooting ability in juvenile and mature cuttings of cherry (*Prunus avium*). J. Hortic. Sci. Biotechnol. 81, 296–302.

Dick, J.Mc.P., Magingo, F., Smith, R.I., McBeath, C., 1999. Rooting ability of *Leucaena leucocephala* stem cuttings. Agrofor. Syst. 42, 149–157.

Dickmann, D.I., 1985. The ideotype concept applied to forest trees. In: Cannell, M.R.G., Jackson, J.E. (Eds.), Attributes of Trees as Crop Plants. Institute of Terrestrial Ecology, Huntington, England, pp. 89–101.

Dickmann, D.I., Gold, M.A., Flore, J.A., 1994. The ideotype concept and the genetic improvement of tree crops. Plant Breed. Rev. 12, 163–193.

Dix, M.E., Bishaw, B., Workman, S.W., Barnhart, M.R., Klopfenstein, N.B., Dix, A.M., 1999. Pest management in energy- and labor-intensive agroforestry systems. In: Buck, L.E., Lassoie, J.P., Fernandes, E.C.M. (Eds.), Agroforestry in Sustainable Agricultural Systems. CRC Press/Lewis, Boca Raton, FL, pp. 131–156.

Djimde M., 1991. Synthese zonal des systemes d'utilisation des terres et des potentialites agroforestieres dans le cadre de la Planification du Reseau Collaboratif de Recherche en Agroforesterie de l'ICRAF pour les Zones Semi-Arides du Burkina Faso, Mali, Niger et Senegal. AFRENA Report No. 48. ICRAF, Nairobi, Kenya.

Donald, C.M., 1968. The breeding of crop ideotypes. Euphytica. 17, 385–403.

Dosunmu, M.I., Johnson, E.C., 1995. Chemical evaluation of the nutritive value and changes in ascorbic acid content during storage of the fruit of 'bitter kola' (*Garcinia kola*). Food Chem. 54, 67–71.

Drechsel, P., Glaser, B., Zech, W., 1991. Effect of four multipurpose tree species on soil amelioration during tree fallow in Central Togo. Agrofor. Syst. 16, 193–202.

Duguma, B., Mollet, M., Tiki Manga, T., 1994. Annual Progress Report. Institute of Agronomic Research (IRA), Cameroon and International Centre for Research in Agroforestry (ICRAF) AFRENA Report.

Dupain, D., 1994. Une Region Traditionnellement Agroforestiere en Mutation: Le PESISIR. CNEARC, Montpellier, France.

Dyson, W.F., 1965. The justification of plantation forestry in the tropics. Turrialba. 15, 135–139.

Dyson, W.O., 1981. Report (1980) to ODA of the UK Government, London, 16 pp.

Dzowela, B.H., 1994. *Acacia angustissima*: a Central America tree that is going places. Agrofor. Today. 6 (3), 13–14.

Dzowela, B.H., Mafongoya, P.L., Hove, L., 1994. SADC-ICRAF Agroforestry project, Zimbabwe. 1994 Progress Report. ICRAF, Nairobi.

Edem, D.O., Eka, O.U., Ifon, E.T., 1984. Chemical evaluation of the value of the fruit of African star apple (*Chrysophyllum albidum*). Food Chem. 14, 303–311.

Edinburgh Center for Tropical Forests, 1993. Multiplying Tropical Trees: Vegetative Propagation and Selection. 5 Video Programmes, ECTF, Edinburgh, Scotland, UK.

Edmunds, W.M., Gaye, C.B., 1997. Naturally high nitrate concentrations in groundwaters from the Sahel. J. Environ. Qual. 26, 1231–1239.

Egbe Enow, A., Eni, K.I., Tchoundjeu, Z., 2013. Phenotypic variation in fruits and nuts of *Cola acuminata* in three populations of the centre region of Cameroon. Int. Res. J. Plant Sci. 4, 236–247.

Ejiofor, M.A.N., unpublished. Nutritional values of Ogbono (*Irvingia gabonensis* var. excelsa, In: Boland, D., Ladipo, D.O., (Eds), Irvingia: Uses, Potential and Domestication, ICRAF, Nairobi, Kenya.

Ejiofor, M.A.N., Onwubuke, S.N., Okafor, J.C., 1987. Developing improved methods of processing and utilization of kernels of *Irvingia gabonensis* (var. *gabonensis* and var. *excelsa*). Int. Tree Crops J. 4, 283–290.

Elevitch, C.R., 2006. Traditional Trees of Pacific Islands: Their Culture, Environment and Use. Permanent Agriculture Resources, Hawaii, 800pp.

Elevitch, C.R., 2011. Speciality Crops for Pacific Islands. Permanent Agriculture Resources, Hawaii, 576pp.

Eliasson, L., Brunes, L., 1980. Light effects on root formation in aspen and willow cuttings. Physiol. Plant. 48, 261–265.

Ellis, F., 1993. Peasant Economics: Farm Households in Agrarian Development. second edition Cambridge University Press, UK, 311pp.

Ellis, F., Allison, E., 2004. Livelihood diversification and natural resources access. Livelihood Support Program Working Paper 9. FAO, Rome, Italy.

Emebiri, L.C., Nwufo, M.I., 1990. Effect of fruit type and storage treatments on the biodeterioration of African Pear (*Dacryodes edulis* (G. Don.) H.J. Lam.). Int. Biodet. 26, 43–50.

Eromosele, I.C., Eromosele, C.O., Kuzhkuzha, D.M., 1991. Evaluation of mineral elements and ascorbic acid contents in some wild plants. Plant Foods Hum. Nutr. 41, 151–154.

Essien, E.U., Esenowo, G.J., Akpanabiatu, M.I., 1995. Lipid composition of lesser known tropical seeds. Plant Foods Hum. Nutr. 48, 135–140.

Estrada-Carmona, N., Hart, A.K., DeClerck, F.A.J., Harvey, C., Milder, J.C., 2014. Integrated landscape management for agriculture, rural livelihoods and ecosystem conservation: an assessment of experience from Latin America and the Caribbean. Landsc. Urban Plan. 129, 1–11.

Eswaran H, Reich P, Beinroth F., 2006. Land degradation: an assessment of the human impact on global land resources. In: 18th World Congress on Soil Science. International Union of Soil Sciences, Philadelphia, USA. < http://cropsconfex.com/crops/wc2006/techprogram/P11487.htm >

European Commission, OECD, United Nations, World Bank (2013) White Cover version on Experimental Ecosystem Accounting in System of Environmental-Economic Accounting 2012 http://unstats.un.org/unsd/envaccounting/eea_White_cover.pdf.

Evenson, R.E., Gollin, D., 2003. Assessing the Impact of the Green Revolution, 1960 to 2000. Science. 300, 758–762.

Ewald, D., Kretzschmar, U., 1996. The influence of micrografting *in vitro* on tissue culture behavior and vegetative propagation of old European larch trees. Plant Cell Tissue Organ Culture. 44, 249–252.

Facheux, C., Tchoundjeu, Z., Foundjem, D., Mbosso, C., 2006. From research to farmer enterprise development in Cameroon: case study of kola nuts. Acta Hortic. (ISHS). 699, 181–188.

Facheux, C., Tchoundjeu, Z., Foundjem-Tita, D., Degrande, A., Mbosso, C., 2007. Optimising the production and marketing of NTFPs. Afr. Crops Sci. Conf. Proc. 8, 1248–1254.

Facheux, C., Gyau, A., Foundjem-Tita, D., Russell, D., Mbosso, C., Franzel, S., et al., 2012. Comparison of three modes of improving benefits to farmers within agroforestry product market chains in Cameroon. Afr. J. Agric. Res. 7, 2336–2343.

Falconer J., 1990. *The Major Significance of 'Minor' Forest Products. The local use and value of forests in the West African humid forest zone. Forests, Trees and People*, Community Forestry Note No. 6, FAO, Rome, Italy

Falconer, J., 1992. *Non-timber Forest Products in Southern Ghana, ODA Forestry Series No 2*. UK Overseas Development Administration, London.

Fandohan, B., Assogbadjo, A., Glèlè Kakaï, R., Kyndt, T., Sinsin, B., 2011. Quantitative morphological descriptors confirm traditionally classified morphotypes of *Tamarindus indica* L. fruits. Genet. Resour. Crop Evol. 58, 299–309.

FAO, 1975. The methodology of conservation of forest genetic resources; report on a pilot study, FAO/UNEP, Misc. 75/8, Rome, 117 pp.

FAO, 1986. *Brise-vent et Rideaux Abris avec Reference Particuliere aux Zone Seches. FAO Conservation Guide*. FAO, Rome.

FAO, 1995. FAO Fertilizer Yearbook Volume 44—1994. Food and Agriculture Organization of the United Nations, Rome.

FAO, 1996. World Food Summit: Synthesis of the Technical Background Documents. Food and Agriculture Organization of the United Nations, Rome.

FAO, 2001. Report on the 12th Session of the Panel of Experts on Forest Genetic Resources. FAO, Rome, Italy, 108 pp.

FAO. 2002. FAOSTAT Database www.fao.org, Rome, Italy.

FAO, 2003. State of the World's Forests. Food and Agriculture Organization, Rome, Italy, 151 pp.

FAO, WFP and IFAD, 2012. *The State of Food Insecurity in the World 2012. Economic Growth is Necessary but Not Sufficient to Accelerate Reduction of Hunger and Malnutrition*. FAO, Rome, 62pp.

Faria, D., Baumgarten, J., 2007. Shade cacao plantations (*Theobroma cacao*) and bat conservation in southern Bahia, Brazil. Biodivers. Conserv. 16, 291–312.

Faria, D., Laps, R.R., Baumgarten, J., Cetra, M., 2006. Bat and bird assemblages from forests and shade cacao plantations in two contrasting landscapes in the Atlantic Forest of southern Bahia, Brazil. Biodivers. Conserv. 15, 587–612.

Faye, M.D., Weber, J.C., Mounkoro, B., Dakouo, J.-M., 2010. Contribution of parkland trees to village livelihoods: a case study from Mali. Dev Pract. 20, 428–434.

Faye, M.D., Weber, J.C., Abasse, T.A., Boureima, M., Larwanou, M., Bationo, A.B., et al., 2011. Farmers' preferences for tree functions and species in the West African Sahel. Forest Tree Livelihoods. 20, 113–136.

Felker, P., Clark, P.R., 1981. Rooting of Mesquite (*Prosopis*) cuttings. J. Range Manage. 34, 466–468.

Ferket, B., 2012. The impacts of improved commercialization of kola nuts on farmer livelihoods in the Western highlands of Cameroon. Master of Science thesis in Conflict and Development, University of Gent, Belgium, 111 pp.

Fish, D., Soria, S., 1978. Water-holding plants (Phytotelmata) as larval habitats for ceratopogonid pollinators of cacao in Bahia, Brazil. Rev. Theobroma. 8, 133–146.

Fish, F., Meshal, I.A., Waterman, P.G., 1978. Alkaloids of *Oricia suaveolens*. Planta Med. 33, 227–231.

Foley, J.A., Ramankutty, N., Brauman, K.A., Cassidy, E.S., Gerber, J.S., Johnston, M., et al., 2011. Solutions for a cultivated planet. Nature. 478, 337—342.

Foma, M., Abdala, T., 1985. Kernel oils of seven plant species of Zaire. J. Am. Oil Chem. Soc. 62, 910—911.

Foster, G.S., Bertolucci, F.L.G., 1994. Clonal development and deployment strategies to enhance gain while minimising risk. In: Leakey, R.R.B., Newton, A.C. (Eds.), Tropical Trees: Potential for Domestication and the Rebuilding of Tropical Forest Resources. HMSO, London, pp. 103—111.

Fouladbash, L., Currie, W.S., 2015. Agroforestry in Liberia: household practices, perceptions and livelihood benefits. Agrofor. Syst. 89, 247—266.

Foundjem-Tita, D., D'Haese, M., Degrande, A., Tchoundjeu, Z., van Damme, P., 2011. Farmers' satisfaction with group market arrangements as a measure of group market performance: a transaction cost analysis of non-timber forest products' producer groups in Cameroon. For. Policy Econ. 13, 545—553.

Foundjem-Tita, D., Degrande, A., D'Haese, M., van Damme, P., Tchoundjeu, Z., Gyau, A., et al., 2012a. Building long-term relationships between producers and trader groups in the non-timber forest product sector in Cameroon. Afr. J. Agric. Res. 7, 230—239.

Foundjem-Tita, D., Tchoundjeu, Z., Speelman, S., D'Haese, M., Degrande, A., Asaah, E., et al., 2012b. Policy and legal frameworks governing trees: incentives or disincentives for smallholder tree planting decisions in Cameroon? Small-scale For. 12, 489—505, DOI 10.1007/s11842-012-9225-z.

Frankel, O.H., Hawkes, I.G., 1975. Crop Genetic Resources for Today and Tomorrow. International Biological Program, 2. Cambridge University Press, Cambridge, 492 pp.

Franks, P.C., 1992. The extension of agroforestry for improving soil fertility in small holder agriculture with a maize-based cropping system: a case study from Malawi. MSc dissertation, University of Edinburgh.

Franzel, S., Jaenicke, H., Janssen, W., 1996. Choosing the Right Trees: Setting Priorities for Multipurpose Tree Improvement. ISNAR Research Report 8. International Service for National Agricultural Research, The Hague, 87 p.

Franzel, S., Denning, G.L., Lilisøe, J.-P., Mercado Jr., A.R., 2004. Scaling up the impact of agroforestry: Lessons from three sites in Africa and Asia. Agrofor. Syst. 61, 329—344.

Franzel, S., Akinnifesi, F.K., Ham, C., 2008. Setting priorities among indigenous fruit tree species in Africa: examples from southern, eastern and western Africa regions. In: Akinnifesi, F.K., Leakey, R.R.B., Ajayi, O.C., Sileshi, G., Tchoundjeu, Z., Matakala, P., et al.,Indigenous Fruit Trees in the Tropics: Domestication, Utilization and Commercialization. CAB International, Wallingford, UK, pp. 1—27. , in association with the World Agroforestry Centre, Nairobi, Kenya.

Franzel, S., Kiptot, E., Lukuyu, B. 2014. Agroforestry: fodder trees. In: Encyclopedia of Agriculture and Food Systems, Vol. 1, Neal van Alfen, editor-in-chief. Elsevier, San Diego, pp. 235—243.

Franzel, S., Degrande, A., Kiptot, E., Kirui, J., Kugonza, J., Preissing, J., et al., 2015. Farmer-to-Farmer Extension. Note 7, GFRAS Good Practice Note for Extension and Advisory Services, Global Forum for Rural Advisory Services, Lindau, Switzerland.

Freeman, O.E., Duguma, L.A., Minang, P.A., 2015. Operationalizing the integrated landscape approach in practice. Ecol. Soc. 20, 24pp. http://dx.doi.org/10.5751/ES-07175-200124.

Frison, E.A., Cherfas, J., Eyzaguirre, P.B., Johns, T., 2004. Biodiversity, nutrition and health: Making a difference to hunger and conservation in the developing world. Seventh Meeting of the Conference of the Parties to the Convention on Biological Diversity (COP7). International Plant Genetic Resources Institute, Rome, Italy.

Fujisaka, S., 1993. A case study of farmer adaptation and adoption of contour hedgerows for soil conservation. Exp. Agric. 29, 97—106.

Fungo, R., 2011. An Analysis of the Nutrition Situation, Agroecosystems and Food Systems of West and Central Africa. Bioversity International, Rome, Italy, 42pp.

Fungo, R., Muyonga, J., Kaaya, A., Okia, C., Tieguhong, J.C., Baidu-Forson, J., 2015. Nutrients and bioactive compounds content of *Baillonella toxisperma*, *Trichoscypha abut* and *Pentaclethra macrophylla* from Cameroon. Food Sci. Nutr. 3, 292—301.

Futuyma, D.J., 1998. Evolutionary Biology. 3rd Edition Sinauer Associates, Sunderland, Massachusetts, 763p.

Gachengo, C.N., 1996. Phosphorus release and availability on addition of organic materials to phosphorus fixing soils. M.Sc. Thesis. *Moi University*, Eldoret, Kenya.

Gacheru, E., Rao, M.R., 2001. Managing Striga infestation on maize using organic and inorganic nutrient sources in western Kenya. Int. J. Pest Manage. 47, 233—239.

Gaiwe, R., Nkulinkiye-Neura, T., Bassene, E., Olschwang, D., Ba, D., Pousset, J.L., 1989. Calcium et mucilage dans les feuilles de *Adansonia digitata* (Boabab). Int. J. Crude Drug Res. 27, 101—104.

Gallina, S., Mandujano, S., Gonzalez-Romero, A., 1996. Conservation of mammalian biodiversity in coffee plantations of Central Veracruz, Mexico. Agrofor. Syst. 33, 13—27.

Garbach, K., Milder, J.C., Montenegro, M., Karp, D.S., DeClerck, F.A.J., 2014. Biodiversity and ecosystem services in agroecosystems. In: van Alfen, N., et al., (Eds.), Encyclopedia of Agriculture and Food Systems, Vol 2. Elsevier Publishers, San Diego, pp. 21—40.

Garnett, T., Godfray, H.C.J., 2012. Sustainable intensification in agriculture: navigating a course through competing food system priorities. Food Climate Research Network and the Oxford Martin Programme on the Future of Food. University of Oxford, UK.

Garnett, T., Appleby, M.C., Balmford, A., Bateman, I.J., Benton, T.G., Bloomer, P., et al., 2013. Sustainable intensification in agriculture: premises and policies. Science. 341, 33—34.

Garrity, D., 2004. World agroforestry and the achievement of the millenium development goals. Agrofor. Syst. 61, 5—17.

Garrity, D., 2012. Agroforestry and the future of global land use. In: Nair, P.K.R., Garrity, D., (Eds.), Agroforestry – The Future of Global Land Use, Adv. Agrofor., 9, 21—27. Springer, Dordrecht, Germany.

Garrity, D.P., 1994. Tree-crop interactions on slopes. In: Ong, C.K., Huxley, P. (Eds.), The Physiological Basis for Tree-Crop Interactions. CABI, Wallingford, UK.

Garrity, D.P., 1996. Tree−soil−crop interactions on slopes. In: Ong, C.K., Huxley, P.A. (Eds.), *Tree−Crop Interactions, a* Physiological *Approach*. CAB International, Wallingford, UK, pp. 299−318.

Garrity, D.P., Akinnifesi, F.K., Ajayi, O.C., Weldesemayat, S.G., Mowo, J.G., Kalinganire, A., et al., 2010. Evergreen agriculture: a robust approach to sustainable food security in Africa. Food Security. 3, 197−214.

Genstat 5 Committee of the Statistics Department, Rothamsted Experimental Station, 1993. Genstat 5, Release 3 Reference Manual. Clarendon Press, Oxford.

GEO, 2007. *Global Environmental Outlook 4: Past, Present and Future Perspectives*. UNEP, Nairobi, Kenya, 572pp.

Gepts, P., 2014. Domestication of Plants. In: van Alfen, N., et al., (Eds.), Encyclopedia of Agriculture and Food Systems, Vol 2. Elsevier Publishers, San Diego, USA, pp. 474−486.

Geurts, I.F., 1982. The Indian jujube or ber (*Zizyphus mauritiana* Lamk.). Aspects related to germplasm conservation. A Preliminary Report, Royal Tropical Institute, Amsterdam, The Netherlands.

Giami, S.Y., Okonkwo, V.I., Akusu, M.O., 1994. Chemical composition and functional properties of raw, heat-treated and partially proteolysed wild mango (*Irvingia gabonensis*) seed flour. Food Chem. 49, 237−243.

Gichuru, M.P., 1991. Residual effects of natural bush, *Cajanus cajan* and *Tephrosia Candida* on the productivity of an acid soil in southeastern Nigeria. Plant Soil. 134, 31−36.

Giller, K.E., Wilson, K.J., 1991. Nitrogen Fixation in Tropical Cropping Systems. CAB International, Wallingford, UK.

Giovannucci D, Scherr S, Neirenberg D, Hebebrand C, Shapiro J, et al., 2012. Food and Agriculture: The Future of Sustainability. A Strategic Input to the Sustainable Development in the 21st Century (SD21) Project. New York: UN Dep. Econ. Soc. Aff. Div. Sustain. Dev. 94pp.

Girma, H., Rao, M.R., Day, R., Ogol, C.K.P.O., 2006. Abundance of insect pests and their effects on biomass yields of single versus multi-species planted fallows. Agrofor. Syst. 93, 93−102.

Glicksman, M., 1996. Tamarind seed gum. In: Glicksman, M. (Ed.), Food Hydrocolloids, Volume III. CRC Press Inc., Boca Raton, FL, pp. 191−202.

Gliessman, S.R., 1998. Agroecology: Ecological Processes in Sustainable Agriculture. Ann Arbor Press, Chelsea.

Gockowski, J., Tonye, J., Baker, D., 1997. Characterization and Diagnosis of Agricultural Systems in the Alternatives to Slash and Burn Forest Margins Benchmark of Southern Cameroon. Report to ASB Program, IITA, Ibadan, Nigeria.

Gockowski J., Baker D., Tonye J., Weise S., Ndoumbe M., Tiki-Manga T., et al., 1998. Characterization and diagnosis of farming systems in the ASB forest margin benchmark of southern Cameroon. *Mimieograph*. Yaoundé, IITA Humid Forest Ecoregional Center.

Gockowski, J., Blaise Nkamleu, G., Wendt, J., 2001. Implications of resource-use intensification for the environment and sustainable technology in the Central African rainforest. In: Lee, D.R., Barrett, C.B. (Eds.), Trade-offs or Synergies? Agricultural Intensification, Economic Development and the Environment. CABI, Wallingford, England, pp. 197−219.

Gockowski, J.J., Dury, S., 1999. The economics of cocoa-fruit agroforests in southern Cameroon. In: Jiménez, F., Beer, J. (Eds.), Multi-strata Agroforestry Systems with Perennial Crops. CATIE, Turrialba, Costa Rica, pp. 239−241.

Godfray, H.C.J., Garnett, T., 2014. Food security and sustainable intensification. Phil. Trans. R. Soc. B. 369, 20120273. Available from: http://dx.doi.org/10.1098/rstb.2012.0273.

Goenster, S., Wiehle, M., Kehlenbeck, K., Jamnadass, R., Gebauer, J., Buerkert, A., 2011. Indigenous fruit trees in homegardens of the Nuba Mountains, Central Sudan: tree diversity and potential for improving the nutrition and income of rural communities. Acta Hortic. 911, 355−364.

Gowda, D.K.S., Venkatesha, M.G., Bhat, P.K., 1995. Preliminary investigations on the incidence of termites on coffee and its shade trees. J. Coffee Res. 25, 30−34.

Grace, J., Fasehun, F.E., Dixon, M., 1980. Boundary layer conductance of the leaves of some tropical timber trees. Plant, Cell Environ. 3, 443−450.

Graham C., Hart D., 1997. Prospects for the Australian Native Bushfood Industry. RIRDC Research Paper No. 97/22, Canberra, Rural Industries Research and Development Corporation.

Grainger, A., 1980. The state of the world's tropical forests. The Ecologist. 10, 6−54.

Greenberg, R., 2000. The conservation value for birds of planted shade cacao plantations in Mexico. Anim. Conserv. 3, 105−112.

Greenberg, R., Bichier, R., Sterling, J., 1997. Bird populations in rustic and planted shade coffee plantations of Eastern Chiapas, Mexico. Biotropica. 29, 501−514.

Greenberg, R., Bichier, P., Angon, A.C., Macvean, C., Perez, R., Cano, E., 2000. The impact of avian insectivory on arthropods and leaf damage in some Guatemalan coffee plantations. Ecology. 81, 1750−1755.

Gross, B., Miller, A., 2014. From field to table: perspectives and potential for fruit domestication. In: Batello, C., Cox, S., Wade, L., Pogna, N., Bozzini, A., Choptiany, J. (Eds.), Proceedings from the FAO expert workshop on perennial crops for food security. FAO, Rome.

Gu, H., Subramanian, S.M., 2012. Socio-Ecological Production Landscapes: Relevance to the Green Economy Agenda. UN University, Institute of Advanced Studies Policy Report.

Guarino, L., 1997. Traditional African Vegetables. International Plant Genetic Resources Institute, Rome, Italy.

Guissou, K.M.L., Kristiansen, T., Lykke, A.M., 2015. Local perceptions of food plants in Eastern Burkina Faso. Ethnobot. Res. Appl. 14, 199−209.

Gulen, H., Arora, R., Kuden, A., Krebs, S.L., Postman, J., 2002. Peroxidase isozyme profiles in compatible and incompatible pear-quine graft combinations. J. Am. Soc. Hortic. Sci. 127, 152−157.

Gunasena, H.P.M., Roshetko, J.M., 2000. Tree Domestication in Southeast Asia: Results of a Regional Study on Institutional Capacity. International Center for Research in Agroforestry (ICRAF), Bogor, Indonesia, 86 pp.

Gutteridge, R.C., 1992. Evaluation of the leaves of a range of tree legumes as a source of nitrogen for crop growth. Exp. Agric. 28, 195–202.

Gyau, A., Takoutsing, B., Degrande, A., Franzel, S., 2012. Producers' motivation for collective action for kola production and marketing in Cameroon. J. Agric. Rural Dev. Trop. Subtrop. 113, 43–50.

Gyau, A., Ngum Faith, A., Foundjem-Tita, D., Ajaga, N., Catacutan, D., 2014. Small-holder farmers' access and rights to land of Njombé in the Littoral region of Cameroon. Afrika Focus. 27, 23–39.

Haglund, E., Ndjeunga, J., Snook, L., Pasternak, D., 2011. Dryland management for improved livelihoods: Farmer managed natural regeneratiom in Niger. J. Environ. Manage. 92, 1696–1705.

Hailemariam, M., Birhane, E., Asfaw, Z., Zewdie, S., 2013. Arbuscular mycorrhizal association of indigenous agroforestry tree species and their infective potential with maize in the rift valley, Ethopia. Agrofor. Syst. 87, 1261–1272.

Hailey, L., 1938. An African Survey. Oxford University Press, London.

Haissig, B.E., 1974. Metabolism during adventitous root primoridum initiation and development. N. Zeal. J. For. Sci. 4, 324–337.

Hall, J.B., Bada, S.O., 1979. The distribution and ecology of Obeche (*Triplochiton scleroxylon*). J. Ecol. 67, 543–564.

Hall, J.B., O'Brien, E.M., Sinclair, F.L., 2000. *Prunus africana*: a monograph. University of Wales, Bangor, UK, 104 pp.

Hall, N.M., Kaya, B., Dick, J., Skiba, U., Niang, A., Tabo, R., 2006. Effect of improved fallow on crop productivity, soil fertility and climate-forcing gas emissions in semi-arid conditions. Biol. Fert. Soils. 42, 224–230.

Hallé, F., Oldeman, R.A.A., Tomlinson, P.B., 1978. Tropical Trees and Forests: An Architectural Analysis. Springer Verlag, Berlin, Germany, p. 441.

Hands, M.R., Harrison, A.F., Bayliss-Smith, T., 1995. Phosphorus dynamics in slash-and -burn and alley cropping systems of the humid tropics. In: Tiessen, H. (Ed.), Phosphorus in the Global Environment. John Wiley & Sons, Chichester, UK, pp. 155–170.

Hansen, J., Eriksen, E.N., 1974. Root formation of pea cuttings in relation to the irradiance of the stockplants. Physiol. Plant. 32, 170–173.

Hansen, J., Stromquist, L.H., Ericsson, A., 1978. Influence of the irradiance on carbohydrate content and rooting of cuttings of pine seedlings *(Pinus sylvestris* L.). Plant Physiol. 61, 975–979.

Harlan, J.R., 1975. Crops and Man. American Society of Agronomy/Crop Science Society of America, Madison, WI.

Haro-Carrión, X., Lozada, T., Navarrete, H., de Koning, G.H.J., 2009. Conservation of vascular epiphyte diversity in shade cacao plantations in the Chocó region of Ecuador. Biotropica. 41, 520–529.

Harris, D.J., 1996. A revision of the Irvingiaceae in Africa. Bull. Jardin Bot. Natl. Belgique. 65, 143–196.

Hartemink, A.E., Buresh, R.J., Jama, B., Janssen, B.H., 1996. Soil nitrate and water dynamics in Sesbania fallow, weed fallows, and maize. Soil. Sci. Soc. Am. J. 60, 568–574.

Hartmann, H.T., Kester, D.E., Davis, F.T., Geneve, R.L., 2002. Plant Propagation: Principles and Practices. seventh ed. Prentice-Hall, Upper Saddle River, NJ, p. 880.

Hartmann, H.T., Kester, D.E., Davies, F.T., Geneve, R., 2010. Hartmann and Kester's Plant Propagation: Principles and Practices. eighth ed. Prentice Hall, New Jersey.

Harvey, C.A., Gonzales, J.G., Somarriba, E., 2006a. Dung beetle and terrestrial mammal diversity in forest, indigenous agroforestry systems and plantain monocultures in Talamanca, Costa Rica. Biodivers. Conserv. 15, 555–585.

Harvey, C.A., Medina, A., Merlo Sánchez, D., Vílchez, S., Hernández, B., et al., 2006b. Patterns of animal diversity associated with different forms of tree cover retained in agricultural landscapes. Ecol. Appl. 16, 1986–1999.

Harvey, C.A., González Villalobos, J.A., 2007. Agroforestry systems conserve species-rich but modified assemblages of tropical birds and bats. Biodivers. Conserv. 16, 2257–2292.

Harvey, C.A., Chacón, M., Donatti, C.I., Garen, E., Hannah, L., Andrade, A., et al., 2014. Climate-smart landscapes: opportunities and challenges for integrating adaptation and mitigation in tropical agriculture. Conserv. Lett. 7, 77–90.

Harwood, C.E. (Ed.), 1992. *Grevillea robusta* in Agroforestry and Forestry. *Proceedings of an International Workshop. International Centre for Research in Agroforestry*, Nairobi, Kenya, 190 pp.

Harwood, C.E., 1997. Domestication of Australian tree species for agroforestry. In: Roshetko, J.M., Evans, D.O. (Eds.), Domestication of Agroforestry Trees in Southeast Asia. Winrock International, Arkansas, USA, pp. 64–72.

Harwood, R.R., 1994. Agronomic alternatives to slash-and-burn in the humid tropics. In: Sanchez, P.A., van Houten, H. (Eds.), *Alternatives to* Slash-and-burn *Agriculture*: Symposium ID-6, 15th World Congress of Soil Science Acapulco, Mexico. International Society of Soil Science, Chapingo, Mexico, pp. 93–106.

Hassan, R., Scholes, R., Ash, N. (Eds.), 2005. *Ecosystems and Human Well-Being, Vol 1: Current State and Trends*. Island Press, for Millennium Ecosystem Assessment, Washington DC.

Hawkins, R.H., Sembiring, H. Suwardjo, D.L., 1990. The potential of alley cropping in the uplands of East and Central Java: a review. Upland Agricultural Research Project, Jeteng. Agency for Agricultural Research and Development, Indonesia.

He, J., Yang, H., Jamnadass, R., Xu, J., Yang, Y., 2011. Decentralization of tree seedling supply systems for afforestation in the west of Yunnan Province, China. Small Scale For. Available from: http://dx.doi.org/10.1007/s11842-011-9176-9.

Hele, A.E., 2001. Muntries production. Agdex 244/11; ISSN1323-0409, Australian Native Produce Industries; Primary Industries and Resources S.A.

Herridge, D., Peoples, M.B., Boddey, R.M., 2008. Global inputs of biological nitrogen fixation in agricultural systems. Plant Soil. 311, 1–18.

Hess, C.E., 1969. Internal and external factors regulating root initiation. In: Whittington, W.J. (Ed.), Root Growth. Plenom Press, New York, pp. 42–52.

Heybroek, A.M., 1978. Primary considerations: multiplication and genetic diversity. 3rd World Consultation on Forest Tree Breeding. Unasylva. 30 (27–33), 49–50.

Hilal, N.S., 1993. Contribution à l'étude du Karité, *Butyrospermum paradoxum* (Gaertn.F.) Hepper Sapotaceae, Monograph of Faculté de Medecine et de Pharmacie, University of Cheikh Anta Diop, Dakar, Senegal, 144p.

Hoad, S.P., Leakey, R.R.B., 1994. Effects of light quality on gas exchange and dry matter partitioning in *Eucalyptus grandis* W. Hill ex Maiden. For. Ecol. Manage. 70, 265–273.

Hoad, S.P., Leakey, R.R.B., 1996. Effects of pre-severance light quality on the vegetative propagation of *Eucalyptus grandis* W. Hill ex Maiden: cutting morphology, gas exchange and carbohydrate status during rooting. Trees. 10, 317–324.

Holding-Anyonge, C., Roshetko, J.M., 2003. Farm-level timber production: orienting farmers towards the market. Unaslyva. 212 (54), 48–56.

Hollingsworth, P., Dawson, I., Goodall-Copestake, W., Richardson, J., Weber, J.C., Sotelo Montes, C., et al., 2005. Do farmers reduce genetic diversity when they domesticate tropical trees? A case study from Amazonia. Mol. Ecol. 14, 497–501.

Holmgren, P., Masakha, E.J., Sjöholm, H., 1994. Not all African land is being degraded: a recent survey of trees on farms in Kenya reveals rapidly increasing forest resources. Ambio. 23, 390–395.

Holtzhausen, L.C., Swart, E., van Rensburg, R., 1990. Propagation of the marula (*Sclerocarya birrea* subsp. *caffra*). Acta Hortic. 275, 323–334.

Homma A.K.O., 1994. Plant extractivism in the Amazon: Limitations and possibilities. In: Clusener-Godt, M., Sachs, I., (Eds.), Extractivism in the Brazilian Amazon: Perspectives on Regional Development, MAB Digest 18, Man and the Biosphere, UNESCO, Paris, France, pp. 34–57.

Horst, W.J., Kühne, R., Kang, B.T., 1995. Nutrient use in *Leucaena leucocephala* and *Cajanus cajan* in maize-cassava alley cropping on Terre de Barre, Benin Republic. In: Kang, B.T., Osiname, A.O., Larbi, A. (Eds.), Alley Farming Research and Development. Alley Farming Network for Tropical Africa, Ibadan, Nigeria, pp. 122–136.

Houghton, R.A., Boone, R.D., Fruci, J.R., 1987. The flux of carbon from terrestrial ecosystems to the atmosphere in 1980 due to changes in land use: geographic distribution of the global flux. Tellus. 39B, 122–139.

Howard, B.H., Blasco, A.B., 1979. Variation of rooting ability within apple hardwood cutting hedges. Report of East Mailing Research Station for 1978, p. 74.

Howland, P., 1975a. Vegetative propagation methods for *Triplochiton scleroxylon* K. Schum. Proceedings of the Symposium on Variation and Breeding Systems of Triplochiton scleroxylon K. Schum. Federal Department of Forest Research, Ibadan, Nigeria, pp. 99–109.

Howland, P., 1975b. Variation in rooting stem cuttings of *Triplochiton scleroxylon* K. Schum, Ibid, pp. 110–124.

Howland, P., 1975c. Current management techniques for raising *Triplochiton scleroxylon* K. Schum. Ibid. pp. 125–129.

Howland, P., Bowen, M.R., 1977. *Triplochiton scleroxylon* K. Schum and other West African tropical hardwoods. West African Hardwoods Improvement Project, Research Report, 1971–1977. Forestry Research Institute of Nigeria, 154 pp.

Howland, P., Bowen, M.R., Ladipo, D.O., Oke, J.B., 1978. The study of clonal variation in *Triplochiton scleroxylon* K. Schum as a basis for selection and improvement. In: Nikles, D.G., Burley, J., et al.,Proc. joint workshop IUFRO working parties S2.02-08 and S2. 03-1, Brisbane, 1977. Commonwealth Forestry Institute, Oxford, UK, pp. 898–904.

Hulse, J., 1996. Flavours, spices and edible gums: opportunities for integrated agroforestry systems. In: Leakey, R.R.B., Temu, A.B., Melnyk, M., Vantomme, P. (Eds.), *Domestication and Commercialization of Non-timber Forest Products in Agroforestry Systems, Non-Wood Forest Products No. 9.* FAO, Rome, Italy, pp. 86–96.

Hulugalle, N.R., Kang, B.T., 1990. Effect of hedgerow species in alley cropping systems on surface soil physical properties of an Oxic Paleustalf in southwestern Nigeria. J. Agric. Sci. 114, 301–307 (Cambridge, UK).

Hulugalle, N.R., Ndi, J.N., 1993. Effects of no-tillage and alley cropping on soil properties and crop yields in a Typic Kandiudult of southern Cameroon. Agrofor. Syst. 22, 207–220.

IAASTD, 2009. Agriculture at a Crossroads: International Assessment of Agricultural Science and Technology for Development Global Report. In: McIntyre, B.D., Herren, H.R., Wakhungu, J., Watson, R.T. (Eds.), Island Press, Washington DC, USA, 590pp.

Ibiyemi, S.A., Abiodun, A., Akanji, S.A., 1988. *Andasonia [sic] digitata Bombax* and *Parkia filicoideae Welw*: fruit pulp for the soft drink industry. Food Chem. 28, 111–116.

IBSRAM, 1994. International Board for Soils Research and Management, Bangkok. Highlights 1993.

ICRAF, 1992a. A Selection of Useful Trees and Shrubs in Kenya. ICRAF, Nairobi, Kenya.

ICRAF, 1992b. Annual Report 1991. International Centre for Research in Agroforestry, Nairobi, Kenya.

ICRAF, 1993. *ICRAF: The Way Ahead. Strategic Plan.* ICRAF, Nairobi, Kenya.

ICRAF, 1994. Annual Report 1993. International Centre for Research in Agroforestry, Nairobi, Kenya.

ICRAF, 1995. Annual Report 1994. International Centre for Research in Agroforestry, Nairobi, Kenya.

ICRAF, 1996. 1995 Annual Report. International Centre for Research in Agroforestry, Nairobi, Kenya.

ICRAF, 1997a. Annual Report 1996. ICRAF, Nairobi, Kenya.

ICRAF, 1997b. ICRAF Medium-Term Plan 1998–2000, ICRAF, PO Box 30677, Nairobi, Kenya, 73p.

ICRAF, 2012. Agroforestry Tree Genetic Resources Strategy 2013–2017, World Agroforestry Centre PO Box 30677-00100 Nairobi, Kenya, 19pp.

ICRAF, 2015. Strategy on Tree-based Energy: Clean and Sustainable Energy for Improving Livelihoods of Poor People. World Agroforestry Centre, Nairobi, Kenya, 25pp.

ICSU, ISSC 2015. Review of the Sustainable Development Goals: The Science Perspective. International Council for Science, Paris, pp 92.

IFAD, 2004. Livestock Services and the Poor. International Fund for Agricultural Development, Rome, Italy.

IFPRI, 1996. Feeding the World, Preventing Poverty and Protecting the Earth: a 2020 Vision. International Food Policy Research Institute, Washington.

IFPRI. 2011. Keynote lecture by Dr Mark Rosegrant to Ag Economic Forum, St Louis, MO (Press release). May 23, 2011

Ikerra, T.W.D., Mnekeni, P.N.S., Singh, B.R., 1994. Effects of added compost and farmyard manure on P release from Minjingu phosphate rock and its uptake by maize. Norwegian J. Agric. Sci. 8, 13−23.

Ingram, J., 1990. The role of trees in maintaining and improving soil productivity—a review of the literature. In: Prinsley, R.T. (Ed.), Agroforestry for Sustainable Production: Economic Implications. Commonwealth Science Council, London, pp. 243−303.

Ingram, V., Ros-Tonen, M.A.F., Dietz, T., 2015. A fine mess: Bricolaged forest governance in Cameroon. Int. J. Commons. 9, 000-000.

IPCC, 2007. Climate Change 2007 − Synthesis Report. Contribution of Working Groups I, II and III to the Fourth Assessment Report of the Intergovernmental Panel on Climate Change. In: Core Writing Team, Pachauri, R.K., Reisinger, A., (Eds.), IPCC, Geneva, Switzerland, 104 pp.

Irvine, F.R., 1961. Woody Plants of Ghana. Oxford University Press, Oxford, UK, 868 pp.

Iyamuremye, F., Dick, R.P., 1996. Organic amendments and phosphorus sorption by soils. Adv. Agron. 56, 139−185.

Jabbar, M.R., Labri, A., Reynolds, L., 1994. Profitability of alley farming with and without fallow in southwest Nigeria. Exp. Agric. 30, 319−327.

Jackson, L.E., Pulleman, M.M., Brussaard, L., Bawa, K.S., Brown, G.G., Cardoso, I.M., et al., 2012. Social-ecological and regional adaptation of agro-biodiversity management across a global set of research regions. Global Environ. Change. 22, 623−639.

Jaenicke, H., 1999. Good Tree Nursery Practices − Practical Guidelines for Research Nurseries. International Center for Research in Agroforestry, Nairobi, Kenya, 95 pp.

Jaenicke, H., Franzel, S., Boland, D., 1995. Towards a method to set priorities amongst multipurpose trees for improvement activities: a case study from west Africa. J. Trop. For. Sci. 7, 490−506.

Jaenicke, H., Simons, A.J., Maghembe, J., Weber, J.C., 2000. Domesticating indigenous fruit trees for agroforestry. Acta Hortic. 523, 45−52.

Jaffee, S., Henson, S., Diaz Rios, L., 2011. Making the Grade: Smallholder Farmers, Emerging Standards, and Development Assistance Programs in Africa (A Research Program Synthesis) (Rep. No. 62324-AFR, The International Bank for Reconstruction and Development/The World Bank).

Jamnadass, R., Lowe, A., Dawson, I.K., 2009. Molecular markers and the management of tropical trees: the case of indigenous fruits. Trop. Plant Biol. 2, 1−12.

Jamnadass, R., Dawson, I.K., Anegbeh, P., Asaah, E., Atangana, A., Cordeiro, N.J., et al., 2010. *Allanblackia*, a new tree crop in Africa for the global food industry: market development, smallholder cultivation and biodiversity management. Forests Trees Livelihoods. 19, 251−268.

Jamnadass, R.H., Dawson, I.K., Franzel, S., Leakey, R.R.B., Mithöfer, D., Akinnifesi, F.K., et al., 2011. Improving livelihoods and nutrition in sub-Saharan Africa through the promotion of indigenous and exotic fruit production in smallholders' agroforestry systems: a review. Int. For. Rev. 13, 338−354.

Jamnadass, R., Langford, K., Anjarwalla, P., Mithöfer, D., 2014. Public−Private partnerships in agroforestry. In: van Alfen, N. (Ed.), Encyclopedia of Agriculture and Food Systems, Vol. 4. Elsevier, San Diego, 544−564pp.

Jamnadass, R., McMullin, S., Iiyama, M., Dawson, I.K., Powell, B., Termote, C., et al., 2015. Understanding the roles of forests and tree-based systems in food provision. In: Vira, B., Wildburger, C., Mansourian, S. (Eds.), Forests, Trees and Landscapes for Food Security and Nutrition, A Global Assessment Report, 33. International Union of Forest Research Organizations, IUFRO World Series, Vienna, Austria, pp. 25−49.

Janick, J., Moore, J.N., 1996. Fruit Breeding: Volume 1. Tree and Tropical Fruits. Wiley, New York, 632 pp.

Jeffree, C.E., Yeoman, M.M., 1983. Development of intercellular connections between opposing cells in a graft union. New Phytol. 93, 481−509.

Johnson, M.D., Levy, N.J., Kellermann, J.L., Robinson, D.E., 2009. Effects of shade and bird exclusion on arthropods and leaf damage on coffee farms in Jamaica's Blue Mountains. Agrofor. Syst. 76, 139−148.

Jones, A.M.P., Murch, S.J., Ragone, D., 2010. Diversity of breadfruit (*Artocarpus altilis*, Moraceae) seasonality: a resource for year-round nutrition. Econ. Bot. 64, 340−351.

Jones, N., 1974. Records and comments regarding the flowering of *Triplochiton scleroxylon* K. Schum. Commonw. For. Rev. 53, 52−56.

Jones, N., 1975. Observations on *Triplochiton scleroxylon* K. Schum flower and fruit development. In: Proceedings of the Symposium on Variation and Breeding of *Triplochiton scleroxylon* K. Schum, Ibadan, Nigeria, 28−37.

Jong, K., Stone, B.C., Soepadmo, E., 1973. Malaysian tropical forests an unexplored genetic reservoir of edible-fruit tree species. In: Soepadmo, E., Singh, K.G. (Eds.), Proceedings of the Symposium on Biological Resources and Natural Development. Malayan Nature Society, Kuala Lumpar, pp. 113−122.

Jonsson, K., Ståhl, L., Högberg, P., 1996. Tree fallows: a comparison between five tropical tree species. Biol. Fert. Soils. 23, 50−56.

Joseph, J.K., 1995. Physico-chemical attributes of wild mango (*Irvingia gabonensis*) seeds. Bioresour. Technol. 53, 179−181.

Joseph, J.K., Aworh, O.C., 1991. Composition, sensory quality and respiration during ripening and storage of edible wild mango (*Irvingia gabonensis*). Int. J. Food Sci. Technol. 26, 337−342.

Juo, A.S.R., Franzluebbers, K., Dabiri, A., Ikhile, B., 1995. Changes in soil properties during long-term fallows and continuous cultivation after forest clearing in Nigeria. Agric. Ecosyst. Environ. 56, 9−18.

Kadzere, I., Watkins, C.B., Merwin, I.A., Akinnifesi, F.K., Saka, J.D.K., Mhango, J., 2006. Fruit variability and relationships between color at harvest and quality during storage of *Uapaca kirkiana* (Muell. Arg.) fruit from natural woodlands. Hortscience. 41, 352−356.

Kaitho, R.J., Tamminga, S., Bruchem, J., 1993. Rumen degradation and in vivo digestibility of dried *Calliandra calothyrsus* leaves. Anim. Feed Sci. Technol. 43, 19−30.

Kalenda, D.T., Missang, C.E., Kinkela, T.T., Krebs, H.C., Renard, C.M.G.C., 2002. New developments in the chemical characterization of the fruit of *Dacryodes edulis* (G. Don) H.J. Lam. Forests Trees Livelihoods. 11, 119−123.

Kalinganire, A., Weber, J.C., Uwamariya, A., Kone, B., 2008. Improving rural livelihoods through domestication of indigenous fruit trees in parklands of the Sahel. In: Akinnifesi, F.K., Leakey, R.R.B., Ajayi, O.C., Sileshi, G., Tchoundjeu, Z., Matacala, P., Kwesiga, F.R. (Eds.), Indigenous Fruit Trees in the Tropics: Domestication, Utilization and Commercialization. CABI, Wallingford, pp. 186−203.

Kalinganire, A., Weber, J.C., Coulibaly, S., 2012. Improved *Ziziphus mauritiana* germplasm for Sahelian smallholder farmers: First steps toward a domestication programme. Forests Trees Livelihoods. 21, 128–137.

Kamatali, P., Teller, E., Vanbelle, M., Collignon, G., Foulon, M., 1992. In situ degradability of organic matter, crude protein and cell wall of various tree forages. Anim. Prod. 55, 29–34.

Kang, B.T., 1993. Alley cropping: past achievements and future directions. Agrofor. Syst. 23, 141–155.

Kang, B.T., Wilson, G.F., Sipkens, L., 1981. Alley cropping maize and *Leucaena leucocephala* Lam. In Southern Nigeria. Plant Soil. 63, 165–179.

Kang, B.T., Reynolds, L., Atta-Krah, R.N., 1990. Alley farming. Adv. Agronomy. 43, 315–359.

Kang, B.T., Akinnifesi, F.K., Ladipo, D.O., 1994. Performance of selected woody agroforestry species grown on Alfisol and Ultisol in the humid lowland of West Africa, and their effects on soil properties. J. Trop. Forest Sci. 7, 303–312.

Kang, B.T., Salako, F.K., Akobundu, I.O., Pleysier, J.L., Chianu, J.N., 1997. Amelioration of a degraded Oxic Paleustalf by leguminous and natural fallows. Land Use Manage. 13, 130–136.

Kapseu, C., Parmentier, M., 1997. Composition en acides gras de quelques huiles végétal du Cameroun. Sciences des Aliments. 17 (3), 85–92.

Kapseu, C., Tchiegang, C., 1996. Composition de l'huile des fruits de deux cultivars de safou au Cameroun. Fruits. 51, 185–191.

Kapseu, C., Avouampo, E., Djeumako, B., 2002. Oil extraction from *Dacryodes edulis* (G. Don) H.J. Lam fruit. Forests Trees Livelihoods. 11, 97–104.

Kärki, L., Tigerstedt, P.M.A., 1985. Definition and exploitation of forest tree ideotypes in Finland. In: Cannell, M.R.G., Jackson, J.E. (Eds.), Attributes of Trees as Crop Plants. Institute of Terrestrial Ecology, Huntington, England, pp. 102–108.

Kengni, E., Tchoundjeu, Z., Tchouanguep, F.M., Mbofung, C.M.F., 2001. Sensory evaluation of *Dacryodes edulis* fruit types. For. Trees Livelihoods. 11, 57–66.

Kengue, J., 1998. Point sur la biologie de la reproduction du safoutier (*Dacryodes edulis* (G. Don) H.J. Lam). In: Kapseu, C., Kayem, G.J. (Eds.), Proceedings of the 2nd International Workshop on African Pear Improvement and Other New Sources of Vegetable Oils. ENSAI, Presses Universitaires de Yaoundé, Cameroon, pp. 97–111.

Kengue, J., 2002. *Safou: Dacryodes edulis G. Don.* Fruits for the Future 3. International Center for Underutilized Crops, Southampton, UK, 147 pp.

Kengue, J., Nya-Ngatchou, J., 1994. Le Safoutier: the African pear. In: Proceedings of the Domestication of the African Pear Workshop, Douala, Cameroon, 4–6 October 1994.

Kengue, J., Singa, E.M., 1998. Preliminary characterization of some African Plum accessions in Barombi-Kang germplasm collection. In: Kapseu, C., Kayem, G.J. (Eds.), Proceedings of the 2nd International Workshop on African Pear Improvement and Other New Sources of Vegetable Oils. ENSAI, Presses Universitaires de Yaoundé, Cameroon, pp. 113–122.

Kengue, J., Anegbeh, P., Waruhiu, A., Avana, M.-L., Kengni, E., Tsobeng, A., et al., 2002a. Domestication du Safoutier (*Dacryodes edulis* (G Don) H.J. Lam): un état des lieux. In: Kengue, J., Kapseu, C., Kayem, G.J. (Eds.), *3ème Séminaire International sur la Valorisation du Safoutier et autres Oléagineux Non-conventionnels, Yaoundé, Cameroun, 3–5 Octobre 2000.* Presses Universitaires d'Afrique, Yaoundé, pp. 60–72.

Kengue, J., Tchuenguem Fohouo, F.N., Adewusi, H.G., 2002b. Towards the improvement of Safou (*Dacryodes edulis*): Population variation and reproductive biology. Forests Trees Livelihoods. 11, 73–84.

Kerkhoven, A.R.W., 1913. Het Tegengaan van Afspoeling door Rationed Tuinaanleg. Soekaboemische Landbouw Vereenniging, Soekaboemi.

Khan, Z.R., Hassanali, A., Overholt, W., Khamis, T.M., Hooper, A.M., Pickett, J.A., et al., 2002. Control of witchweed *Striga hermonthica* by intercropping with Desmodium spp. and the mechanism defined as allelopathic. J. Chem. Ecol. 28, 1871–1885.

Khan, Z.R., Midega, C.A.O., Hassanali, A., Pickett, J.A., Wadhams, L.J., Wanjoya, A., 2006. Management of witchweed, *Striga hermonthica*, and stemborers in sorghum, *Sorghum bicolor*, through intercropping with greenleaf desmodium. *Desmodium intortum*. Int. J. Pest Manage. 52, 297–302.

Khan, Z.R., Midega, C.A.O., Hassanali, A., Pickett, J.A., Wadhams, L.J., 2007. Assessment of different legumes for the control of *Striga hermonthica* in maize and sorghum. Crop Sci. 47, 730–734.

Khasanah, N., Perdana, A., Rahmanullah, A., Manurung, G., Roshetko, J.M., van Noordwijk, M., 2015. Intercropping teak (*Tectona grandis*) and maize (*Zea mays*): bioeconomic trade-off analysis of agroforestry management practices in Gunungkidul, West Java. Agrofor. Syst. 89, 1019–1033.

Kiepe, P., Rao, M.R., 1994. Management of agroforestry for the conservation and utilization of land and water resources. Outlook Agric. 23 (1), 17–25.

Kiers, E.T., Leakey, R.R.B., Izac, A.-M., Heinemann, J.A., Rosenthal, E., Nathan, D., et al., 2008. Agriculture at a crossroads. Science. 320, 320–321.

Kimani, P.G., 2002. Population variation in yield and composition in bark chemical extracts of *Prunus africana*, and its potential for domestication. M. Phil Thesis. Moi University, Kenya, 73 pp.

Kimondo, J.M., Agea, J.G., Okia, C.A., Abohassan, R.A.A., Nghitoolwa Ndeunyema, E.T.N., Woiso, D.A., et al., 2012. Physiochemical and nutritional characterization of *Vitex payos* (Lour.) Merr. (Verbenaceae): an indigenous fruit tree of Eastern Africa. J. Hortic. For. 4, 161–168.

Kindt, R., 2002. Methodology for tree species diversification planning for African ecosystems. PhD thesis. University of Ghent, 327 pp.

Kindt, R., Mutua, A., Muasya, S., Kimotho, J., 2003. Tree Seed Suppliers Directory. World Agroforestry Center, Nairobi, Kenya, 426 pp.

Kindt, R., van Damme, P., Simons, A.J., 2006. Tree diversity in western Kenya: using profiles to characterise richness and evenness. Top. Biodivers. Conserv. 2, 193–210.

Kindt, R., van Damme, P., Simons, A.J., 2006c. Patterns of species richness at varying scales in western Kenya: planning for agroecosystem diversification. Biodivers. Conserv. 15, 3235–3249.

Kinjo, T., Pratt, P.F., 1971. Nitrate adsorption. Soil Sci. Soc. Am. Proc. 35, 722–732.

Kiptot, E., Franzel, S., 2012. Gender and agroforestry in Africa: who benefits? The African perspective. In: Nair, P.K.R., Garrity, D. (Eds.), Agroforestry – The Future of Global Land Use, Adv. Agrofor., 9, 463–496. Springer, Dordrecht, Germany.

Kiptot, E., Franzel, S., 2015. Farmer-to-farmer extension: opportunities for enhancing performance of volunteer farmer trainers in Kenya. Dev. Pract. 25, 503–517.

Koffa S.N., Roshetko J.M., 1999. Farmer-managed germplasm production-diffusion pathways in Lantapan, Philippines. In: Roshetko, J.M., Evans, D. O., (Eds.) Domestication of agroforestry trees in Southeast Asia. Forest Farm. Comm. Tree Res Rep., Special Issue. Winrock International, Morrilton, pp 142–150, 242p

Konig, D., 1992. The potential of agroforestry methods for erosion control in Rwanda. Soil Technol. 5, 167–176.

Krauss, U., 2004. Diseases in tropical agroforestry landscapes: the role of biodiversity. In: Schroth, G., da Fonseca, G.A.B., Harvey, C.A., Gascon, C., Vasconcelos, H.L., Izac, A.-M.N. (Eds.), Agroforestry and Biodiversity Conservation in Tropical Landscapes. Island Press, Washington, DC, pp. 397–412.

Krauss, U., Soberanis, W., 2001. Rehabilitation of diseased cacao fields in Peru through shade regulation and timing of biocontrol measures. Agrofor. Syst. 53, 179–184.

Kubiszewski, I., Costanza, R., Franco, C., Lawn, P., Talberth, J., Jackson, T., et al., 2013. Beyond GDP: Measuring and achieving global genuine progress. Ecol. Econ. 93, 57–68.

Kumar, B.M., Nair, P.K.R., 2004. The enigma of tropical homegardens. Agrofor. Syst. 61, 135–152.

Kumar Rao, J.D.V.K., Tompson, J.A., Sastry, P.V.S.S., Giller, K.E., Day, J.M., 1987. Measurement of N_2-fixation, growth and yield of pigeon pea [*Cajanus cajan* (L.) Millsp.] grown in a vertisol. Biol. Fertil. Soils. 7, 95–100.

Kuznets, S., 1934. National Income, 1929–1932. 73rd US Congress (Senate doc. no. 124 1934), pp. 1–12. www.nber.org/books/kuzn34-1.

Kwamina, E.B., Ockie, J.H.B., Nam, C.N., 2015. A systemic intervention to assess resource impact on the quality of life among women farmers in developing countries: evidence from Ghana. Acad. J. Agric. Res. 3, 15–22.

Kwesiga, F., Chisumpa, S., 1990. Ethnobotanical Survey in the Eastern Province of Zambia. ICRAF, Nairobi.

Kwesiga, F., Coe, R., 1994. The effect of short rotation *Sesbania sesban* planted fallows on maize yield. For. Ecol. Manage. 64, 199–208.

Kwesiga, F., Kamau, I., 1989. *Agroforestry Potential in the Unimodal Upland Plateau of Zambia. AFRENA Report No. 7.* ICRAF, Nairobi.

Kwesiga, F., Phiri, D., Simwanza, C.P., Mwanza, S., 1994. Zambia-ICRAF Research Project No. 87. ICRAF, Nairobi.

Kwesiga, F., Akinnifesi, F.K., Ramadhani, T., Kadzere, I., Saka, J., 2000. Domestication of indigenous fruit trees of the miombo in southern Africa. In: Shumba, E.M., Luseani, E., Hangula, R. (Eds.), Proceedings of a SADC Tree Seed Centre Network Technical meeting, Windhoek, Namibia, 13–14 March 2000. Co-sponsored by CIDA and FAO, pp. 8–24.

Kwesiga, F.R., Grace, J., 1986. The role of the red/far-red ratio in the response of tropical tree seedlings to shade. Ann. Bot.(London). 57, 283–290.

Kwesiga, F.R., Grace, J., Sandford, A.P., 1987. Some photosynthetic characteristics of tropical timber trees as affected by the light regime during growth. Ann. Bot. (London). 58, 23–32.

Kwesiga, F.R., Franzel, S., Place, F., Phiri, D., Simwanza, C.P., 1999. *Sesbania sesban* improved fallows in eastern Zambia: their inception, development and farmer enthusiasm. Agrofor. Syst. 47, 49–66.

Ladipo, D.O., 1981. Branching patterns of the tropical hardwood *Triplochiton scleroxylon* K. Schum. with special reference to the selection of superior clones at an early age. Ph.D. Thesis. University of Edinburgh, p. 248.

Ladipo, D.O., Leakey, R.R.B., Longman, K.A., Last, F.T., 1980. A study of variation in *Triplochiton scleroxylon* K. Schum: some criteria for clonal selection. In: Proceedings of IUFRO. Symposium on Genetic Improvement and Productivity of Fast Growing Tree Species, Sao Paulo, Brazil.

Ladipo, D.O., Grace, J., Sandford, A.P., Leakey, R.R.B., 1984. Clonal variation in photosynthetic and respiration rates and diffusion resistances in the tropical hardwood *Triplochiton scleroxylon* K. Schum. Photosynthetica. 18, 20–27.

Ladipo, D.O., Leakey, R.R.B., Grace, J., 1991a. Clonal variation in apical dominance in young plants of *Triplochiton scleroxylon* K. Schum.: responses to decapitation. Silvae Genet. 40, 135–140.

Ladipo, D.O., Leakey, R.R.B., Grace, J., 1991b. Clonal variation in a four-year-old plantation of *Triplochiton scleroxylon* K. Schum. and its relation to the predictive test for branching habit. Silvae Genet. 40, 130–135.

Ladipo, D.O., Leakey, R.R.B., Grace, J., 1992. Variations in bud activity from decapitated, nursery-grown plants of *Triplochiton scleroxylon* in Nigeria: effects of light, temperature and humidity. For. Ecol. Manage. 50, 287–298.

Ladipo, D.O., Fondoun, J.-M., Ganga, N., 1996. Domestication of bush mango (*Irvingia* spp.): some exploitable intraspecific variations in west and central Africa. In: Leakey, R.R.B., Newton, A.C. (Eds.), Tropical Trees: The Potential for Domestication and the Rebuilding of Forest Resources. HMSO, London, UK, pp. 193–205.

Laird, S.A. (Ed.), 2002. Biodiversity and Traditional Knowledge: Equitable Partnerships in Practice, People and Plants Conservation Series. Earthscan Publication Ltd., London. UK, 504pp.

Lajtha, K., Harrison, A.F., 1995. Strategies of phosphorus acquisition and conservation by plant species and communities. In: Tiessen, H. (Ed.), Phosphorus in the Global Environment. John Wiley & Sons, Chichester, UK, pp. 139–148.

Lal, R., 1989a. Agroforestry systems and soil surface management of a tropical soil: water run-off erosion and nutrient loss. Agrofor. Syst. 8, 97–111.

Lal, R., 1989b. Agroforestry systems and soil surface management of a tropical Alfisol. II. Water runoff, soil erosion, and nutrient loss. Agrofor. Syst. 8, 97–111.

Lal, R., 1989c. Agroforestry systems and soil surface management of a tropical Alfisol. IV. Effect on soil physical and mechanical properties. Agrofor. Syst. 8, 197–215.

Lal, R., 1989d. Agroforestry systems and soil surface management on a tropical Alfisol. V. Water infiltrability, transmissivity and soil water sorptivity. Agrofor. Syst. 8, 217–238.

Lamien N., Sidibe A., Bayala J., 1996. Use and commercialization of non-timber forest products in Western Burkina Faso. In: Leakey, R.R.B., Temu, A.B., Melnyk, M., Vantomme, P. (Eds.), Domestication and Commercialization of Non-timber Forest Products in Agroforestry Systems. Non-Wood Forest Products No. 9. FAO, Rome, Italy, pp 51–64.

Lankoandé, B., Ouédraogo, A., Boussim, J.I., Lykke, A.M., 2015. Phenotypic traits of *Carapa procera* fruits from riparian forests of Burkina Faso, West Africa. J. Hortic. For. 7, 160–167.

Lapeyrie, F., Högberg, P., 1994. Harnessing symbiotic associations: ectomychorrizas. In: Leakey, R.R.B., Newton, A.C. (Eds.), *Tropical Trees: The Potential for Domestication and the Rebuilding of* Forest Resources. HMSO, London, pp. 158–164.

Last, F.T., Dighton, J., Mason, P.A., 1987. Successions for sheathing mychorrizal fungi. Trends Ecol. Evol. 2, 157–161.

Lavelle, P., 1996. Diversity of soil fauna and ecosystem function. Biol. Int. 33, 3–16.

Lavelle, P., Pashanasi, B., 1989. Soil macrofauna and land management in Peruvian Amazonia. Pedobiologia. 33, 283–291.

Lavelle, P., Moreira, F., Spain, A., 2014. Biodiversity: Conserving biodiversity in agroecosystems. In: van Alfen, N., et al., (Eds.), Encyclopedia of Agriculture and Food Systems, Vol 2. Elsevier Publishers, San Diego, pp. 41–60.

Leakey, R.R.B., 1983. Stockplant factors affecting root initiation in cuttings of *Triplochiton scleroxylon* K. Schum., an indigenous hardwood of West Africa. J. Hortic. Sci. 58, 277–290.

Leakey, R.R.B., 1985. The capacity for vegetative propagation in trees. In: Cannell, M.G.R., Jackson, J.E. (Eds.), Attributes of Trees as Crop Plants. Institute of Terrestrial Ecology, Huntingdon, UK, pp. 110–133.

Leakey, R.R.B., 1987. Clonal forestry in the tropics– a review of developments, strategies and opportunities. Commonw. For. Rev. 66, 61–75.

Leakey, R.R.B., 1989. Vegetative propagation methods for tropical trees: rooting leafy softwood cuttings, In: Agroforestry and Mycorrhizal Research for Semi-arid lands of East Africa Workshop, National Museums of Kenya, Sept. 1989, 26–64, Institute of Terrestrial Ecology, Bush Estate, Penicuik, EH26 OQB, Scotland, UK.

Leakey, R.R.B., 1990. *Nauclea diderrichii*: rooting stem cuttings, clonal variation in shoot dominance and branch plagiotropism. Trees. 4, 164–169.

Leakey, R.R.B., 1991. Clonal forestry: towards a strategy. Some guidelines based on experience with tropical trees. In: Jackson, J.E. (Ed.), *Tree Breeding and Improvement*. Royal Forestry Society of England, Wales and Northern Ireland, Tring, England, pp. 27–42.

Leakey, R.R.B., 1992. Enhancement of rooting-ability in *Triplochiton scleroxylon* by injecting stockplants with auxins and a cytokinin. For. Ecol. Manage. 54, 305–313.

Leakey, R.R.B., 1996. Definition of agroforestry revisited. Agroforestry Today. 8 (1), 5–7.

Leakey, R.R.B., 1997. Domestication potential of *Prunus africana* ('Pygeum') in sub-Saharan Africa. In: Kinyua, A.M., Kofi-Tsekpo, W.M., Dangana, L.B. (Eds.), Conservation and Utilization of Medicinal Plants and Wild Relatives of Food Crops. UNESCO, Nairobi, Kenya, pp. 99–106.

Leakey R.R.B., 1998a. The use of biodiversity and implications for agroforestry. In: Proceedings of FORUM BELEM. POEMA, Belém, Brazil, pp. 43–58.

Leakey, R.R.B., 1998b. Agroforestry in the humid lowlands of West Africa: some reflections on future directions for research. Agrofor. Syst. 40, 253–262.

Leakey, R.R.B., 1999a. Potential for novel food products from agroforestry trees. Food Chem. 64, 1–14.

Leakey, R.R.B., 1999b. Agroforestry for biodiversity in farming systems. In: Collins, W.W., Qualset, C.O. (Eds.), Biodiversity in Agroecosystems. CRC Press, New York, pp. 127–145.

Leakey, R.R.B., 2001a. Win:Win landuse strategies for Africa: 1. Building on experience with agroforests in Asia and Latin America. Int. For. Rev. 3, 1–10.

Leakey, R.R.B., 2001b. Win:Win landuse strategies for Africa: 2. capturing economic and environmental benefits with multistrata agroforests. Int. For. Rev. 3, 11–18.

Leakey, R.R.B., 2003. IPUF: to domesticate or not to domesticate? S. Afr. Ethnobot. Newsl. Indig. Plant Use Forum. 1 (1), 40–42.

Leakey, R.R.B., 2004. Physiology of vegetative propagation in trees. In: Burley, J., Evans, J., Youngquist, J.A. (Eds.), Encyclopedia of Forest Sciences. Academic Press, London, pp. 1655–1668.

Leakey, R.R.B., 2005. Domestication potential of Marula (*Sclerocarya birrea* subsp *caffra*) in South Africa and Namibia: 3. Multi-trait selection. Agrofor. Syst. 64, 51–59.

Leakey, R.R.B., 2010. Agroforestry: a delivery mechanism for multi-functional agriculture. In: Kellimore, L.R. (Ed.), Handbook on Agroforestry: Management Practices and Environmental Impact. Environmental Science, Engineering and Technology Series. Nova Science Publishers, New York, pp. 461–471.

Leakey, R.R.B., 2012a. The intensification of agroforestry by tree domestication for enhanced social and economic impact. CAB Rev. Perspect. Agric. Vet. Sci. Nutr. Nat. Resour. 7 (035), 1–3.

Leakey, R R.B., 2012b. Living with the Trees of Life – Towards the Transformation of tropical Agriculture. CABI, Wallingford, UK, 200pp.

Leakey, R.R.B., 2012c. Participatory domestication of indigenous fruit and nut trees: new crops for sustainable agriculture in developing countries. In: Gepts, P.L., Famula, T.R., Bettinger, R.L., Brush, S.B., Damania, A.B., McGuire, P.E., Qualset, C.O. (Eds.), Biodiversity in Agriculture: Domestication, Evolution and Sustainability. Cambridge University Press, Cambridge, pp. 479–501.

Leakey, R.R.B., 2012d. Non-timber forest products – a misnomer? Guest editorial. J. Trop. For. Sci. 24, 145–146.

Leakey, R.R.B., 2012e. Multifunctional agriculture and opportunities for agroforestry: implications of IAASTD. In: Nair, P.K.R., Garrity, D.P. (Eds.), Agroforestry: The Future of Global Land Use. Springer, Dordrecht, pp. 203–214.

Leakey, R.R.B., 2013. Addressing the causes of land degradation, food/nutritional insecurity and poverty: a new approach to agricultural intensification in the tropics and sub-tropics. In: Hoffman, U. (Ed.), Wake Up Before It Is Too Late: Make Agriculture Truly Sustainable Now for Food Security in a Changing Climate, UNCTAD Trade and Environment Review 2013, Chapter 3. UN Publications,, Geneva, pp. 192–198.

Leakey, R.R.B., 2014a. Agroforestry: participatory domestication of trees. In: van Alfen, N. (Ed.), Encyclopedia of Agriculture and Food Systems, Vol. 1. Elsevier, San Diego, pp. 253–269.

Leakey, R.R.B., 2014b. An African solution to the problems of African agriculture, In: Sustainable Natural Resources Management in Africa's Urban Food and Nutrition Equation. Nature & Faune 28(2), 17–20 (Published by FAO Regional Office for Africa).

Leakey, R.R.B., 2014c. Plant cloning: macro-propagation. In: van Alfen, N. (Ed.), Encyclopedia of Agriculture and Food Systems, Vol 4. Elsevier Publishers, San Diego, pp. 349–359, Elsevier, USA.

Leakey, R.R.B., 2014d. Sustainable natural resources management in Africa's urban food and nutrition equation: an African solution to the problems of African agriculture. In: F. Bojang, A. Ndeso-Atanga (Eds.), Sustainable Natural Resources Management in Africa's Urban Food and Nutrition Equation., Nat. Faune J., 28(2), 17–20.

Leakey, R.R.B., 2014e. The role of trees in agroecology and sustainable agriculture in the tropics. Annu. Rev. Phytopathol. 52, 113–133.

Leakey RRB, 2014f. Twelve principles for better food and more food from mature perennial agroecosystems. In: Perennial Crops for Food Security, pp. 282-306, Proceedings of Perennial Crops for Food Security Workshop, Rome, Italy August 28–30. Rome: FAO.

Leakey, R.R.B., in press. The role of trees in agroecology. In: Hunter, D., Guarino, L., Spillane, C., McKeown, P. (Eds.), The Routledge Handbook of Agricultural Biodiversity. Routledge Publishers, London.

Leakey, R.R.B., 2017a. Trees: a keystone role in agroecosystem function: an update. Multifunctional Agriculture. Elsevier, San Diego, CA, pp. 19–20.

Leakey, R.R.B., 2017b. Trees: delivering enhanced crop production and income: an update. Multifunctional Agriculture. Elsevier, San Diego, CA, pp. 65–67.

Leakey, R.R.B., 2017c. Trees: an important source of food and non-food products for farmers: an update. Multifunctional Agriculture. Elsevier, San Diego, CA, pp. 99–100.

Leakey, R.R.B., 2017d. Trees: capturing useful traits in elite cultivars: an update. Multifunctional Agriculture. Elsevier, San Diego, CA, pp. 155–158.

Leakey, R.R.B., 2017e. Trees: skills and understanding essential for domestication: an update. Multifunctional Agriculture. Elsevier, San Diego, CA, pp. 273–279.

Leakey, R.R.B., 2017f. Trees: ensuring that farmers benefit from domestication: an update. Multifunctional Agriculture. Elsevier, San Diego, CA, pp. 345–347.

Leakey, R.R.B., 2017g. Trees: a call to policy makers to meet farmers' needs by combining environmental services with marketable products: an update. Multifunctional Agriculture. Elsevier, San Diego, CA, pp. 369–371.

Leakey, R.R.B., 2017h. Trees: delivering productive and sustainable farming systems: an update. Multifunctional Agriculture. Elsevier, San Diego, CA, pp. 389–391.

Leakey, R.R.B., 2017i. Trees: meeting the social, economic and environmental needs of poor farmers—scoring sustainable development goals: an update. Multifunctional Agriculture. Elsevier, San Diego, CA, pp. 417–420.

Leakey, R.R.B., Akinnifesi, F.K., 2008. Towards a domestication strategy for indigenous fruit trees in the tropics. In: Akinnifesi, F.K., Leakey, R.R.B., Ajayi, O.C., Sileshi, G., Tchoundjeu, Z., Matakala, P., Kwesiga, F. (Eds.), Indigenous Fruit Trees in the Tropics: Domestication, Utilization and Commercialization. CAB International, Wallingford, pp. 28–49.

Leakey, R.R.B., Asaah, E.K., 2013. Underutilised species as the backbone of multifunctional agriculture: the next wave of crop domestication. Acta. Hortic. 979, 293–310.

Leakey, R.R.B., Coutts, M.P., 1989. The dynamics of rooting in *Triplochiton scleroxylon* cuttings: their relation to leaf area, node position, dry weight accumulation, leaf water potential and carbohydrate composition. Tree Physiol. 5, 135–146.

Leakey, R.R.B., Izac, A.-M.N., 1996. Linkages between domestication and commercialization of non-timber forest products: implications for agroforestry. In: Leakey, R.R.B., Temu, A.B., Melnyk, M., Vantomme, P. (Eds.), Domestication and Commercialization of Non-timber Forest Products. Non-Wood Forest Products No. 9. FAO, Rome, Italy, pp. 1–7.

Leakey, R.R.B., Jaenicke, H., 1995. The domestication of indigenous fruit trees: opportunities and challenges for agroforestry. In: Suzuki, K., Sakurai, S., Ishii, K., Norisada, M. (Eds.), Procedings of 4th International BIO-REFOR Workshop. BIO-REFOR, Tokyo, Japan, pp. 15–26.

Leakey, R.R.B., Ladipo, D.O., 1987. Selection for improvement in vegetatively-propagated tropical hardwoods. In: Atkin, R., Abbott, J. (Eds.), Improvement of Vegetatively Propagated Plants. Academic Press, London, pp. 324–336.

Leakey, R.R.B., Ladipo, D.O., 1996. Trading on genetic variation − fruits of *Dacryodes edulis*. Agrofor. Today. 8 (2), 16–17.

Leakey, R.R.B., Last, F.T., 1980. Biology and potential of *Prosopis* species in arid environments, with particular reference to *P. cineraria*. J. Arid Environ. 3, 9–24.

Leakey, R.E., Lewin, R., 1996. The Sixth Extinction: Biodiversity and Its Survival. Weidenfeld and Nicholson, London, 271 pp.

Leakey, R.R.B., Longman, K.A., 1986. Physiological, environmental and genetic variation in apical dominance as determined by decapitation in *Triplochiton scleroxylon*. Tree Physiol. 1, 193–207.

Leakey, R.R.B., Longman, K.A., 1988. Low-tech cloning of tropical trees. Appropr. Technol. 15, 6.

Leakey, R.R.B., Maghembe, J.A., 1994. Domestication of high value trees for agroforestry: an alternative to slash and burn agriculture, ICRAF Position Paper 1, 12pp.

Leakey, R.R.B., Mohammed, H.R.S., 1985. The effects of stem length on root initiation in sequential single-node cuttings of *Triplochiton scleroxylon* K. Schum. J. Hortic. Sci. 60, 431−437.

Leakey, R.R.B., Newton, A.C., 1994a. Domestication of 'Cinderella' species as a start of a woody plant revolution. In: Leakey, R.R.B., Newton, A.C. (Eds.), Tropical Trees: Potential for Domestication and the Rebuilding of Forest Resources. HMSO, London, pp. 3−6.

Leakey, R.R.B., Newton, A.C. (Eds.), 1994b. Tropical Trees: Potential for Domestication, Rebuilding Forest Resources. HMSO, London, 284 pp.

Leakey, R.R.B., Newton, A.C. (Eds.), 1994c. Domestication of Timber and Non-timber Forest Products MAB Digest 17. UNESCO, Paris.

Leakey, R.R.B., Newton, A.C., 1995. *Domestication of Tropical Trees for Timber and Non-timber Forest Products, MAB Digest 17*. UNESCO, Paris, 94 pp.

Leakey, R.R.B., Page, T., 2006. The 'ideotype concept' and its application to the selection of 'AFTP' cultivars. Forest Tree Livelihoods. 16, 5−16.

Leakey, R.R.B., Prabhu, R., 2017. Toward multifunctional agriculture − an African initiative. Multifunctional Agriculture. Elsevier, San Diego, CA, pp. 395−416.

Leakey, R.R.B., Sanchez, P.A., 1997. How many people use agroforestry products? Agrofor. Today. 9 (3), 4−5.

Leakey, R.R.B., Simons, A.J., 1997. The domestication and commercialization of indigenous trees in agroforestry for the alleviation of poverty. Agroforestry Systems. 38, 165−176.

Leakey, R.R.B., Simons, A.J., 2000. When does vegetative propagation provide a viable alternative to propagation by seed in forestry and agroforestry in the tropics and sub-tropics? In: Wolf, H., Arbrecht, J. (Eds.), Problem of Forestry in Tropical and Sub-tropical Countries: The Procurement of Forestry Seed − The Example of Kenya. Ulmer Verlag, Germany, pp. 67−81.

Leakey, R.R.B., Storeton-West, R., 1992. The rooting ability of *Triplochiton scleroxylon* cuttings: the interaction between stockplant irradiance, light quality, and nutrients. For. Ecol. Manage. 49, 133−150.

Leakey, R.R.B., Tchoundjeu, Z., 2001. Diversification of tree crops: domestication of companion crops for poverty reduction and environmental services. Exp. Agric. 37, 279−296.

Leakey, R.R.B., Tomich, T.P., 1999. Domestication of tropical trees: from biology to economics and policy. In: Buck, L.E., Lassoie, J.P., Fernandes, E.C.M. (Eds.), Agroforestry in Sustainable Ecosystems. CRC Press/Lewis Publishers, New York, USA, pp. 319−338.

Leakey, R.R.B., van Damme, P., 2014. The role of tree domestication in value chain development. For. Trees Livelihoods. 23, 116−126.

Leakey, R.R.B., Chapman, V.R., Longman, K.A., 1975. Studies on root initiation and bud outgrowth in nine clones of *Triplochiton scleroxylon* K. Schum. Proceedings of Symposium on Variation and Breeding Systems of *Triplochiton scleroxylon* K. Schum., 1975, Ibadan, Nigeria. Forestry Research Institute of Nigeria, lbadan, Nigeria, pp. 86−92.

Leakey, R.R.B., Ferguson, N.R., Longman, K.A., 1981. Precocious flowering and reproductive biology of *Triplochiton scleroxylon* K. Schum. Commonw. For. Rev. 60, 117−126.

Leakey, R.R.B., Last, F.T., Longman, K.A., Ojo, G.O.A., Oji, N.O., Ladipo, D.O., 1980. Triplochiton scleroxylon: a tropical hardwood for plantation forestry. In: Proceedings of IUFRO Symposium on Genetic Improvement and Productivity of Fast growing Tree Species, Sao Paulo, Brazil.

Leakey, R.R.B., Chapman, V.R., Longman, K.A., 1982a. Physiological studies for tropical tree improvement and conservation. Factors affecting root initiation in cuttings of Triplochiton scleroxylon K. Schum. For. Ecol. Manage. 4, 53−66.

Leakey, R.R.B., Last, F.T., Longman, K.A., 1982b. Domestication of forest trees: a process to secure the productivity and future diversity of tropical ecosystems. Commonw. For. Rev. 61, 33−42.

Leakey, R.R.B., Mesén, J.F., Tchoundjeu, Z., Longman, K.A., Dick, J.Mc.P., Newton, A.C., et al., 1990. Low-technology techniques for the vegetative propagation of tropical tress. Commonw. For. Rev. 69, 247−257.

Leakey, R.R.B., Newton, A.C., Dick, J.Mc.P., 1994. Capture of genetic variation by vegetative propagation: processes determining success. In: Leakey, R.R.B., Newton, A.C. (Eds.), Tropical Trees: The Potential for Domestication and the Rebuilding Forest Resources. HMSO, London, pp. 72−83.

Leakey, R.R.B., Temu, A.B., Melnyk, M., Vantomme, P. (Eds.), 1996. Domestication and Commercialization of Non-timber Forest Products in Agroforestry Systems, Non-Wood Forest Products No. 9. FAO, Rome, Italy.

Leakey, R.R.B., Fondoun, J.-M., Atangana, A., Tchoundjeu, Z., 2000. Quantitative descriptors of variation in the fruits and seeds of *Irvingia gabonensis*. Agrofor. Syst. 50, 47−58.

Leakey, R.R.B., Atangana, A.R., Kengni, E., Waruhiu, A.N., Usoro, C., Tchoundjeu, Z., et al., 2002. Domestication of *Dacryodes edulis* in West and Central Africa: characterization of genetic variation. For. Trees Livelihoods. 12, 57−71.

Leakey, R.R.B., Schreckenberg, K., Tchoundjeu, Z., 2003. The participatory domestication of West African indigenous fruits. Int. For. Rev. 5, 338−347.

Leakey, R.R.B., Tchoundjeu, Z., Smith, R.I., Munro, R.C., Fondoun, J.-M., Kengue, J., et al., 2004. Evidence that subsistence farmers have domesticated indigenous fruits (*Dacryodes edulis* and *Irvingia gabonensis*) in Cameroon and Nigeria. Agrofor. Syst. 60, 101−111.

Leakey, R.R.B., Tchoundjeu, Z., Schreckenberg, K., Shackleton, S., Shackleton, C., 2005a. Agroforestry Tree Products (AFTPs): targeting poverty reduction and enhanced livelihoods. Int. J. Agric. Sustain. 3, 1−23.

Leakey, R.R.B., Shackleton, S., du Plessis, P., 2005b. Domestication potential of Marula (*Sclerocarya birrea* subsp. *caffra*) in South Africa and Namibia: 1. Phenotypic variation in fruit traits. Agrofor. Syst. 64, 25−35.

Leakey, R.R.B., Pate, K., Lombard, C., 2005c. Domestication potential of Marula (*Sclerocarya birrea* subsp *caffra*) in South Africa and Namibia: 2. Phenotypic variation in nut and kernel traits. Agrofor. Syst. 64, 37−49.

Leakey, R.R.B., Greenwell, P., Hall, M.N., Atangana, A.R., Usoro, C., Anegbeh, P.O., et al., 2005d. Domestication of *Irvingia gabonensis*: 4. Tree-to—tree variation in food thickening properties and in fat and protein contents of dika nut. Food Chem. 90, 365−378.

Leakey R.R.B., Tchoundjeu Z., Schreckenberg K., Simons A.J., Shackleton S., Mander M., et al., 2006. Trees and markets for agroforestry tree products: targeting poverty reduction and enhanced livelihoods, In: World Agroforestry and the Future, Proceedings of 25th Anniversary Conference, November 2003, ICRAF, Nairobi, Kenya, pp. 17–28.

Leakey, R.R.B., Fuller, S., Treloar, T., Stevenson, L., Hunter, D., Nevenimo, T., et al., 2008. Characterization of tree-to-tree variation in morphological, nutritional and chemical properties of *Canarium indicum* nuts. Agrofor. Syst. 73, 77–87.

Leakey, R.R.B., Kranjac-Berisavljevic, G., Caron, P., Craufurd, P., Martin, A., McDonald, A., et al., 2009. Impacts of AKST on development and sustainability goals. In: McIntyre, B.D., Herren, H., Wakhungu, J., Watson, R. (Eds.), International Assessment of Agricultural Science and Technology for Development: Global Report. Island Press, New York, USA, pp. 145–253. , Chapter 3.

Leakey, R.R.B., Weber, J.C., Page, T., Cornelius, J.P., Akinnifesi, F.K., et al., 2012. Tree domestication in agroforestry: progress in the second decade. In: Nair, P.K., Garrity, D. (Eds.), *Agroforestry: The Future of Global* Land Use. Springer, New York, pp. 145–173.

Lengkeek, A.G., Jaenicke, H., Dawson, I.K., 2005. Genetic bottlenecks in agroforestry systems: results of tree nursery surveys in East Africa. Agrofor. Syst. 63, 149–155.

Libby, W.J., 1982. What is the safe number of clones per plantation? In: Heybroek, H.M., Stephan, B.R., von Weissenberg, K. (Eds.), Resistance to Diseases and Pests in Forest Trees. PUDOC, Wageningen, The Netherlands, pp. 324–360.

Libby, W.J., Jund, E., 1962. Variance associated with cloning. Heredity. 17, 533–540.

Lipton, M., 1999. Reviving global poverty reduction: What role for genetically modified plants? 1999 Sir John Crawford Memorial Lecture at CGIAR Centres Week, Washington DC (October 28th 1999). CGIAR Secretariat, The World Bank, Washington DC, 41p.

Loach, K., 1977. Leaf water potential and the rooting of cuttings under mist and polythene. Physiol. Plant. 40, 191–197.

Lombard, C., Leakey, R.R.B., 2010. Protecting the rights of farmers and communities while securing long term market access for producers of non-timber forest products: Experience in southern Africa. Forests, Trees Livelihoods. 19, 235–249.

Lompo, F., 1993. Contribution à la Valorisation des phosphates naturels du Burkina Faso: etudes des effets de l'interaction phosphates naturels–materies organiques. These Docteur Ingenieur. Faculte des Sciences et Techniques de L'Université Nationale de Cote d'Ivoire, Abidjan.

Long, S.P., Marshall-Colon, A., Zhu, X.-G., 2015. Meeting the global food demand of the future by engineering crop photosynthesis and yield potential. Cell. 161, 56–66.

Longman, K., Manurung, R.M., Leakey, R.R.B., 1990. Use of small, clonal plants for experiments on factors affecting flowering in tropical trees. In: Bawa, K.S., Hadley, M. (Eds.), *Reproductive Ecology of Tropical Flowering Plants.* Man and the Biosphere Series No. 7. UNESCO and Parthenon Publishing Group, Carnforth, pp. 389–399.

Longman, K.A., 1976a. Conservation and utilization of gene resources by vegetative multiplication of tropical trees. In: Burley, J., Styles, B.T. (Eds.), Tropical Trees: Variation, Breeding and Conservation. Academic Press, London, pp. 19–24.

Longman, K.A., 1976b. Some experimental approaches to the problem of phase change in forest trees. Symp. Juv. Woody Perenn. Acta Hortic. 56, 81–90.

Longman, K.A., 1993c. Rooting Cuttings of Tropical Trees, Tropical Trees: Propagation and Planting Manuals, Vol 1. Commonwealth Science Council, London.

Longman, K.A., Leakey, R.R.B., 1995. La domestication du Samba. Ann. Sci. For. 52, 43–56.

Longman, K.A., Leakey, R.R.B., Howland, P., Bowen, M.R., 1978. Physiological approaches for utilizing and conserving the genetic resources of tropical trees, *Proceedings of World Consultation on Forest Tree Breeding*, 3rd, 1977, Vol. 2. CSIRO, Canberra, Australia, pp. 1043–1054.

Longman, K.A., Leakey, R.R.B., Denne, M.P., 1979a. Genetic and environmental effects on shoot growth and xylem formation in a tropical tree. Ann. Bot. 44, 377–380.

Longman, K.A., Leakey, R.R.B., Howland, P., Bowen, M.R., 1979b. Physiological approaches for utilizing and conserving the genetic resources of tropical trees. In: Proceedings of 3rd World Consultation on Forest Tree Breeding, Canberra, 1977, Vol. 2., 1043–1054.

Loos, J., Abson, D.J., Chappell, M.J., Hanspach, J., Mikulcak, F., Tichit, M., et al., 2014. Putting meaning back into "sustainable intensification". Front. Ecol. Environ. 93, 57–68.

Lovett, P.N., Haq, N., 2000. Evidence for anthropic selection of the sheanut tree (*Vitellaria paradoxa*). Agrofor. Syst. 48, 273–288.

Lowe, A.J., Russell, J.R., Powell, W., Dawson, I.K., 1998. Identification and characterization of nuclear, cleaved amplified polymorphic sequences (CAPS) loci in *Irvingia gabonensis* and *I. wombolu*, indigenous fruit trees of west and central Africa. Mol. Ecol. 7, 1771–1788.

Lowe, A.J., Gillies, A.C.M., Wilson, J., Dawson, I.K., 2000. Conservation genetics of bush mango from central/west Africa: implications from random amplified polymorphic DNA analysis. Mol. Ecol. 9, 831–841.

Lozada, T., de Koning, G.H.J., Marche, R., Klein, A.M., Tscharntke, T., 2007. Tree recovery and seed dispersal by birds: comparing forest, agroforestry and abandoned agroforestry in coastal Ecuador. Perspect. Plant Ecol. Evol. Syst. 8, 131–140.

Lubrano, C., Robin, J.R., Khaiat, A., 1994. Composition en acides gras, stérols et tocophérols d'huiles de pulpe de fruits de six espèces de palmiers de Guyane. Oleagineux. 49, 59–65.

Maas, B., Clough, Y., Tscharntke, T., 2013. Bats and birds increase crop yield in tropical agroforestry landscapes. Ecol. Lett. 16, 1480–1487.

MacDonald, A.M., Bonsor, H.C., Dochartaigh, B.É.Ó., Taylor, R.G., 2012. Quantitative maps of groundwater resources in Africa. Environ. Res. Lett. 7, 024009.

Maclean, A.J., Litsinger, J.A., Moody, K., Watson, A.K., 1992. The impact of alley cropping *Gliricidia sepium* and *Cassia spectabilis* on upland rice and maize production. Agrofor. Syst. 20, 213–219.

Maes, J., Paracchini, M.L., Zulian, G., Dunbar, M.B., Alkemade, R., 2012. Synergies and trade-offs between ecosystem service supply, biodiversity and habitat conservation status in Europe. Biol. Conserv. 155, 1–12.

Mafongoya, P.L., Mpepereki, S., Dzowela, B.H., Mangwayana, E., Makonese, F., 1997. Effect of pruning quality and method of pruning placement on soil microbial composition. Afr. Crop Sci. Proc. 3, 393−398.

Mafongoya, P.L., Kuntashula, E., Sileshi, G., 2006. Managing soil fertility and nutrient cycles through fertilizer trees in southern Africa. In: Uphoff, N., Ball, A.S., Fernandes, E., Herren, H., Husson, O., Laing, M., et al.,Biological Approaches to Sustainable Soil Systems. CRC Press, New York, pp. 273−289.

Maghembe, J.A., 1994. Out of the forest: indigenous fruit trees in southern Africa. Agrofor. Today. 6 (2), 4−6.

Maghembe, J.A., Kwesiga, F., Ngulube, M., Prins, H., Malaya, F.M., 1994. The domestication potential of indigenous fruit trees of the miombo woodlands of southern Africa. In: Leakey, R.R.B., Newton, A.C. (Eds.), Tropical Trees: Potential for Domestication and the Rebuilding of Forest Resources. HMSO, London, pp. 220−229.

Maghembe, J.A., Ntupanyama, Y., Chirwa, P.W., 1995. Improvement of Indigenous Fruit Trees of the Miombo Woodlands of Southern Africa. ICRAF, Nairobi, Kenya, 138pp.

Maghembe, J.A., Simons, A.J., Kwesiga, F., Rarieya, M., 1998. *Selecting Indigenous Trees for Domestication in Southern Africa: Priority Setting with Farmers in Malawi, Tanzania, Zambia and Zimbabwe*. ICRAF, Nairobi, Kenya, 94pp.

Makueti, J.T., Tchoundjeu, Z., Kalinganire, A., Nkongmeneck, B.A., Asaah, E., Tsobeng, A., et al., 2012. Influence de la provenance du géniteur et du type de pollen sur la fructification sous pollinisation contrôlée chez *Dacryodes edulis* (Burseraceae) au Cameroun. Int. J. Biol. Chem. Sci. 2, 1−17.

Mander, M., Cribbins, J., Lewis, F., 2002. The commercial marula industry: a sub-sector analysis. *Report to UK DFID Forestry Research Programme (Project No R7795)*. Institute of Natural Resources, Scottsville, South Africa, p. 79.

Manfredini, S., Vertuani, S., Braccioli, E., Buzzoni, V., 2002. Antioxidant capacity of *Adansonia digitata* fruit pulp and leaves. Acta Phytother. 2, 2−7.

Mangaoang, E.O., Roshetko, J.M., 1999. Establishment and production of *Gliricidia sepium* in farmers' seed orchard in the Uplands of Leyte, Philippines. For. Farm. Comm. Tree Res. Rep. 4, 73−77.

Mansfield, J.E., Bennet, J.G., King, R.B., Lang, D.M., Lawton, R.M., 1975. Land Resources of Northern and Luapula Provinces of Zambia. A Reconnaissance Assessment, Vols. 3, 4 and 5. Ministry of Overseas Development, London.

Maranz, S., Wiesman, Z., 2003. Evidence for indigenous selection and distribution of shea tree, *Vitellaria paradoxa*, and its potential significance to prevailing parkland savanna tree patterns in sub-Saharan Africa north of the equator. J. Biogeogr. 30, 1505−1516.

Maroko, J., Buresh, R.J., Smithson, P.C., 1999. Soil phosphorus fractions in unfertilized fallow−maize systems in two tropical soils. Soil. Sci. Soc. Am. J. 63, 320−326.

Mason, P.A., Musoko, M.O., Last, F.T., 1992. Short-term changes in vesicular-arbuscular mycorrhizal spore populations in *Terminalia* plantations in Cameroon. In: Read, D.J., Lewis, D.H., Fitter, A.H., Alexander, I.J. (Eds.), Mycorrhizas in Ecosystems. CABI, Wallingford, UK, pp. 261−267. 1992.

Matin, M.A., 1989. Carbon economy during rooting of cutting of *Nauclea diderrichii* (de Wild & Th. Dur.) Merill. *M.Phil Thesis*. University of Edinburgh, p. 123.

Matos, J, Sousa, S, Wandelli, E, Perin, R, Arcoverde, M, Fernandes, E, 2003. Performance of big-leaf mahogany in agroforestry systems in the western Amazon region of Brazil. Proceedings of the International Conference on Big-leaf Mahogany: ecology, genetic resources and management. 22−24 October 1996, San Juan, Puerto Rico.

Matthews, R.B., Holden, S.T., Volk, J., Lungu, S., 1992. The potential of alley cropping in improvement of cultivation systems in the high rainfall areas of Zambia. I. Chitemene and Fundikila. Agrofor. Syst. 17, 219−240.

Mbile, P., Tchoundjeu, Z., Degrande, A., Asaah, E., Nkuinkeu, R., 2003. Mapping the biodiversity of 'Cinderella' trees in Cameroon. Biodiversity. 4, 17−21.

Mbile, P., Tchoundjeu, Z., Degrande, A., Avana, M.-L., Tsobeng, C., 2004. Non-mist vegetative propagation by resource-poor, rural farmers of the forest zone of Cameroon: Some technology adaptations to enhance practice. Forest Trees Livelihoods. 14, 43−52.

Mbofung, C.M.F., Silou, T., Mouragadja, I., 2002. Chemical characterization of safou (*Dacryodes edulis*) and evaluation of its potential as an ingredient in nutritious biscuits. Forest Tree Livelihoods. 12, 105−118.

Mbora, A., Lillesø, J-.P.B., 2007. Sources of tree seed and vegetative propagation of trees around Mt. Kenya, Development and Environment No. 9, World Agroforestry Center (ICRAF), Nairobi, Kenya, 58 pp.

Mbosso, C., Degrande, A., Villamor, G.B., van Damme, P., Tchoundjeu, Z., Tsafack, S., 2015. Factors affecting the adoption of agricultural innovation: the case of *Ricinodendron heudelotii* kernel extraction machine in southern Cameroon. Agrofor. Syst. 89, 799−811, http://dx.doi.org/10.1007/s10457-015-9813-y.

Mbosso, M.F.P.C., 1999. Opportunités et contraintes socio-économiques liées à la culture des arbres fruitiers locaux dans les basses terres humides du Cameroun: *Dacryodes edulis* et *Irvingia gabonensis*. Mémoire de fin d'études présenté pour l'obtention du diplome d'Ingenieur Agronome. Université de Dschang, 111p.

McCalla, A.F., Brown, L.R., 1999. Feeding the developing world in the next Millenium: A question of science? Proceedings of Conference on "Ensuring Food Security, Protecting the Environment, Reducing Poverty in Developing Countries. Can Biotechnology Help?", 21−22 October 1999. World Bank, Washington DC, 6p.

McHardy, T., 2002. Inventory of available marula resources on the Makhatini Flats, Maputaland, in the fruiting season of 2002, *Report to UK DFID Forestry Research Programme (Project No R7795)*. Institute of Natural Resources, Scottsville, South Africa, 14pp.

McIntyre, B.D., Herren, H., Wakhungu, J., Watson, R. (Eds.), 2009. International Assessment of Agricultural Knowledge Science and Technology for Development: Global Report. Island Press, New York, NY, 590pp.

McNeely, J.A., 2004. Nature vs. nurture: Managing relationships between forests, agroforestry and wild biodiversity. Agrofor. Syst. 61, 155–165.

McNeely, J.A., Scherr, S.J., 2001. Saving our Species: How the Earth's Biodiversity Depends on Progress in Agriculture. Future Harvest, Washington, DC.

McNeely, J.A., Scherr, S.J., 2003. Ecoagriculture: Strategies to Feed the World and Save Wild Biodiversity. Island Press, Washington DC, 323 pp.

McNeely, J.A., Schroth, G., 2006. Agroforestry and biodiversity conservation: traditional practices, present dynamics, and lessons for the future. Biodivers. Conserv. 15, 549–554.

MEA, 2005. Ecosystems and Human Well-Being, Vol 1. Current State and Trends. In: Hassan, R., Scholes, R., Ash, N. (Eds.), Millennium Ecosystem Assessment. Island Press, Washington DC, USA, 917 pp.

Meadows, D.H., Meadows, G., Randers, J., Behrens III, W.W., 1972. The Limits to Growth. Universe Books, New York.

Medaglia, J.C., 2009. Comments of UPOV. Secretariat of the Convention on Biological Diversity, January 2009. (www.upov.int/en/about/pdf/upov_-comments_medaglia_study_final.pdf).

Mekonnen, K., Buresh, R.J., Jama, B., 1997. Root and inorganic nitrogen distributions in Sesbania fallow, natural fallow and maize. Plant. Soil. 188, 319–327.

Melnyk, M., 1996. Indigenous enterprise for the domestication of trees and the commercialization of their fruits. In: Leakey, R.R.B., Temu, A.B., Melnyk, M., Vantomme, P. (Eds.), Domestication and Commercialization of Non-timber Forest Products in Agroforestry Systems. Non-Wood Forest Products No. 9. FAO, Rome, Italy, pp. 97–103.

Mesén, F.J., Newton, A.C., Leakey, R.R.B., Grace, J., 1997a. Vegetative propagation of *Cordia alliodora* (Ruiz and Pavon) Oken: the effects of IBA concentration, propagation medium and cutting origin. For. Ecol. Manage. 92, 45–54.

Mesén, F.J., Newton, A.C., Leakey, R.R.B., 1997b. The effects of propagation environment and foliar area on the rooting physiology of *Cordia alliodora* (Ruiz and Pavon) Oken cuttings. Trees. 11, 404–411.

Metzner, J.K., 1976. Lamtoronisasi, an experiment in soil conservation. Bull. Indones. Econ. Stud. 12, 103–109.

Mhango, J., Akinnifesi, F.K., Mng'omba, S.A., Sileshi, G., 2008. Effect of growing medium on early growth and survival of *Uapaca kirkiana* Müell Arg. seedlings in Malawi. Afr. J. nol. 7, 2197–2202.

Mialoundama, F., Avana, M.-L., Youmbi, E., Mampouya, P.C., Tchoundjeu, Z., Mbeuyo, M., et al., 2002. Vegetative propagation of *Dacryodes edulis* (G. Don) H.J. Lam by marcots, cuttings and micropropagation. Forests Trees Livelihoods. 12, 85–96.

Michon, G., de Foresta, H., 1995. The Indonesian agroforest model. Forest resource management and biodiversity conservation. In: Halliday, P., Gilmour, D.A. (Eds.), *Conserving Biodiversity Outside Protected Areas: The Role of* Traditional Agro-ecosystems. IUCN, Gland, Switz, pp. 90–106.

Michon, G., de Foresta, H., 1996a. The agroforest model as an alternative to the pure plantation model for domestication and commercialization of NTFP's. In: Leakey, R.R.B., Temu, A.B., Melnyk, M., Vantomme, P. (Eds.), Domestication and Commercialization of Non-timber Forest Products in Agroforestry Systems. Non-Wood Forest Products No. 9. FAO, Rome, Italy, pp. 160–175.

Michon, G., de Foresta, H., 1996b. Agroforests as an alternative to pure plantations for the domestication and commercialization of NTFPs. In: Leakey, R.R.B., Temu, A.B., Melnyk, M. (Eds.), *Domestication and Commercialiation of Non-timber Forest Products for Agroforestry*. NonWood forest products 9. Food and Agriculture Organization of the United Nations, Rome, pp. 160–175.

Minang, P.A., van Noordwijk, M., Freeman, O.E., Mbow, C., de Leeuw, J., Catacutan, D. (Eds.), 2015. Climate-Smart Landscapes: Multifunctionality In Practice. World Agroforestry Centre (ICRAF), Nairobi, Kenya.

Minang, PA, van Noordwijk, M, and Swallow, B, 2012. High-carbon-stock rural development pathways in Asia and Africa: How improved land management can contribute to economic development and climate change mitigation. In: Nair, P.K.R., Garrity, D., (Eds.), Agroforestry − The Future of Global Land Use. Adv. Agrofor., 9, 127–144. Springer, Dordrecht, Germany.

Mitschein, T.A., Miranda, P.S., 1998. POEMA: a proposal for sustainable development in Amazonia. In: Leihner, D.E., Mitschein, T.A. (Eds.), AThird Millenium for Humanity? The Search for Paths of Sustainable Development. Peter Lang, Frankfurt am Main, pp. 329–366.

Mizrahi, Y., Nerd, A., 1996. New crops as a possible solution for the troubled Israeli export market. In: Janick, J., Simon, J.E. (Eds.), Progress in new crops: Proceedings of the third national new crops symposium. American Society for Horticultural Science, pp. 46–64.

Mkonda, A., Lungu, S., Maghembe, J.A., Mafongoya, P.L., 2003. Fruit- and seed germination characteristics of *Strychnos cocculoides*, an indigenous fruit tree from natural populations in Zambia. Agrofor. Syst. 58, 25–31.

Mng'omba, S.A., Du Toit, E.S., Akinnifesi, F.K., Venter, H.M., 2007a. Repeated exposure of jacket plum (*Pappea capensis*) micro-cuttings to indole-3-butyric acid (IBA) improved in vitro rooting capacity. S. Afr. J. Bot. 73, 230–235.

Mng'omba, S.A., Du Toit, E.S., Akinnifesi, F.K., Venter, H.M., 2007b. Early recognition of graft compatibility in *Uapaca kirkiana* fruit tree clones, provenances and species using in vitro callus technique. Hortic. Sci. 43, 732–736.

Mng'omba, S.A., Akinnifesi, F.K., Mkonda, A., Mhango, J., Chilanga, T., Sileshi, G., et al., 2015. Ethnoecological knowledge for identifying elite phenotypes of the indigenous fruit tree, *Uapaca kirkiana* in the Miombo Woodlands of Southern Africa. Agroecol. Sustain. Food Syst. 39, 399–415.

Moe, R., Andersen, A.S., 1988. Stockplant environment and subsequent adventitious rooting. In: Davies, T.D., Haissig, B.E., Sankula, N. (Eds.), Adventitious Root Formation in Cuttings. Dioscorides, Portland, OR, pp. 214–234.

Moguel, P., Toledo, V.M., 1999. Biodiversity conservation in traditional coffee systems of Mexico. Conserv. Biol. 13, 11–21.

Molden, D. (Ed.), 2007. Water for Food, Water for life: A Comprehensive Assessment of Water Management in Agriculture. Earthscan, London and International Water Management Institute, Colombo.

Montagnini, F., Sancho, F., 1990. Impacts of native trees on tropical soils: a study in the Atlantic lowlands of Costa Rica. Ambio. 19, 386–390.

Morgan, D.C., Smith, H., 1976. Linear relationship between phytochrome photoequilibrium and growth in plants under simulated natural radiation. Nature. 262, 210−212.

Muchugi, A.M., 2001. Genetic variation in the threatened medicinal tree *Prunus africana* in Kenya and Cameroon: implications for the genetic management of the species. M. Sc. thesis. Kenyatta University, 90 pp.

Muchugi, A., Lengkeek, A.G., Kadu, C.A.C., Muluvi, G.M., Njagi, E.N.M., Dawson, I.K., 2006. Genetic variation in the threatened medicinal tree *Prunus africana* in Cameroon and Kenya: implications for current management and evolutionary history. S. Afr. J. Bot. 72, 498−506.

Muchugi, A., Muluvi, G.M., Kindt, R., Kadu, C.A.C., Simons, A.J., Jamnadass, R.H., 2008. Genetic structuring of important medicinal species of genus *Warburgia* as revealed by AFLP analysis. Tree Genet. Genomes. 4, 787−795.

Mudge, K.W., Brennan, E.B., 1999. Clonal propagation of multipurpose and fruit trees used in agroforestry. In: Buck, L.E., Lassoie, J.P., Fernandes, E.C.M. (Eds.), Agroforestry in Sustainable Ecosystems. CRC Press/Lewis Publishers, New York, pp. 157−190.

Muinga, R.W., Topps, J.H., Rooke, J.A., Thorpe, W., 1995. The effect of supplementation with *Leucaena leucocephala* and maize bran on voluntary food intake, digestibility, live weight and milk yield of *Bos indicus* × *Bos taurus* dairy cows and rumen fermentation in steers offered *Pennisetum purpureum ad libitum* in the semi-humid tropics. Anim. Sci. 50, 13−23.

Mulawarman, R.J.M., Sasongko, S.M., Iriantono, D., 2003. Tree seed management seed sources, seed collection and seed handling: a field manual for field workers and farmer. World Agroforestry Center-ICRAF and Winrock International, Bogor, Indonesia, 54 pp.

Mulyoutami, E., Roshetko, J.M., Martini, E., Awalina, D., Janudianto, 2015. Gender roles and knowledge in plant species selection and domestication: a case study in South and Southeast Sulawesi. Int. For. Rev. 17, 99−111.

Munjuga, M., Ofori, D.A., Asaah, E., Tchoundjeu, Z., Simons, A., Jamnadass, R., 2011. *Grafting Techniques of* Allanblackia *Species: Extension Guide*. World Agroforestry Centre, Nairobi, Kenya, p. 8.

Muñoz, D., Estrada, A., Naranjo, E., Ochoa, S., 2006. Foraging ecology of howler monkeys in a cacao (*Theobroma cacao*) plantation in Comalcalco, México. Am. J. Primatol. 68, 127−142.

Murdiyarso, D., Hariah, K., Husin, Y.A., Wasrin, U.R., 1996. Greenhouse gas emisions and carbon balance in slash and burn practices. In: van Noordwijk, M., Tomich, T.P., Garrity, D.P., Fagi, A.M. (Eds.), *Alternatives to Slash-and-burn in* Indonesia. AARD, Bogor, Indonesia, pp. 15−38.

Mwakalogho, R.J.M., 1986. A CB study of soil conservation in Malawi. *Cost-benefit Analysis of Soil and Water Conservation Projects. Land Utilization Programme Report 4*. SADCC, Maseru, Lesotho, pp. 35−42.

Mwamba, C.K., 1995. Natural variation in fruits of *Uapaca kirkiana* in Zambia. For. Ecol. Manage. 26, 299−303.

Mwase, W.F., Bjørnstad, Å., Ntupanyama, Y.M., Kwapata, M.B., Bokosi, J.M., 2006a. Phenotypic variation in fruit, seed and seedling traits of nine *Uapaca kirkiana* provenances found in Malawi. S. Afr. For. J. 208, 15−21.

Mwase, W.F., Bjornstad, A., Stedje, S., Bokosi, J.M., Kwapata, M.B., 2006b. Genetic diversity of *Uapaca kirkiana* Muel. Arg. Populations as revealed by amplified fragment length polymorphisms (AFLP). Afr. J. Biotechnol. 5, 1205−1213.

Mwase, W.F., Erik-Lid, S., Bjornstad, A., Stedje, A., Kwapata, M., Bokosi, J., 2010. Application of amplified fragment length polymorphism (AFLPs) for detection of sex-specific markers in diecious *Uapaca kirkiana* Müell. Årg. Afr. J. Biotechnol. 6, 137−142.

Mweta, D.E., Akinnifesi, F.K., Saka, J.D.K., Makumba, W., Chokotho, N., 2007. Green manure from prunings and mineral fertilizer affect phosphorus adsorption and uptake by maize crop in a gliricidia-maize intercropping. Sci. Res. Essay. 2, 446−453.

Myers, N., 1980. The present status and future prospects of tropical moist forests. Environ. Conserv. 7, 101−114.

Nair, P.K.R., 1989. Agroforestry defined. In: Nair, P.K.R. (Ed.), Agroforestry Systems in the Tropics. Kluwer Academic Publishers, Dordrecht, pp. 13−18.

Nair, P.K.R., 1990. *The Prospects for Agroforestry in the Tropics. World Bank Technical Paper Mo. 131*. The World Bank, Washington, DC.

Nair, PKR, 1993b. An Introduction to Agroforestry. Kluwer Academic Publishers, Dordrecht, The Netherlands, with ICRAF, Nairobi, Kenya.

Nair, PKR, 2012b. Climate change mitigation: a low-hanging fruit of agroforestry. In: Nair, P.K.R., Garrity, D., (Eds.), Agroforestry − The future of global land use, Adv. Agrofor., 9, 31−68. Springer, Dordrecht, Germany.

Nair, P.K.R., 2014. Agroforestry: practices and systems. In: van Alfen, N., et al., (Eds.), Encyclopedia of Agriculture and Food Systems, Vol 1. Elsevier Publishers, San Diego, pp. 270−282.

Namkoong, G., Kang, H.C., Brouard, J.S., 1988. Tree Breeding: Principles and Strategies. Springer-Verlag, New York.

Narendra, B.H., Mulyoutami, E., Roshetko, J.M., Tata, H.L., 2013. Prioritizing underutilized tree species for domestication in smallholder systems of West Java. Small-scale For. 12, 519−538.

National Academy of Sciences, 1975. Under-exploited Tropical Plants with Promising Economic Value. N.A.S., Washington D.C., 188 pp.

National Academy of Sciences, 1979. Tropical Legumes: Resources for the Future. N.A.S., Washington D.C., 331 pp.

Navarro, M., Malres, S., Labouisse, J.P., Roupsard, O., 2007. Vanuatu breadfruit project: survey on botanical diversity and traditional uses of *Artocarpus altilis*. Acta Hortic. 757, 81−88.

Naylor, R.L., 2014. The Evolving Sphere of Food Security. Oxford University Press, New York, 416pp.

Ndjouenkeu, R., Goycoolea, F.M., Morris, E.R., Akingbala, J.O., 1996. Rheology of okra (*Hibiscus esculentus* L.) and dika nut (*Irvingia gabonensis*) polysaccharides. Carbohydr. Polym. 29, 263−269.

Ndoye, O, 1995. The markets for non-timber forest products in the humid forest zone of Cameroon and its borders: Structure, conduct, performance and policy implications, Report to CIFOR, Bogor, Indonesia, 86 p.

Ndoye, O, Tchamou, N, unpublished. Utilization and marketing of *Irvingia garbonensis* in the humid forest zone of Cameroon. In: Boland, D., Ladipo, D.O. (Eds.), Proceedings of ICRAF Workshop, May 1994, Ibadan, Nigeria.

Ndoye, O., Ruiz-Perez, M., Ayebe, A., 1997. The Markets of Non-timber Forest Products in the Humid Forest Zone of Cameroon. *Rural Development Forestry Network, Network Paper* 22c. Overseas Development Institute, London, UK, p. 20.

Ndoye, O., Ruiz Pérez, M., Eyebe, A., 1998. The markets of non-timber forest products in the humid forest zone of Cameroon. Rural Development Forestry Network Paper 22c. ODI, London, UK.

Ndoye, O., Awono, A., Schreckenberg, K., Leakey, R.R.B., 2005. Commercializing indigenous fruit for poverty alleviation. A Policy Briefing Note for Governments in the African Humid Tropics Region. Overseas Development Institute, London.

Ndungu, J.N., Boland, D.J., 1994. *Sesbania sesban* collections in Southern Africa: developing a model for co-operation between a CGIAR Centre and NARS. Agrofor. Syst. 27, 129−143.

Ndungu, J., Jaenicke, H., Boland, D., 1995. Considerations for germplasm collection of indigenous fruit trees in the Miombo. In: Maghembe, J.A., Ntupanyama, Y., Chirwa, P.W. (Eds.), Improvement of Indigenous Fruit Trees of the Miombo Woodlands of Southern Africa. International Center for Research in Agroforestry, Nairobi, pp. 1−11.

Nelder, J.A., 1962. New kinds of systematic designs for spacing trials. Biometrics. 18, 283−307.

Nelson, E., Mendoza, G., Regetz, J., Polasky, S., Tallis, H., Cameron, D.R., et al., 2009. Modeling multiple ecosystem services, biodiversity conservation, commodity production and tradeoffs at landscape scales. Front. Ecol. Environ. 7, 4−11.

Nerd, A., Mizrahi, Y., 1993. Domestication and introduction of marula (*Sclerocarya birrea* subsp. *caffra*) as a new crop for the Negev desert of Israel. In: Janick, J., Simon, J.E. (Eds.), New Crops. Wiley, New York, pp. 496−499.

Nestel, D., Dickschen, F., Altieri, M.A., 1993. Diversity patterns of soil Coleoptera in Mexican shaded and unshaded coffee agroecosystems: an indication of habitat perturbation. Biodivers. Conserv. 2, 70−78.

Nestel, P., Bouis, H.E., Meenakshi, J.V., Pfeiffer, W., 2006. Biofortification of staple food crops. J. Nutr. 136, 1064−1067.

Nevenimo, T., Johnston, M., Binifa, J., Gwabu, C., Anjen, J., Leakey, R.R.B., 2008. Domestication potential and marketing of *Canarium indicum* nuts in the Pacific: producer and consumer surveys in Papua New Guinea (East New Britain). Forest Tree Livelihoods. 18, 253−269.

Nevenimo, T., Moxon, J., Wemin, J., Johnston, M., Bunt, C., Leakey, R.R.B., 2007. Domestication potential and marketing of *Canarium indicum* nuts in the Pacific: 1. A literature review. Agrofor. Syst. 69, 117−134.

Newton, A.C., Jones, A.C., 1993. Characterization of microclimate in mist and non-mist propagation systems. J. Hortic. Sci. 68, 421−430.

Newton, A.C., Leakey, R.R.B., Mesén, J.F., 1993. Genetic variation in Mahoganies: its importance, capture and utilization. Biodivers. Conserv. 2, 114−126.

Newton, A.C., Moss, R., Leakey, R.R.B., 1994. The hidden harvest of tropical forests: domestication of non-timber products. Ecodecis. Environ. Policy Mag. 13 (July), 48−52.

Newton, A.C., Dick, J.Mc.P., McBeath, C., Leakey, R.R.B., 1996. The influence of R:FR ratio on the growth, photosynthesis and rooting ability of *Terminalia spinosa* Engl. and *Triplochiton scleroxylon* K. Schum. Ann. Appl. Biol. 128, 541−556.

Nghitoolwa, E., Hall, J.B., Sinclair, F.L., 2003. Population status and gender imbalance of the marula tree, *Sclerocarya birrea* subsp. *caffra*, in northern Namibia. Agrofor. Syst. 59, 289−294.

Ngo Mpeck, M.L., Tchoundjeu, Z., Asaah, E., 2003a. Vegetative propagation of *Pausinystalia johimbe* (K. Schum) by leafy stem cuttings. Propagat. Ornament. Plants. 3, 11−18.

Ngo Mpeck, M.L., Asaah, E., Tchoundjeu, Z., Atangana, A.R., 2003b. Strategies for the domestication of *Ricinodendron heudelotii*: evaluation of variability in natural populations from Cameroon. Agric. Environ. 1, 257−262.

Ngulube, M.R., 1995. Indigenous fruit trees in southern Africa: the potential of *Uapaca kirkiana*. Agrofor. Today. 7 (3-4), 17−18.

Niang, A., Amadalo, B., Gathumbi, S., 1996a. Green manure from the roadside, Miti Ni Maendeleo 2, 10. Maseno Agroforestry Research Centre, Kisumu, Kenya.

Niang, A.I., Ugeziwe, J., Styger, E., Gahamanyi, A., 1996b. Forage potential of eight woody species: Intake and growth rates for local young goats in the highland region of Rwanda. Agrofor. Syst. 34, 171−178.

Nin-Pratt, A., McBride, L., 2014. Agricultural intensification in Ghana: evaluating the optimist's case for a Green Revolution. Food Policy. 48, 153−167.

Njenga, A., Wesseler, J., 1999. Participatory farming systems analysis of tree based farming systems in Southern Malawi. GTZ−ITFSP Project, Nairobi, Kenya, 34 pp.

Njoroge, M, Rao, MR, 1994. Barrier hedgerow intercropping for soil and water conservation on sloping lands. Paper Presented at the 8th International Soil Conservation Organization Conference. Soil and Water Conservation: Challenges and Opportunities. 4−8 December, 1994, New Delhi, India.

Nkonde, C, Jayne, TS, Richardson, RB, Place, F, 2015. Testing the farm-size productivity relationship over a wide range of farm sizes: Should the relationship be a decisive factor in guiding agricultural development and land policies in Zambia. In: Linking Land Tenure and Use for Shared Prosperity, Proc. 2015 World Bank 2015 Conference on Land and Poverty, 23−27 March 2015, Washington DC, USA.

Nkongmeneck, B., 1985. Le genre *Cola* au Cameroun. Revues Sciences et Techniques, Serie de la Science Agronomique. 1 (3), 57−70, Yaounde, Cameroon.

Nono-Wombin, R, Ojiewo, C, Abang, M, Oluoch, MO, 2012. Good Agricultural Practices for African Indigenous Vegetables, Scripta Horticulturae No 15, 248pp.

Nordeide, M.B., Hatløy, A., Følling, M., Lied, E., Oshaug, A., 1996. Nutrient composition and nutritional importance of green leaves and wild food resources in an agricultural district, Koutiala, in Southern Mali. Int. J. Food Sci. Nutr. 47, 455−468.

Norgrove, L., Csuzdi, C., Forzi, F., Canet, M., Gounes, J., 2009. Shifts in soil faunal community structure in shaded cacao agroforests and consequences for ecosystem function in Central Africa. Trop. Ecol. 50, 71−78.

Novick, R.R., Dick, C.W., Lemes, M.R., Navarro, C., Caccone, A., Bermingham, E., 2003. Genetic structure of Mesoamerican populations of Big-leaf mahogany (*Swietenia macrophylla*) inferred from microsatellite analysis. Mol. Ecol. 12, 2885−2893.

Nyamulinda, V., 1991. Erosion sans cultures de versant et transports solides dans les bassin de drainage des hautes terres de Ruhengeri au Rwanda. Reseau Eros. Bull. 11, 38−63.

Nyathi, P., Campbell, B.M., 1993. The acquisition and use of miombo litter by small scale farmers in Masvingo, Zimbabwe. Agrofor. Syst. 22, 43−48.

Nygren, S.E.P., Fernández, M., Harmand, J.-M., Leblanc, H.A., 2012. Symbiotic dinitrogen fixation by trees: an underestimated resource in agroforestry systems? Nutr. Cycl. Agroecosyst. 94, 123−160.

Nyoka, B.I., Mng'omba, S.A., Akinnifesi, F.K., Ajayi, O.C., Sileshi, G., Jamnadass, R., 2011. Agroforestry tree seed production and supply systems in Malawi. Small Scale For. Available from: http://dx.doi.org/10.1007/s11842-011-9159-x.

Nzilani, N.J., 1999. The status of *Prunus africana* in Kakamega Forest and the prospects for its vegetative propagation. M. Phil. thesis. Moi University, Kenya, 78 pp.

O'Neill, G.A., Dawson, I.K., Sotelo Montes, C., Guarino, L., Current, D., Guariguata, M., et al., 2001. Strategies for genetic conservation of trees in the Peruvian Amazon basin. Biodivers. Conserv. 10, 837−850.

Obasi, N.B.B., Okolie, N.P., 1993. Nutritional constituents of the seeds of the African pear, Dacryodes edulis. Food Chem. 46, 297−299.

Obizoba, I.C., Amaechi, N.A., 1993. The effect of processing methods on the chemical composition of baobab (*Adansonia digitata*.L.) pulp and seed. Ecol. Food Nutr. 29, 199−205.

Odetokun, S.M., 1996. The nutritive value of baobab fruit (*Andasonia [sic] digitata*). Riv. Ital. Sost. Gras. 23, 371−373.

OED, 1989. In: Simpson, J.A., Weiner, E.S.C. (Eds.), Oxford English Dictionary, 2nd Edition. Clarendon Press, Oxford.

Ofori, D.A., Newton, A.C., Leakey, R.R.B., Grace, J., 1996a. Vegetative propagation of *Milicia excelsa* Welw. by leafy stem cuttings. Effects of auxin concentration, leaf area and rooting medium. Forest Ecol. Manage. 84, 39−48.

Ofori, D.A., Newton, A.C., Leakey, R.R.B., Cobbinah, J.R., 1996b. Vegetative propagation of *Milicia excelsa* Welw. by root cuttings. J. Trop. Forest Sci. 9, 124−127.

Ofori, D.A., Newton, A.C., Leakey, R.R.B., Grace, J., 1997. Vegetative propagation of *Milicia excelsa* Welw. by leafy stem cuttings. I. Effects of maturation, coppicing, cutting length and position on rooting ability. J. Trop. For. Sci. 10, 115−129.

Ofori, D.A., Gyau, A., Dawson, I.K., Asaah, E., Tchoundjeu, Z., Jamnadass, R., 2014. Developing more productive African agroforestry systems and improving food and nutritional security through tree domestication. Curr. Opin. Environ. Sustain. 6, 123−127.

Ofori, D.A., Asomaning, J.M., Peprah, T., Agyeman, V.K., Anjarwalla, P., Tchoundjeu, Z., et al., 2015. Addressing constraints in propagation of *Allanblackia* species through seed sectioning and air layering. J. Exp. Biol. Agric. Sci. 3, 89−96.

Okafor, J.C., 1974. Varietal delimitation in *Irvingia gabonensis* (Irvingiaceae). Niger. J. For. 4, 80−87.

Okafor, J.C., 1978. Development of forest tree crops for food supplies in Nigeria. Forest Ecol. Manage. 1, 235−247.

Okafor, J.C., 1980. Edible indigenous woody plants in the rural economy of the Nigerian forest zone. Forest Ecol. Manage. 3, 45−55.

Okafor, J.C., 1983. Varietal delimitation in *Dacryodes edulis* (G. Don) H.J. Lam. (Burseraceae). Int. Tree Crops J. 2, 255−265.

Ogbobe, O., 1992. Physiochemical composition and characterization of the seed and oil of *Sclerocarya birrea*. Plant Foods Hum. Nutr. 42, 201−206.

Okafor, J.C., Fernandes, E.C.M., 1987. Compound farms of southeastern Nigeria: a predominant agroforestry homegarden system with crops and small livestock. Agrofor. Syst. 5, 153−168.

Okafor, J.C., Lamb, A., 1994. Fruit trees: diversity and conservation strategies. In: Leakey, R.R.B., Newton, A.C. (Eds.), Tropical Trees: Potential for Domestication and the Rebuilding of Forest Resources. HMSO, London, pp. 34−41.

Oke, O.L., Umoh, I.B., 1978. Lesser known oilseeds. I. Chemical composition. Nutr. Rep. Int. 17, 293−297.

Okigbo, BN, 1981. Plants and agroforestry in land use systems of West Africa - a conceptual framework for planning and establishment of research priorities. In: Proceedings of ICRAF Consultative Meeting on Plant Research and Agroforestry, Nairobi, April 1981.

Okolo, HC, unpublished. Industrial potential of various *Irvingia gabonensis* products such as oil, Ogbono and juice, In: Boland, D., Ladipo, D.O., (Eds), Irvingia: Uses, Potential and Domestication, ICRAF, Nairobi, Kenya.

Okorio, J., Byenkya, S., Wajja, N., Peden, D., 1994. Comparative performance of seventeen upperstorey tree species associated with crops in the highlands of Uganda. Agrofor. Syst. 26, 1−19.

Omode, A.A., Fatoki, O.S., Olaogun, K.A., 1995. Physicochemical properties of some underexploited and nonconventional oilseeds. J. Agric. Food Chem. 43, 2850−2853.

O'Neill, MK (Compiler), 1993. KARI Regional Research Centre, Embu: annual report, March 1992−April 1993. AFRE.NA Report No. 69. Nairobi: ICRAF.

O'Neill, MK, Muriithi, FM, 1994. Local adoption of agroforestry technologies in the Mt Kenya highlands. A.S.A. Abstracts.

Ong, C.K., 1994. The 'dark side' of intercropping: manipulation of soil resources. Ecophysiology of Tropical Intercropping. Institut National de la Recherche Agronomique, Paris.

Ong, C.K., Huxley, P.A. (Eds.), 1996. Tree−crop Interactions, a Physiological Approach. CAB International, Wallingford, UK.

Ong, C.K., Leakey, R.R.B., 1999. Why tree-crop interactions in agroforestry appear at odds with tree-grass interactions in tropical savannahs. Agrofor. Syst. 45, 109−129.

Ong, C.K., Black, C.R., Wilson, J., Muthuri, C., Bayala, J., Jackson, N.A., 2014. Agroforestry: Hydrological Impacts. In: van Alfen, N. (Ed.), Encyclopedia of Agriculture and Food Systems, Vol. 1. Elsevier, San Diego, pp. 244–252.

Onyekwelu, J.C., Olusola, J.A., Stimm, B., Mosandl, R., Agbelade, A.D., 2014. Farm-level tree growth characteristics, fruit phenotypic variation and market potential assessment of three socio-economically important forest fruit tree species. For. Trees Livelihoods. 24, 27–42.

Otani, T., Ae, N., Tanaka, H., 1996. Phosphorus (P) uptake mechanisms of crops grown in soils with low P status. II. Significance of organic acids in root exudates of pigeonpea. Soil Sci. Plant Nutr. 42, 553–560.

Otsyina, R, Asenga, D, 1993. Ngitiri: a traditional Sukuma silvopastoral system. Survey report. Tanzania/ICRAF Agroforestry project, Shinyanga, Tanzania.

Otsyina, R, Hanson J, Akyeampong, E, 1994a. Potential for development and research for leucaena in East Africa. Paper presented at the International Leucaena Workshop, Bogor, Indonesia, 25–29 January 1994.

Otsyina, R., Msangi, R., Gama, B., Nyadzi, G., Madulu, J., 1994b. *SADC-FCRAF Agroforestry Research Project Tumbi, Tabora, Tanzania. Annual Progress Report. AFRENA Report No. 83.* ICRAF, Nairobi.

Page, T., 2003. The domestication and improvement of *Kunzea pomifera* (F. Muell.), Muntries. *PhD Thesis. University of Melbourne*, Australia, p. 402.

Page, T., Potrawiak, A., Berry, A., Tate, H., Tungon, J., Tabi, M., 2010a. Production of sandalwood (*Santalum austrocaledonicum*) for improved smallholder incomes in Vanuatu. Forest Tree Livelihoods. 19, 299–316.

Page, T., Southwell, I., Russell, M., Tate, H., Tungon, J., Sam, C., et al., 2010b. Geographic and phenotypic variation in heartwood and essential oil characters in natural populations of *Santalum austrocaledonicum* in Vanuatu. Chem. Biodivers. 7, 1990–2006.

Pagiola, S., 1994. Soil conservation in a semi-arid region of Kenya: rates of return and adoption by farmers. In: Napier, T.L., Camboni, S.M., El-Swaify, S.A. (Eds.), *Adopting Conservation on the* Farm. Soil and Water Conservation Society, Alkeny, Iowa, pp. 171–187.

Palm, C.A., 1995. Contribution of agroforestry trees to nutrient requirements of intercropped plants. Agrofor. Syst. 30, 105–124.

Palm, C.A., 1996. Contribution of agroforestry trees to nutrient requirements of intercropped plants. Agrofor. Syst. 30, 105–124.

Palm, C.A., Sanchez, P.A., 1991. Nitrogen release from the leaves of some tropical legumes as affected by their lignin and polyphenolic contents. Soil. Biol. Biochem. 23, 83–88.

Palm, C.A., Myers, R.J.K., Nandwa, S.M., 1997. Organic-inorganic nutrient interaction in soil fertility replenishment. In: Buresh, R.J., Sanchez, P.A., Calhoun, F. (Eds.), Replenishing Soil Fertility in Africa. Special Publication 51. Soil Science Society of America, Madison, WI, USA, pp. 193–218.

Palmer, B., Minson, D.J., unpublished. The in sacco digestibility of the tropical shrub legume *Calliandra calothyrsus* as affected by the diet of the fistulated animal, temperature of drying and source of rumen inoculation.

Palmer, B., Schink, A.C., 1992. The effect of drying on the intake and rate of digestion of the shrub legume *Calliandra calothvrsus*. Trop. Grassl. 26, 89–93.

Pandey, D.N., 2002. Carbon sequestration in agroforestry systems. Clim. Policy. 2, 367–377.

Panik, F., 1998. The use of biodiversity and implications for industrial production. In: Leihner, D.E., Mitschein, T.A. (Eds.), A Third Millennium for Humanity? The Search for Paths of Sustainable Development. Peter Lang, Frankfurt am Main, pp. 59–73.

Paningbatan, E.P., 1990. Alley cropping for managing soil erosion on sloping lands. Trans. 14th Int. Congr. Soil Sci. 7, 376–377.

Pardee, G.L., Philpott, S.M., 2011. Cascading indirect effects in a coffee agroecosystem: effects of parasitic phorid flies on ants and the coffee berry borer in a high-shade and low-shade habitat. Environ. Entomol. 40, 581–588.

Parker, I.M., López, I., Petersen, J.J., Anaya, N., Cubilla-Rios, L., Potter, D., 2010. Domestication syndrome in Caimito (*Chrysophyllum cainito* L.): fruit and seed characteristics. Econ. Bot. 64, 161–175.

Patel, S.H., Pinckney, T.C., Jaeger, W.K., 1995. Smallholder wood production and population pressure in East Africa: evidence of an environmental Kuznets curve? Land. Econ. 71, 516–530.

Patterson, H.D., Thompson, R., 1971. Recovery of inter-block information when block sizes are unequal. Biometrika. 58, 545–554.

Pauku, R.L., 2005. Domestication of Indigenous Nuts for Agroforestry in the Solomon Islands. *PhD Thesis. James Cook University*, Cairns, Queensland, Australia, 381p.

Pauku, R.L., Lowe, A., Leakey, R.R.B., 2010. Domestication of indigenous fruit and nut trees for agroforestry in the Solomon islands. Forest Tree Livelihoods. 19, 269–287.

Peay, K.G., Kennedy, P.G., Bruns, T.D., 2011. Rethinking ectomycorrhizal succession: Are root density and hyphal exploration types drivers of spatial and temporal zonation? Fungal Ecol. 4, 233–240.

Pélé, J., Berre, S., 1967. Les aliments d'origine végétale au Caméroun. Cam. Agric., Past. For. ORSTOM, Paris, France, pp. 108–111.

Pelleck, R., 1992. Contour hedgerows and other soil conservation interventions for the hilly terrain of Haiti. In: Tato, K., Hurni, H. (Eds.), Soil Conservation for Survival. Soil and Water Conservation Society, Ankeny, Iowa, pp. 313–320.

Peng, J., Wang, Y., Wu, J., Jing, J., Ye, M., 2011. The contribution of landscape ecology to sustainable land use research. Environ. Dev. Sustain. 13, 953.

Pennington, T.D., Robinson, R.K., 1998. Utilization profile of a new species: *Inga ilta* T.D. Penn. In: Pennington, T.D., Fernandes, E. (Eds.), The Genus Inga: Utilization. Royal Botanic Garden, Kew and ICRAF, Nairobi, Kenya.

Peprah, T., Ofori, D.A., Siaw, D.E.K.A., Addo-Danso, S.D., Cobbinah, J.R., Simons, A.J., et al., 2009. Reproductive biology and characterization of *Allanblackia parviflora* A. Chev. in Ghana. Genet. Resour. Crop Evol. 56, 1037–1044.

Perera, A.N.F., Yaparatne, V.M.K., van Bruchem, J., 1992. Characterisation of protein in some Sri Lankan tree fodders and agro-industrial by-products by nylon bag degradation studies. In: Ibrahim, M.N., de Jong, R., van Bruchem, J., Purnamo, H. (Eds.), Livestock and Feed Development in the Tropics. Agricultural University of Wageningen, Brawijaya University, Malang and Commission of European Communities, Malang, Indonesia, pp. 171−177.

Perfecto, I., Snelling, R., 1995. Biodiversity and the transformation of a tropical agroecosystem: ants in coffee plantations. Ecol. Appl. 5, 1084−1097.

Perfecto, I., Vanderdeer, J., 2008. Biodiversity conservation and tropical agroecosystems: a new conservation paradigm. Ann. N.Y. Acad. Sci. 1134, 173−200.

Perfecto, I., Vandermeer, J., 2010. The agroecological matrix as alternative to the land sparing/agricultural intensification model. Proc. Natl. Acad. Sci. USA. 107, 5786−5791.

Perfecto, I., Rice, R., Greenberg, R., van der Voorst, M.E., 1996. Shade coffee: a disappearing refuge for biodiversity. Bioscience. 46, 598−608.

Perfecto, I., Vandermeer, J., Hanson, P., Cartin, V., 1997. Arthropod biodiversity loss and the transformation of a tropical agroecosystem. Biodivers. Conserv. 6, 935−945.

Perfecto, I., Vandermeer, J.H., Bautista, G.L., Nunez, G.I., Greenberg, R., et al., 2004. Greater predation in shaded coffee farms: the role of resident neotropical birds. Ecology. 85, 2677−2681.

Perfecto, I., Vandermeer, J., Philpott, S.M., 2014. Complex ecological interactions in coffee agroecosystems. Annu. Rev. Ecol. Evol. Syst. 45, 137−158.

Peters, V.E., Greenberg, R., 2013. Fruit supplementation affects birds but not arthropod predation by birds in Costa Rican agroforestry systems. Biotropica. 45, 102−110.

Peters, C.M., Gentry, A.H., Mendelsohn, R.O., 1989a. Valuation of an Amazonian rainforest. Nature. 339, 655−656.

Peters, C.M., Gentry, A., Mendelson, R., 1989b. Valuation of a tropical forest in Peruvian Amazonia. Nature. 339, 655−657.

Phalan, B., Onial, M., Balmford, A., Green, R.E., 2011. Reconciling food production and biodiversity conservation: land sharing and land sparing compared. Science. 333, 1289−1291.

Philpott, S.M., Bichier, P., 2012. Effects of shade tree removal on birds in coffee agroecosystems in Chiapas, Mexico. Agric. Ecosyst. Environ. 149, 171−180.

Phiri, D.M., Coulman, B., Steppler, H.A., Kamara, C.S., Kwesiga, F., 1992. The effect of browse supplementation on maize husk utilization by goats. Agrofor. Syst. 17, 153−158.

Phofeutsile, K., Jaenicke, H., Muok, B., 2002. Propagation and management. In: Hall, J.B., O'Brien, E.M., Sinclair, F.L. (Eds.), Sclerocarya birrea: A Monograph. School of Agricultural and Forest Sciences Publication No. 19, University of Wales, Bangor, UK, pp. 93−106.

Pingali, P.L., 2012. Green revolution: Impacts, limits and the path ahead. PNAS. 109, 12302−12308.

Place, F., 1995. The role of land and tree tenure on the adoption of agroforestry technologies in Zambia, Burundi, Uganda and Malawi: a summary and synthesis. Land Tenure Center, University of Wisconsin, Madison.

Place, F. (Ed.), 1996. Towards improved policy making for natural resources and ecosystem management in sub-Saharan Africa. ICRAF, Nairobi, Kenya.

Place, F., Otsuka, K., 1997. Population density, land tenure and resource management in Uganda. ICRAF and IFPRI, Nairobi, Kenya.

Place, F., Otsuka, K., 2000. Population pressure, land tenure, and tree resource management in Uganda. Land Econ. 76, 233−251.

Poch, T.J., Simonetti, J.A., 2013. Ecosystem services in human-dominated landscapes: insectivory in agroforestry systems. Agrofor. Syst. 87, 871−879.

Poppy, G.M., Chiotha, S., Eigenbrod, F., Harvey, C.A., Honzák, M., Hudson, M.D., et al., 2014. Food security in a perfect storm: using the ecosystem services framework to increase understanding. Phil. Trans. R. Soc. B. 369, 000-000. http://dx.doi.org/10.1098/rstb.2012.0288 369(1639), 20120288.

Poth, M., La Favre, J.S., Focht, D.D., 1986. Quantification by direct ^{15}N dilution of fixed N_2 incorporation into soil by *Cajanus cajan* (pigeon pea). Soil Biol. Biochem. 18, 125−127.

Poulsen, A., Andersen, A.S., 1980. Propagation of *Hedera helix;* influence of irradiance to stockplants, length of internode and topophysis of cutting. Physiol. Plant. 49, 359−365.

Poulton, C., Poole, N., 2001. Poverty and Fruit Tree Research: Issues and Options Paper. DFID Forestry Research Programme, Wye College, Ashford, England, 70p. < www.nrinternational.co.uk/forms2/frpzf0141b.pdf> .

Pramono, A.A., Fauzi, M.A., Widyani, N., Heriansyah, I., Roshetko, J.M., 2011. Management of community teak forests: a field manual for farmers. CIFOR, ICRAF, and FORDA, Bogor.

Prance, G.T., 1994. Amazonian tree diversity and the potential for supply of non-timber forest products. In: Leakey, R.R.B., Newton, A.C. (Eds.), Tropical Trees: The Potential for Domestication and the Rebuilding of Forest Resources. HMSO, London, UK, pp. 7−15.

Prastowo, N.H., Roshetko, J.M., Manurung, G.E.S., Nugraha, E., Tukan, J.M., Harum, F., 2006. Tehnik pembibitan dan perbanyakan vegetatif tanaman buah. ICRAF and Winrock International, Bogor (Indonesia), 100 pp.

Pretorius, V., Rohwer, E., Rapp, A., Holtzhausen, L.C., Mandery, H., 1985. Volatile flavour components of marula juice. Z. Lebensm. Unter.-Forsch. 181, 458−461.

Pretty, J., 2008. Agricultural sustainability: Concepts, principles and evidence. Philos. Trans. R. Soc. B. 363, 447−465.

Pretty, J., 2013. The consumption of a finite planet: well-being, convergence, divergence and the nascent green economy. Environ. Resour. Econ. 55, 475−499.

Pretty, J., Bharucha, Z.P., 2014. Sustainable intensification in agricultural systems. Ann. Bot. 114, 1571−1596.

Pretty, J., Noble, A.D., Bossio, D., Dixon, J., Hine, R.E., Penning de Vries, F.W.T., et al., 2006. Resource-conserving agriculture increases yields in developing countries. Environ. Sci. Technol. 3, 24–43.

Pretty, J., Toulmin, C., Williams, S., 2011. Sustainable intensification in African agriculture. Int. J. Agric. Sustain. 9, 5–24.

Purnomosidhi, P., Suparman, Roshetko, J.M., Mulawarman, 2007. Perbanyakan dan budidaya tanaman buah-buahan: Pedoman Lapang edisi kedua. World Agroforestry Center-ICRAF, SEA Regional Office and Winrock International, Bogor, Indonesia, 51 pp.

Ræbild, A., Larsen, A.S., Jensen, J.S., Ouedraogo, M., de Groote, S., van Damme, P., et al., 2011. Advances in domestication of indigenous fruit trees in the west African Sahel. New Forests. 41, 297–315.

Ragone, D., 1997. Breadfruit, *Artocarpus altilis* (Parkinson) Fosberg, Promoting the Conservation and Use of Underutilized and Neglected Crops Series, vol 10. Institute of Plant Genetics and Crop Plant Research, Gatersleben/International Plant Genetic Resources Institute, Rome, 77 pp.

Raintree, J.B., 1987a. The state of the art of agroforestry diagnosis and design. Agrofor. Syst. 5, 219–250.

Raintree, JB, 1987b. Frontiers of agroforestry diagnosis and design. Washington State University, Agroforestry Consortium. Distinguished Lecture Series 1987.

Rao, M.R., 1994. Agroforestry for sustainable soil management in humid and sub-humid tropical Africa. In: de Pauw, E. (Ed.), *Strategies for the Management of Upland Soils of Humid Tropical Africa. IBSRAM Network Document No. 9*. IBSRAM, Bangkok, Thailand.

Rao, M.R., Kamara, C.S., Kwesiga, F., Duguma, B., 1990. Methodological issues for research on improved fallows. Agrofor. Today. 2 (4), 8–9.

Rao, M.R., Nair, P.K.R., Ong, C.K., 1998. Biophysical interactions in tropical agroforestry. Agrofor. Syst. 53, 3–50.

Rao, MR, Palada, MC, Becker, BN, 2004. Medicinal and aromatic plants in agroforestry systems. In: Nair, P.K.R., Rao, M.R., Buck, L.E. (Eds.), New Vistas in Agroforesty: A Compendium for 1st World Congress of Agroforestry. Agrofor. Syst. 61, 107–122.

Rasmusson, D.C., 1987. An evaluation of ideotype breeding. Crop Sci. 27, 1140–1146.

Read, M.D., Kang, B.T., Wilson, G.F., 1985. Use of *Leucaena leucocephala* (Lam. de Wit) leaves as a nitrogen source for crop production. Fertil. Res. 8, 107–116.

Read, P.E., Preece, J.E., 2014. Cloning plants: micropropagation and tissue culture. In: van Alfen, N. (Ed.), Encyclopedia of Agriculture and Food Systems, Vol. 2. Elsevier, San Diego, pp. 317–336.

Reardon, T., Timmer, C.P., Barrett, C.B., Berdegue, J., 2003. The rise of supermarkets in Africa, Asia and Latin America. Am. J. Agric. Econ. 85, 1140–1146.

Reid, J.S.G., Edwards, M.E., 1995. In: Stephen, A.M. (Ed.), Food Polysaccharides and their Applications. Marcel Dekker Inc, New York, pp. 155–186.

Reitsma, R., Parrish, J.D., McLarney, W., 2001. The role of cacao plantations in maintaining forest avian diversity in southeastern Costa Rica. Agrofor. Syst. 53, 185–193.

Reynolds, L., Bimbuzi, S., 1993. The Leucaena psyllid in Coast Province, Kenya. Nitrogen Fixing Tree Res. Rep. 11, 103.

Reynolds, L., de Leeuw, P., 1994. Myth and manure in nitrogen cycling: a case study of Kaloneni division in Coast Province, Kenya. In: Powell, J.M., Fernandez-Rivera, S., Williams, T.O., Renard, C. (Eds.), Livestock and Sustainable Nutrient Cycling in Mixed Farming Systems of Sub-Saharan Africa. ILCA, Addis Ababa, Ethiopia.

Reynolds, L., Ekurukwe, J.O., 1988. Effect of *Trypanosoma vivax* infection on West African Dwarf sheep at two planes of nutrition. Small Rumin. Res. 1, 175–188.

Reynolds, L., Jabbar, M., 1994. The role of alley farming in African livestock production. Outlook Agric. 23, 105–113.

Reynolds, L, 1994. A review of the biophysical and socio-economic basis of alley farming. Consultants Report for AFNETA, August 1994.

Robbins, S.R.J., Matthews, W.S.A., 1974. Minor forest products; their total value is of major order. Unasylva. 26 (106), 7–14.

Roberts, D.L., Cooper, R.J., Petit, L.J., 2000a. Flock characteristics of ant-following birds in premontane moist forest and coffee agroecosystems. Ecol. Appl. 10, 1414–1425.

Roberts, D.L., Cooper, R.J., Petit, L.J., 2000b. Use of premontane moist forest and shade coffee agroecosystems by army ants in Western Panama. Conserv. Biol. 14, 192–199.

Roche, LA, Wyn Jones, RG, Retallick, SJ, 1989. A comparative study in Africa of agroforestry trees on vertisolic soils under low rainfall. Annual Progress Report to ODA 1988–89, Project R4181, University College of North Wales, 13 pp.

Rockström, J., Steffen, W., Noone, K., Persson, Å., Chapin III, F.S., Lambin, E.F., et al., 2009. Planetary boundaries: exploring the safe operating space for humanity. Ecol. Soc. 14 (2), 32.

Rojas, J., Godoy, C., Hanson, P., Hilje, L., 2001a. A survey of homopteran species (Auchenorrhyncha) from coffee shrubs and poró and laurel trees in shaded coffee plantations in Turrialba, Costa Rica. Rev. Biol. Trop. 49, 1057–1065.

Rojas, J., Godoy, C., Hanson, P., Kleinn, C., Hilje, L., 2001b. Hopper (Homoptera: Auchenorrhyncha) diversity in shaded coffee systems of Turrialba, Costa Rica. Agrofor. Syst. 53, 171–177.

Rosenstock, TS, Mpanda, M, Kimaro, A, Luedeling, E, Kuyah, S, Anyekulu, E, et al., 2015. Science to support climate-smart agricultural development: Concepts and results from the MICCA pilot projects in East Africa, Mitigation of Climate Change in Agriculture Series, 10, FAO Rome 47pp.

Roshetko, JM, Evans, DO, 1999. Domestication of Agroforestry Trees in Southeast Asia. Farm, Forest and Community Tree Research Reports, Special Issue, Winrock International (FACT Net), Morrilton, Arkansas, 242p.

Roshetko, J.M., Suarna, M., Nitis, M.I., Lana, K., Sukanten, W., Puger, A.W., 1999c. *Gliricidia sepium* seed production under indigenous lopping and planting systems in a seasonally dry area of Bali, Indonesia. For. Farm Comm. Tree Res. Rep. 4, 83–88.

Roshetko, J.M., Mulawarman, Purnomosidhi, P., 2004a. *Gmelina arborea*: a viable species for smallholder tree farming in Indonesia? New Forest. 28, 207−215.

Roshetko, J.M., Purnomosidhi, P., Mulawarman, 2004b. Farmer Demonstration Trials (FDTs): promoting tree planting and farmer innovation in Indonesia. In: Gonsalves, J., Becker, T., Braun, A., Caminade, J., Campilan, D., de Chavez, H., Fajber, E., Kapiriri, M., Vernooy, R. (Eds.), Participatory Research and Development for Sustainable Agriculture and Natural Resource Management: A Sourcebook. International Potato Center-Users' Perspectives With Agricultural Research and Development/ International Development Research Center/International Fund for Agricultural Development, Laguna/Ottawa/Rome, pp. 384−392.

Roshetko, J.M., Lasco, R.D., Delos Angeles, M.D., 2007. Smallholder agroforestry systems for carbon storage. Mitig. Adapt. Strat. Global Change. 12, 219−242.

Roshetko, J.M., Mulawarman, Dianarto, A., 2008. Tree seed procurement-diffusion pathways in Wonogiri and Ponorogo, Java. Small Scale For. 7, 333−352.

Roshetko, J.M., Tolentino Jr., E.L., Carandang, W.M., Bertomeu, M., Tabbada, A.U., Manurung, G., et al., 2010. Tree nursery sourcebook. Options in support of sustainable development. World Agroforestry Center (ICRAF), SEA Regional Office and Winrock International, Bogor, Indonesia, 52 pp.

Rowell, D., 2014. Soil Science: Methods and Applications. Routledge, New York, USA.

Royal Society, 2009. Reaping the Benefits: Science and the Sustainable Intensification of Global Agriculture. The Royal Society, London.

Ruiz Pérez, M., Ndoye, O., Eyebe, A., 1999. Marketing of nonwood forest products in the humid forest zone of Cameroon. Unasylva. 50 (198), 12−19.

Russell, J.R., Weber, J.C., Booth, A., Powell, W., Sotelo Montes, C., Dawson, I.K., 1999. Genetic variation of riverine populations of *Calycophyllum spruceanum* in the Peruvian Amazon Basin, revealed by AFLP analysis. Mol. Ecol. 8, 199−204.

Saka, J.D.K., 1995. The nutritional value of edible indigenous fruits: present research status and future directions. In: Maghembe, J.A., Ntupanyama, Y., Chirwa, P.W. (Eds.), Improvement of Indigenous Fruit Trees of the Miombo Woodlands of Southern Africa. ICRAF, Nairobi, Kenya, pp. 50−57. , PO Box 30677.

Saka, J.D.K., Msonthi, J.D., Maghembe, J.A., 1994. Nutritional value of edible fruits of indigenous wild trees in Malawi. For. Ecol. Manage. 64, 245−248.

Saka, J.D.K., Kadzere, I., Ndabikunze, B.K., Akinnifesi, F.K., Tiisekwa, B.P.M., 2008. Product development: nutritional value, processing and utilization of indigenous fruits from the miombo ecosystem. In: Akinnifesi, F.K., Leakey, R.R.B., Ajayi, O.C., Sileshi, G., Tchoundjeu, Z., Matakala, P., Kwesiga, F. (Eds.), Indigenous Fruit Trees in the Tropics: Domestication, Utilization and Commercialization. CAB International, Wallingford, pp. 288−309.

Sanchez, A.C., de Smedt, S., Haq, N., Samson, R., 2011. Comparative study on baobab fruit morphological variation between western and south-eastern Africa: opportunities for domestication. Genet. Resour. Crop Evol. 58 (8), 1143−1156.

Sanchez, P.A., 1976. Properties and Management of Soils in the Tropics. Wiley, New York, p. 618.

Sanchez, P.A., 1994. Alternatives to slash-and-burn: a pragmatic approach for mitigating tropical deforestation. In: Anderson, J.R. (Ed.), Agricultural Technology, Policy Issues for the International Community. CAB International, Wallingford, UK, pp. 451−480.

Sanchez, P.A., 1995. Science in agroforestry. Agrofor. Syst. 30, 5−55.

Sanchez, P.A., 2002. Soil fertility and hunger in Africa. Science. 192, 2019−2020.

Sanchez, P.A., Leakey, R.R.B., 1997. Land use transformation in Africa: three determinants for balancing food security with natural resource utilization. Eur. J. Agronomy. 7, 15−23.

Sanchez, PA, Miller, RH, 1986. Organic matter and soil fertility management in acid soils of the tropics. In: Transactions of the 13th International Congress on Soil Science, Hamburg 6, 609−625.

Sanchez, P.A., Palm, C.A., 1996. Nutrient cycling and agroforestry in Africa. Unasylva. 185 (47), 24−28.

Sanchez, P.A., Palm, C.A., Davey, C.B., Szott, L.T., Russell, C.E., 1985. Trees as soil improvers in the humid tropics? In: Cannell, M.G.R., Jackson, J.E. (Eds.), *Trees as* Crop Plants. Institute of Terrestrial Ecology, Huntingdon, UK, pp. 327−358.

Sanchez, PA, Woomer, PL, Palm, CA, 1994. Agroforestry approaches for rehabilitating degraded lands after tropical deforestation. In: Rehabilitation of degraded forest lands in the tropics−technical approach. JIRCAS International Symposium Series 1, pp. 108−119. Tsukuba, Japan: JIRCAS.

Sanchez, P.A., Izac, A.-M., Valencia, I.M., Pieri, C., 1996. Soil fertility replenishment in Africa. In: Breth, S.A. (Ed.), Achieving Greater Impact from Research Investments in Africa. Sasakawa Africa Association, Mexico City, pp. 200−208.

Sanchez, P.A., Buresh, R.J., Leakey, R.R.B., 1997a. Trees, soils and food security. Philos. Trans. R. Soc. Lond. B. 352, 949−961.

Sanchez, P.A., Shepherd, K.D., Soule, M.J., Place, F.M., Buresh, R.J., Izac, A.-M.N., et al., 1997b. Soil fertility replenishment in Africa as an investment in natural resource capital. In: Buresh, R.J., Sanchez, P.A., Calhoun, F. (Eds.), Replenishing Soil Fertility in Africa. Soil Sci. Soc. Amer., Madison, USA, pp. 1−46. , ASA-SSSA Special Publication 51.

Sanginga, N., Bowen, G.D., Danso, S.K.A,, 1990. Assessment of genetic variability for N-fixation between and within provenances of *Leucaena leucocephala* and *Acacia albida* estimated by ^{15}N labelling techniques. Plant. Soil. 127, 169−178.

Sanginga, N., Manrique, K., Hardarson, G., 1991. Variation in nodulation and N-fixation by the *Gliricidia sepium-Rhiobium* spp. symbiosis in a calcareous soil. Biol. Fert. Soils. 11, 273−278.

Sanginga, N., Danso, S.K.A., Zapata, F., Bowen, G.D., 1994. Field validation of intraspecific variation in phosphorus use efficiency and nitrogen fixation by provenances of *Gliricidia sepium* grown in low P soils. Appl. Soil Ecol. 1, 127−138.

Sanginga, N., Vanlauwe, B., Danso, S.K.A., 1995. Management of biological N-fixation in alley cropping systems: Estimation and contribution to N balance. Plant. Soil. 174, 119–141.

Sanou, H., Picard, N., Lovett, P.N., Dembélé, M., Korbo, A., Diarisso, D., et al., 2006. Phenotypic variation of agromorphological traits of the shea tree, *Vitellaria paradoxa* C.F. Gaertn., in Mali. Genet. Resour. Crop Evol. 53, 145–161.

Santilli, J, 2015. Agroforestry and the Law: the impact of legal instruments on agroforestry systems, Final Report to World Agroforestry Centre, Nairobi, Kenya, 86pp.

Sawadogo, K., Bezard, J., 1982. Etude de la structure glyceridique du beurre de karité. Oleagineux. 37, 69–74.

Sayer, J., Sunderland, T., Ghazoul, J., Pfund, J.L., Sheil, D., Meijaard, E., et al., 2013. Ten principles for a landscape approach to reconciling agriculture, conservation and other competing land uses. Proc. Natl. Acad. Sci. USA. 110, 8349–8356.

Schäfer, G., McGill, A.E.J., 1986. Flavour profiling of juice of the Marula (*Sclerocarya birrea* subsp. *caffra*) as an index for cultivar selection. Acta Hortic. 194, 215–222.

Scherr, S.J., 2000. A downward spiral? Research evidence on the relationship between poverty and natural resource degradation. Food Policy. 25, 479–498.

Scherr, S.J., Buck, L., Willemen, L., Milder, J.C., 2014. Ecoagriculture: integrated landscape management for people, food and nature. In: van Alfen, N., et al., (Eds.), Encyclopedia of Agriculture and Food Systems, Vol 3. Elsevier Publishers, San Diego, pp. 1–17.

Scherr, SJ, Hazell, PA, 1994. Sustainable agricultural development strategies in fragile lands. Environment and production technology division discussion paper 1. Washington, DC: IFPRI.

Schippers, R.R., 2000a. African Indigenous Vegetables: An Overview of the Cultivated Species. University Greenwich, Kent, UK, 214 pp.

Schippers, RR, 2000b. African Indigenous Vegetables: An Overview of the Cultivated Species. Natural Resources Institute/ACP-EU Technical Center for Agricultural and Rural Cooperation.

Schippers, R.R., Budd, L., 1997. African Indigenous Vegetables. IPGRI, and England: Natural Resources Institute, Rome.

Schreckenberg K, Leakey RRB, Kengue J, (Eds.), 2002a. A Fruit Tree with a Future: *Dacryodes edulis* (Safou, the African Plum) – Special Issue of For. Trees Livelihoods, 12, 1–152.

Schreckenberg, K., Degrande, A., Mbosso, C., Boli Baboulé, Z., Boyd, C., Enyong, L., et al., 2002b. The social and economic importance of *Dacryodes edulis* (G.Don) H.J. Lam in southern Cameroon. Forests, Trees Livelihoods. 12, 15–40.

Schreckenberg, K., Awono, A., Degrande, A., Mbosso, C., Ndoye, O., Tchoundjeu, Z., 2006a. Domesticating indigenous fruit trees as a contribution to poverty reduction. Forests Trees Livelihoods. 16, 35–51.

Schreckenberg, K., Marshall, E., Newton, A., Te Velde, D.W., Rushton, J., Edouard, F., 2006b. Commercialisation of non-timber forests products: what determines success. ODI Forestry Briefing. Overseas Development Institute, London, UK.

Schroeder, P., 1993. Agroforestry systems: integrated land use to store and conserve carbon. Clim. Res. 3, 53–60.

Schroeder, P., 1994. Carbon storage benefits of agroforestry systems. Agrofor. Syst. 27, 89–97.

Schroth, G., do Socorro Souza da Mota, M., 2014. Agroforestry: Complex Multistrata Agriculture. In: van Alfen, N. (Ed.), Encyclopedia of Agriculture and Food Systems, Vol. 1. Elsevier, San Diego, pp. 195–207.

Schroth, G., Kolbe, D., Pity, B., Zech, W., 1995. Searching for criteria for the selection of efficient tree species for fallow improvement, with special reference to carbon and nitrogen. Fert. Res. 42, 297–314.

Schroth, G., Krauss, U., Gasparotto, L., Duarte Aguilar, J.A., Vohland, K., 2000. Pests and diseases in agroforestry systems of the humid tropics. Agrofor. Syst. 50, 199–241.

Schroth, G., Salazar, E., da Silva, J.P., 2001. Soil nitrogen mineralization under tree crops and a legume cover crop in multi-strata agroforestry in central Amazonia: spatial and temporal patterns. Expl. Agric. 37, 253–267.

Schroth, G., Da Fonseca, G.A.B., Harvey, C.A., Gascon, C., Vasconcelos, H.L., Izac, A.-M.N., 2004. Agroforestry and Biodiversity Conservation in Tropical Landscapes. Island Press, Washington, DC.

Scott, J.C., 1998. Seeing Like a State: How Certain Schemes to Improve the Human Condition Have Failed. Yale University Press, MA.

Sebastian, K. (Ed.), 2014. Atlas of African Agriculture Research and Development – Revealing Agriculture's Place in Africa. IFPRI, Washington DC, USA, 108pp.

Sekercioglu, C.H., 2012. Bird functional diversity and ecosystem services in tropical forests, agroforests and agricultural areas. J. Ornithol. 153 (S1), 153–161.

Serageldin I, Persley GJ, 2000. Promethean Science: Agricultural Biotechnology, the Environment, and the Poor. CGIAR, 1818 H St, Washington, DC 20433, USA, 41 pp.

Shackleton, C.M., Botha, J., Emanuel, P.L., Ndlovu, S., 2002a. Inventory of marula (*Sclerocarya birrea* subsp. *caffra*) stocks and fruit yields in communal and protected areas of the Bushbuckridge lowveld, Limpopo Province, South Africa. Report to UK DFID Forestry Research Programme (Project No R7795). Rhodes University, Grahamstown, South Africa, p. 17.

Shackleton, C.M., Dzerefos, C.M., Shackleton, S.E., Mathabela, F.R., 2000. The use of and trade in indigenous edible fruits in the Bushbuckridge savanna region, South Africa. Ecol. Food Nutr. 39, 225–245.

Shackleton, C.M., Botha, J., Emanuel, P.L., 2003. Productivity and abundance of *Sclerocarya birrea* subsp. caffra in and around rural settlements and protected areas of the Bushbuckridge lowveld, South Africa. For. Trees Livelihoods. 13, 217–232.

Shackleton, S.E., 2002. The Informal Marula Beer Traders of Bushbuckridge, Limpopo Province, South Africa, Report to UK DFID Forestry Research Programme (Project No R7795). Rhodes University, Grahamstown, South Africa, p. 25.

Shackleton, S.E., Shackleton, C.M., Cunningham, A.B., Lombard, C., Sullivan, C.A., Netshiluvhi, T.R., 2002. Knowledge on *Sclerocarya birrea* in South and Southern Africa: a summary. Part 1. Taxonomy, ecology and role in rural livelihoods. South. Afr. For. J. 194, 27−41.

Shackleton, S, Wynberg, R, Sullivan, C, Shackleton, C, Leakey, R, Mander, M, et al., 2003. Marula commercialisation for sustainable and equitable livelihoods: synthesis of a southern African case study. Winners and Losers: Final Technical Report to DFID (FRP Project R7795), vol. 4, Appendix 3.5, 57 pp. (see www.nerc-wallingford.uk/research/winners/literature.html).

Shackleton, S.E., Shackleton, C.M., 2005. Contribution of marula (*Sclerocarya bir*re*a*) fruit and fruit products to rural livelihoods in the Bushbuckridge district, South Africa: balancing domestic needs and commercialization. For. Trees Livelihoods. 15, 3−24.

Shackleton, S., Shanley, P., Ndoye, O., 2007. Invisible but viable: recognizing localmarkets for non-timber forest products. Int. For. Rev. 9, 697−712.

Shackleton, S., Ziervogel, G., Sallu, S., Gill, T., Tschakert, P., 2015. Why is socially-just climate change adaptation in sub-Saharan Africa so challenging? A review of barriers identified from empirical cases. WIREs Clim. Change. Available from: http://dx.doi.org/10.1002/wcc.335.

Shackleton, S.E., Paumgarten, F., Kassa, H., Husselman, M., Zida, M., 2011. Opportunities for enhancing poor women's socio-economic empowerment in the value chains of three African non-timber forest products (NTFPs), Int. For. Rev, 13. pp. 136−151.

Shames, S., Clarvis, M.H., Kissinger, G., 2014. Financing Strategies for Integrated Landscape Investment: Synthesis Report. In: Shames, S. (Ed.), Financing Strategies for Integrated Landscape Investment. EcoAgriculture Partners, on behalf of the Landscapes for People, Food and Nature Initiative, Washington, DC.

Shanks, E., Carter, J., 1994. The Organization of Small-Scale Nurseries. Overseas Development Institute, London, UK, 144 pp.

Shepherd, K.D., Ohlsson, E., Okaleba, J.R., Ndufa, J.K., David, S., 1994. A static model for nutrient flow on mixed farms in the highlands of western Kenya to explore the possible impact of improved management. In: Powell, J.M., Fernandez-Rivera, S., Williams, T.O., Renard, C. (Eds.), Livestock and Sustainable Nutrient Cycling in Mixed Farming Systems of Sub-Saharan Africa. ILCA, Addis Ababa.

Shiembo, P.N., Newton, A.C., Leakey, R.R.B., 1996a. Vegetative propagation of *Irvingia gabonensis* Baill. a West African fruit tree. For. Ecol. Manage. 87, 185−192.

Shiembo, P.N., Newton, A.C., Leakey, R.R.B., 1996b. Vegetative propagation of *Gnetum africanum* Welw., a leafy vegetable from West Africa. J. Hortic. Sci. 71 (1), 149−155.

Shiembo, P.N., Newton, A.C., Leakey, R.R.B., 1997. Vegetative propagation of *Ricinodendron heulelotii* (Baill) Pierre ex Pax, a West African fruit tree. J. Trop. Forest Sci. 9, 514−525.

Siaw, D.E.K.A., Kang, B.T., Okali, D.U.U., 1991. Alley cropping with *Leucaena leucocephala* (Lam.) de Wit and *Acioa barteri* (Hook. f.) Engl. Agrofor. Syst. 14, 219−231.

Sibuea, T.T.H., 1995. Short Notes on the Sumatran Rhino (*Dicerorhinus sumatrensis*) in the Agroforest Areas (damar garden) in Krui, Lampung. AWB, Bogor, Indonesia.

Sibuea, T.T.H., Herdimansyah, D., 1993. The variety of mammal species in the agroforest areas of Krui (Lampung), Muara Bungo (Jambi) and Maninjau (West Sumatra). Internal Report OSTROM-Himbio. Universitas Padjajaran, Bandung, Indonesia, 62 pp.

Siddique, K.M.H., Sedgeley, R.H., 1987. Canopy development modifies the water economy of chickpea (*Cicer arietinum* L.), in south western Australia. Aust. J. Agric. Res. 37, 599−561.

Sidibé, M., Scheuring, J.F., Tembely, D., Sidibé, M.M., Hofman, P., Frigg, M., 1996. Baobab − homegrown vitamin C for Africa. Agrofor. Today. 8 (2), 13−15.

Siebert, S.F., 2002. From shade- to sun-grown perennial crops in Sulawesi, Indonesia: implications for biodiversity conservation and soil fertility. Biodivers. Conserv. 11, 1889−1902.

Sigaud, P., Hald, S., Dawson, I., Ouedraogo, A., 1998. FAO/IPGRI/ICRAF workshop on the conservation, management, sustainable utilization and enhancement of forest genetic resources in dry-zone sub-saharan Africa. Forest Genetic Resources No. 26. FAO, Rome, pp. 9−12.

Sileshi, G., Mafongoya, P.L., 2006. Long-term effect of legume-improved fallows on soil invertebrates and maize yield in eastern Zambia. Agric. Ecosyst. Environ. 115, 69−78.

Sileshi, G., Akinnifesi, F.K., Ajayi, O.C., Place, F., 2008a. Meta-analysis of maize yield response to planted fallow and green manure legumes in sub-Saharan Africa. Plant Soil. 307, 1−19.

Sileshi, G., Chintu, R., Mafongoya, P.L., Akinnifesi, F.K., 2008b. Mixed-species legume fallows affect faunal abundance and richness and N cycling compared to single species in maize-fallow rotations. Soil Biol. Biochem. 40, 3065−3075.

Sileshi, G., Schroth, G., Rao, M.R., Girma, H., 2008c. Weeds, diseases, insect pests and tri-trophic interactions in tropical agroforestry. In: Batish, D. R., Kohli, R.K., Jose, S., Singh, H.P. (Eds.), Ecological Basis of Agroforestry. CRC Press, Boca Raton, pp. 73−94.

Sileshi, G., Akinnifesi, F.K., Ajayi, O.C., Place, F., 2008d. Meta-analysis of maize yield response to planted fallow and green manure legumes in sub-Saharan Africa. Plant Soil. 307, 1−19.

Sileshi, G.W., Akinnifesi, F.K., Ajayi, O.C., Muys, B., 2011. Integration of legume trees in maize-based cropping systems improves rainfall use efficiency and crop yield stability. Agric. Water Manage. 98, 1364−1372.

Sileshi, G.W., Mafongoya, P., Akinnifesi, F.K., Phiri, E., Chirwa, P., Beedy, T., et al., 2014. Agroforestry: Fertilizer trees. In: van Alfen, N. (Ed.), Encyclopedia of Agriculture and Food Systems, Vol. 1. Elsevier, San Diego, pp. 222−234.

Silou, T., 1996. Le safoutier (*Dacryodes edulis*): un arbre mal connu. Fruits. 51, 47−60.

Silou, T., Rocquelin, G., Gallon, G., Molangui, T., 2000. Contribution á la caracterization des safous (*Dacryodes edulis*) d'Afrique Centrale II. Composition chimique et characteristiques nutritionnelles des safous du district de Biko (Congo-Brazzaville): Variation inter-arbre. Rev. Ital. della Sostanze Grasse. 77, 85−89.

Silva, W.G., Amelotti, G., 1983. Composition of the fatty substances from the fruit of *Guilielma speciosa* (Pupunha). Riv. Ital. Sost. Gras. 60, 767—770.

Sim, B.L., Jones, N., 1985. Improvement of *Gmelina arborea* in "Sabah Softwoods" plantations. In: Barnes, R.D., Gibson, G.L. (Eds.), Provenance and Genetic Improvement Strategies in Tropical Forest Trees. Commonwealth Forestry Institute, Forest Research Centre, Oxford, UK and Harare, Zimbabwe, pp. 604—609.

Simbo, D.J., de Smedt, S., van den Bilcke, N., de Meulenaer, B., van Camp, J., Uytterhoeven, V., et al., 2013. Opportunities for domesticating the African baobab (*Adansonia digitata* L.): multi-trait fruit selection. Agrofor. Syst. 87, 493—505.

Simitu, P., Jamnadass, R., Kindt, R., Kungu, J., Kimiywe, J., 2008. Consumption of dryland indigenous fruits to improve livelihoods in Kenya: the case of Mwingi district. Acta Hortic. 806, 93—98.

Simmonds, N.W. (Ed.), 1976. Evolution of Crop Plants. Longman, London and New York.

Simmonds, N.W., 1979. Principles of Crop Improvement. Longman, London, p. 408.

Simons, A.J., 1996a. ICRAF's strategy for domestication of indigenous tree species. In: Leakey, R.R.B., Temu, A.B., Melnyk, M., Vantomme, P. (Eds.), Domestication and Commercialization of Non-timber Forest Products in Agroforestry Systems, Non-wood Forest Products, vol 9. FAO, Rome, pp. 8—22.

Simons, A.J., 1996b. Delivery of improvement for agroforestry trees. In: Dieters, M.J., Matheson, A.C., Nikles, D.G., Harwood, C.E., Walker, S.M. (Eds.), Tree Improvement for Sustainable Tropical Forestry, Vol 2. Queensland Forestry Research Institute, Gympie, pp. 391—400.

Simons, A.J., 2003. Concepts and principles of tree domestication. In: Simons, A.J., Beniest, J. (Eds.), Tree Domestication in Agroforestry. World Agroforestry Center, Nairobi, Kenya, 244 pp.

Simons, A.J., Leakey, R.R.B., 2004. Tree domestication in tropical agroforestry. Agrofor. Syst. 61, 167—181.

Simons, A.J., Stewart, J.L., 1994. Gliricidia sepium. In: Gutteridge, R.C., Shelton, H.M. (Eds.), Fodder Tree Legumes in Tropical Agriculture. CAB International, Wallingford, UK, pp. 30—48.

Simons, A.J., MacQueen, D.J., Stewart, J.L., 1994. Strategic concepts in the domestication of nonindustrial trees. In: Leakey, R.R.B., Newton, A.C. (Eds.), Tropical Trees: Potential for Domestication and the Rebuilding of Forest Resources. HMSO, London, pp. 91—102.

Simons, A.J., Dawson, I., Tchoundjeu, Z., Duguma, B., 1998. Passing problems: prostate and *Prunus*. J. Am. Bot. Council. 43, 49—53.

Simons, A.J., Jaenicke, H., Tchoundjeu, Z., Dawson, I., Kindt, R., Oginosako, Z., et al., 2000a. The future of trees is on farms: tree domestication in Africa. Forests and Society: The Role of Research. XXI IUFRO World Congress, Kuala Lumpur, Malaysia, pp. 752—760.

Simons, A.J., Tchoundjeu, Z., Munjuga, M., Were, J., Dawson, I., Ruigu, S., et al., 2000b. Domestication Strategy. In: Hall, J.B., O'Brien, E.M., Sinclair, F.L. (Eds.), *Prunus africana*: A Monograph. UNCW, Bangor, Wales, pp. 39—43.

Singh, R.P., Ong, C.K., Saharen, N., 1989. Above and below ground interactions in alley cropping in semi-arid India. Agrofor. Syst. 9, 259—274.

Singh, Y., Khind, C.S., Singh, B., 1991. Efficient management of leguminous green manures in wetland rice. Adv. Agronomy. 45, 135—189.

Smaling, E., 1993. An agroecological framework for integrated nutrient management with special reference to Kenya. *Ph.D. thesis*. Agricultural University, Wageningen, The Netherlands.

Smartt, J., Haq, N., 1997. Domestication, Production and Utilization of New Crops. International Center for Underutilized Crops, Southampton.

Smith, J.W., Larbi, A., Jabbar, M.A., Akinlade, J., 1994. Voluntary Intake by Sheep and Goats of *Gliricidia sepium* Fed in Three Forms and at Three Levels of Supplementation to a Basal Diet of *Panicum maximum*. ILCA, Ibadan (Internal document.).

Snapp, S., 1995. Improving fertiliser efficiency with small additions of high quality organic inputs. In: Waddington, S.R. (Ed.), *Report on the First Meeting of the Network Working Group. Soil Fertility Network for Maize-based Farming Systems in Selected Countries of Southern Africa*. Rockefeller Foundation/ CIMMYT, Harare, Zimbabwe, pp. 60—65.

Soemwarto, O., 1987. Homegardens: a traditional agroforestry system with a promising future. In: Steppler, H.A., Nair, P.K.R. (Eds.), Agroforestry: A Decade of Development. ICRAF, Nairobi, pp. 157—170.

Solera, CR, 1993. Determinants of Competition between Hedgerow and Alley Species in a Contour Intercropping System. PhD thesis, University of the Philippines, Los Banos.

Soloviev, P., Niang, T.D., Gaye, A., Totte, A., 2004. Variability of fruit physicochemical characters for three harvested woody species in Senegal: *Adansonia digitata, Balanites aegyptiaca* and *Tamarindus indica*. Fruits. 59, 109—119.

Somarriba, E., Harvey, C.A., Samper, M., Anthony, F., Gonzalez, J., et al., 2004. Biodiversity conservation in neotropical coffee (*Coffea arabica*) plantations. In: Schroth, G., da Fonseca, G.A.B., Harvey, C.A., Gascon, C., Vasconcelos, H.L., Izac, A.-M.N. (Eds.), Agroforestry and Biodiversity Conservation in Tropical Landscapes. Island Press, Washington, DC, pp. 198—226.

Sonwa, D.J., Nkongmeneck, B.A., Weise, S.F., Tchatat, M., Adesina, A.A., et al., 2007. Diversity of plants in cocoa agroforests in the humid forest zone of Southern Cameroon. Biodivers. Conserv. 16, 2385—2400.

Sotelo-Montes, C., Weber, J., 1997. Prioritization of tree species for agroforestry systems in the lowland amazon forests of Peru. Agroforesteria en las Amecicas. 4 (14), 12—17.

Sotelo Montes, C., Weber, J.C., 2009. Genetic variation in wood density and correlations with tree growth in *Prosopis africana* from Burkina Faso and Niger. Ann. For. Sci. 66, 13.

Sotelo Montes, C., Hernández, R., Beaulieu, J., Weber, J.C., 2006. Genetic variation and correlations between growth and wood density of *Calycophyllum spruceanum* Benth. at an early age in the Peruvian Amazon. Silvae Genet. 55, 217—228.

Sotelo Montes, C., Hernández, R.E., Beaulieu, J., Weber, J.C., 2008. Genetic variation in wood color and its correlations with tree growth and wood density of *Calycophyllum spruceanum* at an early age in the Peruvian Amazon. New For. 35, 57—73.

Sotelo Montes, C., Silva, D.A., Garcia, R.A., Muñiz, G.I.B., Weber, J.C., 2011. Calorific value of *Prosopis africana* and *Balanites aegyptiaca* wood: relationships with tree growth, wood density and rainfall gradients in the West African Sahel. Biomass Bioenergy. 35, 346−353.

Spears, J.S., 1980. Can the wet tropical forest survive? Commonw. For. Rev. 58, 165−180.

Sperber, C.F., Azevedo, C.O., Muscardi, D.C., Szinwelski, N., Almeida, S., 2012. Drivers of parasitoid wasp community composition and cocoa agro-forestry practice in Bahia State, Brazil. In: Kaonga, M.L. (Ed.), Agroforestry for Biodiversity and Ecosystem Services. InTech, Rijeka, pp. 45−64.

Sporn, S.G., Bos, M.M., Hoffstätter-Müncheberg, M., Kessler, M., Gradstein, S.R., 2009. Microclimate determines community composition but not richness of epiphytic understory bryophytes of rainforest and cacao agroforests in Indonesia. Funct. Plant Biol. 36, 171−179.

Sprent, J., 2009. Legume Nodulation: A Global Perspective. Wiley-Blackwell, Chichester, UK.

Stadlmayr, B., Charrondière, U.R., Eisenwagen, S., Jamnadass, R., Kehlenbeck, K., 2013. Nutrient composition of selected indigenous fruits from sub-Saharan Africa. J. Sci. Food Agric. 2013 (93), 2627−2636.

Staver, C.F., Guharay, F., Monterroso, D., Muschler, R., 2001. Designing pest-suppressive multi-strata perennial crop systems: shade-grown coffee in Central America as a case study. Agrofor. Syst. 53, 151−170.

Stearns, S.C., Hockstra, R.F., 2000. Evolution: An Introduction. Oxford University Press, Oxford, England, 381 p.

Stenchly, K., Clough, Y., Buchori, D., Tscharntke, T., 2011. Spider web guilds in cacao agroforestry: comparing tree, plot and landscape-scale management. Divers. Distrib. 17, 748−756.

Stewart, J.L., Dunsdon, A.J., Hellín, J.J., Hughes, C.E., 1992. Wood Biomass Estimation of Central American Dry-Zone Species. Oxford Forestry Institute, Oxford, UK, 83 pp.

Stewart, J.L., Allison, G.E., Simons, A.J., 1996. *Gliricidia sepium*: Genetic Resources for Farmers. Oxford Forestry Institute, Oxford, UK, 125 pp.

Stomgaard, P., 1985. Biomass, growth and burning of woodland in a shifting cultivation area of South Central Africa. For. Ecol. Manage. 12, 163−178.

Stoorvogel, JJ, Smaling, EMA, JanssenB. H, 1993. Calculating soil nutrient balances in Africa at different scales. I: supra-national scale. In: Smaling, E. (Eds.), An Agro-ecological Framework for Integrated Nutrient Management with Special Reference to Kenya. Doctoral thesis, Agricultural University, Wageningen, The Netherlands, pp. 75−90.

Sufi, N.A., Kaputo, M.T., 1977. Identification and determination of free sugars in Masuku fruit (*Uapaca kirkiana*). Zam. J. Sci. Tech. 2, 23−25.

Sullivan, C.A., 2003. Forest use by Amerindians in Guyana − implications for development policy. In: Barker, D., McGregor, D. (Eds.), Resources, Planning and Environmental Management in a Changing Caribbean. UWI Press, Kingston.

Sullivan, C.A., O'Regan, D.P., Shackleton, S., Ousman, S., Shackleton, C., Mander, M., et al., 2003. *Winners and Losers in Forest Product Commercialisation*, Final Report to DFID Forestry Research Programme (R7795). Centre for Ecology and Hydrology, Wallingford, UK, www.ceh-wallingford.ac.uk/research/winners/literature.html.

Sunderland, T.C.H., Ndoye, O. (Eds.), 2004. Forest Products, Livelihoods and Conservation: Case Studies of Non-Timber Forest Product Systems − Volume 2 − Africa. CIFOR, Indonesia.

Sunderland, T.C.H., Clark, L.E., Vantomme, P., 1999.) *Non-Wood Forest Products of Central Africa, Current Research Issues and Prospects for Conservation and Development*. FAO, Rome.

Swallow, BM, Villarreal, M, Kwesiga, F, Holding, AC, Agumya, A, Thangata, P, 2007. Agroforestry and its role in mitigating the impact of HIV/AIDS. In: World Agroforestry and the Future: Proceedings of 25th Anniversary Conference, 1−5 November 2003, Nairobi. ICRAF, Nairobi.

Swallow, B.M., Kallesoe, M.K., Iftikhar, U.A., van Noordwijk, M., Bracer, C., 2009. Compensation and rewards for environmental services in the developing world: framing pan-tropical analysis and comparison. Ecol. Soc. 14 (2), 26.

Swaminathan, MS, 2012. Agroforestry for an ever-green revolution. In: Nair, P.K.R., Garrity, D., (Eds.), Agroforestry − The Future of Global Land Use, Adv. Agrofor., 9, 7−10. Springer, Dordrecht, Germany.

Swift, M.J., Anderson, J.M., 1994. Biodiversity and ecosystem function in agricultural systems. Biodivers. Ecosyst. Funct. 99, 15−42.

Szott, L.T., Fernandes, E.C.M., Sanchez, P.A., 1991. Soil-plant interactions in agroforestry systems. For. Ecol. Manage. 45, 127−152.

Tabuna, H, 1999. Le marche des produits forestiers non ligneux de l'Afrique Centrale en France et en Belgique. Produits, acteurs, circuits de distribution et debouches actuels. CIFOR Occasonal Paper No 19. CIFOR, Indonesia. (3 x e acute in marche and debouches).

Tacio, H.D., 1991. The SALT system: agroforestry for sloping lands. Agrofor. Today. 3 (1), 12−13.

Takoutsing, B., Tchoundjeu, Z., Degrande, A., Asaah, E., Tsobeng, A., 2014. Scaling-up sustainable land management practices through the concept of the rural resource centre: reconciling farmers' interests with research agendas. J. Agric. Educ. Ext. 5, 463−483.

Tata epse Ngome, PI 2015. The contribution of fruit from trees to improve household food security in the context of deforestation in Cameroon. PhD thesis, Rhodes University, Grahamstown, 195 pp.

Taylor, F.W., Kwerepe, B., 1995. Towards domestication of some indigenous fruit trees in Botswana. In: Maghembe, J.A., Ntupanyama, Y., Chirwa, P.W. (Eds.), Improvement of Indigenous Fruit Trees of the Miombo Woodlands of Southern Africa. ICRAF, Nairobi, Kenya, pp. 113−134. , PO Box 30677.

Taylor, F.W., Mateke, S.M., Butterworth, K.J., 1996. A holistic approach to the domestication and commercialisation of non-timber forest products. In: Leakey, R.R.B., Temu, A.B., Melnyk, M., Vantomme, P. (Eds.), *Domestication and Commercialization of Non-timber Forest Products in Agroforestry Systems*. FAO Non-Wood Forest Products No. 9. FAO, Rome, pp. 8−22.

Tchatchou, TH, 1989. Influence des facteurs extrinseques et intrinseques sur la rhizogenese des boutures uninodales et unifoliées de *Triplochiton scleroxylon* (K. Schum.). Thesis, ENSA Centre Universitaire de Dschang, Cameroon, 81 pp.

Tchoundjeu, Z., 1989. Vegetative propagation of the tropical hardwoods *Khaya ivorensis* A. Chev. and *Lovoa trichiloides* Harms., *Ph.D. thesis*. University of Edinburgh, p. 270.

Tchoundjeu, Z., 1996. Vegetative propagation of sahelian agroforestry tree species: *Prosopis africana* and *Bauhinia rufescens*. In: Dieters, M.J., Matheson, A.C., Nickles, D.G., Harwood, C.E., Walker, S.M. (Eds.), Tree Improvement for Sustainable Tropical Forestry. Queensland Forestry Research Institute, Caloundra, pp. 416−419.

Tchoundjeu, Z., Atangana, A.R., 2006. Ndjanssang: *Ricinodendron heudelotii* (Baill). International Centre for Underutilised Crops, Southampton, UK, 20 pp.

Tchoundjeu, Z., Leakey, R.R.B., 2000. Vegetative propagation of African Mahogany: effects of stockplant flushing cycle, auxin and leaf area on carbohydrate and nutrient dynamics of cuttings. J. Trop. For. Sci. 12, 77−91.

Tchoundjeu, Z., Weber, J.C., Guarino, L., 1997. Germplasm collection of threatened, valuable multipurpose tree species: the case of *Prosopis africana* in SALWA. Agrofor. Syst. 39, 91−100.

Tchoundjeu, Z., Duguma, B., Fondoun, J.-M., Kengue, J., 1998. Strategy for the domestication of indigenous fruit trees of West Africa: case of *Irvingia gabonensis* in southern Cameroon. Cam. J. Biol. Biochem. Sci. 4, 21−28.

Tchoundjeu, Z., Avana, M.L., Leakey, R.R.B., Simons, A.J., Asaah, E., Duguma, B., et al., 2002a. Vegetative propagation of *Prunus africana*: effects of rooting medium, auxin concentrations and leaf area. Agrofor. Syst. 54, 183−192.

Tchoundjeu, Z., Kengue, J., Leakey, R.R.B., 2002b. Domestication of *Dacryodes edulis*: state-of-the art. Forests Trees Livelihoods. 12, 3−14.

Tchoundjeu, Z., Ngo Mpeck, M.L., Asaah, E., Amougou, A., 2004. A role of vegetative propagation in the domestication of *Pausinystalia johimbe* (K. Schum), a highly threatened medicinal species of West and Central Africa. Forest Ecol. Manage. 188, 175−183.

Tchoundjeu, Z., Degrande, A., Leakey, R.R.B., Schreckenberg, K., 2005. Participatory domestication of indigenous trees for improved livelihoods and a better environment. A policy briefing note for governments in the African humid tropics region. Overseas Development Institute, London (2 pp).

Tchoundjeu, Z., Asaah, E., Anegbeh, P.O., Degrande, A., Mbile, P., Facheux, C., et al., 2006. Putting participatory domestication into practice in West and Central Africa. Forests, Trees Livelihoods. 16, 53−70.

Tchoundjeu, Z., Degrande, A., Leakey, R.R.B., Simons, A.J., Nimino, G., Kemajou, E., et al., 2010a. Impact of participatory tree domestication on farmer livelihoods in west and central Africa. Forest Tree Livelihoods. 19, 219−234.

Tchoundjeu, Z., Tsobeng, A.C., Asaah, E., Anegbeh, P., 2010b. Domestication of *Irvingia gabonensis* (Aubry Lecomte) by air layering. J. Hortic. For. 2, 171−179.

Téhé, H., 1986. Utilizations des resources forestières chez les Guérés et les Oubis Côte d'Ivoire). Banco, (Côte d'Ivoire). 4, 26−30.

ten Kate, K., Laird, S., 1999. The Commercial Use of Biodiversity: Access to Genetic Resources and Benefit-Sharing. Earthscan, London.

Thijssen, H.J.C., Muriithi, F.M., Nyaata, O.Z., Mwangi, J.N., Aiyelaagbe, I.O.O., Mugendi, D.N., 1993. *Report on an Ethnobotanical Survey of Woody Perennials in the Coffee Zone of Embu District, Kenya. AFRENA Report No. 62*. ICRAF, Nairobi.

Thiollay, J.-M., 1995. The role of traditional agroforests in the conservation of rain forest bird diversity in Sumatra. Conserv. Biol. 9, 335−353.

Thiong'o, M., Kingori, S., Jaenicke, H., 2002. The taste of the wild: variation in the nutritional quality of marula fruits and opportunities for domestication. Acta Hortic. 575, 237−244.

Thomas, E., Jalonen, R., Loo, J., Boshier, D., Gallo, L., Cavers, S., et al., 2014. Genetic considerations in ecosystem restoration using native tree species. For. Ecol. Manage. 333, 66−75.

Tian, G., Brussard, L., Kang, B.T., 1995. An index for assessing the quality of plant residues and evaluating their effects on soil and crop use in the subhumid tropics. Appl. Soil Ecol. 2, 25−32.

Tian, G., Salako, F.K., Ishida, F., Zhang, J., 2001. Biological restoration of a degraded Alfisol in humid tropics using planted woody fallow: synthesis of 8-year-results. In: Scott, D.E., Mohtar, R.H., Steinhardt, G.C. (Eds.), Sustaining the Global Farm. USDA-ARS, West Lafayette, IN, pp. 333−337.

Tiffen, M., Mortimer, M., Gichuki, F., 1994. More People, Less Erosion: Environmental Recovery in Kenya. John Wiley & Sons, Chichester, UK.

Tindale NB, 1981. Desert Aborigines and the southern coastal peoples: some comparisons. In: Keast, A., (Eds.), Ecological Biogeography of Australia. The Hague, Junk, p. 1879.

Tisdall, J.M., Oades, J.M., 1982. Organic matter and water-stable aggregates in soils. J. Soil Sci. 33, 141−163.

Tittonell, P., 2014. Ecological intensification of agriculture − sustainable by nature. Curr. Opin. Environ. Sustain. 8, 53−61.

Tittonell, P., Giller, K.E., 2013. When yield gaps are poverty traps: the paradigm of ecological intensification in African smallholder agriculture. Field Crops Res. 143, 76−90.

Toensmeier, E., 2016. The Carbon Farming Solution: A Global Toolkit of Perennial Crops and Regenerative Agricultural Practices for Climate Change Mitigation and Food Security. Chelsea Green Publishing, White River Junction, Vermont.

Tolkamp, B.J., Brouwer, B.O., 1993. Statistical review of digestion in goats compared with other ruminants. Small Rumin. Res. 11, 107−124.

Tomich, T.P., van Noordwijk, M., Budidarsono, S., Gillison, A., Kusumanto, T., Stolle, F., et al., 2001. Agricultural intensification, deforestation and the environment: Assessing trade offs in Sumatra, Indonesia. In: Lee, D.R., Barrett, C.B. (Eds.), Trade offs or Synergies? Agricultural Intensification, Economic Development and the Environment. CABI Publishing, Wallingford, pp. 221−244.

Torquebiau, E., 1992. Are tropical agroforestry home gardens sustainable? Agric. Ecosyst. Environ. 41, 189−207.

Torquebiau, E., Cholet, N., Ferguson, W., Letourmy, P., 2013. Designing an index to reveal the potential of multipurpose landscapes in Southern Africa. Land. 2, 705−725.

Tracy, M.D., 1996. Pejibaye flour, a hopeful option. Bol. Retadar. 23, 3.

Tscharntke, T., Clough, Y., Bhagwat, S.A., Buchori, D., Faust, H., et al., 2011. Multifunctional shade-tree management in tropical agroforestry landscapes—a review: multifunctional shade-tree management. J. Appl. Ecol. 48 (3), 619−629.

Tscharntke, T., Clough, Y., Wanger, T.C., Jackson, L., Motzke, I., et al., 2012a. Global food security, biodiversity conservation and the future of agricultural intensification. Biol. Conserv. 151, 53−59.

Tscharntke, T., Clough, Y., Jackson, L.E., Motzke, I., Perfecto, I., Vandermeer, J., et al., 2012b. Global food security, biodiversity conservation and the future of agricultural intensification. Biol. Conserv. 151, 53−59.

Tsobeng, A., Tchoundjeu, Z., Degrande, A., Asaah, E., Bertin, T., van Damme, P., 2015. Phenotypic variation in *Pentaclethra macrophylla* Benth (Fabaceae) from the humid lowlands of Cameroon. Afrika Focus. 28, 47−61.

Tsobeng, A., Ofori, D., Tchoundjeu, Z., Asaah, E., van Damme, P., 2016. Improving growth of stockplants and rooting ability of leafy stem cuttings of *Allanblackia floribunda* Oliver (Clusiaceae) using different NPK fertilizers and periods of application. New For. Available from: http://dx.doi.org/10.1007/s11056-015-9517-1.

Tsobeng, A, Asaah, E, Leakey, RRB, Tchoundjeu, Z, van Damme, P, Ofori, D, et al. In press. Effects of perseverance irradiance on the growth of *Allanblackia floribunda* stockplants and on the subsequent rooting capacity of leafy stem cuttings, in press. Trees 0: 000-000.

Tukan, C.M.J., Roshetko, J.M., Budidarsono, S., Manurung, G.S., 2006. Market chain improvement: linking farmers to markets in Nanggung, West Java, Indonesia. Acta Hortic. 699, 429−438.

Tyndall, B, 1996. The Anatomy of Innovative Adoption: The Case of Successful Agroforestry in East Africa. Ph.D. Thesis, Colorado State University, 212pp.

Ukafor, V., 2002. Variability studies in fruit and tree characteristics of *Dacryodes edulis* in south eastern Nigeria. MSc Thesis. Rivers State University of Science and Technology, Port Harcourt, Nigeria.

Umoro Umati, U., Okiy, A., 1987. Characteristics and composition of the pulp oil and cake of the African pear, *Dacryodes edulis* (*G.Don*) H.J.Lam. J. Sci.Food Agric. 38, 67−72.

UNEP, 1992. In: Tolba, M.K., El-Kholy, O.A. (Eds.), *The World Environment, 1972−1992*. Chapman and Hall, London.

UNFPA, 1992. *A World in Balance. State of the World Population, 1992*. United Nations Fund for Population Activities, New York.

United Nations, 2015. Transforming our world: the 2030 Agenda for Sustainable Development, Resolution adopted by the General Assembly on 25 September 2015, UN General Assembly, Seventieth Session (October 21 2015), Agenda items 15 and 116, New York, USA.

Unruh, J.D., Houghton, R.A., Lefebvre, P.A., 1993. Carbon storage in agroforestry: an estimate for sub-Saharan Africa. Clim. Res. 3, 39−52.

UPOV, 1991. International Convention for the Protection of New Varieties of Plants, of December 2, 1961, as revised at Geneva in 1972, 1978 and on March 19, 1991.

Uzo, JO, 1980. Yield and harvest predictions of some indigenous perennial fruits, roots and leafy vegetables in tropical West Africa. In: Proceedings of International Symposium on the Current Problems of Fruits and Vegetables (Tropical and Sub-tropical), Laguna, 24−26 March 1980, 22p.

van Bael, S.A., Bichier, P., Greenberg, R., 2007a. Bird predation on insects reduces damage to the foliage of cocoa trees (*Theobroma cacao*) in western Panama. J. Trop. Ecol. 23, 715−719.

van Bael, S.A., Bichier, P., Ochoa, I., Greenberg, R., 2007b. Bird diversity in cacao farms and forest fragments of western Panama. Biodivers. Conserv. 16, 2245−2256.

van Bael, S.A., Philpott, S.M., Greenberg, R., Bichier, P., Barbier, N.A., et al., 2008. Birds as predators in tropical agroforestry systems. Ecology. 89, 928−934.

van Damme, P., Termote, C., 2008. African botanical heritage for new crop development. Afrika Focus. 21, 45−64.

van Damme, P., Kindt, R., 2011. Ethnobotanical methods. In: Dawson, I., Harwood, C., Jamnadass, R., Beniest, J. (Eds.), Agroforestry Tree Domestication: a Primer. ICRAF, Nairobi, pp. 28−35.

van Damme, P., Termote, C., 2008. African botanical heritage for new crop development. Afrika Focus. 21, 45−64.

van den Beldt, R.J., Napompeth, B., 1992. Leucaena psyllid comes to Africa; time to learn some lessons from the Asian experience. Agrofor. Today. 4, 11−12.

van den Bilcke, N., Alaerts, K., Ghaffaripour, S., Simbo, D.J., Samson, R., 2014. Physico-chemical properties of tamarind (*Tamarindus indica* L.) fruits from Mali: selection of elite trees for domestication. Genet. Resour. Crop Evol. 61, 537−553.

van Noordwijk, M., 1999. Nutrient cycling in ecosystems versus nutrient budgets in agricultural systems. In: Smaling, E., Oenema, O., Fresco, L. (Eds.), Nutrient Cycles and Nutrient Budgets in Global Agro-Ecosystems. CAB International, Wallingford, UK, pp. 1−26.

van Noordwijk, M., 2014. Climate change: agricultural mitigation. In: van Alfen, N. (Ed.), Encyclopedia of Agriculture and Food Systems, Vol. 2. Elsevier, San Diego, pp. 220−231.

van Noordwijk, M., Brussaard, L., 2014. Minimizing the ecological footprint of food: closing yield and efficiency gaps simultaneously? Curr. Opin. Environ. Sustain. 8, 62−70.

van Noordwijk, M., Widianto, M.H., Hairiah, K., 1991. Old tree root channels in acid soils in the humid tropics: important for crop root penetration, water infiltration and nitrogen management. Plant Soil 134, 37−44.

van Noordwijk M, van Schaik CP, de Foresta H, Tomich TP, 1995. Segregate or integrate nature and agriculture for biodiversity conservation. In: Proceedings of Biodiversity Forum, 4−5 November 1995, Jakarta, Indonesia.

van Noordwijk, M., Lawson, G., Soumare, A., Groot, J.J.R., Hairiah, K., 1996. Root distribution of trees and crops: competition and/or complementarity. In: Ong, C.K., Huxley, P.A. (Eds.), Tree−Crop Interactions: A Physiological Approach. CAB International, Wallingford, UK, pp. 319−364.

van Noordwijk, M., Hoang, M.H., Neufeldt, H., Öborn, I., Hoang, M.H., Neufeldt, H., et al., 2011a. How trees and people can co-adapt to climate change: Reducing vulnerability in multifunctional landscapes. World Agroforestry Center, Nairobi, Kenya, 134pp.

van Noordwijk, M., Hoang, M.H., Neufeldt, H., Öborn, I., Yatich, T. (Eds.), 2011b. How trees and people can co-adapt to climate change: reducing vulnerability through multifunctional agroforestry landscapes. World Agroforestry Centre, Nairobi, Kenya, 131pp.

van Noordwijk, M., Leimona, B., Jindal, R., Villamor, G., Vardhan, M., Namirembe, S., et al., 2012a. Payments for environmental services: evolution toward efficient and fair incentives for multifunctional landscapes. Annu. Rev. Environ. Resour. 37, 389—420.

van Noordwijk, M, Tata, HL, Xu, J, Dewi, S, Minang, P, 2012b. Segregate or integrate for multifunctionality and sustained change through landscape agroforestry involving rubber in Indonesia and China In: Nair, P.K.R., Garrity, D., (Eds.), Agroforestry — The Future of Global Land Use. Adv. Agrofor., 9, 69—104. Springer, Dordrecht, Germany.

Vanhove, W., van Damme, P., 2013. On-farm conservation of cherimoya (*Annona cherimola* Mill.) germplasm diversity: a value chain perspective. Trop. Conserv. Sci. 6, 158—180.

Vaughan, C., Ramírez, O., Herrera, G., Guries, R., 2007. Spatial ecology and conservation of two sloth species in a cacao landscape in Limón, Costa Rica. Biodivers. Conserv. 16, 2293—2310.

Vedeld, P., Angelsen, A., Sjaastad, E., Kobugabe Berg, G., 2004. Counting on the Environment: Forest Incomes and the Rural Poor. Environmental Economics Series No 98. World Bank, Washington DC, USA, 95 pp.

Veierskov, B., 1978. A relationship between length of basis and adventitious root formation in pea cuttings. Physiol. Plant. 42, 146—150.

Villachica, H, 1996. Frutales y Hortalizas Promisorios de la Amazonia, Tratado de Cooperacion Amazonica, Lima, Peru, 367p.

Villarreal, M., Holding, A.C., Swallow, B.M., Kwesiga, F., 2006. The challenge of HIV/AIDS: Where does agroforestry fit in? In: Garrity, D., Okono, A., Grayson, M., Parrott, S. (Eds.), World Agroforestry into the Future. World Agroforestry Center., Nairobi, pp. 181—192.

Vivien, J., Faure, J.J., 1985. Arbres des Forêts Denses d'Afrique Centrale. Agence de Coopération Culturelle et Technique, Paris.

Vivien, J., Faure, J.J., 1996. Fruitiers Sauvages d'Afrique (Espèces du Cameroon). Le Centre Technique de Coopération Agricole et Rurale. Wageningen, The Netherlands, 415pp.

von Holdt, B.M., Gray, M.M., Wayne, R.K., 2012. Genome-wide approaches for the study of dog domestication. In: Gepts, P., Famula, T.R., Bettinger, R.L., Brush, S.B., Damania, A.B., McGuire, P.E., Qualset, C.O. (Eds.), Biodiversity in Agriculture: Domestication, Evolution and Sustainability. Cambridge University Press, Cambridge, pp. 275—298.

von Teichmann, I, 1982. Notes on the distribution, morphology, importance and uses of the indigenous Anacardiaceae: 1. The distribution and morphology of *Sclerocarya birrea* (the Marula). Trees in South Africa: October-December, 2—7.

von Teichmann, I, 1983. Notes on the distribution, morphology, importance and uses of the indigenous Anacardiaceae: 2. The importance and uses of *Sclerocarya birrea* (the Marula). Trees in South Africa: April - September, 35—41.

Vosti, S., Reardon, T., 1997. Sustainability, Growth and Poverty Alleviation: A Policy and Agroecological Perspective. The John Hopkins University Press, Baltimore, USA.

Vosti, S., Witcover, J., 1995. Non-timber tree product market research workshop - an overview. IFPRI, Washington, D.C.

Vutilolo, I.V.N., Tyagi, A.P., Thomson, L.A.J., Heads, M., 2005. Comparison of performance of whitewood *Endospermum medullosum* L. (S. Smith) provenances and families in Vanuatu. S. Pac. J. Nat. Sci. 23, 37—42.

Wahyuni, S., Yulianti, E.S., Komara, W., Yates, N.G., Obst, J.M., Lawry, J.B., 1982. The performance of ongole cattle offered either grass, sun-dried *Leucaena leucocephala* or varying proportions of each. Trop. Anim. Prod. 7, 275—283.

Wald, W., 1940. A note on the analysis of variance with unequal class frequencies. Ann. Math. Stat. 11, 96.

Walker, B.H., Holling, C.S., Carpenter, S.R., Kinzig, A., 2004. Resilience, adaptability and transformation of social-ecological systems. Ecol. Soc. 9 (2), 5 [online].

Wambugu, C., Franzel, S., Cordero, J., Stewart, J., 2006. Fodder shrubs for dairy farmers in East Africa: making extension decisions and putting them into practice. World Agroforestry centre, Nairobi. Kenya.

Wanger, T.C., Saro, A., Iskandar, D.T., Brook, B.W., Sodhi, N.S., et al., 2009. Conservation value of cacao agroforestry for amphibians and reptiles in South-East Asia: combining correlative models with follow-up field experiments. J. Appl. Ecol. 46, 823—832.

Wanger, T.C., Iskandar, D.T., Motzke, I., Brook, B.W., Sodhi, N.S., et al., 2010a. Land-use change affects community composition and traits of tropical amphibians and reptiles in Sulawesi (Indonesia). Conserv. Biol. 24, 795—802.

Wanger, T.C., Wielgoss, A.C., Motzke, I., Clough, Y., Brook, B.W., et al., 2010b. Endemic predators, invasive prey, and native diversity. Proc. R. Soc. B. 278 (1706), 690—694.

Wambugu, C., Place, F., Franzel, S., 2011. Research, development and scaling up the adoption of fodder shrub innovations in east Africa. Int. J. Agric. Sustain. 9, 100—109.

Waring, R.H., Cleary, B.D., 1967. Plant moisture stress: evaluation by pressure bomb. Science. 155, 1248—1254.

Warrington, I.J., Rook, D.A., Morgan, D.C., Turnbull, H.L., 1988. The influence of simulated shadelight and daylight on growth, development and photosynthesis of *Pinus radiata, Agathis australis* and *Dacrydium cupressinum*. Plant Cell Environ. 11, 343—356.

Waruhiu, A.N., 1999. Characterization of fruit traits towards domestication of an indigenous fruit tree of west and central Africa: A case study of *Dacryodes edulis* in Cameroon, *MSc Thesis*. University of Edinburgh, p. 58.

Waruhiu, A.N., Kengue, J., Atangana, A.R., Tchoundjeu, Z., Leakey, R.R.B., 2004. Domestication of *Dacryodes edulis*: 2. Phenotypic variation of fruit traits in 200 trees from four populations in the humid lowlands of Cameroon. Food Agric. Environ. 2, 340—346.

Waterman, P.G., Meshal, I.A., Hall, J.B., Swaine, M.D., 1978. Biochemical systematics and ecology of the Toddalioideae in the central part of the West African forest zone. Biochem. Syst. Ecol. 6, 239—245.

Watson, G.A., 1983. Development of mixed tree and food crop systems in the humid tropics: a response to population pressure and deforestation. Exp. Agric. 19, 311—332.

Watson, G.A., 1990. Tree crops and farming systems development in the humid tropics. Exp. Agric. 26, 143−159.

Watt, J.M., Breyer-Brandwijk, M.G., 1962. Medicinal and poisonous plants of southern and eastern Africa. Livingstone, Edinburgh & London.

Weber, J.C., Sotelo Montes, C., 2005. Variation and correlations among stem growth and wood traits of *Calycophyllum spruceanum* Benth. from the Peruvian Amazon. Silvae Genet. 54, 31−41.

Weber, J.C., Sotelo Montes, C., 2008. Variation in tree growth and wood density of *Guazuma crinite* Mart. in the Peruvian Amazon. New For. 36, 29−52.

Weber, J.C., Sotelo Montes, C., 2010. Correlations and clines in tree growth and wood density of *Balanites aegyptiaca* (L.) Delile provenances in Niger. New For. 39, 39−49.

Weber, J.C., Labarta Chávarri, R.L., Sotelo Montes, C., Brodie, A.W., Cromwell, E., Schreckenberg, K., et al., 1997. Farmers' use and management of tree germplasm: case studies from the Peruvian Amazon Basin. In: Simons, A.J., Kindt, R., Place, F. (Eds.), Policy Aspects of Tree Germplasm Demand and Supply. The World Agroforestry Center (ICRAF), Nairobi, pp. 57−63.

Weber, J.C., Sotelo Montes, C., Vidaurre, H., Dawson, I.K., Simons, A.J., 2001. Participatory domestication of agroforestry trees: an example from the Peruvian Amazon. Dev. Pract. 11, 425−433.

Weber, J.C., Larwanou, M., Abasse, T.A., Kalinganire, A., 2008. Growth and survival of *Prosopis africana* provenances tested in Niger and related to rainfall gradients in the West African Sahel. For. Ecol. Manage. 256, 585−592.

Weber, J.C., Sotelo Montes, C., Ugarte, J., Simons, A.J., 2009. Phenotypic selection of *Calycophyllum spruceanum* on farms in the Peruvian Amazon: evaluating a low-intensity selection strategy. Silvae Genet. 58, 172−179.

Weber, J.C., Sotelo Montes, C., Cornelius, J., Ugarte, J., 2011. Genetic variation in tree growth, stem form and mortality of *Guazuma crinita* in slower- and faster-growing plantations in the Peruvian Amazon. Silvae Genet. 60, 70−78.

Weber, J.C., Sotelo Montes, C., Kalingare, A., Abasse, T., Larwanou, M., 2015. Genetic variation and clines in growth and survival of *Prosopis Africana* from Burkina Faso and Niger: Comparing results from a nursery test and long-term field test in Niger. Euphytica. Available from: http://dx.doi.org/10.1007/s/10681-015-1413-4.

Wedding, B.B., White, R.D., Grauf, S., Tilse, B., Gadek, P.A., 2008. Near infrared spectroscopy as a rapid, non-invasive method for sandalwood oil determination. Sandalwood Res. Newsl. 23, 1−4.

Weinert, I.A.G., van Wyk, P.J., Holtzhausen, L.C., 1990. Marula. In: Nagy, S., Shaw, P.E., Wardowski, W.F. (Eds.), Fruits of Tropical and Subtropical origin: Composition, Properties and Uses. Florida Science Source Inc., Lake Alfred, FL, pp. 88−115.

Weist, M., Tscharntke, T., Sinaga, M.H., Maryanto, I., Clough, Y., 2010. Effect of distance to forest and habitat characteristics on endemic versus introduced rat species in agroforest landscapes of Central Sulawesi, Indonesia. Mamm. Biol. 75, 567−571.

Welford, L., Le Breton, G., 2008. Bridging the gap: PhytoTrade Africa's experience of certification of natural products. Forests Trees Livelihoods. 18, 69−79.

Weston, P., Hong, R., Kaboré, C., Kull, C.A., 2015. Farmer-managed natural regeneration enhances rural livelihoods in dryland West Africa. Environ. Manage. 0, 000-000 http://dx.doi.org/10.1007/s00267-015-0469-1.

Wheatley, C, Woods, EJ, Setyadjit, 2004. The Benefits of Supply Chain Practice in Developing Countries − Conclusions from an International Workshop held in Indonesia. ACIAR Proceedings No. 119e.

Whitmore, T.C., 1972. *Tree Flora of Malaya, I*. Longmans, Kuala Lumpur.

Whittome, MPB, 1994. The Adoption of Alley Farming in Nigeria and Benin. On-farm experience of IITA and ILCA. PhD thesis, University of Cambridge.

WHO, 2014. Household air pollution and health. *WHO Fact Sheet 292*. WHO, Geneva.

Wielgoss, A., Tscharntke, T., Buchori, D., Fiala, B., Clough, Y., 2010. Temperature and a dominant dolichoderine ant species affect ant diversity in Indonesian cacao plantations. Agric. Ecosyst. Environ. 135, 253−259.

Wiersum, K.F., 1984. Surface erosion under various tropical agroforestry systems. In: O'Loughlin, C.L., Pearce, A.J. (Eds.), Symposium on Effects of Forest Land Use on Erosion and Slope Stability. East-West Centre, Honolulu, Hawai, USA, pp. 231−239.

Wiersum, K.F., 1985. Effects of various vegetation layers in an *Acacia auriculiformis* forest plantation on surface erosion in Java, Indonesia. In: El-Swaify, S.A., Moldenhaur, W.C., Lo, A. (Eds.), Soil Erosion and Conservation. Soil Conservation Society of North America, Ankeny, Iowa, USA, pp. 78−89.

Wiersum, K.F., 1996. Domestication of valuable tree species in agroforestry systems: Evolutionary stages from gathering to breeding. In: Leakey, R.R.B., Temu, A.B., Melnyk, M., Vantomme, P. (Eds.), Domestication and Commercialization of Non-timber Forest Products, Non-Wood Forest Products, 9. FAO, Rome, pp. 147−158.

Wiggins, S., 2014. African agricultural development: lessons and challenges. J. Agric. Econ. 65, 529−556.

Wightman, K., 1999. Good Tree Nursery Practices − Practical Guidelines for Community Nurseries. International Center for Research in Agroforestry, Nairobi, Kenya, 95 pp.

Wildin, J.H., 1994. Beef production from broadacre *Leucaena* in central Queensland. In: Gutteridge, R.C., Shelton, H.M. (Eds.), Forage Tree Legumes in Tropical Agriculture. CABI, Wallingford, UK, pp. 352−356.

Wilkinson, JA 2006. Baobab Dried Fruit Pulp Novel Food Application, http://www.acnfp.gov.uk/assess.

Wilkinson, J.A., Hall, M., 2007. Baobab fruit: the upside down tree that could turn around the drinks industry. Soft Drinks Int. Bot. (April).26−28.

Wilson, J., 1978. Some physiological responses *of Acer pseudoplatanus L.* to wind at different levels of soil water, and the anatomical features of abrasive leaf damage. Ph.D. Thesis. University of Edinburgh, p. 282.

Wilson, J., Munro, R.C., Ingleby, K., Mason, P.A., Jefwa, J., et al., 1991. Agroforestry in semi-arid lands of Kenya: role of mycorrhizal inoculation and water retaining polymer. For. Ecol. Manag. 45, 153−163.

Wilson, J., Ingleby, K., Mason, P.A., Ibrahim, K., Lawson, G.J., 1992. Long-term changes in vesicular-arbuscular mycorrhizal spore populations in *Terminalia* plantations in Côte d'Ivoire. In: Read, D.J., Lewis, D.H., Fitter, A.H., Alexander, I.J. (Eds.), Mycorrhizas in Ecosystems. CABI, Wallingford, UK, pp. 268−275. , 1992.

Winrock International, 1992. Assessment of Animal Agriculture in Sub-Saharan Africa. Winrock International Institute for Agricultural Development, Arkansas, USA.

Woods, EJ 2004. Supply Chain Management: Understanding the Concept and Its Implications in Developing Countries. ACIAR Proceedings No. 119e.

Woomer, P.L., Palm, C.A., Qureshi, J.N., Kotto-Same, J., 1997. Carbon sequestration and organic resource management in African smallholder agriculture. In: Lal, R., Kimble, J.M., Follett, R.F., Stewart, B.A. (Eds.), Management of Carbon Sequestration in Soil. CRC Press, Boca Raton, FL, pp. 58−78. , USA Advances in Soil Science.

Woomer, P, Bajah, O, Atta-Krah, AN, Sanginga, N, 1995. Analysis and interpretation of Alley Farming Network data from tropical Africa. Paper presented at the International Alley Farming Conference, 14−18 September, 1992, IITA, Ibadan, Nigeria.

World Agroforestry Centre, 2007. Criteria and indicators for environmental service compensation, and reward mechanisms: realistic, voluntary, conditional and pro-poor. ICRAF Working Paper 37, World Agroforestry Centre, Bogor, Indonesia.

World Bank, 2004. Sustaining Forest: A Development Strategy. World Bank, Washington DC.

World Bank, 2007. World Development Report 2008: Agriculture for Development. The World Bank, Washington DC.

WRI, 1992. World Resources 1992−93. World Resources Institute, New York, USA, Oxford: Oxford University Press.

Wu, J., 2013. Landscape sustainability science: Ecosystem services and human well-being in changing landscapes. Landsc. Ecol. 28, 999−1023.

Wunderle, J., Latte, S., 1998. Avian resource use in Dominican shade coffee plantations. Wilson Bull. 110, 271−281.

Wynberg, R., Cribbins, J., Leakey, R.R.B., Lombard, C., Mander, M., Shackleton, S.E., et al., 2002. A summary of knowledge on marula (*Sclerocarya birrea* subsp. *caffra*) with emphasis on its importance as a non-timber forest product in South and southern Africa. 2. Commercial use, tenure and policy, domestication, intellectual property rights and benefit-sharing. S. Afr. For. J. 196, 67−77.

Wynberg, R.P., Laird, S.A., Shackleton, S., Mander, M., Shackleton, C., du Plessis, P., et al., 2003. Marula policy brief. Marula commercialization for sustainable and equitable livelihoods. Forests Trees Livelihoods. 13, 203−215.

Yaap, B., Campbell, B.M., 2012. Assessing Design of Integrated Conservation and Development Projects. In: Sunderland, T.C.H., Sayer, J.A., Hoang, M.H. (Eds.), Evidence-Based Conservation: Lessons from the Lower Mekong. Routledge, Abingdon, UK, p. 227.

Yazzie, D., VanderJagt, D.J., Pastuszyn, A., Okolo, A., Glew, R.H., 1994. The amino acid and mineral content of baobab (*Adansonia digitata L.*) leaves. J. Food Comp. Anal. 7, 189−193.

Youmbi, E., Clair-Maczulajtys, Bory, G., 1989. Variations de la composition chimique des fruits de *Dacryodes edulis* (DON) LAM. Fruits. 44, 149−153.

Young, A.M., 1982. Effects of shade cover and availability of midge breeding sites on pollinating midge populations and fruit set in two cocoa farms. J. Appl. Ecol. 19, 47−63.

Young, A.M., 1983. Seasonal differences in abundance and distribution of cocoa-pollinating midges in relation to flowering and fruit set between shady and sunny habitats of the Lola Cocoa farm, Costa Rica. J. Appl. Ecol. 20, 801−831.

Young, A., 1989. Agroforestry for Soil Conservation. CAB International and ICRAF, Wallingford, UK.

Zerega, N.J.C., Ragone, D., Motley, T.J., 2004. Complex origins of breadfruit (*Artocarpus altilis*, Moraceae): implications for human migrations in Oceania. Am. J. Bot. 91, 760−766.

Zobel, B.J., Talbert, J.T., 1984. Applied Forest Tree Improvement. John Wiley & Sons, New York, 505 pp.

Zomer, R.J., Trabucco, A., Coe, R., Place, F., 2009. Trees on farm: Analysis of global extent and geographic patterns of agroforestry. ICRAF Working Paper No 89. World Agroforestry Centre, Nairobi, Kenya, 63pp.

Zomer, RJ, Trabucco, A, Coe, R, Place, F, van Noordwijk, M, Xu, JC, 2014. Trees on farms: an update and reanalysis of agroforestry's global extent and socio-ecological characteristics. Working Paper 179. Bogor, Indonesia: World Agroforestry Centre (ICRAF) Southeast Asia Regional Program. DOI: 10.5716/ WP14064.PDF.

Index

Note: Page numbers followed by "*f*" and "*t*" refer to figures and tables, respectively.

Printed in the United States
By Bookmasters